Molecular Biology of Ornamental Plants

Molecular Biology of Ornamental Plants

Editors

Aiping Song
Yu Chen

MDPI

Basel • Beijing • Wuhan • Barcelona • Belgrade • Novi Sad • Cluj • Manchester

Editors
Aiping Song
Nanjing Agricultural University
Nanjing, China

Yu Chen
Nanjing Agricultural University
Nanjing, China

Editorial Office
MDPI
St. Alban-Anlage 66
4052 Basel, Switzerland

This is a reprint of articles from the Special Issue published online in the open access journal *Plants* (ISSN 2223-7747) (available at: https://www.mdpi.com/journal/plants/special_issues/SI_Ornamental).

For citation purposes, cite each article independently as indicated on the article page online and as indicated below:

Lastname, A.A.; Lastname, B.B. Article Title. *Journal Name* **Year**, *Volume Number*, Page Range.

ISBN 978-3-0365-9368-5 (Hbk)
ISBN 978-3-0365-9369-2 (PDF)
doi.org/10.3390/books978-3-0365-9369-2

Contents

About the Editors . . . . **vii**

Aiping Song and Yu Chen
Molecular Biology of Ornamental Plants
Reprinted from: *Plants* **2023**, *12*, 3493, doi:10.3390/plants12193493 . . . . **1**

Aidi Zhang, Yuhong Xiong, Jing Fang, Xiaohan Jiang, Tengfei Wang, Kangchen Liu, et al.
Diversity and Functional Evolution of Terpene Synthases in Rosaceae
Reprinted from: *Plants* **2022**, *11*, 736, doi:10.3390/plants11060736 . . . . **5**

Yao Zhang, Xi Liu, Yuying Li, Xiongfang Liu, Hong Ma, Suping Qu and Zhenghong Li
Basic Characteristics of Flower Transcriptome Data and Derived Novel EST-SSR Markers of *Luculia yunnanensis*, an Endangered Species Endemic to Yunnan, Southwestern China
Reprinted from: *Plants* **2022**, *11*, 1204, doi:10.3390/plants11091204 . . . . **25**

Na Yuan, Shujing Wei, Hans Peter Comes, Sisheng Luo, Ruisen Lu and Yingxiong Qiu
A Comparative Study of Genetic Responses to Short- and Long-Term Habitat Fragmentation in a Distylous Herb *Hedyotis chyrsotricha* (Rubiaceae)
Reprinted from: *Plants* **2022**, *11*, 1800, doi:10.3390/plants11141800 . . . . **39**

Riwen Fei, Shixin Guan, Siyang Duan, Jiayuan Ge, Tianyi Sun and Xiaomei Sun
Elucidating Biological Functions of 9-*cis*-Epoxycarotenoid Dioxygenase Genes Involved in Seed Dormancy in *Paeonia lactiflora*
Reprinted from: *Plants* **2023**, *12*, 710, doi:10.3390/plants12040710 . . . . **55**

Zhiquan Wang, Longjie Ni, Liangqin Liu, Haiyan Yuan and Chunsun Gu
IlAP2, an AP2/ERF Superfamily Gene, Mediates Cadmium Tolerance by Interacting with *IlMT2a* in *Iris lactea* var. *chinensis*
Reprinted from: *Plants* **2023**, *12*, 823, doi:10.3390/plants12040823 . . . . **69**

Nanyan Zhu and Chunhua Zhou
Transcriptomic Analysis Reveals the Regulatory Mechanism of Color Diversity in *Rhododendron pulchrum* Sweet (Ericaceae)
Reprinted from: *Plants* **2023**, *12*, 2656, doi:10.3390/plants12142656 . . . . **81**

Lingling Zhang, Weimin Fang, Fadi Chen and Aiping Song
The Role of Transcription Factors in the Regulation of Plant Shoot Branching
Reprinted from: *Plants* **2022**, *11*, 1997, doi:10.3390/plants11151997 . . . . **99**

Yang Yang, Yunfeng Miao, Shiwei Zhong, Qiu Fang, Yiguang Wang, Bin Dong and Hongbo Zhao
Genome-Wide Identification and Expression Analysis of *XTH* Gene Family during Flower-Opening Stages in *Osmanthus fragrans*
Reprinted from: *Plants* **2022**, *11*, 1015, doi:10.3390/plants11081015 . . . . **113**

Xiaohan Cao, Wenyang Wan, Huimin Mao, Dandan Yin, Xianhui Deng, Huan Yan and Liping Ren
Genome-Wide Identification and Expression Analysis of Dof Transcription Factors in Lotus (*Nelumbo nucifera* Gaertn.)
Reprinted from: *Plants* **2022**, *11*, 2057, doi:10.3390/plants11152057 . . . . **131**

Cuihua Gu, Linxue Shang, Guozhe Zhang, Qun Wang, Qingqing Ma, Sidan Hong, et al.
Identification and Expression Analysis of NAC Gene Family in Weeping Trait of *Lagerstroemia indica*
Reprinted from: *Plants* **2022**, *11*, 2168, doi:10.3390/plants11162168 . . . . **147**

Jin Ma, Guozhe Zhang, Yacheng Ye, Linxue Shang, Sidan Hong, Qingqing Ma, et al.
Genome-Wide Identification and Expression Analysis of HSF Transcription Factors in Alfalfa (*Medicago sativa*) under Abiotic Stress
Reprinted from: *Plants* **2022**, *11*, 2763, doi:10.3390/plants11202763 . . . . **163**

Zhenhao Guo, Lisi He, Xiaobo Sun, Chang Li, Jiale Su, Huimin Zhou and Xiaoqing Liu
Genome-Wide Analysis of the *Rhododendron* AP2/ERF Gene Family: Identification and Expression Profiles in Response to Cold, Salt and Drought Stress
Reprinted from: *Plants* **2023**, *12*, 994, doi:10.3390/plants12050994 . . . . **183**

Dina Liu, Chunsun Gu, Zekai Fu and Zhiquan Wang
Genome-Wide Identification and Analysis of *MYB* Transcription Factor Family in *Hibiscus hamabo*
Reprinted from: *Plants* **2023**, *12*, 1429, doi:10.3390/plants12071429 . . . . **201**

Xiangshang Song, Yaping Kou, Mingao Duan, Bo Feng, Xiaoyun Yu, Ruidong Jia, et al.
Genome-Wide Identification of the Rose *SWEET* Gene Family and Their Different Expression Profiles in Cold Response between Two Rose Species
Reprinted from: *Plants* **2023**, *12*, 1474, doi:10.3390/plants12071474 . . . . **211**

Hongli Liu, Jing Liu, Peng Chen, Xin Zhang, Ke Wang, Jiuxing Lu and Yonghua Li
Selection and Validation of Optimal RT-qPCR Reference Genes for the Normalization of Gene Expression under Different Experimental Conditions in *Lindera megaphylla*
Reprinted from: *Plants* **2023**, *12*, 2185, doi:10.3390/plants12112185 . . . . **231**

Guanghao Cheng, Xiaochun Shu, Zhong Wang, Ning Wang and Fengjiao Zhang
Establishing a Virus-Induced Gene Silencing System in *Lycoris chinensis*
Reprinted from: *Plants* **2023**, *12*, 2458, doi:10.3390/plants12132458 . . . . **253**

About the Editors

Aiping Song

Aiping Song, Ph.D., is an Associate Professor in the Department of Ornamental Horticulture, College of Horticulture, Nanjing Agricultural University. He has investigated many aspects of ornamental plant genomics and molecular biology, emphasizing the role of stress response, development, and plant–pathogen interaction. His project management experience includes the National Natural Science Foundation of China, Natural Science Fund of Jiangsu Province and wide collaboration with other research groups. He is on the Editorial Board of *BMC Plant Biology, Int J Mol Sci, Plants* and several other plant-science-related journals. He has published over 100 research articles with an H index of 32.

Yu Chen

Yu Chen, Ph.D., is an Associate Professor at the College of Agro-Grassland Science, Nanjing Agricultural University. He has conducted research in many areas, including turfgrass breeding for quality and stress resistance, and molecular mechanisms of stress tolerance in turfgrass and forage. His project management experience has included work with the National Natural Science Foundation of China, Natural Science Fund of Jiangsu Province, and wide collaboration with other research groups. He has published over 30 research articles.

plants

MDPI

Editorial

Molecular Biology of Ornamental Plants

Aiping Song [1,2,*] and Yu Chen [3]

1 State Key Laboratory of Crop Genetics & Germplasm Enhancement and Utilization, Key Laboratory of Landscaping, Key Laboratory of Flower Biology and Germplasm Innovation (South), Ministry of Agriculture and Rural Affairs, Key Laboratory of Biology of Ornamental Plants in East China, National Forestry and Grassland Administration, College of Horticulture, Nanjing Agricultural University, Nanjing 210095, China
2 Zhongshan Biological Breeding Laboratory, No.50 Zhongling Street, Nanjing 210014, China
3 College of Agro-Grassland Science, Nanjing Agricultural University, Nanjing 210095, China; cyu801027@njau.edu.cn
* Correspondence: aiping_song@njau.edu.cn

1. Summary

Relative to model plants, ornamental plants have many special characteristics, such as their flower color and shape, and a floral fragrance. Furthermore, these traits are unique and complex to different plants. With the development of sequencing technology and omics tools, core regulatory networks and genes related to specific traits were discovered in a variety of ornamental plants. These research methods have enriched the exploration of the regulatory mechanisms of ornamental traits and provided elite candidate genes for directional breeding. Our Special Issue "Molecular Biology of Ornamental Plants" covers 13 ornamental plant species, including 7 woody flowers, 3 perennial flowers, 1 aquatic flower and 2 herbs. The research topics include flower color, abiotic stress, species diversity, plant development and research protocol innovation. The research methods integrate the latest methods of bioinformatics and molecular biology. The following is a brief introduction to the main progresses in this Special Issue.

2. Main Progresses in this Special Issue

2.1. Molecular Evolution and Species Diversity

Zhang et al. investigated the TPS gene family in eight sequenced Rosaceae species through classification, chromosomal location, orthologous relationships and duplication analysis. The results showed that the distribution of the TPS gene family among Rosaceae species revealed an assortment of family numbers and functions, and further observed that lineage-specific expansions of the TPSs accompanied by frequent domain loss were widely observed within different TPS clades [1]. Zhang et al. conducted the transcriptome sequencing of *Luculia yunnanensis*. Then, the MISA (Microsatellite) tool was used to identify SSR loci from all unigenes, and a total of 15,384 SSRs were identified. Furthermore, 17 of the synthesized EST-SSR primers can be used for subsequent population genetic diversity analyses and molecular marker-assisted breeding [2]. Yuan et al. compared the population structure and genetic diversity of a distylous herb, *Hedyotis chyrsotricha* (Rubiaceae), in two contrasting island systems of southeast China. RAD-Seq data emphasized that the matrix of water in the island system may facilitate the seed (fruit) dispersal of *H. chrysotricha*, thus maintaining population connectivity and providing ongoing resilience to the lasting effects of habitat fragmentation [3].

2.2. Gene Function Research

Fei et al. identified and cloned two genes from the NCED family in *Paeonia lactiflora*, which are related to ABA synthesis in the herbaceous peony, named *PlNCED1* and *PlNCED2*. Further studies confirmed the function of the *PlNCED* gene in inducing seed dormancy and hindering seed germination [4]. In iris plants, Wang et al. confirmed the

Citation: Song, A.; Chen, Y. Molecular Biology of Ornamental Plants. *Plants* **2023**, *12*, 3493. https://doi.org/10.3390/plants12193493

Received: 16 September 2023
Accepted: 27 September 2023
Published: 7 October 2023

function of the AP2/ERF superfamily member IlAP2, which interacts with IlMT2a, in enhancing Cd tolerance [5]. Zhu et al. used transcriptome approaches to study the molecular mechanisms affecting the petal coloration of four varieties of *Rhododendron pulchrum*. The results showed that there were differentially expressed genes among the four different petal samples, including signal transduction-related genes, anthocyanin biosynthesis genes and transcription factors. The development of a red color may be related to a higher expression of *DFR* genes, which promotes the accumulation of anthocyanin [6]. Zhang et al. were involved in regulating plant branching and development to provide candidate genes for improving crop architecture through gene editing or directed breeding [7].

2.3. Gene Family Identification

Yang et al. identified 38 *OfXTH* genes belonging to the four main phylogenetic groups in *Osmanthus fragrans*. The gene structure, chromosomal location, synteny relationship and cis-acting elements and expression patterns were analyzed on a genome-wide scale. The expression patterns showed that most *OfXTHs* were closely associated with the flower-opening period of *O. fragrans*, and five of them were involved in the regulation of flower opening by responding to ambient temperature changes [8]. Cao et al. identified twenty-nine *NnDofs* in lotuses, of which the physicochemical properties vary; however, all of which contain conserved zinc finger structures. A promoter analysis, RNA-seq atlas and qRT-PCR were performed to demonstrate the potential function of *NnDofs* in the salt tolerance of lotuses [9]. Gu et al. identified 21 *LiNAC* genes from the transcriptome data in *Lagerstroemia indica*, and the physicochemical characteristics of amino acids, their subcellular localization, transmembrane structure, GO and KEGG enrichment, and expression patterns were examined. A further qRT-PCR indicated that these *LiNAC* genes may be involved in the regulation of weeping traits in *L. indica* [10]. Ma et al. identified 16 *MsHSF* genes in *Medicago sativa*, and the physicochemical properties, subcellular localization, synteny analysis, GO annotation and enrichment, and the protein interaction networks of amino acids were also examined. Similarly, the RNA-Seq results found that the *MsHSF* gene family plays a key role in abiotic stress [11]. Guo et al. performed a phylogenetic analysis of 120 AP2/ERF genes in the *Rhododendron simsii* genome; cis-acting elements involving plant growth regulators, the response to abiotic stress and MYB binding sites were detected in the upstream sequences of *RsAP2* genes. In addition, twenty *RsAP2* genes were selected for a qRT-PCR, and the results showed that most of the *RsAP2* genes responded to these abiotic stresses [12]. Liu et al. analyzed 204 *HhMYB* family members in *Hibiscus hamabo* using RNA-seq and a qRT-PCR, and found that *HhMYB* was involved in plant responses to salt and drought stress to varying degrees [13]. Song et al. identified and analyzed 25 *SWEET* genes in *Rosa chinensis* 'Old Blush', and through a transcriptome analysis, *SWEET2a* and *SWEET10c* were found to be candidate genes involved in the rose's ability to tolerate cold conditions [14].

2.4. Research Protocol Innovation

Continuous updates in technology and tools promote further studies of ornamental plants. Liu et al. selected 14 candidate genes from the transcriptome database of *Lindera megaphylla* for additional qRT-PCRs under different conditions to further verify the reliability of selected reference genes above [15]. Cheng et al. developed an efficient virus-induced gene silencing (VIGS) system using the leaf tip needle injection method, which could effectively silence genes in *Lycoris chinensis*, providing a powerful tool for gene function studies [16].

Author Contributions: A.S. and Y.C. wrote the manuscript. All authors have read and agreed to the published version of the manuscript.

Funding: This research was funded by the National Natural Science Foundation of China (32172609), the Natural Science Fund of Jiangsu Province (BK20190076) and a project funded by the Priority Academic Program Development of Jiangsu Higher Education Institution.

Conflicts of Interest: The authors declare that the research was conducted in the absence of any commercial or financial relationships that could be construed as potential conflicts of interest.

References

1. Zhang, A.D.; Xiong, Y.H.; Fang, J.; Jiang, X.H.; Wang, T.F.; Liu, K.C.; Peng, H.X.; Zhang, X.J. Diversity and Functional Evolution of Terpene Synthases in Rosaceae. *Plants* **2022**, *11*, 736. [CrossRef] [PubMed]
2. Zhang, Y.; Liu, X.; Li, Y.Y.; Liu, X.F.; Ma, H.; Qu, S.P.; Li, Z.H. Basic Characteristics of Flower Transcriptome Data and Derived Novel EST-SSR Markers of Luculia yunnanensis, an Endangered Species Endemic to Yunnan, Southwestern China. *Plants* **2022**, *11*, 1204. [CrossRef] [PubMed]
3. Yuan, N.; Wei, S.J.; Comes, H.P.; Luo, S.S.; Lu, R.S.; Qiu, Y.X. A Comparative Study of Genetic Responses to Short- and Long-Term Habitat Fragmentation in a Distylous Herb Hedyotis chyrsotricha (Rubiaceae). *Plants* **2022**, *11*, 1800. [CrossRef] [PubMed]
4. Fei, R.W.; Guan, S.X.; Duan, S.Y.; Ge, J.Y.; Sun, T.Y.; Sun, X.M. Elucidating Biological Functions of 9-cis-Epoxycarotenoid Dioxygenase Genes Involved in Seed Dormancy in Paeonia lactiflora. *Plants* **2023**, *12*, 710. [CrossRef] [PubMed]
5. Wang, Z.Q.; Ni, L.J.; Liu, L.Q.; Yuan, H.Y.; Gu, C.S. IlAP2, an AP2/ERF Superfamily Gene, Mediates Cadmium Tolerance by Interacting with IlMT2a in Iris lactea var. chinensis. *Plants* **2023**, *12*, 823. [CrossRef] [PubMed]
6. Zhu, N.Y.; Zhou, C.H. Transcriptomic Analysis Reveals the Regulatory Mechanism of Color Diversity in Rhododendron pulchrum Sweet (Ericaceae). *Plants* **2023**, *12*, 2656. [CrossRef] [PubMed]
7. Zhang, L.L.; Fang, W.M.; Chen, F.D.; Song, A.P. The Role of Transcription Factors in the Regulation of Plant Shoot Branching. *Plants* **2022**, *11*, 1997. [CrossRef] [PubMed]
8. Yang, Y.; Miao, Y.F.; Zhong, S.W.; Fang, Q.; Wang, Y.G.; Dong, B.; Zhao, H.B. Genome-Wide Identification and Expression Analysis of XTH Gene Family during Flower-Opening Stages in Osmanthus fragrans. *Plants* **2022**, *11*, 1015. [CrossRef] [PubMed]
9. Cao, X.H.; Wan, W.Y.; Mao, H.M.; Yin, D.D.; Deng, X.H.; Yan, H.; Ren, L.P. Genome-Wide Identification and Expression Analysis of Dof Transcription Factors in Lotus (Nelumbo nucifera Gaertn.). *Plants* **2022**, *11*, 2057. [CrossRef] [PubMed]
10. Gu, C.H.; Shang, L.X.; Zhang, G.Z.; Wang, Q.; Ma, Q.Q.; Hong, S.D.; Zhao, Y.; Yang, L.Y. Identification and Expression Analysis of NAC Gene Family in Weeping Trait of Lagerstroemia indica. *Plants* **2022**, *11*, 2168. [CrossRef] [PubMed]
11. Ma, J.; Zhang, G.Z.; Ye, Y.C.; Shang, L.X.; Hong, S.D.; Ma, Q.Q.; Zhao, Y.; Gu, C.H. Genome-Wide Identification and Expression Analysis of HSF Transcription Factors in Alfalfa (Medicago sativa) under Abiotic Stress. *Plants* **2022**, *11*, 2763. [CrossRef] [PubMed]
12. Guo, Z.H.; He, L.S.; Sun, X.B.; Li, C.; Su, J.L.; Zhou, H.M.; Liu, X.Q. Genome-Wide Analysis of the Rhododendron AP2/ERF Gene Family: Identification and Expression Profiles in Response to Cold, Salt and Drought Stress. *Plants* **2023**, *12*, 994. [CrossRef] [PubMed]
13. Liu, D.A.; Gu, C.S.; Fu, Z.K.; Wang, Z.Q. Genome-Wide Identification and Analysis of MYB Transcription Factor Family in Hibiscus hamabo. *Plants* **2023**, *12*, 1429. [CrossRef]
14. Song, X.S.; Kou, Y.P.; Duan, M.G.; Feng, B.; Yu, X.Y.; Jia, R.D.; Zhao, X.; Ge, H.; Yang, S.H. Genome-Wide Identification of the Rose SWEET Gene Family and Their Different Expression Profiles in Cold Response between Two Rose Species. *Plants* **2023**, *12*, 1474. [CrossRef]
15. Liu, H.L.; Liu, J.; Chen, P.; Zhang, X.; Wang, K.; Lu, J.X.; Li, Y.H. Selection and Validation of Optimal RT-qPCR Reference Genes for the Normalization of Gene Expression under Different Experimental Conditions in Lindera megaphylla. *Plants* **2023**, *12*, 2185. [CrossRef] [PubMed]
16. Cheng, G.H.; Shu, X.C.; Wang, Z.; Wang, N.; Zhang, F.J. Establishing a Virus-Induced Gene Silencing System in Lycoris chinensis. *Plants* **2023**, *12*, 2458. [CrossRef]

plants

Article

Diversity and Functional Evolution of Terpene Synthases in Rosaceae

Aidi Zhang [1,2], Yuhong Xiong [1,2,3], Jing Fang [1,2,3], Xiaohan Jiang [1,2,3], Tengfei Wang [1,2,3], Kangchen Liu [1,2,3], Huixiang Peng [1,2,3] and Xiujun Zhang [1,2,*]

1 Key Laboratory of Plant Germplasm Enhancement and Specialty Agriculture, Wuhan Botanical Garden, Chinese Academy of Sciences, Wuhan 430000, China; zhangaidi@wbgcas.cn (A.Z.); xiongyuhong20@mails.ucas.ac.cn (Y.X.); fangjing17@mails.ucas.ac.cn (J.F.); jiangxiaohan16@mails.ucas.ac.cn (X.J.); wangtengfei17@mails.ucas.ac.cn (T.W.); liukangchen18@mails.ucas.ac.cn (K.L.); penghuixiang20@mails.ucas.ac.cn (H.P.)
2 Center of Economic Botany, Core Botanical Gardens, Chinese Academy of Sciences, Wuhan 430074, China
3 University of Chinese Academy of Sciences, Beijing 100049, China
* Correspondence: zhangxj@wbgcas.cn; Tel.: +86-27-87700844

Abstract: Terpenes are organic compounds and play important roles in plant development and stress response. Terpene synthases (TPSs) are the key enzymes for the biosynthesis of terpenes. For Rosaceae species, terpene composition represents a critical quality attribute, but limited information is available regarding the evolution and expansion occurring in the terpene synthases gene family. Here, we selected eight Rosaceae species with sequenced and annotated genomes for the identification of TPSs, including three *Prunoideae*, three *Maloideae*, and two *Rosoideae* species. Our data showed that the TPS gene family in the Rosaceae species displayed a diversity of family numbers and functions among different subfamilies. Lineage and species-specific expansion of the TPSs accompanied by frequent domain loss was widely observed within different TPS clades, which might have contributed to speciation or environmental adaptation in Rosaceae. In contrast to *Maloideae* and *Rosoideae* species, *Prunoideae* species owned less TPSs, with the evolution of *Prunoideae* species, TPSs were expanded in modern peach. Both tandem and segmental duplication significantly contributed to TPSs expansion. Ka/Ks calculations revealed that *TPSs* genes mainly evolved under purifying selection except for several pairs, where the divergent time indicated TPS-e clade was diverged relatively anciently. Gene function classification of TPSs further demonstrated the function diversity among clades and species. Moreover, based on already published RNA-Seq data from NCBI, the expression of most TPSs in *Malus domestica*, *Prunus persica*, and *Fragaria vesca* displayed tissue specificity and distinct expression patterns either in tissues or expression abundance between species and TPS clades. Certain putative TPS-like proteins lacking both domains were detected to be highly expressed, indicating the underlying functional or regulatory potentials. The result provided insight into the TPS family evolution and genetic information that would help to improve Rosaceae species quality.

Keywords: terpene synthases; Rosaceae; expansion; evolution

Citation: Zhang, A.; Xiong, Y.; Fang, J.; Jiang, X.; Wang, T.; Liu, K.; Peng, H.; Zhang, X. Diversity and Functional Evolution of Terpene Synthases in Rosaceae. *Plants* **2022**, *11*, 736. https://doi.org/10.3390/plants11060736

Academic Editors: Wolfgang Friedt and Igor Jerković

Received: 31 December 2021
Accepted: 4 March 2022
Published: 10 March 2022

1. Introduction

The Rosaceae family consists of more than 2500 species in more than 90 genera and the family is divided into four subfamilies based on fruits: *Spiraeoideae* (*Spirea* subfamily), *Rosoideae* (rose subfamily), *Prunoideae* (plum subfamily), and *Maloideae* (apple subfamily). A number of species in the Rosaceae family are of economic importance as food crops, such as peaches, apples, almonds, cherries, pears, raspberries, and strawberries. Some species in the Rosaceae family are grown as ornamentals, such as the spiraea and rose. For Rosaceae, the inner quality of fruits or flowers is mainly determined by aroma and flavor. The aroma components are mainly composed of volatile products and terpenoids are important components of the volatile products such as linalool, (E)-β–damarone, and β–ionone [1].

Terpenoid compounds have been characterized in almond, apple, and peach [2–5]. Volatile compounds in almond (plum subfamily) mainly consist of fatty acid-derived volatiles, several monoterpenes, sesquiterpenes, and phenylpropanoids, and significant differences in volatile composition were observed between different tissues and various varieties [3]. Apple (apple subfamily) fruit produce more than 300 volatile organic compounds (VOCs), including alcohols, aldehyde esters, and ketones; the specific VOC composition in apple depends on several factors, including cultivar, climacteric ethylene production levels, maturity, and environmental conditions [4,5]. More than 100 volatile chemicals have been identified in peach (plum subfamily) fruit, in which linalool is a key odorant that affects fruit aroma and consumer preference; over-expression of *terpene synthase gene 3* led to linalool accumulation [2]. Thus, small changes in volatile content have the potential to affect fruit flavor quality, there is an emphasized interest to regulate fruit flavor related volatiles, and epigenetic regulation of terpenoids is also a control strategy during fruit ripening [6,7]. To facilitate the breeding of Rosaceae species with desirable sensory qualities and improve their qualities under stress conditions, a better understanding of the genetic determinants of aroma and flavor in general and terpenes in particular is required.

Terpenoids represent the largest group of natural products and make up diverse secondary metabolites. Terpenoid composition not only represents a critical attribute in determining the quality of horticultural food products, such as taste and aroma, but also functions widely in plant development and defense, such as attracting pollinators, defending against herbivores, and acting as anti-bacterial agents [8–10]. Plant terpenoids (isoprene-C5, monoterpenes-C10, sesquiterpenes-C15, diterpenes-C20, and polyterpenoids-C5xn) are some compounds derived from isomeric 5-carbon building blocks isopentenyl diphosphate (IPP) and dimethylallyl diphosphate (DMAPP) [11,12]. About 50,000 terpenoid metabolites including monoterpenes, sesquiterpenes, and diterpenes have been identified in higher plants, liverworts, and fungi [11,12]. In general, for a better adaptation to a local ecological niche, each species typically synthesizes only a small fraction of terpenoid metabolites [9,13]. In plants, geranyl diphosphate (GPP), farnesyl diphosphate (FPP), and geranylgeranyl diphosphate (GGPP) are the precursors for monoterpenes, sesquiterpenes, and diterpenes, respectively; terpene synthase genes (TPSs) are responsible for converting them into a multitude of cyclic and acyclic terpenoids [12]. The characteristic catalytic function of TPSs is to generate multiple terpenoid products using one substrate, thus collectively contributing to numerous different structures of plant terpenoids in addition to modifying enzymes [12]. Generally, the TPS family is characterized by two large domains including the N-terminal domain (PF01397) and the C-terminal metal cofactor binding domain (PF03936) [14]. The N-terminal domain possesses a conserved RRX8W (R, arginine, W, tryptophan, and X, alternative amino acid) motif and the C-terminal domain contains two highly conserved aspartate-rich motifs, the DDxxD motif and the NSE/DTE motif. The DDxxD motif is involved in the coordination of divalent ion(s), water molecules, and the stabilization of the active site, and is found among all functional TPSs [3]. The NSE/DTE motif flanks the entrance of the active site and function in binding a trinuclear magnesium cluster [14,15]. TPSs are split into seven subgroups based on their amino acid sequence relatedness, namely TPS a-g. The majority of TPSs in most plants fall into one or two clades and the TPS-d clade was only encoded in gymnosperms [15,16]. Different TPS clades differ considerably in catalysate and sequence. As the largest clade, the TPS-a clade mainly encodes sesquiterpenes. TPS-b and TPS-g clades are clustered closely to TPS-a; the TPS-b clade contains the conserved R(R)X8W motif and usually encodes monoterpenes, while the TPS-g clade lacks the conserved R(R)X8W motif, and functions in producing mono- and sesquiterpenes [17]. The TPS-c clade is characterized by the "DXDD" motif but not the "DDXXD" that was detected in other clades, and mainly functions in producing diterpene products [17]. The TPS-d clade is only encoded in gymnosperms, and function in producing mono- and sesquiterpene products. The TPS-e/f clades are clustered closely, and mainly encode diterpene products [17]. TPS-a/b clades lack the N-terminal γ domain characteristic of diterpene synthases found in clades c, e, and f. In general, the biosynthesis of isoprene, monoterpenes, and diterpenes occurs in the plastid and the biosynthesis of sesquiterpenes occurs in the cytosol [17].

So far, the TPS gene family members have been characterized in many plant species. The sizes of TPS families in the majority of sequenced plants genomes range from 1 to 100. The TPS families probably evolved through duplication of genes followed by functional divergence [13]. The bryophyte *Physcomitrella patens* has a single TPS gene [13]. In *Arabidopsis thaliana, Vitis vinifera, Ocimum sanctum, Daucus carota*, tomato, and *Camellia sinensis*, 32, 69, 81, 19, 44, and 80 TPSs have been identified, respectively [13,18–20]. However, not all of the TPSs were functional. In recent years, the genomes of Rosaceae species such as peach, plum, apple, pear, and strawberry have been sequenced, which has significantly promoted the related studies of this family. Although abundant terpenes were characterized in different tissues of Rosaceae species [3,4,21], a comprehensive study on TPSs has not been reported in Rosaceae.

In this study, we mainly focused on three subfamilies in Rosaceae that are of economic importance as food crops, and did not select the *Spirea* subfamily, as the *Spirea* subfamily is only composed of ornamental flowers; thus, we excluded it from this study. For each subfamily, we chose two representative species with available sequenced genomes of high quality (chromosome scale). Based on the annotated genomes, we carried out the identification, characterization, and metabolite pathway mapping predictions of all TPSs encoded by eight Rosaceae species, including three Prunoideae species, three Maloideae species, and two Rosoideae species. We classified these TPSs into putative TPS-like proteins containing either PF01397 or PF03936 and complete ones containing both of them, respectively. The analysis of phylogeny, gene structure, and expression patterns were conducted with a special focus on their family number distribution and tissue expression patterns among subfamilies and different TPS clades. The findings revealed the diversity and functional evolution of TPSs in Rosaceae. The results provide a foundation for the exploration of TPSs to improve the understanding of the evolution and biosynthesis of terpenoids in Rosaceae.

2. Results

2.1. Identification of TPS Family Members in Rosaceae

To explore the distribution of TPSs among Rosaceae species and the evolutionary trajectory in the subfamily, we selected eight Rosaceae species for the identification of TPSs, including three Prunoideae species (*P. persica, P. mira, P. mume*), three Maloideae species (*P. betulifolia, M. domestica, M. baccata*) and two Rosoideae species (*F. vesca, R. chinensis*) (Table 1). BLASTP and HMM searches were performed against their entire protein sequences, and these two approaches produced the similar number of hits, indicating the relative conservation of the TPS family. We merged the hits together and verified them for the existence of Pfam domains PF03936 (metal-binding domain) and PF01397 (N-terminal TPS domain). Pfam domain distribution in the TPSs of eight Rosaceae species is listed in Table S1. As a result, hundreds of complete TPSs that contained both domains were identified. A recent study used this similar approach to detect TPSs in peach [2], and it detected 38 full-length TPSs with both domains; another study detected cultivated apple contained 55 putative TPS genes [4], the similar amount of TPSs as our results indicates the approach used in our study is reliable. For each Rosoideae species, we found that a certain ratio (64.29–0%) of putative TPSs was composed of both domains. The reference information and family number distribution of TPSs is listed in Table 1. The average family number of all TPSs in the Rosoideae species is the highest (65–76), followed by Maloideae (48–56) and Prunoideae (10–45). The Prunoideae species showed varied numbers of TPSs; there were only 10 putative TPSs identified in *P. mira*. In contrast, 30 and 45 putative TPSs were detected in *P. mume* and *P. persica*, respectively, whereas for the complete TPSs with both domains, 38, 36, and 43 TPSs were identified in *P. persica, M. domestica*, and *F. vesca*, respectively. *M. domestica* displayed a lower ratio of complete TPSs, with less than 64.29%, while the ratio in Prunoideae was up to 90%. All the putative TPSs were renamed numerically with the abbreviation of species names as a prefix (Table S2); only complete TPSs that contained both domains were used for the subsequent analysis.

Table 1. Summary of genome information and TPSs of sequenced Rosaceae species used in this study.

Subfamily	Species	Release Version *	The Prefix of Gene Symbol	Total Genes	Numbers of TPSs	Numbers of TPSs with Both Domains	Percent of TPSs with Both Domains	Numbers of Already Know TPSs
Prunoideae	*Prunus persica*	GDR, v2.0	Pr.pLo	26,873	45	38	84.44%	38 (both domains) [2]
	Prunus mira	GDR, v1.0	Pr.mir	26,958	10	9	90.00%	
	Prunus mume	NCBI, v1.0	Pr.mum	28,638	30	27	90.00%	
Maloideae	*Pyrus betulifolia*	GDR, v1.0	Py.bet	59,552	48	37	77.08%	
	Malus x domestica	GDR, HFTH1 v1.0	Ma.dom	44,677	56	36	64.29%	55 (all) [4]
	Malus baccata	CNGB, v1.0	Ma.bac	45,931	48	33	68.75%	
Rosoideae	*Fragaria vesca*	GDR, v4.0	Fr.ves	35,914	65	43	66.15%	
	Rosa chinensis	GDR, v1.0	Ro.chi	39,669	76	57	75.00%	

* GDR, Genome Database for Rosaceae; CNGB, China National GeneBank DataBase; NCBI, National Center for Biotechnology Information. Species used for the phylogenetic tree are highlighted in bold.

2.2. TPS Classification and Motif/Domain Annotation

All the TPSs containing both PF01397 and PF03936 domains were subjected to classification and motif annotation. TPSs from the three representative Rosaceae species (*P. persica*, *M. domestica*, and *F. vesca*) were chosen for the visualization of the classification and domain/motif distribution in TPSs. After removing putative TPSs lacking both domains, a total of 117 TPSs in the three species were used for phylogenetic construction (Figures 1 and S1). The phylogenetic topology revealed that all the TPSs were divided into seven known clades TPS a-g. TPS-b and TPS-g clustered with the TPS-a clade and forms a large branch. TPS-e and TPS-f formed sister clades and clustered close to the TPS-c clade. No TPSs clustered with the TPS-d clade, which was only encoded in gymnosperms. The conserved motifs were constructed using the online MEME software and three conserved motifs (motifs 1, 2, 3) were detected in nearly all the TPSs (109, 96, and 102). However, the frequency and distribution of these motifs varied among TPSs. For example, *Ma.dom-TPS11* only contained motif 2, and lost motifs 1 and 3, whereas *Fr.ves-TPS28* contains an extra copy of motif 3. The motif distribution of TPSs in the eight Rosaceae species is listed in Table S1; the significant E-value indicates the reliability of the identified motifs. For different clades, we found that the motif composition of TPS-a/b/g clades is more conservative than that of TPS-c/e/f, as shown in Figure 2B; nearly all TPSs in these clades contain all of the three motifs. In contrast, many TPSs from TPS-c/e/f lost motif 2, indicating the differences of motif distribution among clades. The conserved domain (CD) annotation used by the CD-search tool in NCBI revealed the discrepancy of domain annotation among different clades. CD domains Terpene_cyclase_plant_C1 (accession: cd00684) and Isoprenoid_Biosyn_C1 (accession: cd00385), which both belong to superfamily Isoprenoid_Biosyn_C1 superfamily (accession: cl00210), were annotated in clades of TPS-a, TPS-b, and TPS-g. PLN02279 super family (ent-kaur-16-ene synthase) was annotated in TPS-e/f clades. PLN02592 superfamily (ent-copalyl diphosphate synthase) was annotated in TPS-c clade. Pfam domain annotation results verified that each full-length TPS is characterized by two conserved domains with PF01397 (N-terminal) and PF03936 (C-terminal). The protein lengths of TPSs ranged from 232 AA to 1726 AA, showing a wide distribution of TPS lengths. One of the conserved aspartate-rich motifs in the C-terminal domain that is involved in the coordination of divalent ions, water molecules, and the stabilization of the active site was characterized based on motif sequences alignment (Figure 2C). The conservation of amino acid composition varied among different TPS clades. The TPS-c subfamily is characterized by the "DXDD" motif but not the "DDXXD" motif that was detected in other clades.

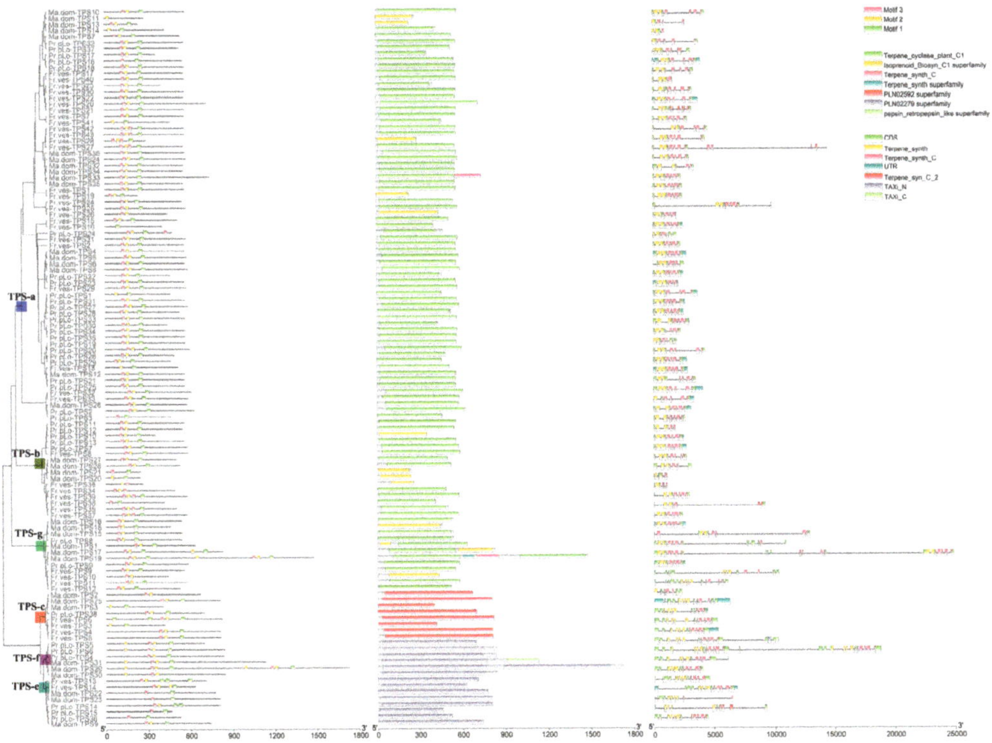

Figure 1. Phylogenetic relationship and distribution of motif/domain of TPSs in three Rosaceae species (*P. persica*, *M. domestica*, *F. vesca*). The phylogenetic tree is shown on the left panel, while conserved motifs, conserved domains, and Pfam domains are shown on the right three panels. The phylogenetic tree from full-length amino acid sequences was constructed using the MEGA with maximum likelihood (ML) method. The conserved motifs were assessed using the online MEME software. The conserved domain was annotated based on the conserved domain database in NCBI, whereas the gene structure and domains were annotated by using the PfamScan tool. The conserved motifs and domains are shaded in different colors. The root nodes of TPS-a, TPS-g, TPS-b, TPS-c, TPS-e, and TPS-f clades are indicated by blue, green, yellow-green, red, benzo, and purple, respectively.

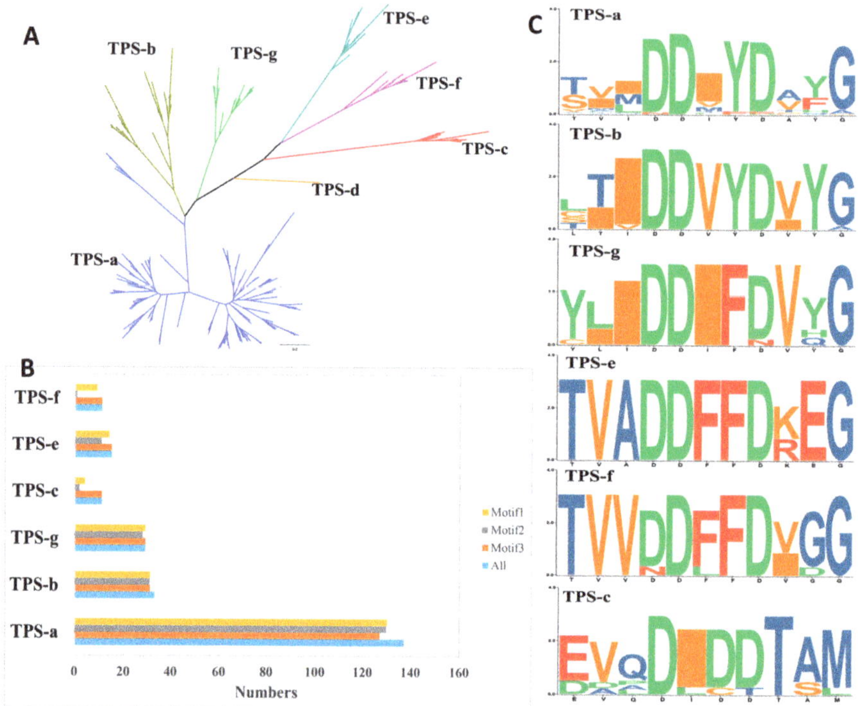

Figure 2. The unrooted phylogenetic tree of TPSs and motifs comparison between different TPS clades. (**A**) The maximum-likelihood phylogenetic tree of the TPS proteins in three Rosaceae species (*P. persica*, *M. domestica*, *F. vesca*). The TPS-a, TPS-g, TPS-b, TPS-c, TPS-e, and TPS-f clades are shaded in blue, green, yellow-green, red, benzo, purple, respectively. (**B**) The frequency of different motifs among different TPS clades. (**C**) The seqLogo of 'DDxxD' motif in the C-terminal domain of different TPS clades, the bit score represents the information content for each position in the sequence.

2.3. Varied Family Number of TPSs among Different Clades in Rosaceae

To explore the evolution of TPSs in Rosaceae, we chose two representative species from each subfamily in Rosaceae for phylogenetic tree construction, including two Prunoideae species (*P. persica*, *P. mume*), two Maloideae species (*M. domestica*, *P. betulifolia*), and two Rosoideae species (*F. vesca*, *R. chinensis*). The detailed distribution of family numbers of different TPS clades is listed in Table 2. The TPS-a clade is the major determinant of family size of individual species; it is the largest group with more TPS copies, followed by TPS-b and TPS-g, TPS-c/e/f have a relatively small family size. On the whole, Rosoideae species owned more TPS copies, followed by Maloideae species, and Prunoideae species. The family number of TPSs from *R. chinensis* is up to 57, of which 37 are TPS-a members. The six Rosaceae species that we chose nearly contained all five TPS clades; however, no TPS-f member was detected in *F. vesca*. The family number distribution of different TPS clades also varied among three Rosaceae subfamilies. For example, more than two TPS-c gene copies existed in the Maloideae and Rosoideae subfamilies, while only one copy was detected in Prunoideae species. The diversity of family number of TPS-f is also obvious; both *M. domestica* and *P. betulifolia* owned three copies, while for Rosoideae species, only one copy was encoded in *R. chinensis*. For Prunoideae species, one copy of TPSs from the TPS-f clade was detected in *P. mira* and *P. mume*, while three copies were detected in *P. persica*, indicating the recent expansion of the TPS-f clade in modern peach. The phylogenetic relationship revealed that lineage-specific expansions of the

TPSs were widely observed within different TPS clades. These expansions are denoted in the phylogenetic tree (Figures 3 and S2), the clustering relationship indicated that many expansions occurred after the split of sister lineages, such as the split of Maloideae and Prunoideae. Two Rosoideae species (*F. vesca*, *R. chinensis*) owned more lineage-specific expansions compared with the other two Rosaceae subfamilies, especially in the branch of TPS-a clade. On the other hand, in the lineage-specific expansion branch, except that many copies were shared by sister species, species-specific expansions were also detected widely. *R. chinensis* was also the most striking species with more copies of TPSs arising by species-specific expansion.

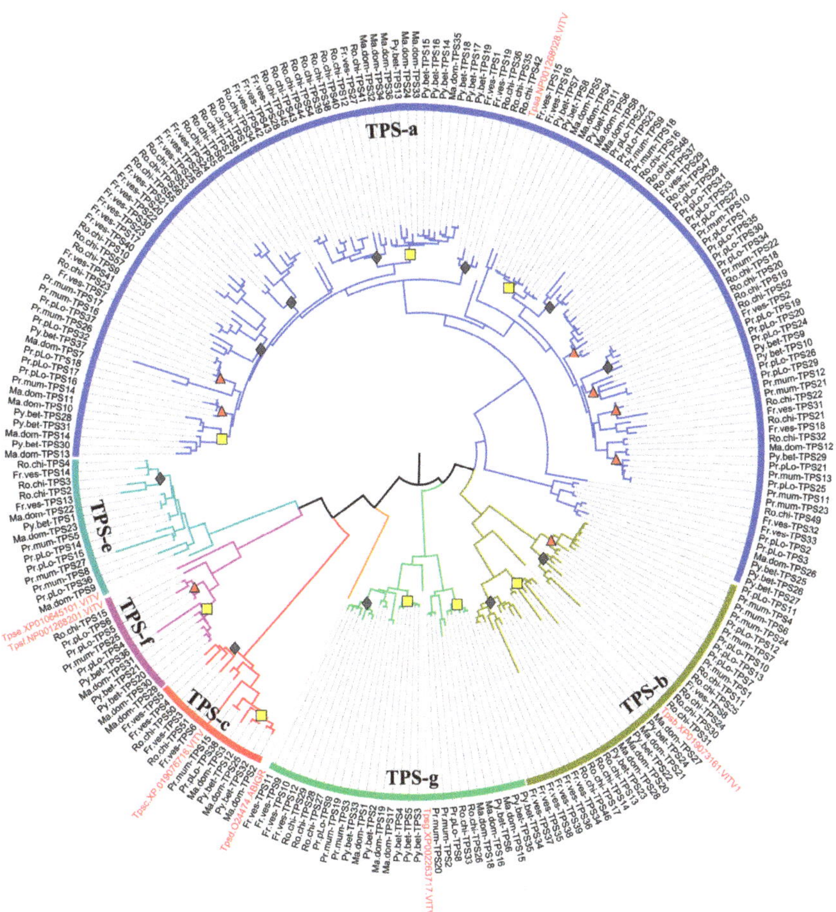

Figure 3. Phylogenetic tree of complete TPSs from six representative Rosaceae species. The TPS members were identified from six species including two Prunoideae species (*P. persica*, *P. mume*), two Maloideae species (*M. domestica*, *P. betulifolia*), and two Rosoideae species (*F. vesca*, *R. chinensis*); only those TPSs containing both PF01397 and PF03936 domains were used. The phylogenetic tree from full-length amino acid sequences was constructed using the MEGA with maximum likelihood (ML) method. Representative sequences of TPSs from *Vitis vinifera* were used as outgroups. The branches of TPS-a, TPS-g, TPS-b, TPS-c, TPS-e, and TPS-f clades are indicated by blue, green, yellow-green, red, benzo, and purple colors, respectively. The lineage-specific expansion of TPSs in Prunoideae, Maloideae, Rosoideae are indicated by yellow squares, grey diamonds, red triangles, respectively. The enlarged phylogenetic tree with bootstrap values is shown in Supplementary Figure S2.

Table 2. Numbers of complete TPSs with both domains from Rosaceae species that were used for the phylogenetic tree of Figure 3.

TPSs Clade	TPS-a	TPS-b	TPS-g	TPS-c	TPS-d	TPS-e	TPS-f	Total
Motif	DDXYD	DDVYD	DDIFD	DIDDT	DXDD, DXXD	DDFFD	DDFFD	
Prunus persica	24	5	2	1	0	3	3	38
Prunus mira	0	0	4	1	0	3	1	9
Prunus mume	13	5	4	1	0	3	1	27
Pyrus betulifolia	20	3	8	2	0	1	3	37
Malus x domestica	17	4	6	3	0	3	3	36
Rosa chinensis	37	9	5	2	0	3	1	57
Fragaria vesca	26	7	4	4	0	2	0	43

2.4. Chromosomal Location, Synteny Analysis of TPS Family Members in Rosaceae

The chromosomal location maps of the TPSs in Rosaceae species were constructed by TBtools, as shown in Figures S3–S9, which indicates the diversity of distribution on chromosomes. For *P. persica*, TPSs were mainly distributed on three chromosomes (chromosomes 3, 4, and 8). Most TPSs (39) were distributed on the 4th chromosome, while five and one TPSs were located on the 3rd and 8th chromosomes, respectively. For the Rosoideae and Maloideae species, TPSs were widely distributed on the 8th and 10th chromosomes, respectively. On the whole, plants from the same subfamily shared a similar chromosome distribution of TPSs. From the physical location maps, we observed that the tandem arrays of TPSs are quite extensive, as in *P. persica*, about 30 TPSs cluster across a stretch of 1051 kb on the 4th chromosome, in *M. domestica*, six TPSs occur in a 200 kb stretch on 10th chromosome. These tandem arrays are likely the consequence of duplication by unequal crossover; genes in tandem arrays are typically highly homologous to each other. For *P. persica*, *M. domestica*, and *F. vesca*, synteny analysis within genomes was conducted to determine their duplication events, as shown in Figure 4, which indicated that genome segmental and tandem duplication were together the driving force for the expansion of the TPS gene family in Rosaceae. For *M. domestica* and *P. persica*, segmental and tandem duplication both contributed to family expansion; for *F. vesca*, tandem duplication played the most important role in the family expansion. MCScanX was used to identify possible collinear blocks between genomes in Rosaceae; the syntenic map between them was constructed, as shown in Figure 5, which showed plenty of syntenic relationships resulting from genome duplication and recombination.

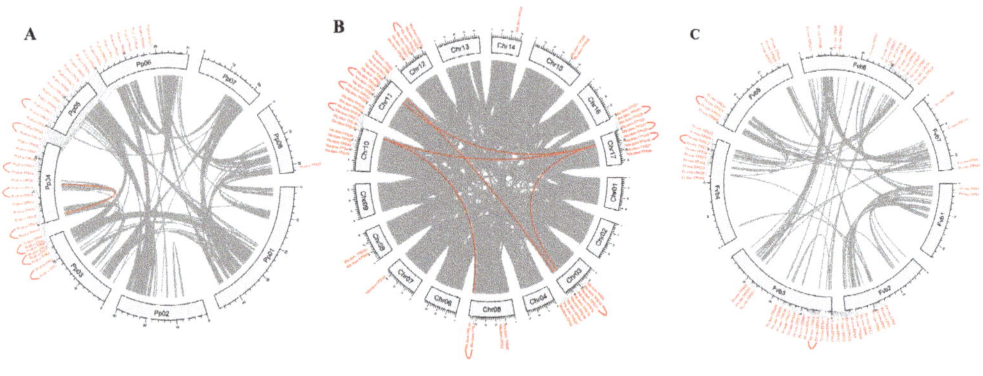

Figure 4. Inter-chromosomal relationships of TPSs in *P. persica* (**A**), *M. domestica* (**B**), and *F. vesca* (**C**). Grey lines in the circle indicate the collinear blocks in the *P. persica* or *M. domestica* genome. The red lines in the circle highlight the segmental duplicated TPS gene pairs, while the red curve outside the circle indicates the tandemly duplicated TPS gene pairs produced by MCSscanX.

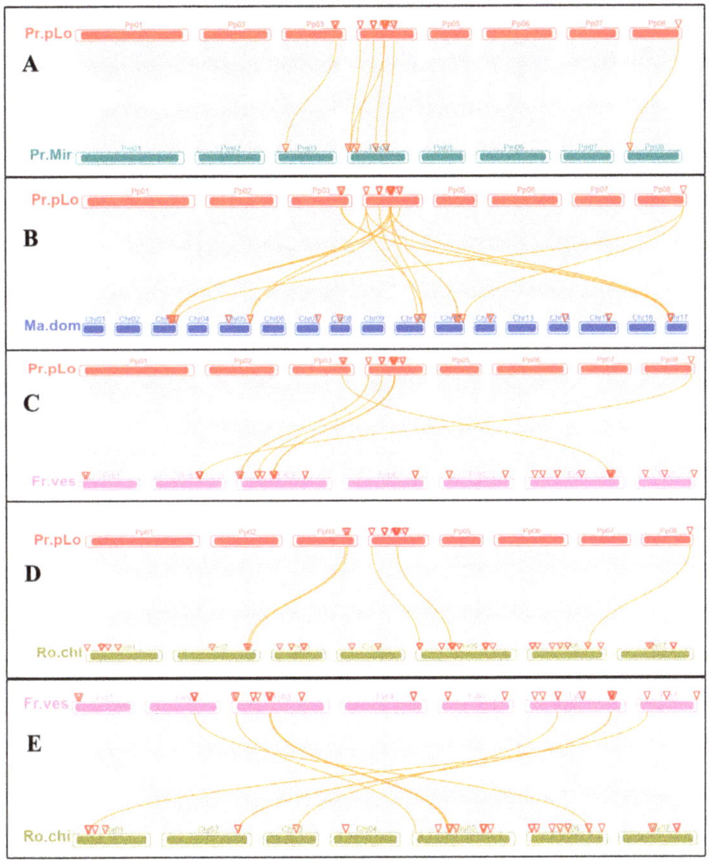

Figure 5. Synteny analysis of TPSs among Rosaceae species. Grey lines in the background indicate the collinear blocks between different genomes, while the yellow lines highlight the syntenic TPS gene pairs. The chromosome is indicated by different colored boxes and labeled by Pr.pLo (*P. persica*), Pr. Mir (*P. mira*), Ma.dom (*M. domestica*), Fr.ves (*F. vesca*), Ro.chi (*R. chinensis*). Collinear relationships between *P. persica-P. mira*, *P. persica-M. domestica*, *P. persica-F. vesca*, *P. persica-R. chinensis*, *F. vesca-R. chinensis* are shown in (**A–E**), respectively.

2.5. Ka/Ks Ratios of TPS Family Members in Rosaceae

In addition, based on the phylogenetic relationship in Figure 3, we assumed TPSs pairs that derived from recent duplication as paralogs. There were two types of paralogs: one is the "between-species paralog" and the other is "within-species paralog" [22]. In this study, we only used the latter type of paralogs. As a result, a total of 82 TPS paralogs from recent duplication were found in the six Rosaceae species (*P. persica*, *P. mume*, *M. domestica*, *P. betulifolia*, *F. vesca*, *R. chinensis*). To explore the selection pressure in the evolution of TPSs, the Ka/Ks values were calculated for the six Rosaceae species (Table S3); a Ka/Ks value of less than one implies purifying selection, Ka/Ks = 1 represents neutral selection, and Ka/Ks > 1 indicates positive selection. The results showed that the Ka/Ks values of TPS paralogs were mostly less than one, suggesting that these genes evolved under purifying selection. There were five gene paralogs with Ka/Ks values greater than one, including three pairs of the TPS-a clade (Ro.chi-TPS18/Ro.chi-TPS20, Ma.dom-TPS14/Ma.dom-TPS13, and Ma.dom-TPS6/Ma.dom-TPS8), one pair from the TPS-b clade (Fr.ves-TPS11/Fr.ves-TPS9), and one pair from the TPS-c clade (Ma.dom-TPS29/Ma.dom-TPS30), which indicates that they were evolved under positive selection. Based on Ks values, the divergence time was calculated, which showed that five gene paralogs were diverged less than 5.5 Mya, especially for Ro.chi-TPS18 and Ro.chi-TPS20, and their divergence time was estimated to be around 0.55 Mya, indicating these gene paralogs were diverged recently. We observed that the divergence time varied among different clades. The divergence time of paralogs from TPS-e were all more than 11.4 Mya, which is higher than other clades, indicating the TPS-e clade diverged relatively anciently.

2.6. Function Diversity of TPSs in Rosaceae

For the six representative plants shown in Figure 3, we predicted their function based on blast searching against Uniprot and KEGG pathway databases; their subcellular localization was also predicted using TargetP [23] and pLoc-mPlant (www.jci-bioinfo.cn/pLoc-mPlant/, accessed on 24 August 2021). Detailed information is listed in Tables 3 and S2. Subcellular localization analysis indicated that TPSs were substantially localized to cytoplasm and chloroplasts, and only a few TPSs were localized to mitochondria. TPSs from the TPS-a clade were mostly located in cytoplasm and chloroplasts, and demonstrated varied function diversity. Most of the characterized TPSs (~95%) in the TPS-a clade are involved in sesquiterpenoid and triterpenoid biosynthesis, and monoterpene biosynthesis pathway, using geranyl diphosphate GPP and farnesyl diphosphate (FPP) as substrate. Whereas for *M. domestica* and *P. betulifolia*, several TPSs from TPS-a clade are involved in diterpenoid biosynthesis that used geranylgeranyl diphosphate (GGPP) as substrate. The finding suggested both the cytosolic mevalonic acid (MVA) pathway and the plastidic methylerythritol phosphate (MEP) pathway coexist in the TPS-a clade. All characterized TPSs in TPS-b clade are involved in either a sesquiterpenoid or monoterpene biosynthesis pathway. The members from the TPS-g clade mainly function in producing acyclic mono- and sesquiterpenoid products. TPS-c and TPS-e clades mainly participate in diterpenoid biosynthesis. Surprisingly, despite the close relationship between TPS-f and TPS-e/c clades, we found that TPS-f members are involved in sesquiterpenoid and triterpenoid biosynthesis, and monoterpene biosynthesis pathway. Furthermore, we observed that TPSs from the same clade participated in different pathways, such as, for TPS-e clade, TPSs were all predicted as ent-kaur-16-ene synthases with exceptions in the *M. domestica* that involved in other pathways. For the Prunoideae subfamily, the function of TPSs among different clades was relatively conserved in three plants, and more TPS copies from the TPS-a clade involved in the monoterpene biosynthesis pathway were detected in *P. persica*. However, the functional diversity of TPSs is more obvious for Maloideae and Rosoideae species, such as the TPS-a clade in *P. betulifolia*. The above findings indicated that the TPS family in Rosaceae species possesses remarkable flexibility to evolve enzymes substrate specificity, and different clades expand in different lineages by gene duplication and divergence. It can be expected that proteins with altered subcellular localization and new substrate specificities would have evolved.

Table 3. Putative functions of TPSs in six Rosaceae species.

TPSs Clade	TPS-a	TPS-b	TPS-g	TPS-c	TPS-e	TPS-f
Catalytic type	**a. monoTPS** 1, (−)-alpha-pinene synthase. C/P 2, S-linalool synthase **b. SesquiTPS** 1, (−)-germacrene D synthase. C/P 2, (3S, 6E)-nerolidol synthase. P 3, (E, E)-alpha-farnesene synthase. P **c. DiTPS** 1, Ent-kaur-16-ene synthase 2, Ent-copalyl diphosphate synthase. P 3, Copal-8-ol diphosphate hydratase. P	**a. monoTPS** 1, (−)-alpha-pinene synthase. C/P 3, Tricyclene synthase EBOS. C/P **b. SesquiTPS** 1, (−)-germacrene D synthase. C/P 2, (3S, 6E)-nerolidol synthase. P 3, (E, E)-alpha-farnesene synthase. P	**a. monoTPS** 1, (−)-alpha-pinene synthase. C/P 2, S-linalool synthase 3, Tricyclene synthase EBOS. C/P **b. SesquiTPS** 2, (3S, 6E)-nerolidol synthase. P **c. DiTPS** 2, Ent-copalyl diphosphate synthase. P	**a. monoTPS** 1, (−)-alpha-pinene synthase. C/P 3, Tricyclene synthase EBOS. C/P **b. SesquiTPS** 1, (−)-germacrene D synthase. C/P 2, (3S, 6E)-nerolidol synthase. P **c. DiTPS** 2, Ent-copalyl diphosphate synthase. P 3, Copal-8-ol diphosphate hydratase. P	**a. monoTPS** 1, (−)-alpha-pinene synthase. C/P **b. SesquiTPS** 2, (3S, 6E)-nerolidol synthase. P 3, (E, E)-alpha-farnesene synthase. P **c. DiTPS** 1, Ent-kaur-16-ene synthase	**a. monoTPS** 2, S-linalool synthase 3, Tricyclene synthase EBOS. C/P **b. SesquiTPS** 1, (−)-germacrene D synthase. C/P **c. DiTPS** 1, Ent-kaur-16-ene synthase
Prunus persica	a1 (19) *; b1 (5);	b3 (5);	b2 (2);	c2 (1);	c1 (3);	a2 (3);
Prunus mume	a1 (6); b1 (4);	b3 (5);	b2 (4);	c2 (1);	c1 (3);	a2 (1);
Prunus mira	b1 (1);	/	b2 (4);	c2 (1);	c1 (3);	a2 (1);
Fragaria vesca	a1 (8); b1 (18);	a3 (6); b3 (1);	b2 (4);	c2 (3), c3 (1);	c1 (2);	/
Rosa chinensis	a1 (10); b1 (16), b2 (5), b3 (5); c1 (1);	a1 (2), a3 (2); b1 (3), b3 (2);	a1 (3), a2 (1); a3 (1);	b1 (2);	c1 (3);	b1 (1);
Pyrus betulifolia	a1 (8), a2 (3); b1 (8), b2 (2), b3 (2); c2 (2), c3 (1);	b1 (3);	a1 (2), a3 (1); b2 (5),	a1(1), a3 (1);	c1 (1);	a3 (1);
Malus x domestica	a1 (4); b1 (4), b2 (2); c1 (5), c2 (2);	a3 (2); b2 (1), b3 (1);	a1 (1); b1 (3), b2 (1); c2 (1);	b1 (1), b2 (1); c2 (1);	a1 (1); b2 (1), b3 (1);	a3 (1); c1 (1);

Notes: *, a1 represents the catalytic type of TPS listed in the second row; number in the bracket "(19)" represents family number of this TPS in the corresponding species.

2.7. Tissue-Specific Expression of TPSs in Rosaceae

For the three representative Rosaceae species (*P. persica*, *M. domestica*, and *F. vesca*), based on RNA-seq data of different tissue development (ripe fruit, immature fruit, leaf), each tissue consisted of two replicates. We examined the tissue expression profiles of TPSs, as listed in Table S4, and observed that TPSs in the three Rosaceae species demonstrated tissue-specific expression (see Figures 6 and S10). There were about 31, 26, and 41 TPSs expressed in at least one tissue for *P. persica*, *M. domestica*, *F. vesca*, respectively; surprisingly, 11 of them were putative TPSs without both domains, such as *Fr.ves-TPS61*, *Fr.ves-TP2*, *Pr.pLo-TPS42*, *Ma.dom-TPS42*, etc. Whereas many TPSs with both domains were not expressed in any of the three tissues, these findings indicated the complexity of TPSs expression. In ripe fruits, expressed TPSs were only detected in strawberries; for *M. domestica* and *P. persica*, no TPSs were expressed in ripe fruits. In addition, we observed that most TPSs were exclusively or highly expressed in one tissue, as shown in Figure 7, such as *Pr.pLo-TPS13*, *Ma.dom-TPS28*, and *Fr.ves-TPS9* genes; they were highly or merely expressed in immature fruit or leaf. We further examined the expression pattern between different TPS clades; for the three plants, different TPSs from TPS-a clade were widely distributed in different tissues, indicating the functional diversity of TPS-a clades. However, the expression from other TPS clades varied among species, such as TPS-b clade; its members in *M. domestica*, *Ma.dom-TPS28*, were only highly expressed in immature fruit, but its counterparts in *P. persica*, *Fr.ves-TPS11 and Fr.ves-TPS13*, displayed an opposite expression pattern despite their high homology. *Fr.ves-TPS11* was highly expressed in leaf, while *Fr.ves-TPS13* was highly expressed in immature fruit, indicating the function differentiation after the recent duplication. Similarly, for TPS-g clade, its members (*Fr.ves-TPS9-12*) in *F. vesca* and members (*Pr.pLo-TPS9, Pr.pLo-TPS9*) in *P. persica* were only highly expressed in leaf, while the counterparts in *M. domestica* (*Ma.dom-TPS1, Ma.dom-TPS15-18*) showed different tissue expression specificity. In addition, we found that the TPS-f clade showed a larger variation between three species. *F. vesca* lost the TPS-f genes, *M. domestica* had two copies that displayed a low expression level, while *P. persica* owned three copies, which is higher than all other species. Two copies were relatively highly expressed in leaf, and one copy was expressed in immature fruit. They were predicted as s-linalool synthases; their product s-linalool is assumed to be an important flavor component in *P. persica* [2]. On the whole, the above finding indicated that TPSs in Rosaceae also underwent extensive expression differentiation after the split of sister lineages and gene duplication.

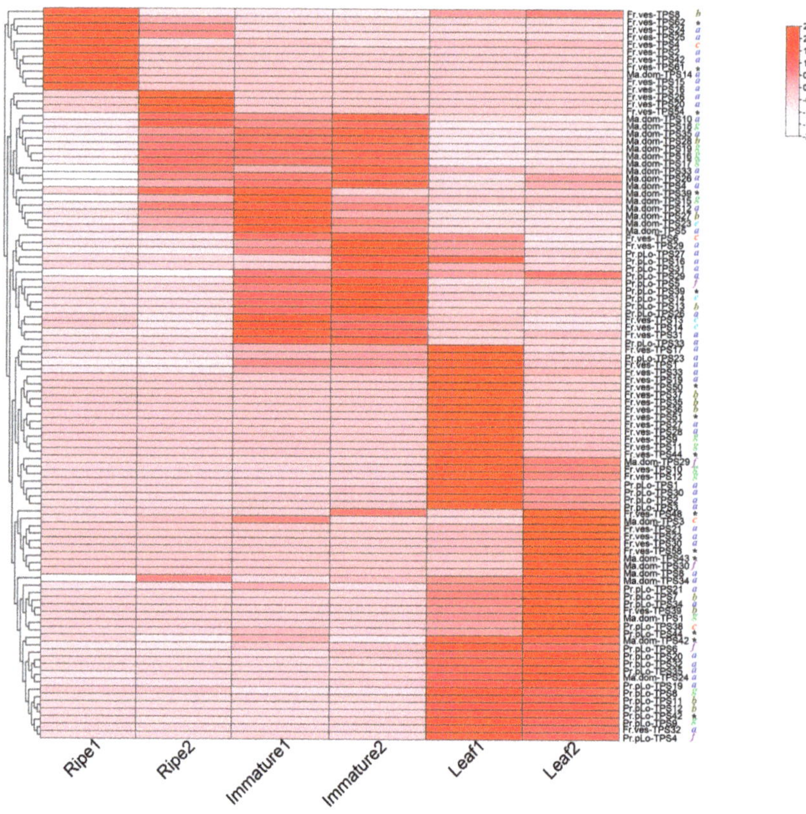

Figure 6. Expression pattern of expressed TPSs in the three Rosaceae species (*P. persica, M. domestica, F. vesca*). Heat mapping of TPSs gene expression in three Rosaceae species. The x-axis represents different samples (ripe fruit, immature fruit, and leaf), the y-axis represents TPSs. There are two replicates for each tissue. The rows and columns were clustered based on row-scale normalized expression values. TPSs clades were shown on the right by different colors; putative TPSs without both domains are indicated by black stars.

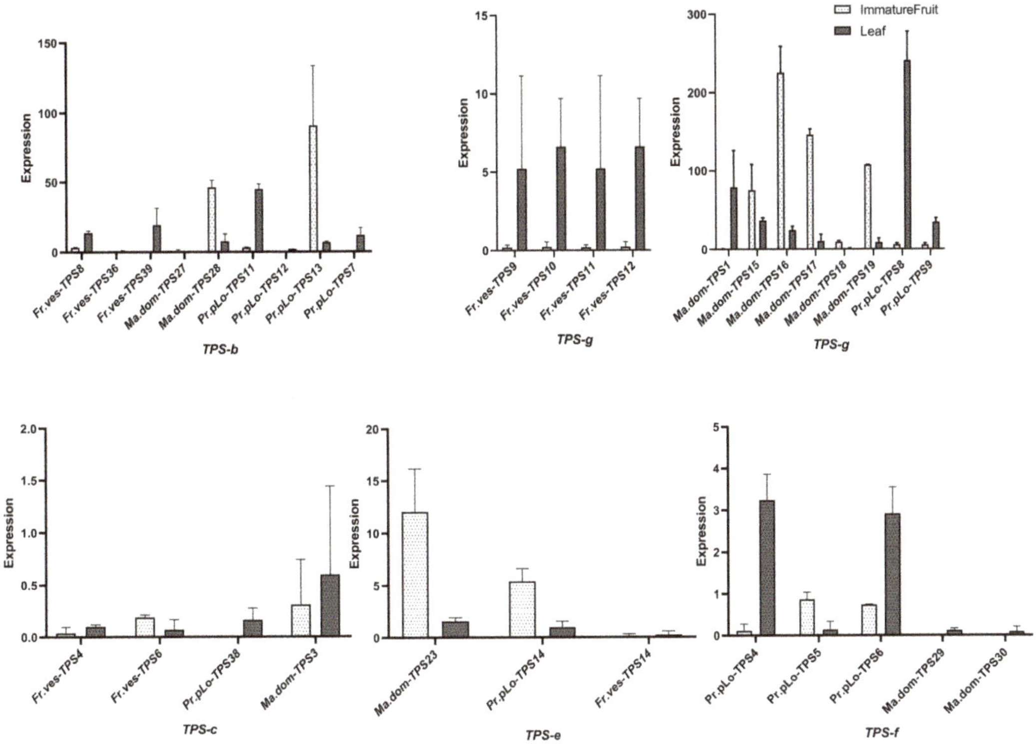

Figure 7. The expression level of TPSs from clades b, c, g, e, and f in the three Rosaceae species (*P. persica, M. domestica, F. vesca*). Tissues (immature fruit, leaf) are indicated by different colors.

3. Discussion

In the plant kingdom, terpenes are traditionally classified as secondary metabolites. Thousands of terpenes have been found, and have proven to play significant roles not only in resistance against stress conditions but also in flavor formation [19,20,24,25]. However, each species is capable of synthesizing only a small fraction, whose synthesis has evolved in plants as a result of selection for increased fitness via better adaptation to the local ecological niche of each species [12]. Terpene synthases are responsible for the synthesis of the various terpene molecules [12]; plant TPS gene families are a medium-sized group, and display varied numbers of TPS families among different species [16]. The Rosaceae family has significant economic value, including the fruit crops and ornamental flowers; however, comprehensive molecular evolutionary and function analysis of TPSs remain elusive. In this study, we screened for the TPS family from eight Rosaceae species. We identified TPSs by detecting both domains and either single domain separately, thus minimizing the chance of missing putative TPSs. We found this family in Rosaceae is a midsized family, as identified in a previous study [13], ranging from 10 TPSs in *P. mira* to 76 in *R. chinensis*. Domain loss for either N-terminal or C-terminal occurred frequently in Rosaceae species.

All the Rosaceae TPSs in this study were divided into seven known clades, TPS a–g. The family numbers of different TPS clades varied among three Rosaceae subfamilies; for example, more than two TPS-c gene copies existed in Maloideae and Rosoideae subfamilies, while only one copy was detected in Prunoideae species. The average number of TPSs in Prunoideae species is lower than that of Maloideae and Rosoideae; no recent WGD except a triplicated arrangement could limit the expansion of TPSs in Prunoideae. Additionally, fewer TPSs in the early Prunoideae species *P. mira*, but more TPSs in modern peach *P. persica*, further revealed the evident evolutionary plasticity of the TPS family. Lineage- and species-specific expansions of the TPSs were widely observed within different TPS clades in Rosaceae. The varied family number and differentiation of TPSs in Rosaceae may play roles in the specialization of essential traits and species differentiation. It has been proposed that lineage-specific genes have a greater chance to contribute to phenotypic variations because their roles are not essential. Further synteny analysis showed that segmental and tandem duplications were both the driving force for the expansion of the TPS gene family in Rosaceae; for *M. domestica* and *P. persica*, segmental and tandem duplication contributed to family expansion; and for *F. vesca*, tandem duplication played the most important role in the family expansion. Ka/Ks calculations further revealed that TPSs genes mainly evolved under purifying selection, except for several pairs; the divergent time indicated TPS-e clade was diverged relatively anciently.

Since closely related enzymes differ in their product profiles, subcellular localization, or substrates, the prediction strategy based on sequence similarity usually cannot accurately describe the specialized function of TPSs, and only roughly obtained their involved pathways. In our study, we still predicted their function based on blast searching against Uniprot and KEGG pathway databases, the results revealed the functional diversity between different TPS clades and species. For the three Prunoideae species, the functional classification of TPSs among different clades was relatively conservative except for the family number variation. Whereas for the other two subfamilies, the putative function of TPSs demonstrated wide diversity, such as TPS-a clade in *P. betulifolia*, up to seven different types of synthases were predicted. In addition, we observed that early diverged TPS-e clade is conserved in function and most of their members were all predicted as ent-kaur-16-ene synthases with exceptions in *M. domestica*. Despite TPS-e and TPS-f being sister clades that clustered together, we found that their function had undergone differentiation, in contrast to TPS-e that mainly participated in diterpenoid biosynthesis, most TPS-f members are involved in the monoterpene biosynthesis pathway. The expansion of TPS-f in *P. persica* and their product, S-linalool synthase, is the essential aromatic substance in peach fruits [21]. Our findings indicated that the TPS family in Rosaceae species possesses remarkable function diversity; different clades expand in different lineages by gene duplication and divergence. The generation of altered subcellular localization, and new substrate specificities of TPSs, is a dynamic process that specialized the trait differentiation. However, experimental data like metabonomics, enzyme assays, are also needed to verify these observations.

The expression profiling results on the *TPS* gene family in three Rosaceae species (*P. persica*, *M. domestica*, and *F. vesca*) showed that most identified TPSs were expressed in at least one tissue. Most of the TPSs were specifically expressed in one certain tissue; the expressed TPSs in ripe fruits are rare. For each TPS clade, the expression pattern also varied among species, such as TPS-f genes, which demonstrated a high expression in *P. persica*, but a lower expression in *M. domestica*. We also found that many paralogs exhibited divergent expression patterns either in tissues or expression abundance, suggesting that expression divergence might significantly contribute to gene survival and function differentiation after gene expansion. It was worth noting that among the expressed TPSs, certain putative TPSs did not have both domains. In contrast, many TPSs with both domains were not expressed, and the finding suggested that in the complexity of TPSs expression, not all of the complete TPSs were functional, and some functional ones may lose activity in either one domain. These putative TPSs without both domains were assumed to be triggered by partial duplication and assumed to be pseudogenes for the loss of original function. A previous study found that a total of 12% of the pseudogenes still contained detectable open reading frames and were effectively expressed. The generated transcripts may contribute to the synthesis of small interfering RNA species that regulate parent transcripts [26]. Fast-evolving families involved in ubiquitination and secondary metabolism families always contain the highest number of pseudogenes [26,27]. Hence, the functionality of the expressed "fragmental" TPSs in our study still needs further investigation.

4. Material and Methods

4.1. Identification of TPSs in Rosaceae

A total of eight Rosaceae species are included in the identification of TPSs, namely three Prunoideae, three Maloideae, and two Rosoideae species. Their genomes have been completely sequenced and annotated. The genome files of the Rosaceae species were mostly downloaded from the NCBI (https://www.ncbi.nlm.nih.gov, accessed on 2 August 2021) and GDR (https://www.Rosaceae.org, accessed on 2 August 2021), the versions of genomes were all the recently released and chromosome-scale, detailed genome information was summarized in Table 1. Six representative sequences of TPS-a, TPS-b, TPS-c, TPS-e, TPS-f, and TPS-g from *Vitis vinifera* from a previous study [28], and one representative sequence of TPS-d from *Abies grandis*, were used as queries to search the corresponding subject protein sequences of each Rosaceae species. Two different methods were used to identify TPSs in Rosaceae species. First, we implemented BLASTP searches of the complete genome with an E-value cut-off of 0.00001 to reduce false positives, and the second method was Hidden Markove Model (HMM) profiles of TPS domains in these Rosaceae species by using HMMER software with an E-value cut off of 0.001 [29]. The redundant sequences were removed by manual inspection. Subsequently, we verified all sequences by checking the existence of Pfam domains PF03936 (metal-binding domain) and PF01397 (N-terminal TPS domain) using PfamScan tools with default parameters [30], Pfam-A was used as the searching database. PfamScan tools search the whole sequences against the Pfam database, and annotate the sequence blocks as known domains, only the significant domains are retained, it can simultaneously predicted the different domains in one protein. Ultimately, genes containing at least one TPS domain were confirmed as members of the TPS gene family and named in numerical order.

4.2. Motif Annotation, Subcellular Localization, and Physical Localization

The conserved motifs were predicted using the online MEME software with the following settings: the motif discovery mode was classic, site distribution was zero or one occurrence per sequence (zoops), the background was a 0-order background model, the maximum number of different motifs was 20, minimum motif width was 6, and maximum motif width was 50 [31]. A shuffling was also performed prior to the MEME/MAST analysis to validate the identified motifs. The conserved domain was annotated based on the conserved domain database (CDD v3.19) in NCBI. TargetP and pLoc-mPlant (www.jci-bioinfo.cn/pLoc-mPlant/, accessed on 24 August 2021) were used to predict the subcellular localization of TPS proteins [32]. For each species, we got the information of the TPSs on the corresponding chromosome according to the annotation documents and drew a sketch map of the gene's physical location using TBTools [33]. Protein functions were also predicted based on blast searching against Uniprot and KEGG pathway databases under default parameters.

4.3. Phylogenetic Tree Construction and TPS Classification

The full-length sequences of TPSs protein sequence with both domains from Rosaceae species were used to perform sequence alignment and phylogenetic tree construction using MEGA7 program [34]. The ClustalW method was used for sequence alignment under default parameters, the Maximum Likelihood and Neighbor-joining methods based on the Jones-Taylor-Thornton (JTT) matrix-based model were used for phylogenetic tree construction, a Bootstrap method was used for the phylogeny test with 500 replications, rates among sites was Gamma distributed (G), and Partial deletion was used for gaps/missing data treatment, with site coverage cutoff at 90%. The produced tree was further embellished by the FigTree program (http://tree.bio.ed.ac.uk/, accessed on 2 September 2021). TPS members were classified by their clustering relation with the query sequence mentioned above. For each TPS clade, their TPS member's sequences were further aligned by ClustalW method, and a sequence logo for the visualization of the highly conserved aspartate-rich motif in C-terminal domain was produced by TBTools [33].

4.4. Ka/Ks Analysis of TPS Family Members in Rosaceae

We chose six representative Rosaceae species for selection pressure analysis, including *P. persica*, *P. mume*, *M. domestica*, *P. betulifolia*, *F. vesca*, and *R. chinensis*. For each species, its TPSs paralogs were determined based on their phylogenetic relationship. If they belong to the same clade and derive from a single gene that was duplicated recently, we assumed they were paralogs. Sometimes those paralogs that arose from a duplication after the speciation event are called "within-species" paralogs [22]. The CDS and protein sequences of each TPS pair were used to compute the Ka (non-synonymous rates) and Ks (synonymous rates) by Ka/Ks calculator (http://services.cbu.uib.no/tools/kaks, accessed on 2 Octomber 2021). The date (T) of the duplication events was estimated by the formula $T = \text{Ks}/2\lambda$, where λ represents the estimated clock-like rate of synonymous substitution; in dicots, it was 1.5×10^{-8} substitutions/synonymous site/year [35].

4.5. Synteny Analysis and Detection of Tandemly/Segmentally Duplicated TPSs

To identify the synteny of TPS family genes among species, we performed all-to-all BLASTP between the genome of *P. persica* and other four species (*P. mira*, *M. domestica*, *F. vesca*, *R. chinensis*). The collinearity analysis was also performed between *F. vesca* and *R. chinensis*. For the three representative species (*F. vesca*, *M. domestica*, *P. persica*), we also performed self-blast by comparing protein-coding genes against their genome using BLASTP with an E-value cut-off of 0.00001. All BLASTP hits were used as input for software MCScanX (Multiple Collinearity Scan toolkit) [36] to identify possible collinear blocks within and between genomes of different species. Based on the self-blast results, we detected the tandemly/segmentally duplicated TPSs for each species. In addition to the tandem duplication that was determined by MCScanX, paralogues that were either adjacent or separated by ≤ 5 genes along a chromosome were also assigned as tandem duplicates. If paralogues were within known genomic duplication blocks, they were considered to be duplicated through segmental duplication. All intra/inter-genomic synteny relationships were visualized with TBtools [33].

4.6. Expression Analysis of TPSs in Rosaceae

For the three representative plants (*F. vesca*, *M. domestica*, *P. persica*), RNA-seq data of different tissues and development stages (ripe fruit, immature fruit, leaf) was retrieved from the fruitENCODE project [37]; each tissue consisted of two replicates. The fruitENCODE project aims to generate a comprehensive annotation of functional elements in seven climacteric fruit species (apple, banana, melon, papaya, peach, pear, and tomato) with sequenced reference genomes. The data were deposited in the SRA database of NCBI with accession number PRJNA381300. Transcriptome analysis was implemented by the protocol in a previous study [38]. The clean reads of RNA-seq data from each sample were mapped against the genome reference with HISAT2 [39]; each SAM file was converted into a BAM file, and sorted. Duplicates were removed with SAMtools [40,41]. Further transcript assembly and quantification of the read alignments were performed using Stringtie [42]. Gene expression levels were measured by FPKM (fragments per kilobase of transcript per million mapped reads) and normalized with the row-scale method. Heatmaps with all samples were plotted using the "HeatMap" function in TBtools.

5. Conclusions

In conclusion, this study investigated the TPS gene family in eight sequenced Rosaceae species through classification, chromosomal location, orthologous relationships, and duplication analysis. The distribution of the TPS gene family among Rosaceae species revealed a diversity of family number and function; lineage-specific expansion of the TPSs accompanied by frequent domain loss were widely observed within different TPS clades. We further provided their tissue-specific expression pattern in *F. vesca*, *M. domestica*, and *P. persica*, revealing the expression differentiation of TPSs between paralogs/species. The findings revealed the evolution of TPSs in Rosaceae and will be highly useful for further genetic improvement of Rosaceae species.

Supplementary Materials: The following supporting information can be downloaded at https://www.mdpi.com/article/10.3390/plants11060736/s1. Table S1: Pfam domain distribution in the TPSs of eight Rosaceae species (sheet a); motif distribution in the TPSs of three Rosaceae species (sheet b); the alternative splicing transcripts of putative TPSs in *F. vesca*, *M. domestica*, and *P. persica* (sheet c). Table S2: Information of all TPSs analyzed in this study, (XLSX); Table S3: Ka/Ks ratio of TPSs in Rosaceae; Table S4: Expression values of TPSs in *F. vesca*, *M. domestica*, and *P. persica*. Figure S1. A Neighbor-joining phylogenetic tree of TPSs of the TPS proteins in three Rosaceae species (*P. persica, M. domestica, F. vesca*). Figure S2. A maximum-likelihood phylogenetic tree of six representative Rosaceae species TPS gene family. The branches of TPS-a, TPS-g, TPS-b, TPS-c, TPS-e and TPS-f clades are colored in blue, yellow-green, green, red, purple and cyan, respectively. Figure S3. Chromosomal locations of TPS family members in peach 'Lovell' genome. The gene location visualize package of TBTools was used to exhibit chromosomal locations of the TPS genes. The number to the left of each chromosome represented the size of the chromosome in bp. Figure S4. Chromosomal locations of TPS family members in *Prunus mume* genome. The gene location visualize package of TBTools was used to exhibit chromosomal locations of the TPS genes. The number to the left of each chromosome represented the size of the chromosome in bp. Figure S5. Chromosomal locations of TPS family members in *Prunus mira* genome. The gene location visualize package of TBTools was used to exhibit chromosomal locations of the TPS genes. The number to the left of each chromosome represented the size of the chromosome in bp. Figure S6. Chromosomal locations of TPS family members in *Malus x domestica* genome. The gene location visualize package of TBTools was used to exhibit chromosomal locations of the TPS genes. The number to the left of each chromosome represented the size of the chromosome in bp. Figure S7. Chromosomal locations of TPS family members in *Pyrus betulifolia* genome. The gene location visualize package of TBTools was used to exhibit chromosomal locations of the TPS genes. The number to the left of each chromosome represented the size of the chromosome in bp. Figure S8. Chromosomal locations of TPS family members in *Fragaria vesca* genome. The gene location visualize package of TBTools was used to exhibit chromosomal locations of the TPS genes. The number to the left of each chromosome represented the size of the chromosome in bp. Figure S9. Chromosomal locations of TPS family members in *Rosa chinensis* genome. The gene location visualize package of TBTools was used to exhibit chromosomal locations of the TPS genes. The number to the left of each chromosome represented the size of the chromosome in bp. Figure S10. Expression pattern of TPS genes in three Rosaceae plants (*P. persica, M. domestica, F. vesca*). The x-axis represents different samples (ripe fruit, immature fruit and leaf), the y-axis represents TPS genes. There are two replicates for each tissue. The phylogenetic tree is shown on the left panel, the root nodes of TPS-a, TPS-g, TPS-b, TPS-c and TPS-e, TPS-f clades are indicated by blue, green, yellowgreen, red, benzo, and purple, respectively.

Author Contributions: Conceptualization, X.Z. and A.Z.; methodology, X.J. and A.Z.; software, A.Z.; validation, Y.X., A.Z. and J.F.; formal analysis, T.W., K.L.; investigation, A.Z.; resources, Y.X.; data curation, A.Z.; writing—original draft preparation, A.Z.; writing—review and editing, A.Z., X.Z. and H.P.; visualization, X.Z.; supervision, X.Z.; project administration, X.Z.; funding acquisition, X.Z. All authors have read and agreed to the published version of the manuscript.

Funding: This research was funded by the National Natural Science Foundation of China, grant number 32070682, the National Science & Technology Innovation Zone Project, grant numbers 1716315XJ00200303 and 1816315XJ00100216.

Institutional Review Board Statement: Not applicable.

Informed Consent Statement: Not applicable.

Data Availability Statement: All raw reads used in this work were deposited in NCBI Bio-Project with the accession number PRJNA381300.

Acknowledgments: We would like to thank the members of the Bioinformatics Group of Wuhan Botanical Garden, Chinese Academy of Sciences, China for the discussion and suggestions to improve the manuscript.

Conflicts of Interest: The authors declare no conflict of interest.

Abbreviations

HMM	Hidden Markove Model
TPSs	terpene synthase genes
GPP	geranyl diphosphate
FPP	farnesyl diphosphate
GGPP	geranylgeranyl diphosphate
VOCs	volatile organic compounds

References

1. Wang, Y.J.; Chen, F.; Fang, J.B.; Yang, C.X.; Zhao, J.B.; Jiang, Q.; Li, S.H. Effects of Germplasm Origin and Fruit Character on Volatile Composition of Peaches and Nectarines. *Flavor Health Benefits Small Fruits* **2010**, *1035*, 95–117.
2. Wei, C.Y.; Liu, H.R.; Cao, X.M.; Zhang, M.L.; Li, X.; Chen, K.S.; Zhang, B. Synthesis of flavour-related linalool is regulated by PpbHLH1 and associated with changes in DNA methylation during peach fruit ripening. *Plant Biotechnol. J.* **2021**, *19*, 2082–2096. [CrossRef] [PubMed]
3. Nawade, B.; Yahyaa, M.; Reuveny, H.; Shaltiel-Harpaz, L.; Eisenbach, O.; Faigenboim, A.; Bar-Yaakov, I.; Holland, D.; Ibdah, M. Profiling of volatile terpenes from almond (*Prunus dulcis*) young fruits and characterization of seven terpene synthase genes. *Plant Sci.* **2019**, *287*, 110187. [CrossRef] [PubMed]
4. Nieuwenhuizen, N.J.; Green, S.A.; Chen, X.; Bailleul, E.J.; Matich, A.J.; Wang, M.Y.; Atkinson, R.G. Functional genomics reveals that a compact terpene synthase gene family can account for terpene volatile production in apple. *Plant Physiol.* **2013**, *161*, 787–804. [CrossRef]
5. Liu, X.J.; Hao, N.N.; Feng, R.F.; Meng, Z.P.; Li, Y.N.; Zhao, Z.Y. Transcriptome and metabolite profiling analyses provide insight into volatile compounds of the apple cultivar 'Ruixue' and its parents during fruit development. *Bmc Plant Biol.* **2021**, *21*, 1–16. [CrossRef]
6. Klee, H.J.; Tieman, D.M. The genetics of fruit flavour preferences. *Nat. Rev. Genet.* **2018**, *19*, 347–356. [CrossRef]
7. Espino-Diaz, M.; Sepulveda, D.R.; Gonzalez-Aguilar, G.; Olivas, G.I. Biochemistry of Apple Aroma: A Review. *Food Technol. Biotechnol.* **2016**, *54*, 375–397. [CrossRef]
8. Zhang, X.; Niu, M.; Teixeira da Silva, J.A.; Zhang, Y.; Yuan, Y.; Jia, Y.; Xiao, Y.; Li, Y.; Fang, L.; Zeng, S.; et al. Identification and functional characterization of three new terpene synthase genes involved in chemical defense and abiotic stresses in Santalum album. *BMC Plant Biol.* **2019**, *19*, 115. [CrossRef]
9. Block, A.K.; Vaughan, M.M.; Schmelz, E.A.; Christensen, S.A. Biosynthesis and function of terpenoid defense compounds in maize (*Zea mays*). *Planta* **2019**, *249*, 21–30. [CrossRef]
10. Huang, X.Z.; Xiao, Y.T.; Kollner, T.G.; Jing, W.X.; Kou, J.F.; Chen, J.Y.; Liu, D.F.; Gu, S.H.; Wu, J.X.; Zhang, Y.J.; et al. The terpene synthase gene family in Gossypium hirsutum harbors a linalool synthase GhTPS12 implicated in direct defence responses against herbivores. *Plant Cell Environ.* **2018**, *41*, 261–274. [CrossRef]
11. Zhou, F.; Pichersky, E. More is better: The diversity of terpene metabolism in plants. *Curr. Opin. Plant Biol.* **2020**, *55*, 1–10. [CrossRef] [PubMed]
12. Kulheim, C.; Padovan, A.; Hefer, C.; Krause, S.T.; Kollner, T.G.; Myburg, A.A.; Degenhardt, J.; Foley, W.J. The Eucalyptus terpene synthase gene family. *BMC Genom.* **2015**, *16*, 450. [CrossRef] [PubMed]
13. Chen, F.; Tholl, D.; Bohlmann, J.; Pichersky, E. The family of terpene synthases in plants: A mid-size family of genes for specialized metabolism that is highly diversified throughout the kingdom. *Plant J.* **2011**, *66*, 212–229. [CrossRef] [PubMed]
14. Gao, Y.; Honzatko, R.B.; Peters, R.J. Terpenoid synthase structures: A so far incomplete view of complex catalysis. *Nat. Prod. Rep.* **2012**, *29*, 1153–1175. [CrossRef] [PubMed]
15. Hillwig, M.L.; Xu, M.; Toyomasu, T.; Tiernan, M.S.; Wei, G.; Cui, G.; Huang, L.; Peters, R.J. Domain loss has independently occurred multiple times in plant terpene synthase evolution. *Plant J.* **2011**, *68*, 1051–1060. [CrossRef] [PubMed]
16. Jiang, S.Y.; Jin, J.; Sarojam, R.; Ramachandran, S. A Comprehensive Survey on the Terpene Synthase Gene Family Provides New Insight into Its Evolutionary Patterns. *Genome Biol. Evol.* **2019**, *11*, 2078–2098. [CrossRef]
17. Karunanithi, P.S.; Zerbe, P. Terpene Synthases as Metabolic Gatekeepers in the Evolution of Plant Terpenoid Chemical Diversity. *Front. Plant Sci.* **2019**, *10*, 1166. [CrossRef]
18. Kumar, Y.; Khan, F.; Rastogi, S.; Shasany, A.K. Genome-wide detection of terpene synthase genes in holy basil (*Ocimum sanctum* L.). *PLoS ONE* **2018**, *13*, e0207097. [CrossRef]
19. Muchlinski, A.; Ibdah, M.; Ellison, S.; Yahyaa, M.; Nawade, B.; Laliberte, S.; Senalik, D.; Simon, P.; Whitehead, S.R.; Tholl, D. Diversity and function of terpene synthases in the production of carrot aroma and flavor compounds. *Sci. Rep.* **2020**, *10*, 9989. [CrossRef]
20. Zhou, H.C.; Shamala, L.F.; Yi, X.K.; Yan, Z.; Wei, S. Analysis of Terpene Synthase Family Genes in Camellia sinensis with an Emphasis on Abiotic Stress Conditions. *Sci. Rep.* **2020**, *10*, 933. [CrossRef]
21. Wang, Y.J.; Yang, C.X.; Li, S.H.; Yang, L.; Wang, Y.N.; Zhao, J.B.; Jiang, Q. Volatile characteristics of 50 peaches and nectarines evaluated by HP-SPME with GC-MS. *Food Chem.* **2009**, *116*, 356–364. [CrossRef]

22. Sonnhammer, E.L.; Koonin, E.V. Orthology, paralogy and proposed classification for paralog subtypes. *Trends Genet.* **2002**, *18*, 619–620. [CrossRef]
23. Boos, F.; Muhlhaus, T.; Herrmann, J.M. Detection of Internal Matrix Targeting Signal-like Sequences (iMTS-Ls) in Mitochondrial Precursor Proteins Using the TargetP Prediction Tool. *Bio-Protocol* **2018**, *8*, e2474. [CrossRef]
24. Lee, G.W.; Lee, S.; Chung, M.S.; Jeong, Y.S.; Chung, B.Y. Rice terpene synthase 20 (OsTPS20) plays an important role in producing terpene volatiles in response to abiotic stresses. *Protoplasma* **2015**, *252*, 997–1007. [CrossRef] [PubMed]
25. Yoshitomi, K.; Taniguchi, S.; Tanaka, K.; Uji, Y.; Akimitsu, K.; Gomi, K. Rice terpene synthase 24 (OsTPS24) encodes a jasmonate-responsive monoterpene synthase that produces an antibacterial gamma-terpinene against rice pathogen. *J. Plant Physiol.* **2016**, *191*, 120–126. [CrossRef] [PubMed]
26. Xie, J.; Li, Y.; Liu, X.; Zhao, Y.; Li, B.; Ingvarsson, P.K.; Zhang, D. Evolutionary Origins of Pseudogenes and Their Association with Regulatory Sequences in Plants. *Plant Cell* **2019**, *31*, 563–578. [CrossRef]
27. Mascagni, F.; Usai, G.; Cavallini, A.; Porceddu, A. Structural characterization and duplication modes of pseudogenes in plants. *Sci. Rep.* **2021**, *11*, 5292. [CrossRef]
28. Zhang, A.; Zhou, H.; Jiang, X.; Han, Y.; Zhang, X. The Draft Genome of a Flat Peach (Prunus persica L. cv. '124 Pan') Provides Insights into Its Good Fruit Flavor Traits. *Plants* **2021**, *10*, 538. [CrossRef]
29. Potter, S.C.; Luciani, A.; Eddy, S.R.; Park, Y.; Lopez, R.; Finn, R.D. HMMER web server: 2018 update. *Nucleic Acids Res.* **2018**, *46*, W200–W204. [CrossRef]
30. Li, W.Z.; Cowley, A.; Uludag, M.; Gur, T.; McWilliam, H.; Squizzato, S.; Park, Y.M.; Buso, N.; Lopez, R. The EMBL-EBI bioinformatics web and programmatic tools framework. *Nucleic Acids Res.* **2015**, *43*, W580–W584. [CrossRef]
31. Bailey, T.L.; Johnson, J.; Grant, C.E.; Noble, W.S. The MEME Suite. *Nucleic Acids Res.* **2015**, *43*, W39–W49. [CrossRef]
32. Emanuelsson, O.; Brunak, S.; von Heijne, G.; Nielsen, H. Locating proteins in the cell using TargetP, SignalP and related tools. *Nat. Protoc.* **2007**, *2*, 953–971. [CrossRef] [PubMed]
33. Chen, C.; Chen, H.; Zhang, Y.; Thomas, H.R.; Frank, M.H.; He, Y.; Xia, R. TBtools: An Integrative Toolkit Developed for Interactive Analyses of Big Biological Data. *Mol. Plant* **2020**, *13*, 1194–1202. [CrossRef] [PubMed]
34. Kumar, S.; Stecher, G.; Tamura, K. MEGA7: Molecular Evolutionary Genetics Analysis Version 7.0 for Bigger Datasets. *Mol. Biol. Evol.* **2016**, *33*, 1870–1874. [CrossRef] [PubMed]
35. Blanc, G.; Wolfe, K.H. Functional divergence of duplicated genes formed by polyploidy during Arabidopsis evolution. *Plant Cell* **2004**, *16*, 1679–1691. [CrossRef] [PubMed]
36. Wang, Y.; Li, J.; Paterson, A.H. MCScanX-transposed: Detecting transposed gene duplications based on multiple colinearity scans. *Bioinformatics* **2013**, *29*, 1458–1460. [CrossRef]
37. Lu, P.; Yu, S.; Zhu, N.; Chen, Y.R.; Zhou, B.; Pan, Y.; Tzeng, D.; Fabi, J.P.; Argyris, J.; Garcia-Mas, J.; et al. Genome encode analyses reveal the basis of convergent evolution of fleshy fruit ripening. *Nat. Plants* **2018**, *4*, 784–791. [CrossRef]
38. Pertea, M.; Kim, D.; Pertea, G.M.; Leek, J.T.; Salzberg, S.L. Transcript-level expression analysis of RNA-seq experiments with HISAT, StringTie and Ballgown. *Nat. Protoc.* **2016**, *11*, 1650–1667. [CrossRef]
39. Kim, D.; Langmead, B.; Salzberg, S.L. HISAT: A fast spliced aligner with low memory requirements. *Nat. Methods* **2015**, *12*, 357–360. [CrossRef]
40. Li, H.; Handsaker, B.; Wysoker, A.; Fennell, T.; Ruan, J.; Homer, N.; Marth, G.; Abecasis, G.; Durbin, R.; Genome Project Data Processing, S. The Sequence Alignment/Map format and SAMtools. *Bioinformatics* **2009**, *25*, 2078–2079. [CrossRef]
41. Danecek, P.; Bonfield, J.K.; Liddle, J.; Marshall, J.; Ohan, V.; Pollard, M.O.; Whitwham, A.; Keane, T.; McCarthy, S.A.; Davies, R.M.; et al. Twelve years of SAMtools and BCFtools. *Gigascience* **2021**, *10*, giab008. [CrossRef] [PubMed]
42. Pertea, M.; Pertea, G.M.; Antonescu, C.M.; Chang, T.C.; Mendell, J.T.; Salzberg, S.L. StringTie enables improved reconstruction of a transcriptome from RNA-seq reads. *Nat. Biotechnol.* **2015**, *33*, 290–295. [CrossRef] [PubMed]

plants MDPI

Article

Basic Characteristics of Flower Transcriptome Data and Derived Novel EST-SSR Markers of *Luculia yunnanensis*, an Endangered Species Endemic to Yunnan, Southwestern China

Yao Zhang [1,2], Xi Liu [3], Yuying Li [4], Xiongfang Liu [1,2], Hong Ma [1,5,*], Suping Qu [6,*] and Zhenghong Li [1,*]

1 Institute of Highland Forest Science, Chinese Academy of Forestry, Kunming 650224, China; zhangyao949@163.com (Y.Z.); liuxiongfang@caf.ac.cn (X.L.)
2 College of Forestry, Nanjing Forestry University, Nanjing 210037, China
3 School of Geography and Ecotourism, Southwest Forestry University, Kunming 650224, China; lx13320441881@163.com
4 College of Grassland Science, Shanxi Agricultural University, Jinzhong 030801, China; liyuy_ing@163.com
5 Key Laboratory of Breeding and Utilization of Resource Insects, National Forestry and Grassland Administration, Kunming 650224, China
6 Flower Research Institute, Yunnan Academy of Agricultural Sciences, Kunming 650205, China
* Correspondence: hortscience@163.com (H.M.); qsp@yaas.org.cn (S.Q.); lzh4949@163.com (Z.L.)

Abstract: *Luculia yunnanensis* (Rubiaceae), an evergreen shrub or small tree, is endemic to China and confined to Nujiang Prefecture, Yunnan Province. This plant is of high ornamental value owing to its attractive pink flowers, sweet fragrance, and long flowering period. Due to the influence of climate change and human factors, the distribution range of *L. yunnanensis* has exhibited a significant shrinking trend, and it has become a vulnerable species that is in urgent need of conservation and rational utilization research. In this study, the flower transcriptome sequencing of *L. yunnanensis* was conducted using an Illumina HiSeq platform. We designed and developed a series of EST-SSR primers based on the flower transcriptome data of *L. yunnanensis*. The results showed that 98,389 unigenes were obtained from the *L. yunnanensis* flower transcriptome, all of which were aligned with sequences in public databases. Nr, Nt, Pfam, KOG/COG, Swiss-Prot, KEGG, and GO annotated 31,859, 13,853, 22,684, 10,947, 21,416, 9722, and 23,390 unigenes, respectively. The MISA (Microsatellite) tool was used to identify SSR loci from all unigenes, and a total of 15,384 SSRs were identified. Repeat motifs were given priority with mononucleotides, dinucleotides, and trinucleotides. The 81 primer pairs were synthesized randomly, of which 44 pairs showed effective amplification. A total of 17 primers showed stable amplification, and rich polymorphism was observed in 6 populations. We concluded via genetic diversity analysis that the average effective number of alleles (Ne), Shannon's information index (I), and polymorphism information content (PIC) were 1.925, 0.837, and 0.403, respectively. In conclusion, 17 EST-SSR primers can be used for subsequent population genetic diversity analysis and molecular-marker-assisted breeding, which is of great significance for formulating resource conservation and utilization strategies for *L. yunnanensis*.

Keywords: *Luculia yunnanensis*; transcriptome; EST-SSR markers; polymorphism; conservation genetics

Citation: Zhang, Y.; Liu, X.; Li, Y.; Liu, X.; Ma, H.; Qu, S.; Li, Z. Basic Characteristics of Flower Transcriptome Data and Derived Novel EST-SSR Markers of *Luculia yunnanensis*, an Endangered Species Endemic to Yunnan, Southwestern China. *Plants* **2022**, *11*, 1204. https://doi.org/10.3390/plants11091204

Academic Editor: Hye Ryun Woo

Received: 25 March 2022
Accepted: 28 April 2022
Published: 29 April 2022

1. Introduction

As an endangered species endemic to Yunnan Province, Southwestern China, *Luculia yunnanensis* is an evergreen shrub or small tree that belongs to the genus *Luculia*, of the family Rubiaceae. This species mainly grows on limestone mountains, secondary shrubby woodland, and open slopes at altitudes of between 1200 and 3200 m. It possesses high ornamental value due to its attractive plant shape, long flowering period, striking pink flowers, sweet fragrance, and the fact that it is evergreen. There is thus great value in its development and utilization.

Our previous field investigations of *L. yunnanensis* and its recorded distribution showed an obvious shrinking trend in its distribution range, which currently only comprises Lushui, Fugong, and Gongshan counties in Nujiang Prefecture, Yunnan Province. The distribution range has been reduced from seven counties in five prefectures or cities to three counties in one prefecture, and the administrative area of the distribution area is 10,465.48 km^2 (only 20% of its historical range). According to the IUCN threat level criteria, *L. yunnanensis* belongs to the vulnerable species category [1]; thus, there is an urgent need for relevant biological conservation research.

The assessment of genetic diversity is one of the most significant parts of the research on the conservation of rare and endangered plants [2]. Considered to be an important component that could be seen as a determination of the viability of a species, it may affect species' adaptive capacity and evolutionary potential, and is often regarded as an indirect warning of the potential extinction of endangered species [3,4]. A decrease in genetic diversity in a species may lead to the degradation of its adaptability to environmental changes, leading to a decline in its evolutionary ability [5,6]. Thus, only by clarifying the genetic diversity of threatened species and taking various corresponding factors into consideration can we propose feasible strategies regarding their protection and utilization.

DNA molecular markers are usually used to estimate the genetic diversity of plants, and a sufficient number of polymorphic primers is the prerequisite for molecular genetic diversity analysis. As one of the most commonly used molecular marker techniques, EST-SSR has the advantages of simple operation, good stability, high accuracy, rich variability, good repeatability, and interspecific transmissibility [7–10]. Compared with genomic SSR markers, EST-SSR markers have low development cost and can be applied to species without reference genomes, and the application of EST-SSR markers has greatly promoted the use of molecular marker techniques in plant research [10–12], as well as being widely used in fields such as plant genetics and breeding, conservation, and the development of germplasm [13–16]. To date, the development of EST-SSR primers for *L. yunnanensis* has not been reported. Zhou [17] made use of a modified biotin–streptavidin capture method to develop 13 pairs of SSR primers suitable for analyzing the genetic diversity of *Luculia*. These 13 pairs of primers can effectively distinguish between *Luculia pinceana* and *L. yunnanensis*. With the same method, Ma [18] developed 24 pairs of SSR primers, and 11 pairs of them were polymorphic in two populations of *L. yunnanensis*. However, this number of markers is far from meeting the needs of follow-up studies on *L. yunnanensis*. Therefore, in this study we designed a series of EST-SSR primers based on the flower transcriptome data of *L. yunnanensis*, and developed a set of polymorphic EST-SSR primers that can be used for subsequent population genetic diversity analysis and molecular-marker-assisted breeding, which are of great significance for formulating resource conservation and utilization strategies for *L. yunnanensis*.

2. Results and Discussion

2.1. Functional Annotation and Classification

The transcriptome sequencing was conducted using an Illumina HiSeq platform (Beijing Novogene Biological Information Technology Co., Ltd., Beijing, China); clean reads were used for de novo assembly using Trinity with the default parameters to obtain 140,042 transcripts, and TGICL version 2.1 software was used to cluster the transcripts into 98,389 unigenes. There were 18,521 unigenes (18.82% of the total unigenes) with a length of more than 1 Kb and high assembly integrity (Table 1). To learn more about the features and functions of these unigenes, we aligned them with the sequences of public databases, including Nr, Nt, Pfam, KOG/COG, Swiss-Prot, KEGG, and GO (see Table 2). It can be seen from the results that at least some of the unigenes were not annotated in all seven databases, which demonstrates that there are unknown unigene sequences. This is consistent with the fact that the annotation of the unigenes of *Glycyrrhiza uralensis* [19], *Luculia gratissima* [20], and *Prunella vulgaris* [21] did not reach 100%. This result may be caused by the length of some unigenes being so short that the annotation information is incomplete, or by the specific unigenes of this species not being fully recognized and the relevant information not being included in the database. The specific reasons for this need to be further studied via sequence analysis and gene expression verification.

Table 1. Length distribution after assembly in the flower transcriptome of *L. yunnanensis*.

Transcript Length Interval	200–500 bp	500–1 kbp	1 k–2 kbp	>2 kbp	Total
Number of transcripts	71,181 (50.83%)	25,812 (18.43%)	23,641 (16.88%)	19,408 (13.86%)	140,042
Number of unigenes	62,121 (63.14%)	17,747 (18.04%)	10,870 (11.05%)	7651 (7.78%)	98,389

Table 2. The statistics of the unigenes' annotation in the flower transcriptome of *L. yunnanensis*.

Annotation	Number of Annotated Unigenes	Percentage of Annotated Unigenes (%)	Percentage of Total Unigenes (%)
Annotated in Nr	31,859	32.38	22.75
Annotated in Nt	13,853	14.07	9.89
Annotated in KO	9722	9.88	6.94
Annotated in Swiss-Prot	21,416	21.76	15.29
Annotated in Pfam	22,684	23.05	16.20
Annotated in GO	23,390	23.77	16.70
Annotated in KOG	10,947	11.12	7.82
Annotated in all databases	4273	4.34	3.05
Annotated in at least one database	36,497	37.09	26.06
Total annotated unigenes	98,389	100	70.26

According to the species classification from the result of BLAST with Nr (Figure 1A), *Coffea canephora* (Rubiaceae, *Coffea*) had the highest matching rate, followed by *Vitis vinifera* (Vitaceae, *Vitis*), with the next highest being *Nicotiana tomentosiformis* (Solanaceae, *Nicotiana*). As can be seen from Figure 1B, 38.5% of the unigenes can be fully matched to Nr (the smaller the E-value, the higher the degree of matching and the similarity), and the degree of matching in Nr is relatively high. Figure 1C shows that 11.0% of the unigenes had more than 95% similarity, and 54% of the unigenes had 80–95% similarity (the higher the similarity, the higher the confidence).

All unigenes were assessed for GO assignments based on the Nr annotations. The unigenes were categorized into biological process (BP), cellular component (CC), and molecular function (MF) to reveal the gene function classification (Figure 2). Within these functional groups, "metabolic process" was the dominant group among biological processes (282,588), which contained 144,606 unigenes, and second was "cellular process" (62,514). Among cellular components (83,109), "cell part" (24,619) contained the largest number of unigenes, followed by "organelle" (19,336). With a total of only 57,203 unigenes in the molecular function category, the two largest groups were "binding" and "catalytic activity", which included 41,251 and 11,311 unigenes, respectively. This result was similar to those for *Lycium barbarum* [22], *Rhododendron fortune* [23], and *Elaeagnus mollis* Diels [24], which showed that metabolic processes and cellular processes contained the largest numbers of unigenes in all subcategories. The genes successfully annotated by KOG were classified into 26 KOG groups, as shown in Figure 3. The dominant group was general function prediction (1936), followed by the posttranslational modification, protein turnover, and chaperones (1459), while the smallest group was cell motility (only 4). The KOG classification results of the three species mentioned above are still consistent with the results of our study. In addition, the group of unknown functions contained 443 unigenes, accounting for 3.6% of the total annotation information in the *L. yunnanensis* flower transcriptome. Studies on *Castanea henryi* (Skan) Rehd. et Wils. [25], *Pinus yunnanensis* Franch. [26], *Rhododendron longipedicellatum* Lei Cai et Y. P. Ma [27], and *Phyllanthus emblica* [28] have produced similar results. We speculate that this may be caused by insufficient annotation information. By comparing the KEGG database, the unigenes can be classified into 22 KEGG pathways according to the signaling pathways involved. Of these pathways, the most represented was translation (957), followed by carbohydrate metabolism (918) and signal transduction (910) (Figure 4).

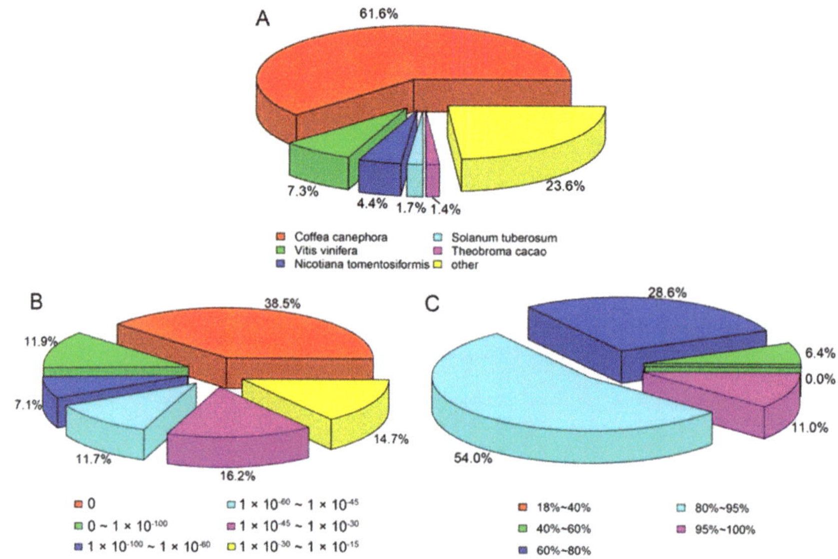

Figure 1. Characterization of assembled *L. yunnanensis* unigenes using the Nr database: (**A**) Species distribution for the assembled unigenes. (**B**) E-value distribution for the assembled unigenes. (**C**) Similarity distribution for the assembled unigenes.

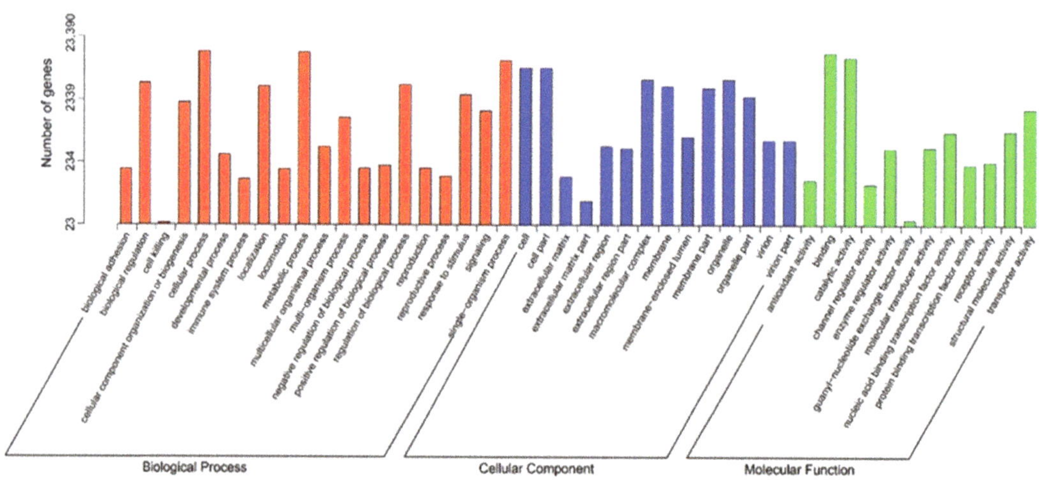

Figure 2. GO classification of assembled unigenes of the *L. yunnanensis* flower transcriptome. The *x*-axis indicates the subgroups in GO annotation; the *y*-axis indicates the number of genes in each category.

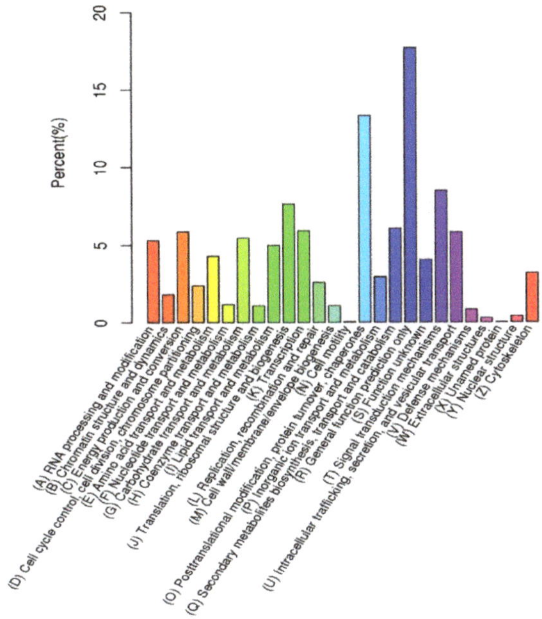

Figure 3. KOG classification of assembled unigenes of the *L. yunnanensis* flower transcriptome. The x-axis indicates the 26 groups in KOG annotation; the y-axis indicates the percentage of the number of genes in each group relative to the total number of annotated genes.

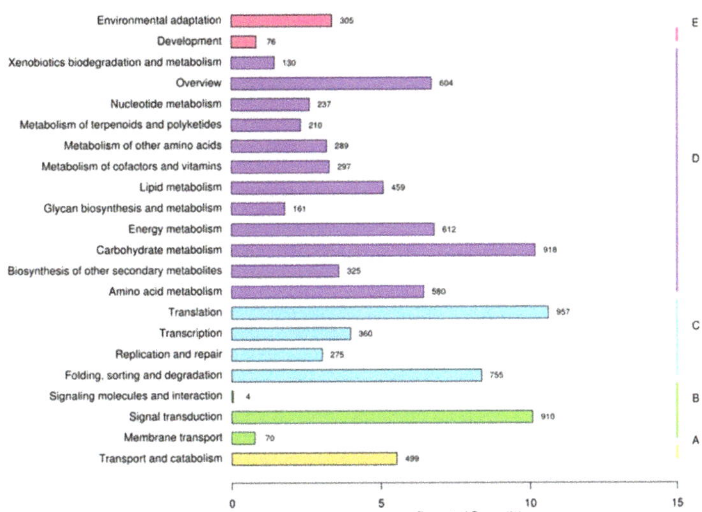

Figure 4. KEGG metabolic pathways of assembled unigenes of the *L. yunnanensis* flower transcriptome: (**A**) cellular processes; (**B**) environmental information processing; (**C**) genetic information processing; (**D**) metabolism; (**E**) organismal systems. The x-axis indicates the number of genes in each metabolic pathway and the ratios of the number of genes to total number of annotated genes; the y-axis indicates the names of the KEGG metabolic pathways.

The unigenes were roughly divided into 3 functional categories and 56 subcategories according to GO function, among which metabolic process, cellular process, cell, cell part, organelle, binding, and catalytic activity were highly enriched. In other words, the expression levels of genes related to cellular activity, metabolic activity, and catalytic activity was high, indicating that *L. yunnanensis* has strong metabolic capacity. We hypothesized that this may be related to the continuous cell proliferation of the meristem during flower development and the vigorous metabolic activities in the flower organs of *L. yunnanensis*. Xia et al. [29] analyzed the flower transcriptome of *Camellia sinensis*, and their results and hypotheses were similar to ours. KEGG functional annotation analysis showed that unigenes could be grouped into 5 categories, among which the pathways related to metabolism and genetic information processing accounted for the highest proportion, and the number of genes related to metabolism-related pathways was the largest. This was further proof that there was strong metabolic activity in *L. yunnanensis* during this period. Additionally, the metabolic pathways of amino acid metabolism, carbohydrate metabolism, lipid metabolism, biosynthesis of other secondary metabolites, and environmental adaptation were involved in the KEGG functional annotation analysis. These data provide a molecular basis for further research on resistance mechanisms, and allow us to explore the genes related to the flowering regulation and environmental adaptability of *L. yunnanensis*.

2.2. Identification of SSRs

The MISA (Microsatellite) tool was used to identify SSR loci in the transcriptome data, and a total of 15,384 SSRs were identified from 98,389 unigenes; the SSR frequency in the transcriptome was 15.63%, and the mean distance of the SSRs in the unigenes was 6.39 kb. It is well known that the distribution frequency of SSR loci varies greatly between different species, which may be related to the genome size of species, the organ and period of the sampled plant used for transcriptome sequencing, the SSR development methods, and the screening criteria [30]. Species with higher frequencies include, for instance, hybrid *Cymbidium* (58.64%) [31], *Morus alba* L. (45.56%) [32], and *Bougainvillea glabra* (44.91%) [33]. Species with lower frequencies include *Paeonia suffruticosa* (6.4%) [34], *Pinus elliottii* (4.80%) [35], and *Chrysanthemum morifolium* (2.84%) [36]. Kumpatal [37] found that the distribution frequency of SSR loci in dicotyledons ranged from 2.65% to 16.82%, with an average of 9.73%. The distribution frequency of SSR loci in the flower transcriptome of *L. yunnanensis* is consistent with the distribution frequency range of SSR loci in dicotyledons, and higher than the average. Therefore, the flower transcriptome data of *L. yunnanensis* had abundant SSR loci, and could be used for the development of SSR markers.

There were abundant SSR types in the flower transcriptome of *L. yunnanensis*, with the distribution of mononucleotide to hexanucleotide repeat types, but the proportion of different repeat types was significantly different. The number of mononucleotide repeat motifs found to be the most frequent was 8889, accounting for 60.29% of the total repeat motifs. Second, dinucleotide and trinucleotide repeat motifs were 3340 and 2309, accounting for 22.65% and 15.66%, respectively, of the total repeat motifs. The tetranucleotide, pentanucleotide, and hexanucleotide repeat motifs accounted for only 158, 23, and 25, respectively. In this study, the number of short-repeat motifs (mononucleotide, dinucleotide, and trinucleotide) was more than that of long-repeat motifs, which is consistent with the distribution in most species, and supports the view that long-repeat motifs have high variability [38]. As shown in Table 3, the mono-, di-, tri-, tetra-, penta-, and hexanucleotide motifs with the highest frequency were A/T, AG/CT, AAG/CTT, AAAG/CTTT, CCTTC/GAAGG, and GAGAC/GTCTC, respectively. The frequency of mononucleotide repeats was the highest. This result was similar to those for *Ricinus communis* [39], *Glycyrrhiza uralensis* Fisch [40], *Arachis hypogaea* L. [41], and *Elaeagnus mollis* Diels [24], which showed that the A/T repeat motif accounted for the largest proportion of SSRs. The proportion of 12–21 bp repeats in SSRs under different repeat types and repeat times was more than 95%, among which the proportion of 5 times and trinucleotide repeat motifs was the largest (22.25%), followed by that of 6 times and dinucleotide repeat motifs (16.48%) (Table 4).

Table 3. Frequency of different repeat motifs in the SSRs of the *L. yunnanensis* flower transcriptome.

Repeat Types	No.	Frequency (%)	Maximum Repeat Motif (Number and %)
Mononucleotide repeat (p1)	8889	60.29	A/T (8816, 99.18%)
Dinucleotide repeat (p2)	3340	22.65	AG/CT (2174, 65.09%)
Trinucleotide repeat (p3)	2309	15.66	AAG/CTT (428, 18.54%)
Tetranucleotide repeat (p4)	158	1.07	AAAG/CTTT (22, 13.92%)
Pentanucleotide repeat (p5)	23	0.16	CCTTC/GAAGG (2, 8.70%) GAGAC/GTCTC (2, 8.70%)
Hexanucleotide repeat (p6)	25	0.16	CCTTC/GAAGG (2, 8.70%) GAGAC/GTCTC (2, 8.70%)

Table 4. Number of SSRs of different repeat motifs on different repeat times in the flower transcriptome of *L. yunnanensis*.

Repeat Types	Repeat Times						
	5	6	7	8	9	10	>10
Dinucleotide repeat	—	965	688	558	591	417	121
Trinucleotide repeat	1303	604	374	23	5	—	
Tetranucleotide repeat	132	25	—	—	—	1	—
Pentanucleotide repeat	14	6	2	—	1	—	—
Hexanucleotide repeat	7	9	5	4	—	—	—

The polymorphism level of SSR markers is an important basis for evaluating their availability, and the length of SSRs is an important factor affecting their polymorphism. The results showed that the polymorphism was high when the length of SSRs was more than or equal to 20 bp, medium when the length was 12–19 bp, and extremely low when the length was less than 12 bp [42]. Therefore, more than 95% SSRs were had a moderate-to-high level of polymorphism in this study. In addition, lower-grade SSR motifs (i.e., mononucleotide, dinucleotide, and trinucleotide) tend to produce polymorphism more easily than higher-order SSR motifs [43,44]. Due to the large number of mononucleotides, dinucleotides, and trinucleotides in this study, it can be predicted that SSRs obtained from the *L. yunnanensis* flower transcriptome have a high potential for polymorphism, and this leads to a high application value in the study of its molecular markers.

2.3. SSR Primer Screening and Verification

A total of 9195 primer pairs were designed using Primer 3.0, and 81 primer pairs were randomly selected to amplify across the genomic DNA template of *L. yunnanensis*. Of the 81 primer pairs, 44 of them successfully amplified fragments from *L. yunnanensis* genomic DNA, with a success rate of 54.32%. Therefore, the primers developed and designed in this study have a certain applicability to the population of *L. yunnanensis*. According to the preliminary screening results, 10 individuals were randomly selected from each population (a total of 60 individuals from 6 populations) to further verify the polymorphism of the 44 primer pairs that could amplify bands. Among them, 17 primer pairs were found to be polymorphic in *L. yunnanensis*, producing clear amplicons of the expected size, which were suitable for the detection of SSR loci by capillary electrophoresis. The specific information of 17 EST-SSR primers is shown in Table 5. A dendrogram clustered the 6 populations of *L. yunnanensis* into 2 clusters (Figure 5). One cluster (I) included 2 populations (GDX and GCG) of Gongshan County and one population (FMM) of Fugong County. The other cluster (II) included 2 populations (LCM and LLO) of Lushui County and one population (FPK) of Fugong County. Geographically, FMM is close to Gongshan County and FPK is close to Lushui County, which is consistent with the results of cluster analysis.

Table 5. The information of 17 EST-SSR primers of *L. yunnanensis*.

Primer ID	Primer Sequence (5′-3′)	Repeat Motif	*Ta* (°C)	Size (bp)	Na	Ne	*I*	PIC
N3	CAAATTCGCGCACCAAAACG GCTAGAGAGAAAAGGGGCCG	(TGGCGT)5	52	261–279	4	2.538	1.061	0.542
N6	ACTGCGTACCTCTCCCTCTT TCTCTCTCTCTCGGACGGAC	(TAATT)5	58	212–228	5	1.933	0.890	0.426
N9	GACCCCAAGTTGGCTGATCA AGGGCACTTCTGTCATTTCGA	(TGTTAC)8	60	127–187	6	2.212	0.923	0.446
N10	CTGGTGCACGAGGATTGAGT GAAGAGTGCCATGGAAACTGC	(TCAATT)7	60	194–224	2	1.427	0.476	0.255
Z13	CCTCCCATAGCAGCAGCAAT AGTAGTATTAATAATGGCTGGAGGT	(ACC)5	54	117–132	5	1.699	0.833	0.382
N22	CGCTTCTGTGTTCGAAACCA CAAAGCTTCCCGTCAACAGC	(ACAAC)7	60	165–180	4	2.575	1.091	0.549
N24	CCCACCGAGCAATACCCAAA ACCTTCTCTGTACTCTGCCT	(GAAA)5	56	268–280	4	1.344	0.555	0.245
Z31	GCAATCCTACTCGTGCTGGT AGCCAAGACTCGGCAGAAAA	(GGC)6	54	228–240	5	2.583	1.239	0.580
Z32	TGCACTCCATAAAAGAAGAAAACACA TGCAGTAACTTCGTGCCCTT	(TATT)5	53	114–122	3	1.405	0.544	0.264
Z33	CCCAACCCACCACACAAGT AGAGAGGAGGATCGAGGACG	(TCT)6	54	255–270	5	1.749	0.887	0.404
Z36	TCGGGTCCTAGGGCTTTCTT GGCCCTCCTTGAGCATTGAT	(CTTT)5	54	211–219	3	1.766	0.768	0.390
Z38	ACCCAAGGAACTCTGTCTCT ACACTTTCGTCGTCCTTAGGT	(AAT)6	53	109–118	4	1.522	0.684	0.321
N41	GCCAGAAGGATAGCTTTCGC GGTTTGTGGTGGTTTTTGGGA	(TCCT)5	53	197–221	6	1.298	0.547	0.223
Z48	AGGAAGGGCTTGTTTTTAAGGT GAGCCAATGACGATCCAGCT	(AG)8	52	215–225	6	1.724	0.914	0.400
Z50	TCTGCTGCATCCAATGTACTGT CCTGCCATAGGTGCCCATTT	(GT)8	54	144–154	5	2.690	1.205	0.576
Z54	AGTAAGTGGGTGGAGGTGGT AGGGGCTGATTCTCTAGCGA	(TTGA)5	59	212–216	2	1.910	0.670	0.363
X70	AGCTGGAAACTAAAGGTGGAGG CTCAGTCTGTCAGGCCTGTG	(ATC)5	58.5	244–250	3	2.355	0.944	0.493

Na = number of different alleles; Ne = number of effective alleles; *I* = Shannon's information index; PIC = polymorphism information content; *Ta* = annealing temperature.

The SSR marker diversity index was estimated using POPGENE 1.32; the effective number of alleles (Ne) and Shannon's information index (I) per locus varied from 1.298 to 2.690 (average of 1.925), and from 0.476 to 1.239 (average of 0.837), respectively. Using CERVUS 3.0 (Field Genetics, London, UK), we calculated that the allelic polymorphism information content (PIC) for each SSR locus ranged from 0.223 to 0.580, with an average of 0.403. Botstein [45] put forward a criterion that primers are highly polymorphic when PIC > 0.5, moderately polymorphic when 0.5 > PIC > 0.25, and low-grade polymorphic when PIC < 0.25. Most of the 17 EST-SSR primer pairs were moderately or highly polymorphic, indicating that these primers could be used for genetic diversity analysis and fingerprint construction of *L. yunnanensis*.

Today, plant diversity is gradually decreasing on a global scale, leading to a related decline in ecosystem service function [46]. Many rare species, such as *L. yunnanensis*, have seldom been studied, and lack protection due to a lack of social awareness. Thus, the numbers and distribution ranges of the species are shrinking. Using transcriptome sequencing technology to study the genetic diversity of *L. yunnanensis* can quickly and efficiently develop mass EST-SSR markers. The application of these markers is conducive to quickly supplementing the molecular biological information of this species, and provides a theoretical basis and scientific evidence for research on the genetic resources, conservation, and utilization of *L. yunnanensis*. In addition, due to the high ornamental value of *L. yunnanensis*, we can refer to Ding's [47] study on *Chrysanthemum*, in which the development of SSR markers for genes related to the regulation of flower shape and color were studied, and these SSR markers were applied to the classification of more than 100 *Chrysanthemum* germplasm resources with different flower colors and shapes. In the same way, we could also screen the SSR loci of important ornamental traits such as the flower shape and color of *L. yunnanensis*, which could lay the foundation for breeding new varieties of *L. yunnanensis*.

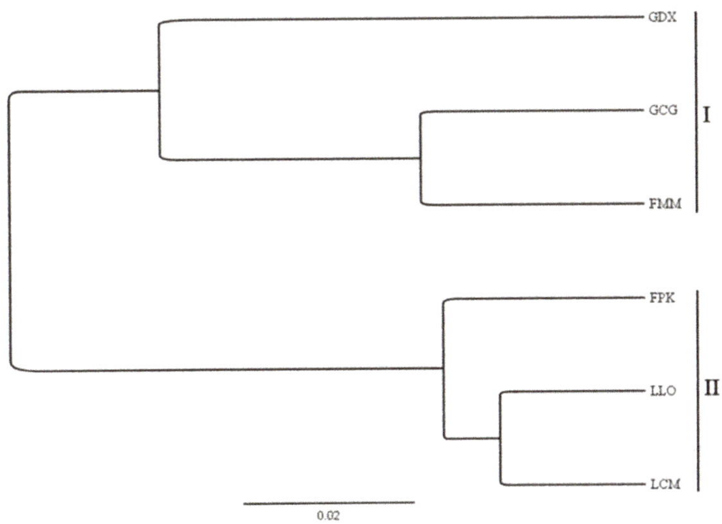

Figure 5. Cluster analysis of 6 *L. yunnanensis* populations based on 17 EST-SSR markers.

3. Materials and Methods

3.1. Plant Materials

We picked the experimental materials from Nujiang Lisu Autonomous Prefecture, Yunnan Province. Inflorescences of *L. yunnanensis* were collected, bottled, and brought back to Kunming for subsequent use. The blooming flowers of five individuals collected within LLO (Table 6) were frozen immediately in liquid nitrogen and stored at −80 °C for use as transcriptome sequencing materials. The plant samples were mixed, and RNA was extracted using an RNA extraction kit from TransGen (Beijing, China). Transcriptomic sequencing was performed at Beijing Novogene Biological Information Technology Co., Ltd., Beijing, China. Based on the results of previous investigations into the resource status of *L. yunnanensis*, 6 populations were selected as primer polymorphism detection materials in regions with a dense distribution and adequate quantity, and where sampling was convenient and relatively representative. These were GDX, GCG, FMM, FPK, LLO, and LCM (Table 6).

3.2. De Novo Assembly, Functional Annotation, and Classification

Transcriptome sequencing was carried using an Illumina HiSeq 4000 platform (Beijing Novogene Biological Information Technology Co., Ltd., Beijing, China); clean reads were used for de novo assembly using Trinity (version: trinityrnaseq-r20131110) with the default parameters to obtain transcripts, and TGICL version 2.1 software was used to cluster the transcripts into unigenes [48,49]. For functional prediction and classification, all unigenes were compared against public databases, including NCBI non-redundant protein sequences (Nr), Nt, Protein Family (Pfam), Eukaryotic Orthologous Groups of Proteins (KOG/COG), Swiss-Prot, the KEGG Orthology database (KO), and Gene Ontology (GO). BLAST [50] was used for unigene annotation in the Nt, Nr, COG, and Swiss-Prot databases. The

KEGG Automatic Annotation Server [51] was used for KEGG annotation of unigenes. Blast2GO [52] and Nr annotation results were used for the GO annotation.

Table 6. The location and habitat of the 6 *L. yunnanensis* populations.

Code	GDX	GCG	FMM	FPK	LLO	LCM
Location	Dulongjiang township, Gongshan County	Cikai town, Gongshan County	Maji township, Fugong County	Pihe township, Fugong County	Luobenzhuo township, Lushui County	Chenggan township, Lushui County
Altitude (m)	1324–1788	1325–1778	1354–1455	1738–1752	2029–2153	1664–1736
Longitude and latitude (°)	98.3188 −98.3466 /27.6915 −27.9160	98.6088 −98.8334 /27.5254 −27.7472	98.8388 −98.8453 /27.3632 −27.3651	98.9102 −98.9106 /26.4799 −27.4801	98.8116 −98.8172 /26.4780 −26.4812	98.8253 −98.8312 /26.3540 −26.3563

3.3. Detection of EST-SSR Markers and Designing of Primers

SSR loci in the transcriptome were identified using the microsatellite identification tool MISA (http://misaweb.ipk-gatersleben.de/) (accessed on 3 December 2020) [53]. The parameters of identifying the mono-, di-, tri-, tetra-, penta-, and hexanucleotide motifs were set to minimum numbers of 10, 6, 5, 5, 5, and 5 repeat units, respectively. Primer pairs were designed using Primer 3.0 software [54], according to the following parameters: the length range of the optimal primer sizes was 18–23 bp, the optimal annealing temperature was from 55–65 °C, the PCR product size range was 80–300 bp, and the GC content was 40–60%.

3.4. DNA Extraction and Quality Check

Total genomic DNA was extracted from the silica-dried leaves of *L. yunnanensis* with a plant genomic DNA extraction kit containing RNase A (TransGen, Beijing, China), following the manufacturer's protocol. The integrity and quality of the genomic DNA was checked using 0.8% agarose gel electrophoresis. The DNA purity and concentration were determined using a NanoDrop 2000 spectrophotometer (Thermo Fisher Scientific, Wilmington, DE, USA), and samples were stored at −20 °C until used for SSR amplification.

3.5. EST-SSR Amplification and Data Analysis

The 81 pairs of randomly selected SSR primers were used for preliminary screening, and 10 individuals per population were used to test the primers amplified with clear bands in order to further determine the polymorphism of the SSR primers. PCR amplification was carried out in a 10 μL reaction mixture containing 10–30 ng of genomic DNA, 0.15 mmol/L of each dNTP, 1 μL of 10×Taq buffer, 0.6 mmol/L each of reverse and forward primer, and 1 unit of Taq DNA polymerase (TransGen, Beijing, China). The PCR amplification was carried out using a Bio-Rad C1000 Touch ™ Thermal Cycler (Bio-Rad Laboratories, CA, USA), as follows: 5 min of initial denaturation at 94 °C, followed by 30 cycles of denaturation for 30 s at 94 °C, a temperature gradient for annealing from 50 °C to 60 °C for 30 s, and extension at 72 °C for 1 min, with a final 10 min at 72 °C for final extension. The PCR products were resolved by electrophoresis in 1.5% agarose gels to determine whether amplification was successful. PCR fluorescent tagging and capillary electrophoresis were performed to further screen polymorphisms. The 5′ end of each forward primer was labeled with FAM or HEX fluorescent dyes (Thermo Fisher Scientific, Wilmington, DE, USA). The PCR amplifications were performed using the PCR conditions mentioned above. The multiplex DNA products labeled with the above-mentioned fluorescent dyes were analyzed on an ABI 3730xl DNA Analyzer with a GeneScan 500 LIZ Size Standard (Thermo Fisher Scientific, Wilmington, DE, USA), and allele sizes were assessed using GeneMapper 4.1 (Thermo Fisher Scientific, Wilmington, DE, USA). The parameters of genetic diversity—including the number of alleles (Na), effective number of alleles (Ne), and Shannon's information index (I)—of these SSR loci were estimated using POPGENE 1.32 [55], and the allelic polymorphism information content (PIC) for each SSR locus was calculated using CERVUS 3.0 (Field Genetics, London, UK) [56]. Cluster analysis of 6 populations on the basis of shared allele distance (DAS) was performed using POPULATIONS 1.2.30 [57], with the unweighted pair group method with arithmetic mean (UPGMA) method. The clustering tree was visualized and edited using Interactive Tree of Life (iTOL) version 3 [58]. The primers with good repeatability, clear bands, and high polymorphism were selected as candidate primers for molecular genetic diversity research on *L. yunnanensis*.

4. Conclusions

The *L. yunnanensis* transcriptome was obtained using an Illumina HiSeq platform. In total, 98,389 unigenes were generated, of which 98,389 unigenes were aligned with the sequences of public databases. Nr, Nt, Pfam, KOG/COG, Swiss-Prot, KEGG, and GO annotated 31,859, 13,853, 22,684, 10,947, 21,416, 9722, and 23,390 unigenes, respectively. EST-SSR specific primer pairs were successfully designed. In addition, MISA was used to identify SSR loci from all unigenes, and a total of 15,384 SSRs was identified. The repeat motif was given priority with mononucleotides, dinucleotides, and trinucleotides. Furthermore, the 81 primer pairs were synthesized randomly, of which 44 pairs showed effective amplification, 17 primers showed stable amplification, and rich polymorphism was observed in 6 populations. This sequence resource and these potential EST-SSR markers can be used for subsequent population genetic diversity analysis and marker-assisted breeding, which is of great significance for formulating resource conservation and utilization strategies for *L. yunnanensis*.

Author Contributions: Z.L., H.M. and Y.Z. conceived and designed the research; X.L. (Xi Liu), X.L. (Xiongfang Liu) and Y.L. performed the experiments; Y.Z. and Y.L. analyzed the data and wrote the paper; Y.Z., H.M. and S.Q. revised the manuscript; Z.L. supervised the project. All authors have read and agreed to the published version of the manuscript.

Funding: This study was supported by the Major Science and Technology Special Program of Yunnan Science and Technology Department (Grant No. 202102AE090052), and the Ten Thousand Talent Program of Yunnan Province (Grant No. YNWRQNBJ-2019-010).

Data Availability Statement: Not applicable.

Conflicts of Interest: The authors declare no conflict of interest.

References

1. Li, Y.Y. Studies on Conservation Genetics of Chinese Endemic Species *Luculia yunnanensis* Hu. Doctoral Dissertation, Chinese Academy of Forestry, Beijing, China, 2017.
2. Allendorf, F.; Ryman, N. The role of genetics in population viability analysis. In *Population Viability Analysis*; Beissinger, S.R., McCullough, D.R., Eds.; The University of Chicago Press: Chicago, IL, USA, 2000; pp. 50–85.
3. Holsinger, K.E.; Gottlieb, L.D. Conservation of rare and endangered plants: Principles and prospects. In *Genetics and Conservation of Rare Plants*; Falk, D.A., Holsinger, K.E., Eds.; Oxford University Press: New York, NY, USA, 1991; pp. 195–208.
4. Lande, R. Risks of population extinction from demographic and environmental stochasticity and random catastrophes. *Am. Nat.* **1993**, *142*, 911–927. [CrossRef] [PubMed]
5. Soule, M.; Simberloff, D. What do genetics and ecology tell us about the design of nature reserves? *Biol. Conserv.* **1986**, *35*, 19–40. [CrossRef]
6. Kaljund, K.; Jaaska, V. No loss of genetic diversity in small and isolated populations of *Medicago sativa* subsp. *falcata*. *Biochem. Syst. Ecol.* **2010**, *38*, 510–520. [CrossRef]
7. Varshney, R.K.; Graner, A.; Sorrells, M.E. Genetic microsatellitemarkers in plants: Features and applications. *Trends Biotechnol.* **2005**, *23*, 48–55. [CrossRef] [PubMed]
8. Kalia, R.K.; Rai, M.K.; Kalia, S.; Singh, R.; Dhawan, A.K. Microsatellite markers: An overview of the recent progress in plants. *Euphytica* **2011**, *177*, 309–334. [CrossRef]
9. Liu, G.; Xie, Y.J.; Zhang, D.Q.; Chen, H.P. Analysis of SSR loci and development of SSR primers in Eucalyptus. *J. For. Res.* **2018**, *29*, 273–282. [CrossRef]
10. Liu, Y.; Fang, X.M.; Tang, T.; Wang, Y.D.; Wu, Y.H.; Luo, J.Y.; Wu, H.T.; Wang, Y.Q.; Zhang, J.; Ruan, R.W.; et al. Inflorescence transcriptome sequencing and development of new EST-SSR markers in common buckwheat (*Fagopyrum esculentum*). *Plants* **2022**, *11*, 742. [CrossRef]

11. Ellis, J.R.; Burke, J.M. EST-SSRs as a resource for population genetic analyses. *Heredity* **2007**, *99*, 125–132. [CrossRef]
12. Yang, X.; Yang, N.; Yuan, Q.; Zhao, G.J.; Li, G.L.; Jia, L.M.; Chen, Z. Development and application of EST-SSR molecular markers in *Pinus bungeana*. *J. Beijing For. Univ.* **2021**, *43*, 1–11.
13. Xiang, X.Y.; Zhang, Z.X.; Wang, Z.G.; Zhang, X.P.; Wu, G.L. Transcriptome sequencing and development of EST-SSR markers in *Pinus dabeshanensis*, an endangered conifer endemic to China. *Mol. Breed.* **2015**, *35*, 1–10. [CrossRef]
14. Li, X.Y.; Guo, Q.; Dong, L.; Sun, Y.H.; Niu, D.S.; Liu, J.P.; Wang, H.S.; Li, Y. EST-SSR analysis of genetic diversity of *Robinia pseudoacacia* clones in Jixian county, Shanxi province of northern China. *J. Beijing For. Univ.* **2019**, *41*, 39–48.
15. Dehkordi, M.K.; Beigzadeh, T.; Sorkheh, K. Novel in silico EST-SSR markers and bioinformatic approaches to detect genetic variation among peach (*Prunus persica* L.) germplasm. *J. For. Res.* **2020**, *31*, 1359–1370. [CrossRef]
16. Wani, M.S.; Sharma, V.; Gupta, R.C.; Munshi, A.H. Development and characterization of SSR markers in Himalayan species Betula utilis. *J. For. Res.* **2020**, *31*, 1453–1460. [CrossRef]
17. Zhou, X.J.; Dong, Y.; Zhao, J.J.; Huang, L.; Ren, X.P.; Chen, Y.N.; Huang, S.M.; Liao, B.S.; Lei, Y.; Yan, L.Y.; et al. Genomic survey sequencing for development and validation of single-locus SSR markers in peanut (*Arachis hypogaea* L.). *BMC Genom.* **2016**, *17*, 420. [CrossRef]
18. Ma, H.; Wang, L.; Wan, Y.M.; Li, H.Z.; Li, Z.H.; Liu, X.X.; Liang, N.; Li, W.J. A set of novel microsatellite markers developed for *Luculia yunnanensis* (Rubiaceae), an endangered plant endemic to Yunnan, China. *Int. J. Mol. Sci.* **2012**, *13*, 534–539. [CrossRef] [PubMed]
19. Ramilowski, J.A.; Sawai, S.; Seki, H.; Mochida, K.; Yoshida, T.; Sakurai, T.; Muranaka, T.; Saito, K.; Daub, C.O. *Glycyrrhiza uralensis* transcriptome landscape and study of phytochemicals. *Plant Cell Physiol.* **2013**, *54*, 697–710. [CrossRef]
20. Wan, Y.M. Mechanism on photoperiodic control of flowering in *Luculia gratissima* 'Xiangfei'. Doctoral Dissertation, Chinese Academy of Forestry, Beijing, China, 2019.
21. Zhu, Y.H.; Zhang, M.J.; Li, L.; Zhao, L.; Dong, C.M. Transcriptome analysis of *Prunella vulgaris* and identification of putative genes involved in second metabolism biosynthesis. *Zhong Cao Yao* **2019**, *50*, 1220–1226.
22. Chen, C.L.; Xu, M.L.; Wang, C.P.; Qiao, G.X.; Wang, W.W.; Tan, Z.Y.; Wu, T.T.; Zhang, Z.S. Characterization of the *Lycium barbarum* fruit transcriptome and development of EST-SSR markers. *PLoS ONE* **2017**, *12*, e0187738. [CrossRef]
23. Xu, Q.W.; Lou, X.Z.; Yang, B.; Lin, E.P.; Tong, Z.K. Transcriptome sequencing and analysis of *Rhododendron fortune*. *J. Zhejiang AF Univ.* **2019**, *36*, 1190–1198.
24. Liu, Y.L.; Li, S.Q.; Wang, Y.Y.; Liu, P.Y.; Han, W.J. *De novo* assembly of the seed transcriptome and search for potential EST-SSR markers for an endangered, economically important tree species: *Elaeagnus mollis* Diels. *J. For. Res.* **2020**, *31*, 69–77. [CrossRef]
25. Zhang, L.; Fan, X.M.; Lin, Q.; Xiang, H.; Wang, W.; Yuan, D.Y. Transcriptome Analysis for developing kernel and expression analysis of starch and sucrose metabolism-related genes in *Castanea henryi*. *J. Plant Genet. Resour.* **2015**, *16*, 603–611.
26. Cai, N.H.; Deng, L.L.; Xu, Y.L.; Xu, Y.; Zhou, L.; Wang, D.W.; Tian, B.; He, C.Z.; Duan, A.A. Transcriptome analysis for *Pinus yunnanensis* based on high throughput sequencing. *Bull. Bot. Res.* **2016**, *36*, 75–83.
27. Li, T.Q.; Liu, X.F.; Wan, Y.M.; Li, Z.H.; Qi, G.H.; Li, Y.Y.; Liu, X.X.; He, R.; Ma, Y.; Ma, H. Transcriptome analysis for *Rhododendron longipedicellatum* (plant species with extremely small populations) based on high throughput sequencing. *Bull. Bot. Res.* **2017**, *37*, 825–834.
28. Liu, X.F.; Li, T.Q.; Li, Z.H.; Wan, Y.M.; Liu, X.X.; Zhang, X.; An, J.; Ma, H. Transcriptome analysis for *Phyllanthus emblica* distributed in dry-hot valleys in Yunnan, China. *Fore. Res.* **2018**, *31*, 1–8.
29. Xia, L.F.; Zhu, X.Z.; Liang, M.Z.; Ma, W.; Song, W.X.; Tian, Y.P.; Zhou, M.; Chen, L.B. Transcriptome analysis of flower from tea plant [*Camellia sinensis* (L.) O. Kuntz] via RNA-Seq technology. *Southwest China J. Agric. Sci.* **2016**, *29*, 2058–2062.
30. Biswas, M.K.; Chai, L.J.; Mayer, C.; Xu, Q.; Guo, W.W.; Deng, X.X. Exploiting BAC−end sequences for the mining, characterization and utility of new short sequences repeat (SSR) markers in *Citrus*. *Mol. Biol. Rep.* **2012**, *39*, 5373–5386. [CrossRef] [PubMed]
31. Zhong, H.Q.; Lin, R.Y.; Lin, B.; Wu, J.S.; Fan, R.H.; Ye, X.X. Analysis on SSR information in transcriptome and development of EST-SSR markers for hybrid cymbidium. *Chin. J. Cell. Biol.* **2020**, *42*, 286–295.
32. Zhao, T.; Chang, S.X.; Leng, Q.Y.; Xu, S.S.; Yin, J.M.; Niu, J.H. Development of SSR molecular markers based on transcriptome sequencing of *Bougainvillea*. *Mol. Plant Breed.* **2019**, *17*, 4331–4341.
33. Wang, H.; Xie, Y.; Sun, Z.C.; Gao, Y.J.; Zhang, D.H.; Song, Y.X.; Gao, Y.X. Analysis on SSR locis in mulberry fruit transcriptome. *Mol. Plant Breed.* **2020**, *18*, 208–215.
34. Wu, J.; Cai, C.; Cheng, F.; Cui, H.; Zhou, H. Characterisation and development of EST-SSR markers in tree peony using transcriptome sequences. *Mol. Breed.* **2014**, *34*, 1853–1866. [CrossRef]
35. Yi, M.; Zhang, L.; Lei, L.; Cheng, Z.S.; Sun, S.W.; Lai, M. Analysis of SSR information in transcriptome and development of EST-SSR molecular markers in *Pinus elliottii* Engelm. *J. Nanjing. For. Univ.* **2020**, *44*, 75–83.
36. Feng, S.; He, R.; Lu, J.; Jiang, M.; Shen, X.; Yan, J.; Wang, Z.; Wang, H. Development of SSR markers and assessment of genetic diversity in medicinal *Chrysanthemum morifolium* cultivars. *Front. Genet.* **2016**, *7*, 113. [CrossRef] [PubMed]
37. Kumpatal, S.P.; Mukhopadhyay, S. Mining and survey of simple sequence repeats in expressed sequence tags of dicotyledonous species. *Genome* **2005**, *48*, 985–998. [CrossRef] [PubMed]
38. Li, Q.; Li, B.; Guo, S.X. SSR information in transcriptome of *Dendrobium nobile*. *China J. Chin. Mater. Med.* **2017**, *42*, 63–69.
39. Zou, M.L.; Xia, Z.Q.; Chen, J.R.; Wang, W.Q. Analysis of SSR information and annotation in EST resource of Caster Bean (*Ricinus communis*). *Chin. J. Trop. Crop.* **2012**, *33*, 2138–2143.

40. Liu, Y.L.; Zhang, P.F.; Song, M.L.; Hou, J.L.; Mei, Q.; Wang, W.Q.; Liu, C.S. Transcriptome analysis and development of SSR molecular markers in *Glycyrrhiza uralensis* Fisch. *PLoS ONE* **2015**, *10*, e0143017. [CrossRef] [PubMed]
41. Zhou, W.; Wang, H.; Li, D.Z.; Yang, J.B.; Zhou, W. Isolation and characterization of 13 microsatellite loci from *Luculia pinceana* (Rubiaceae), a typical distylous species. *Hortscience* **2010**, *45*, 840–841. [CrossRef]
42. Temnykh, S.; De Clerck, G.; Lukashova, A.; Lipovich, L.; Cartinhour, S.; Mc Couch, S. Computational and experimental analysis of microsatellites in rice (*Oryza sativa* L.): Frequency, length variation, transposon associations, and genetic marker potential. *Genome Res.* **2001**, *11*, 1441–1452. [CrossRef]
43. Dreisigacker, S.; Zhang, P.; Warburton, M.L.; Van Ginkel, M.; Hoisington, D.; Bohn, M.; Melchinger, A.E. SSR and pedigree analyses of genetic diversity among CIMMYT wheat lines targeted to different megaenvironments. *Crop. Sci.* **2004**, *44*, 381–388. [CrossRef]
44. Liu, S.S.; Qiao, Z.Q.; Zeng, H.J.; Li, Y.X.; Cai, N.; Liu, X.M.; Wang, X.M. Analysis on SSR loci in transcriptome and development of EST-SSR molecular markers in *Lonicera macranthoides*. *Mol. Plant Breed.* **2021**, *19*, 3015–3021.
45. Botstein, D.; White, R.L.; Skolnick, M.; Davis, R.W. Construction of a genetic linkage map in man using restriction fragment length polymorphisms. *Am. J. Hum. Genet.* **1980**, *32*, 314–331. [PubMed]
46. Li, D.Z.; Yang, X.Y.; Hugh, W.P. Problems and challenges in germplasm preservation. *Bull. Chin. Acad. Sci.* **2010**, *25*, 533–540.
47. Ding, H.X. Development and Application of SSR Markers in *Chrysanthemum* Based on Full-Length Transcriptome Sequence. Master Dissertation, Henan University, Kaifeng, China, 2020.
48. Pertea, G.; Huang, X.Q.; Liang, F.; Antonescu, V.; Sultana, R.; Karamycheva, S.; Lee, Y.D.; White, J.; Cheung, F.; Parvizi, B.; et al. TIGR Gene Indices clustering tools (TGICL): A software system for fast clustering of large EST datasets. *Bioinformatics* **2003**, *19*, 651–652. [CrossRef]
49. Grabherr, G.M.; Haas, B.J.; Yassour, M.; Levin, J.Z.; Thompson, D.A.; Amit, I.; Adiconis, X.; Fan, L.; Raychowdhury, R.; Zeng, Q.D.; et al. Full-length transcriptome assembly from RNA-Seq data without a reference genome. *Nat. Biotechnol.* **2011**, *29*, 644–652. [CrossRef] [PubMed]
50. Altschul, S.F.; Gish, W.; Miller, W.; Myers, E.W.; Lipman, D.J. Basic local alignment search tool. *J. Mol. Biol.* **1990**, *215*, 403–410. [CrossRef]
51. Moriya, Y.; Itoh, M.; Okuda, S.; Yoshizawa, A.C.; Kanehisa, M. KAAS: An automatic genome annotation and pathway reconstruction server. *Nucleic. Acids. Res.* **2007**, *35*, W182–W185. [CrossRef]
52. Götz, S.; García-Gómez, J.M.; Terol, J.; Williams, T.D.; Nagaraj, S.H.; Nueda, M.J.; Robles, M.; Talón, M.; Dopazo, J.; Conesa, A. High-throughput functional annotation and data mining with the Blast2GO suite. *Nucleic. Acids. Res.* **2008**, *36*, 3420–3435. [CrossRef]
53. Beier, S.; Thiel, T.; Munch, T.; Scholz, U.; Mascher, M. MISA-web: A web server for microsatellite prediction. *Bioinformatics* **2017**, *33*, 2583–2585. [CrossRef]
54. Rozen, S.; Skaletsky, H. Primer3 on the WWW for general users and for biologist programmers. *Methods Mol. Biol.* **2000**, *132*, 365–386.
55. Yeh, F.C.; Yang, R.C.; Boyle, T.; Ye, Z.H.; Mao, J.X. *POPGENE, the User-Friendly Shareware for Population Genetic Analysis*; Molecular Biology and Biotechnology Centre, University of Alberta: Edmonton, AB, Canada, 1997; Volume 10, pp. 295–301.
56. Kalinowski, S.T.; Taper, M.L.; Marshall, T.C. Revising how the computer program CERVUS accommodates genotyping error increases success in paternity assignment. *Mol. Ecol.* **2007**, *16*, 1099–1106. [CrossRef]
57. Populations 1.2.30: Population Genetic Software (Individuals or Populations Distances, Phylogenetic Trees). Available online: http://bioinformatics.org/~{}tryphon/populations/ (accessed on 13 April 2022).
58. Letunic, I.; Bork, P. Interactive tree of life (iTOL) v3: An online tool for the display and annotation of phylogenetic and other trees. *Nucleic Acids Res.* **2016**, *44*, W242–W245. [CrossRef] [PubMed]

plants MDPI

Article

A Comparative Study of Genetic Responses to Short- and Long-Term Habitat Fragmentation in a Distylous Herb *Hedyotis chyrsotricha* (Rubiaceae)

Na Yuan [1,†], Shujing Wei [2,†], Hans Peter Comes [3], Sisheng Luo [2], Ruisen Lu [4,*] and Yingxiong Qiu [5,*]

1 Provincial Key Laboratory of Agrobiology, Institute of Crop Germplasm and Biotechnology, Jiangsu Academy of Agricultural Sciences, Nanjing 210014, China; thefuries@163.com
2 Guangdong Academy of Forestry, Guangzhou 510520, China; weishujing2003@163.com (S.W.); sisheng8009@126.com (S.L.)
3 Department of Environment & Biodiversity, Salzburg University, A-5020 Salzburg, Austria; hans-peter.comes@plus.ac.at
4 Institute of Botany, Jiangsu Province and Chinese Academy of Sciences, Nanjing 210014, China
5 Wuhan Botanical Garden, Chinese Academy of Sciences, Wuhan 430074, China
* Correspondence: lurs@cnbg.net (R.L.); qiuyingxiong@wbgcas.cn (Y.Q.)
† These authors contributed equally to this work.

Citation: Yuan, N.; Wei, S.; Comes, H.P.; Luo, S.; Lu, R.; Qiu, Y. A Comparative Study of Genetic Responses to Short- and Long-Term Habitat Fragmentation in a Distylous Herb *Hedyotis chyrsotricha* (Rubiaceae). *Plants* **2022**, *11*, 1800. https://doi.org/10.3390/plants11141800

Academic Editors: Aiping Song and Yu Chen

Received: 23 May 2022
Accepted: 5 July 2022
Published: 7 July 2022

Abstract: The genetic effects of habitat fragmentation are complex and are influenced by both species traits and landscape features. For plants with strong seed or pollen dispersal capabilities, the question of whether the genetic erosion of an isolated population becomes stronger or is counterbalanced by sufficient gene flow across landscapes as the timescales of fragmentation increase has been less studied. In this study, we compared the population structure and genetic diversity of a distylous herb, *Hedyotis chyrsotricha* (Rubiaceae), in two contrasting island systems of southeast China. Based on RAD-Seq data, our results showed that populations from the artificially created Thousand-Island Lake (TIL) harbored significantly higher levels of genetic diversity than those from the Holocene-dated Zhoushan Archipelago (ZA) (π = 0.247 vs. 0.208, H_O = 0.307 vs. 0.256, H_E = 0.228 vs. 0.190), while genetic differences between island and mainland populations were significant in neither the TIL region nor the ZA region. A certain level of population substructure was found in TIL populations, and the level of gene flow among TIL populations was also lower than in ZA populations (m = 0.019 vs. 0.027). Overall, our comparative study revealed that genetic erosion has not become much stronger for the island populations of either the TIL or ZA regions. Our results emphasized that the matrix of water in the island system may facilitate the seed (fruit) dispersal of *H. chrysotricha*, thus maintaining population connectivity and providing ongoing resilience to the effects of habitat fragmentation over thousands of years.

Keywords: habitat fragmentation; RAD-Seq; *Hedyotis chrysotricha*; genetic diversity; population structure; gene flow

1. Introduction

Over the past 200 years, human activities have altered more than 80% of the terrestrial world, leading to increasing habitat reduction and fragmentation of almost all ecosystems [1–3]. Theoretically, habitat fragmentation has severe negative impacts on community structures, population distributions and abundances, patterns of genetic variation, gene flow, etc. [4,5]. These changes, especially genetic alterations, can reduce the individual fitness and evolutionary potential of populations, further impeding the long-term persistence of species and increasing the risk of local extinction [6,7].

Detecting the genetic effects of habitat fragmentation on natural populations is challenging in practice, as there are many confounding factors involved [2]. For instance, life-history traits such as life span, pollination, and seed dispersal, are essential in determining the magnitude of the plant's response to habitat fragmentation. Short-lived herbs experience more generations and may show a faster decrease in genetic diversity than longer-lived trees or shrubs over the same timescale of fragmentation [8]. In addition, plants with abiotic-mediated pollination and seed dispersal are expected to be less susceptible to habitat fragmentation than those with animal-mediated pollination and seed dispersal [9,10]. For example, Wang et al. (2011) [11] observed no significant difference in genetic diversity between pre- and post-fragmentation cohorts of a wind-pollinated tree *Castanopsis sclerophylla* (Lindl. & Paxton) Schottky in a recently fragmented island system. Due to the disappearance of a genetic barrier, the genetic structure of post-fragmentation cohorts was significantly weakened by increased wind speeds and easier pollen movement over water. In contrast, our previous study on the distylous herb *Hedyotis chrysotricha* (Palib.) Merr in the Thousand-Island Lake (TIL) region revealed that 53 years of fragmentation has led to the loss of c. 17.4% of the initial genetic diversity in the island populations [12]. *Hedyotis chrysotricha* is insect-pollinated, with seeds that are dispersed by wind or (and) water [13,14]. Although patterns of gene flow have not been greatly modified among the *H. chrysotricha* populations, it is uncertain whether extensive seed flow can counteract the negative effects of genetic drift to reach a new dynamic equilibrium and assist the long-term persistence of remnant populations.

The various island systems in southeastern China provide natural laboratories for testing the genetic responses of plants to habitat fragmentation. For example, a long-established hotspot for testing the predictions of population genetics theory on recent anthropogenic habitat fragmentation is the TIL region in southeast China (Chun'an County, Hangzhou City, Zhejiang Province, China). This artificial lake (c. 573 km^2), with about 1078 islands of different sizes (0.25–1320 ha) and shapes (c. 83 km^2 in total), was formed in 1959 after the construction of the Xinanjiang River hydroelectric power station [15]. In contrast, the Zhoushan Archipelago (ZA), located in the East China Sea off the northern coast of Zhejiang Province, is an ideal candidate for examining the genetic responses to long-term habitat fragmentation. It was originally an extended part of the nearby continent and was formed by rising sea levels during the early Holocene, c. 7000–9000 years ago. It consists of 1339 islands and 3306 reefs, covering about 1440 km^2 in total [16,17]. The two island systems differ greatly in spatial scale and temporal origin, making them ideal systems for comparative studies detecting both short-term and long-term responses of species to habitat fragmentation. For example, Yuan et al. (2015) [18] compared the patterns of nuclear microsatellite (SSR) variation and the population structure of an evergreen shrub, *Loropetalum chinense* (R. Br.) Oliver in these two island systems. The results showed that this species can mitigate the negative effects of millennia-old natural habit fragmentation via water-facilitated seed dispersal. Similar results were also observed in a comparative genetic study of the perennial vine *Actinidia chinensis* Planch. [19]. For long-lived species, the occurrence of gene flow among populations has great potential to mitigate the genetic erosion caused by both recent and historical habitat fragmentation [20]. However, similar comparative genetic studies to determine whether the same is true for short-lived, heterostylous plants, e.g., *H. chrysotricha*, are still lacking.

Hedyotis chrysotricha is a short-lived, subtropical forest understory herb of southeast China, with a wide distribution in both the TIL and ZA regions [21,22]. This insect-pollinated species has two floral morphs, i.e., long-styled vs. short-styled, which differ reciprocally in the placement of stigmas and anthers to enhance outcrossing. In an ideal population, an equal morph ratio is expected for the availability of compatible pollen and reproductive success [23]. However, skewed morph ratios have been observed in small and fragmented populations of many heterostylous species as a result of demographic stochasticity [24,25]. Morph types differ not only in herkogamy, but also in self- and intramorph compatibility levels, etc. This may further affect levels of inbreeding and population genetic diversity, and the ability to cope with the effects of habitat fragmentation [26–28]. For instance, van Rossum and Triest (2008) [29] found that pin and thrum individuals of distylous

Primula veris differed in fine-scale spatial genetic structure patterns at a small scale, which may be explained by partial self-compatibility of the pin morph combined with restricted seed dispersal and pollinator behavior. In our previous genetic study [13], we surveyed populations of *H. chrysotricha* from the recently fragmented TIL region only, using nuclear SSR markers. Although SSRs are highly polymorphic markers, a limited number of markers might result in low power for addressing landscape genetic questions at large spatial and temporal scales [30–32]. With the decreasing cost of next-generation sequencing, restriction site-associated DNA sequencing (RAD-Seq), which enables the fast identification of thousands of single nucleotide polymorphisms (SNPs) in non-model organisms without any prior information, has become a cost-effective approach in phylogeographic and population genetic studies [33]. In this study, we employed genome-wide SNPs derived from RAD-Seq data to examine and compare short-term vs. long-term effects of habitat fragmentation on *H. chrysotricha* in the TIL vs. ZA regions. Specifically, we aimed to: (1) investigate the effects of short-term vs. long-term habitat fragmentation on the species' population genetic variation (diversity, inbreeding) patterns; (2) examine the genetic differentiation and genetic structure of *H. chrysotricha* populations at small and large spatial scales; and (3) evaluate patterns of gene flow and demographic changes of *H. chrysotricha* populations in the two island systems. This comparative study will increase our understanding of the genetic responses of short-lived plants with strong seed dispersal capabilities to fragmentation and provide an insight into the management and conservation of subtropical forest dwellers in fragmented habitats at both small and large spatial–temporal scales.

2. Results

2.1. RAD-Seq Data and SNP Filtering

Approximately 389.4 billion (389,411,194,200) raw reads were produced from 256 individuals of *H. chrysotricha* sampled across the TIL and ZA regions. After quality control, filtering, and trimming, an average of 1421.5 million and 1638.9 million high-quality reads were retained for the TIL and ZA samples, respectively (Table S1). A catalog containing 9,131,292 loci was constructed, and 9,067,359 loci were genotyped by GSTACKS using the data sets of all *H. chrysotricha* samples. The mean, minimum, and maximum values for effective per-sample coverage were 26.9×, 16.8×, and 53.0×, respectively. After filtering loci of low quality (minor allele frequency < 0.01; missing rate > 0.5), 185 and 188 unlinked SNPs were eventually identified in TIL and ZA populations, respectively, and used for all subsequent analyses (Tables S2 and S3).

2.2. Population Variation

Based on genome-wide SNP markers, we detected significantly higher levels of genetic diversity in *H. chrysotricha* from the TIL region than from the ZA region (π = 0.247 vs. 0.208, $p < 0.01$; H_O = 0.307 vs. 0.256, $p < 0.01$; H_E = 0.228 vs. 0.190, $p < 0.01$; Tables 1 and 2). Within each region, there were no major differences in genetic diversity between island and mainland populations (TIL island vs. mainland populations: π = 0.257 vs. 0.237, H_O = 0.313 vs. 0.302, H_E = 0.237 vs. 0.219; ZA island vs. mainland populations: π = 0.208 vs. 0.207, H_O = 0.256 vs. 0.255, H_E = 0.189 vs. 0.192). F_{IS} values were all negative in all *H. chrysotricha* populations, with a mean value of −0.078 and −0.066 in the TIL and ZA regions, respectively (Tables 1 and 2).

Table 1. Genetic characteristics of island and mainland populations of *Hedyotis chrysotricha* in the Thousand-Island Lake (TIL) region.

TIL	Population ID	Island Size (ha)	Latitude	Longitude	Sample Size	π	H_O	H_E	F_{IS}
Island	IP01	1320	29.5340	118.8818	10	0.236	0.312	0.219	−0.119
	IP02	47.98	29.5254	118.9395	8	0.254	0.316	0.231	−0.080
	IP03	27.49	29.5365	118.9402	9	0.250	0.308	0.231	−0.063
	IP04	1.33	29.5625	118.8990	10	0.255	0.326	0.236	−0.095
	IP05	0.86	29.5487	118.9154	10	0.274	0.317	0.254	−0.053
	IP06	2.56	29.5019	118.8319	10	0.270	0.320	0.250	−0.040
	IP07	2.17	29.4862	118.9098	10	0.252	0.308	0.234	−0.055
	IP08	8.98	29.4949	118.8610	10	0.263	0.294	0.244	−0.008
Mainland	WM01	-	29.5789	118.8671	10	0.240	0.302	0.223	−0.063
	WM02	-	29.5563	118.8154	10	0.227	0.306	0.211	−0.136
	WM03	-	29.4894	118.6907	10	0.221	0.300	0.206	−0.126
	WM04	-	29.4056	118.6367	8	0.219	0.296	0.200	−0.127
	EM01	-	29.3867	118.7484	10	0.235	0.308	0.218	−0.117
	EM02	-	29.4104	118.8441	10	0.256	0.306	0.237	−0.035
	EM03	-	29.4615	118.9056	10	0.221	0.296	0.206	−0.121
	EM04	-	29.4816	119.0097	10	0.275	0.302	0.255	−0.017
Mean (All)						0.247	0.307	0.228	−0.078

Abbreviations: π, nucleotide diversity; H_O, observed heterozygosity; H_E, expected heterozygosity; F_{IS}, within-population inbreeding coefficient.

Table 2. Genetic characteristics of island and mainland populations of *Hedyotis chrysotricha* in the Zhoushan Archipelago (ZA) region.

ZS	Population ID	Island Size (ha)	Latitude	Longitude	Sample Size	π	H_O	H_E	F_{IS}
Island	ZI01	8635	30.0399	121.8784	4	0.221	0.256	0.181	−0.053
	ZI02	50,265	30.1021	121.1167	10	0.209	0.261	0.194	−0.075
	ZI03	50,265	30.0040	122.2927	10	0.206	0.263	0.191	−0.085
	ZI04	4170	29.8266	122.3036	10	0.208	0.253	0.194	−0.064
	ZI05	6182	29.8573	122.3857	10	0.198	0.250	0.184	−0.068
	ZI06	1185	29.9790	122.3782	8	0.213	0.254	0.194	−0.053
	ZI07	11,512	30.2637	122.1850	10	0.202	0.255	0.187	−0.077
Mainland	ZM01	-	29.8350	121.8706	10	0.195	0.243	0.181	−0.058
	ZM02	-	29.8490	121.7392	10	0.215	0.258	0.200	−0.064
	ZM03	-	30.0183	121.5079	10	0.206	0.257	0.191	−0.074
	ZM04	-	30.0685	121.4286	9	0.213	0.260	0.197	−0.055
Mean (All)						0.208	0.256	0.190	−0.066

Abbreviations: π, nucleotide diversity; H_O, observed heterozygosity; H_E, expected heterozygosity; F_{IS}, within-population inbreeding coefficient.

Pairwise F_{ST} values between the TIL populations ranged from 0.001 to 0.393, with about 13% being significant ($p < 0.05$) after sequential Bonferroni correction (Table 3). Among the significant pairwise comparisons, the great majority (11 out of 16) involved population EM04. Pairwise F_{ST} values between the ZA populations ranged from 0.001 to 0.436, with about 24% being significant ($p < 0.05$) after sequential Bonferroni correction (Table 4). Here, a large proportion of significantly high F_{ST} values (6 out of 13) involved population ZI05. The average F_{ST} value across populations was slightly higher in the TIL ($F_{ST} = 0.103$) region than in the ZA ($F_{ST} = 0.092$) region, yet this difference was only marginally significant ($p = 0.046$). No IBD pattern was found in either region (TIL: $r = 0.181$, $p = 0.174$; ZA: $r = 0.072$, $p = 0.242$) (Figure S1).

Table 3. Pairwise F_{ST} values among the 16 populations of *Hedyotis chrysotricha* in the Thousand-Island Lake (TIL) region.

	IP01	IP02	IP03	IP04	IP05	IP06	IP07	IP08	WM01	WM02	WM03	WM04	EM01	EM02	EM03	EM04
IP01																
IP02	0.028															
IP03	0.015	0.090														
IP04	0.022	0.010	0.018													
IP05	0.200	0.006	0.066	**0.185**												
IP06	0.093	0.007	0.006	0.004	0.073											
IP07	0.031	0.029	0.050	0.017	**0.160**	0.031										
IP08	0.092	0.004	0.019	0.008	0.077	0.085	0.001									
WM01	0.047	0.079	0.085	0.002	0.058	0.015	0.019	0.020								
WM02	0.003	0.047	0.008	0.026	0.226	**0.105**	0.013	0.089	0.024							
WM03	0.092	0.002	0.039	0.030	0.201	**0.094**	0.018	0.085	0.047	0.014						
WM04	0.083	0.014	0.024	0.038	0.172	0.004	0.044	0.002	0.051	0.028	0.064					
EM01	0.069	0.047	0.008	0.004	0.211	0.085	0.030	0.083	0.032	0.076	0.049	0.083				
EM02	0.011	0.060	0.079	0.048	0.102	0.056	0.067	0.073	0.049	0.014	0.018	0.058	0.007			
EM03	0.028	0.038	0.012	0.040	0.223	**0.104**	0.034	0.093	0.021	0.030	0.001	0.007	0.021	0.018		
EM04	**0.370**	**0.238**	**0.251**	**0.285**	**0.204**	0.108	**0.287**	0.066	**0.290**	**0.393**	**0.366**	**0.288**	0.376	0.196	**0.386**	

Values in bold are significantly different from zero after sequential Bonferroni correction.

Table 4. Pairwise F_{ST} values among the 11 populations of *Hedyotis chrysotricha* in the Zhoushan Archipelago (ZA) region.

	ZI01	ZI02	ZI03	ZI04	ZI05	ZI06	ZI07	ZM01	ZM02	ZM03	ZM04
ZI01											
ZI02	0.147										
ZI03	0.157	0.075									
ZI04	0.096	0.066	0.013								
ZI05	0.199	**0.303**	**0.270**	0.067							
ZI06	0.162	0.082	0.089	0.057	**0.348**						
ZI07	0.055	0.053	0.036	0.094	**0.315**	0.014					
ZM01	0.139	0.031	0.065	0.053	**0.352**	0.087	0.010				
ZM02	0.014	0.034	0.056	0.078	**0.143**	0.085	**0.174**	**0.158**			
ZM03	0.002	0.059	0.038	**0.167**	**0.436**	0.011	0.051	0.001	**0.242**		
ZM04	0.125	0.073	0.096	0.063	**0.362**	0.123	0.006	0.083	**0.128**	0.001	

Values in bold are significantly different from zero after sequential Bonferroni correction.

2.3. Population Genetic Structure

Based on a STRUCTURE analysis, the most likely number of genetic groups was $K = 3$ for the TIL populations and $K = 2$ for the ZA populations. In the TIL region, Cluster I ('dark blue') was present at high frequency (80% of all local samples) in seven of eight island populations, whereas the great majority of individuals (89%) from the western mainland populations and the remaining island population (IP08) were assigned to Cluster II ('red') (Figure 1 and Figure S2). In the eastern mainland populations, half of the individuals from EM04 were assigned to Cluster III ('green'); the remaining populations contained variable numbers of individuals belonging to both Cluster I and Cluster II (Figure 1). In the ZA region, nearly all populations (excepting ZI04) consisted of variable numbers of individuals with a probability of ≥ 0.5 of belonging to both Cluster I ('blue') and Cluster II ('orange'), when $K = 2$ (Figure 2). At $K = 3$–5, we also observed that the population ZM02 was assigned to a separate cluster (Figure S3), suggesting potential genetic differentiation between this mainland population and the others.

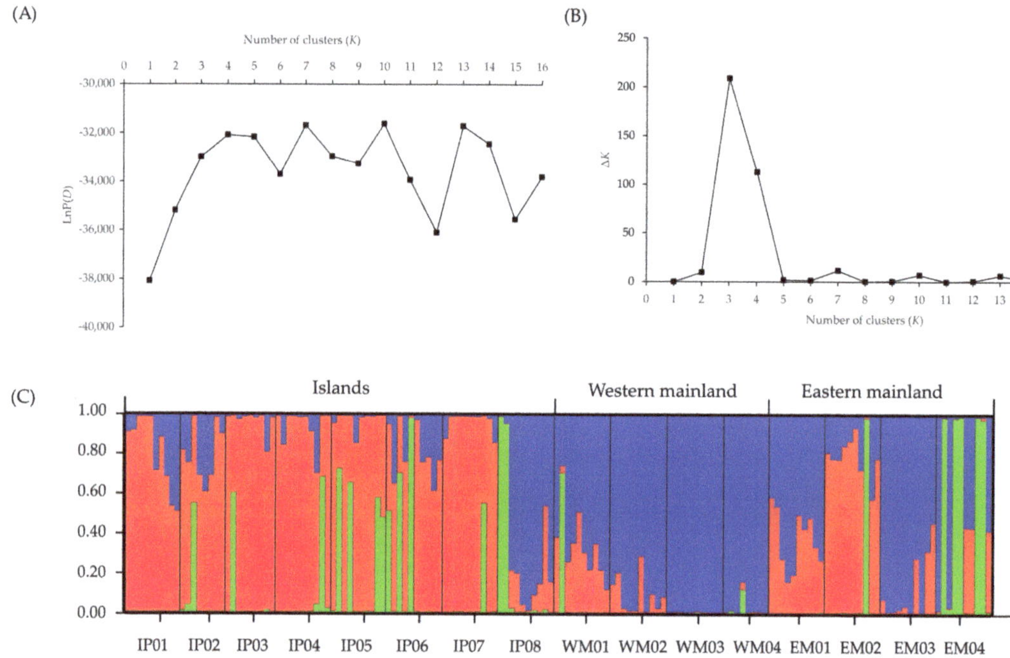

Figure 1. STRUCTURE analysis of 155 individuals (16 populations) of *Hedyotis chrysotricha* from the Thousand-Island Lake (TIL) region, based on RAD-Seq-derived single nucleotide polymorphism (SNP) data. (**A**) Plots of the mean posterior probability [LnP(D)] values of each K and (**B**) the corresponding ΔK statistics. (**C**) Histogram of the STRUCTURE analysis for the model with $K = 3$ (showing the highest ΔK). A vertical bar represents a single individual, and each color corresponds to a suggested cluster (Cluster I: dark blue; Cluster II: red; Cluster III: green). The x-axis corresponds to population codes. The y-axis presents the estimated membership coefficient (Q) for each individual in the three clusters.

2.4. Contemporary Gene Flow

In *H. chrysotricha* populations from the TIL region, the proportion of individuals that originated from within the same site ranged from 68% (IP04) to 81% (WM02), with an average value of 72% (see BAYESASS, Table S4). In contrast, rates of gene flow (or 'migration', m) between populations were low to moderate, and in the majority of populations no more than 5% of the individuals were exchanged with other populations (Table S4). Similarly, in the ZA region, populations largely consisted of individuals originating from the same site (73% on average) (see BAYESASS, Table S5). Nevertheless, the average rate of interpopulation gene flow in the ZA region ($m_{ZA} = 0.027$) was significantly higher than in the TIL region ($m_{TIL} = 0.019$; $p = 0.003$). The DIVMIGRATE method also suggested that the relative migration rate was significantly higher in the ZS region than in the TIL region ($rm_{ZA} = 0.349$, $rm_{TIL} = 0.252$, $p < 0.01$) (Tables S6 and S7). In addition, we observed significant asymmetric gene flow patterns in 10 out of the 16 populations in the TIL region, and 4 out of the 11 populations in the ZA region (Figures 3 and 4).

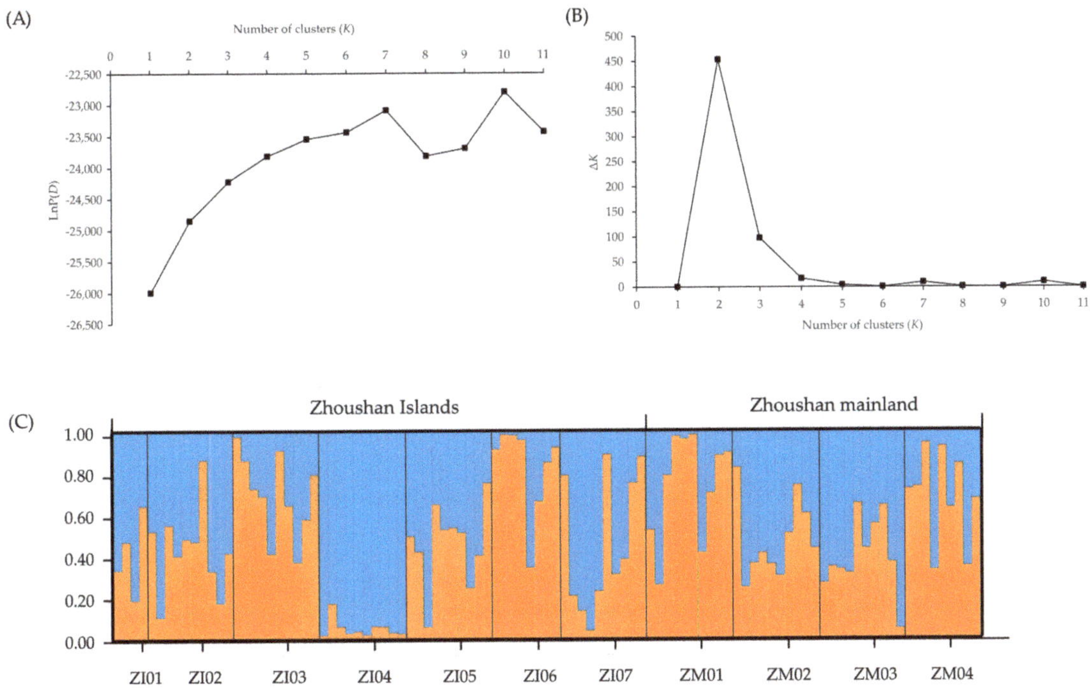

Figure 2. STRUCTURE analysis of 101 individuals (11 populations) of *Hedyotis chrysotricha* from the Zhoushan Archipelago (ZA) region, based on RAD-Seq-derived single nucleotide polymorphism (SNP) data. (**A**) Plots of the mean posterior probability [LnP(D)] values of each K and (**B**) the corresponding ΔK statistics. (**C**) Histogram of the STRUCTURE analysis for the model with $K = 2$ (showing the highest ΔK). A vertical bar represents a single individual, and each color corresponds to a suggested cluster (Cluster I: blue; Cluster II: orange). The x-axis corresponds to population codes. The y-axis presents the estimated membership coefficient (Q) for each individual in the two clusters.

2.5. Demographic (Bottleneck) Analyses

Irrespective of the mutation model assumed (IAM, infinite allele model; SMM, stepwise mutation model; or TPM, two-phase mutation model), we found a significant excess of heterozygosity ($p < 0.05$) for one island (IP05) and one mainland population (EM04) in the TIL region, and one island population (ZI01) in the ZA region (Table 5), indicating that these populations may have experienced a recent bottleneck. However, a significant excess of heterozygosity was only found under the IAM model for one TIL population (IP08) and five ZA populations (ZI02, ZI07, ZM02, ZM03, ZM04; Table 5).

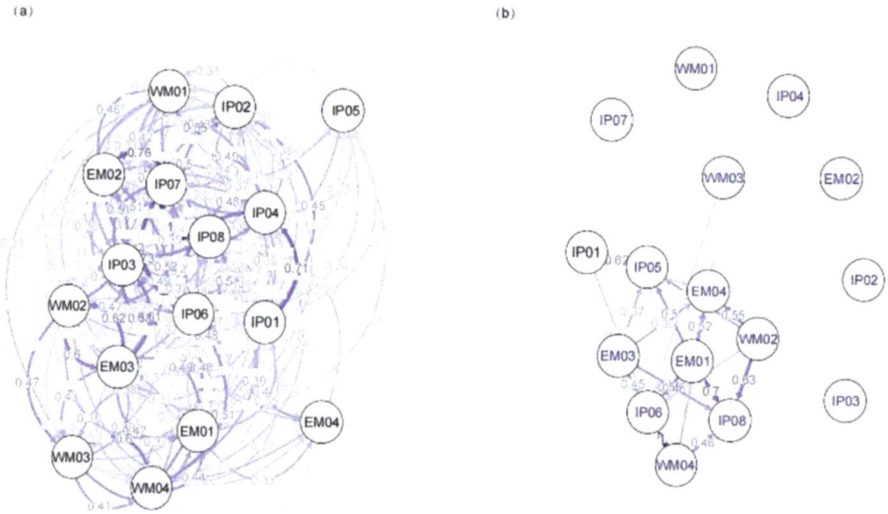

Figure 3. (**a**) The migration network among the 16 *Hedyotis chrysotricha* populations in the Thousand-Island Lake (TIL) region based on G_{ST} (coefficient of gene differentiation) values and (**b**) significant relative migration relationships among populations (after 1000 bootstraps). Arrow thickness and color tone are proportional to the magnitude of the gene flow.

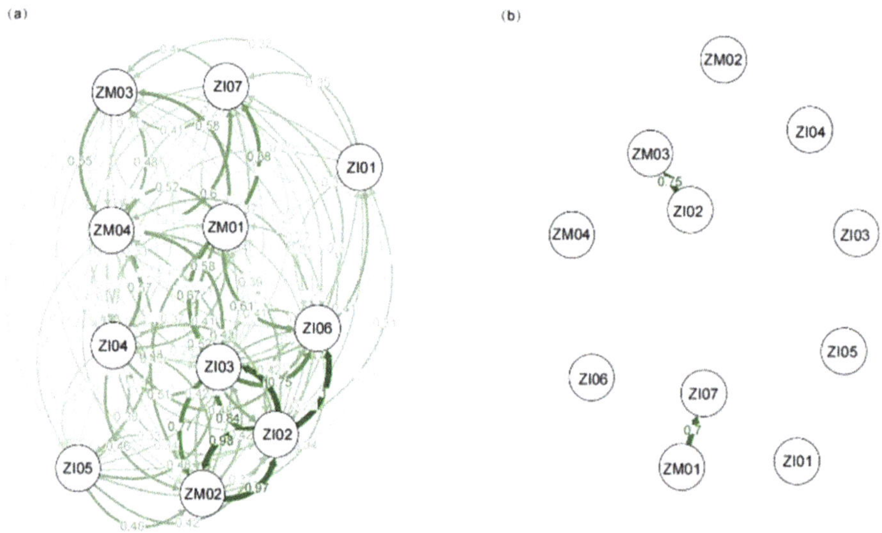

Figure 4. (**a**) The migration network among the 11 *Hedyotis chrysotricha* populations in the Zhoushan archipelago (ZA) region based on G_{ST} (coefficient of gene differentiation) values and (**b**) significant relative migration relationships among populations (after 1000 bootstraps). Arrow thickness and color tone are proportional to the magnitude of the gene flow.

Table 5. Bottleneck analyses for a total of 27 populations of *Hedyotis chrysotricha*, including 16 from the Thousand-Island Lake (TIL) region and 11 from the Zhoushan Archipelago (ZA) region. The *p*-values are shown for the Wilcoxon signed rank tests, as evaluated under the infinite alleles model (IAM), the two-phase mutation model (TPM), and the stepwise mutation model (SMM), respectively.

TIL Region	*p* (Wilcoxon Signed Rank Test)			ZS Region	*p* (Wilcoxon Signed Rank Test)		
Population	**IAM**	**TPM**	**SMM**	**Population**	**IAM**	**TPM**	**SMM**
IP01	0.691	0.984	0.864	ZI01	**0.000**	**0.001**	**0.007**
IP02	0.949	0.999	0.997	ZI02	**0.028**	0.216	0.915
IP03	0.801	0.995	0.958	ZI03	0.202	0.428	0.977
IP04	0.982	0.999	0.997	ZI04	0.175	0.628	0.977
IP05	**0.000**	**0.003**	**0.000**	ZI05	0.137	0.469	0.958
IP06	0.262	0.973	0.786	ZI06	0.211	0.611	0.966
IP07	0.859	0.999	0.959	ZI07	**0.019**	0.242	0.844
IP08	**0.017**	0.637	0.220	ZM01	0.331	0.694	0.981
WM01	0.992	0.999	0.999	ZM02	**0.021**	0.203	0.835
WM02	0.436	0.989	0.786	ZM03	**0.021**	0.206	0.889
WM03	0.654	0.996	0.860	ZM04	**0.034**	0.331	0.909
WM04	0.942	0.999	0.996				
EM01	0.358	0.983	0.660				
EM02	0.873	0.999	0.981				
EM03	0.744	0.996	0.926				
EM04	**0.000**	**0.001**	**0.000**				

Bold values indicate statistical significance ($p < 0.05$).

3. Discussion

Habitat fragmentation changes continuous habitats into small and isolated patches, and populations inhabiting these habitats are often considered to have low genetic diversity [8,34]. In our previous study [12], we found that island populations of *H. chrysotricha* in the TIL region had significantly lower mean genetic diversity than those from the mainland, based on nuclear SSR markers, suggesting that anthropogenic habitat fragmentation can lead to significant loss of genetic diversity over a few decades. In the present study, we further tested whether genetic erosion increases with an increase in spatio-temporal scales of fragmentation. Based on RAD-Seq-derived SNP markers, we observed that the average estimates of genetic diversity for ZA populations were significantly lower than for TIL populations ($\pi = 0.247$ vs. 0.208, $p < 0.01$; $H_O = 0.307$ vs. 0.256, $p < 0.01$; $H_E = 0.228$ vs. 0.190, $p < 0.01$; Tables 1 and 2), and we found no major differences in genetic diversity between island and mainland populations in either the TIL or the ZA region. Although it remains unknown how much of the initial genetic variation has been lost in each region since the formation of islands, these results may indicate that *H. chrysotricha* populations have retained a considerable proportion of their initial genetic diversity over either ~60 years or thousands of years of habitat fragmentation.

Generally, the genetic effects of habitat fragmentation on plants vary, depending on their life-history traits [2]. In this study, the lack of a significant effect of habitat fragmentation on the genetic diversity of *H. chrysotricha* may be explained by demographic factors and/or strong gene dispersal capacity. As a distylous plant species, *H. chrysotricha* has two contrasting flower morphs that reciprocally differ in the spatial separation of stigmas and anthers, which in turn prevents selfing and intramorph mating. Morph bias is expected to reinforce genetic drift effects and inbreeding by reducing mating opportunities [35,36]. We also found that no biased morph ratio existed in either island or mainland populations, and overall levels of inbreeding in both island regions were negative and close to zero, suggesting a sufficient within-population heterozygosity (Tables 1 and 2). In addition, most of the populations showed no genetic signatures of recent bottlenecks (Table 5). Therefore, it is likely that population sizes of *H. chrysotricha* have remained sufficiently large to prevent loss of genetic diversity via inbreeding and genetic drift after fragmentation [37,38].

The gene dispersal ability via pollen or seed is another key characteristic that determines the potential of a plant species to counteract the negative effects of habitat fragmentation [8]. *Hedyotis chrysotricha* is an insect-pollinated plant; its flower visitors include solitary bees (Halictidae and Andrenidae), Diptera (Syrphidae and Bombyliidae), honey bees, small lepidopterans, and/or thrips [13,39]. Due to the limited foraging distances of these pollinators, they seem unlikely to be able to maintain significant levels of gene flow and population connectivity across fragmented habitats. In contrast, the indehiscent fruit of *H. chrysotricha* contains several very small seeds (1.5–2 × 2–2.5 mm), whose dispersal should be easily facilitated by wind [21]. In our previous study, we detected a moderate level of gene flow among 18 *H. chrysotricha* populations in the TIL region [12]. Based on high-quality SNP markers, our genetic analyses here have further verified the undisturbed population connectivity of *H. chrysotricha* populations in the TIL region, with an average of c. 28% individuals in each population being exchanged with other sites (Table S2). Given the relatively large spatial and temporal scales of habitat fragmentation in the ZA region (Figure 5), one might expect high population differentiation and low levels of gene flow between island populations. However, both BAYESASS and DIVMIGRATE analyses suggested that the average level of gene flow among populations in the ZA region was higher than in the TIL region, although the results from these two methods may not be compared directly. Similarly, populations in the ZA region exhibited marginally lower levels of genetic differentiation than those in TIL region (F_{ST} = 0.092 vs. 0.103). A previous study showed that water greatly facilitated the seed dispersal of a typical wind disperser across a fragmented landscape [40]. By referencing the population genetic study of the maritime *Hedyotis strigulosa* var. *parviflora* (Hook. et Arn.) Yamazaki [14], we speculated that water flow may also facilitate the fruit dispersal of *H. chrysotricha*. Although the hydration of *H. chrysotricha* could remain speculative due to the lack of detailed morphological and experimental evidence, the genetic evidence presented here further verified that the species has sufficient fruit (seed) dispersal capabilities to maintain moderate-to-high levels of ongoing gene flow and population connectivity across fragmented landscapes at large spatial and temporal scales.

Corresponding to the above results, little evidence of strong genetic structuring was found in *H. chrysotricha* within the two island systems. Based on SNP markers, the STRUCTURE analysis grouped the TIL populations basically into three clusters (Figure 1) and the ZA populations into two clusters (Figure 2). In the TIL region, most of the island populations were genetically separated from western mainland populations, which is consistent with the results of our previous nSSR analysis, whereas the genetic composition of eastern mainland TIL populations was slightly different from the former study, consisting of a mixture of all three clusters [12]. In the ZA region, the genetic divergence between the mainland and island populations was not as high as expected, and most of the island populations still feature a high percentage of genetic admixtures, even after thousands of years of isolation (Figure 2). The higher levels of population substructure in the TIL region (three clusters) than in the ZA region (two clusters) may be attributed to their contrasting landscape histories. For example, before water flooded the TIL region, mountains (e.g., the Baiji Mountains in the western part of the TIL region) may have acted as a stronger dispersal barrier than water for gene exchange between populations inhabiting the middle and western former hilltops. In the case of the ZA region, which separated from the nearby mainland 7000–9000 years ago due to a rise in sea level, water has been instrumental in facilitating gene exchange between populations for a much longer time than in the TIL region. When combined with the lack of isolation by distance in each island system, these results further suggested that population connectivity in *H. chrysotricha* has not been greatly modified in either island system, regardless of their significant differences in spatial–temporal characteristics.

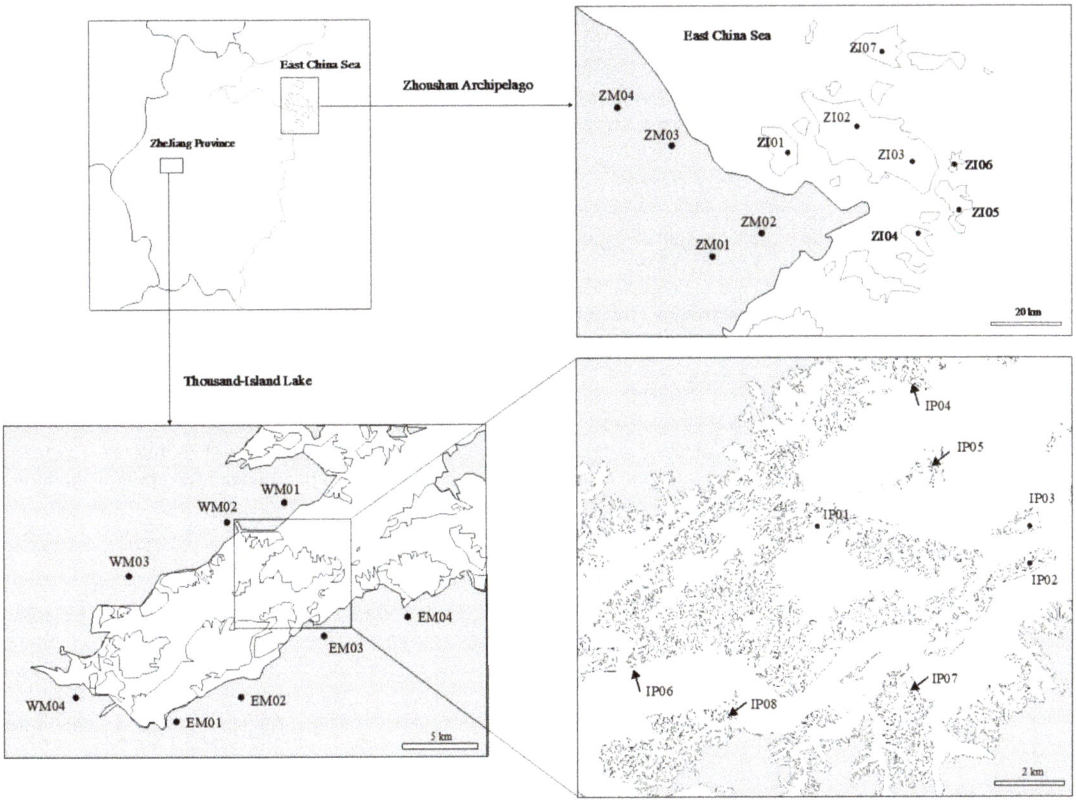

Figure 5. Sample localities of *Hedyotis chrysotricha* in the Thousand-Island Lake (TIL) region and the Zhoushan Archipelago (ZA) region of southeast China. Population codes are identified in Tables 1 and 2.

4. Materials and Methods

4.1. Sample Collection and DNA Extraction

We collected a total of 160 *H. chrysotricha* individuals from eight islands and eight adjacent mainland populations in the TIL region, and 110 individuals from seven islands and four adjacent mainland populations in the ZA region (Tables 1 and 2; Figure 5). In each population, almost equal numbers of individuals with long-styled (L-morph) or short-styled (S-morph) flowers were collected separately. We extracted the total genomic DNA from the dry leaf material of each sample using DNA Plantzol Reagent (Invitrogen, Carlsbad, CA, USA), following the manufacturer's protocol. We then checked the quality of the DNA using 1% agarose gel electrophoresis and measured the DNA concentration using a NanoDrop 2000 spectrophotometer (Thermo Scientific, Wilmington, DE, USA). Finally, 256 of the 270 DNA samples met the minimum quality requirements and were retained for subsequent sequencing and genotyping.

4.2. RAD Sequencing and Data Processing

RAD libraries were prepared using the restriction enzyme *Eco*RI and sample-specific barcodes. Samples in the libraries were pooled and sequenced on an Illumina HiSeq 2500 to generate 150 bp pair-end reads, at the Novogene Bioinformatics Institute (Beijing, China). Standard quality control (QC) pipelines were used to process the raw sequencing data. Reads with > 5% of unknown nucleotides, low-quality reads (quality value ≤ 5), and reads that did not contain sample-specific barcodes were discarded. After filtering procedures, the quality-filtered reads data for each individual were then assembled into de novo loci, and SNP calling was performed using the following core programs: USTACKS, CSTACKS, SSTACKS, GSTACKS, and POPULATIONS, implemented in the STACKS pipeline v2.2 [41]. Briefly, all sequences were firstly processed in USTACKS to produce consensus sequences of RAD tags. The program USTACKS takes a set of short-read sequences as input and aligns them into exactly matching stacks. The minimum depth of coverage required for creating a stack was set at three (–m: 3) sequences, and the maximum distance allowed between stacks was set at four (–M: 4) nucleotides. The program CSTACKS was used to build a catalog of consensus loci containing all the loci from all the *H. chrysotricha* samples and merging all alleles together, with the number of mismatches allowed between sample loci set to four (–n: 4). Next, sets of stacks (i.e., putative loci) created in USTACKS were compared against the catalog using SSTACKS. Then, GSTACKS was used to align the reads to the locus and to call SNPs. Finally, the catalog of reads was filtered using the POPULATIONS program to produce a data set for downstream analyses. To identify high-credibility SNPs, polymorphic RAD loci that were present in at least 75% of the individuals in each population were retained (–r: 0.75). Potential homologs were excluded by removing loci showing heterozygosity of >0.5 within samples [42], and one SNP per RAD locus was kept, to reduce the impact of linkage disequilibrium.

4.3. Genetic Diversity and Population Structure

We calculated the observed heterozygosity (H_O), expected heterozygosity (H_E), nucleotide diversity (π), and inbreeding coefficient (F_{IS}) for each population using the POPULATIONS program in STACKS. Pairwise *F*-statistics (F_{ST}) (1000 permutations) among populations were calculated using ARLEQUIN v3.5.2 [43]. Population structure was analyzed using the program STRUCTURE v2.3.5 [44]. The number of genetic clusters (*K*) was set from 1 to 16 for TIL populations and from 1 to 11 for ZA populations, corresponding to the number of sampled populations in each region. Ten independent iterations were conducted for each *K*, with a burn-in of 10,000 and 100,000 Markov chain Monte Carlo replicates, by assuming the admixture model and independent allele frequencies. The optimal *K* value was determined from the ΔK values calculated by STRUCTURE HARVESTER [45]. A Mantel test for genetic differentiation [$F_{ST}/(1 - F_{ST})$] against geographic distance (log10 transformed) was performed in GENALEX v6.1 [46] for each region, to test the pattern of isolation by distance (IBD). Statistical significance was determined with 1000 permutations.

4.4. Gene Flow and Demographic History Analyses

For each region, we estimated the levels of interpopulation gene flow using the program BAYESASS v3.03 [47]. First, we ran BAYESASS with the default delta values for allelic frequency, migration rates, and inbreeding coefficients. Then, subsequent runs were adjusted with different delta values to ensure that the acceptance rate ranged between 40 and 60% for each parameter [47]. We performed 10 independent runs (1×10^7 iterations with a burn-in of 10^6 generations), each with a different initial seed. Model convergence was assessed by the program TRACER v1.5 [48]. In addition, we also used the DIVMIGRATE function from the DIVERSITY package v1.9.90 in R [49] to calculate the relative direction of the gene flow between island and mainland populations, using Nei's G_{ST} method. To test for asymmetric flow (significantly higher in one direction than the other), 95% confidence intervals were calculated from 1000 bootstrap replicates [50].

We used the program BOTTLENECK v1.2.02 [51] to determine whether *H. chrysotricha* populations underwent significant reductions in effective population size (N_e). We used Wilcoxon's signed rank test, which examines whether populations exhibit a greater level of heterozygosity than predicted in a population at mutation-drift equilibrium, to detect bottlenecks occurring over approximately the last 2–4 N_e generations. For each *H. chrysotricha* population, we performed 10,000 simulations for each of three mutation models (IAM, infinite allele model; SMM, stepwise mutation model; and TPM, two-phase mutation model, with 95% single-step and 5% multi-step mutations). Statistical significance was set at the 0.05 level.

5. Conclusions

Based on high-resolution SNPs, we compared the genetic response of *H. chrysotricha* to short-term and long-term habitat fragmentation. The levels of genetic diversity and the population substructure of *H. chrysotricha* populations were both higher in the TIL region than in the ZA region, while gene flow was higher and less asymmetric among ZA populations. The differences in genetic diversity between island and mainland populations in both regions were not significant, suggesting that genetic erosion of island populations does not increase as the spatial and temporal scales of fragmentation increase. Population connectivity in *H. chrysotricha* has also not been greatly modified by either short-term or long-term habitat fragmentation. The strong fruit (seed) dispersal capacity of *H. chrysotricha*, facilitated by the matrix of water in the island system, could buffer against the negative effects of habitat fragmentation and maintain the population connectivity over thousands of years.

Supplementary Materials: The following supporting information can be downloaded at: https://www.mdpi.com/article/10.3390/plants11141800/s1. Table S1: Raw statistical data for *Hedyotis chrysotricha* samples from the Thousand-Island Lake (TIL) and Zhoushan Archipelago (ZA) regions, generated by Illumina sequencing; Table S2: Genotype information for the 185 filtered SNPs from the TIL dataset; Table S3: Genotype information for the 188 filtered SNPs from the ZA dataset; Table S4: Migration rates (the fraction of individuals in population i (pop i) from population j (pop j)) and confidence intervals between populations of *H. chrysotricha* from the TIL region. Bold values represent the proportion of non-migrants within a population, and italic values indicate asymmetrical migration rates between populations; Table S5: Migration rates (the fraction of individuals in population i (pop i) from population j (pop j)) and confidence intervals between populations of *H. chrysotricha* from the ZA region. Bold values represent the proportion of non-migrants within a population, and values in italics indicate asymmetrical migration rates between populations; Figure S1: Analysis of isolation by distance based on $F_{ST}/(1-F_{ST})$ and the geographic distance (log10 transformed) in (A) the Thousand-Island Lake (TIL) region and (B) the Zhoushan Archipelago (ZA) region; Figure S2: Histogram of the STRUCTURE analysis in the TIL region for the model with $K = 2$, 4, and 5. A vertical bar represents a single individual, and each color corresponds to a suggested cluster; Figure S3: Histogram of the STRUCTURE analysis in the ZA region for the model with $K = 3$–5. A vertical bar represents a single individual, and each color corresponds to a suggested cluster.

Author Contributions: N.Y. and Y.Q. conceived the study; N.Y., R.L. and S.L. collected the samples. N.Y. extracted the DNA samples. N.Y., S.W. and R.L. analyzed the genetic data and wrote the manuscript. H.P.C., R.L. and Y.Q. revised the manuscript. All authors have read and agreed to the published version of the manuscript.

Funding: This research was funded by the Chinese National Natural Science Foundation (No. 31800312), the Jiangsu Innovative and Enterpreneurial Talent Programme (JSSCBS20211311), the Natural Science Foundation of Guangdong Province, China (No. 2021A1515010946), and the Forestry Science and Technology Innovation Project of Guangdong Province, China (No. 2020KJCX003).

Institutional Review Board Statement: Not applicable.

Informed Consent Statement: Not applicable.

Data Availability Statement: SNPs in VCF files across the Thousand-Island Lake (TIL) and the Zhoushan Archipelago (ZA) populations were deposited in Mendeley Data, V1, https://doi.org/10.17632/wsx8swxd27.1 (accessed on 22 May 2022).

Acknowledgments: Special thanks go to the anonymous referees for providing valuable and constructive comments to improve the manuscript.

Conflicts of Interest: The authors declare no conflict of interest.

References

1. Fischer, J.; Lindenmayer, D.B. Landscape modification and habitat fragmentation: A synthesis. *Glob. Ecol. Biogeogr.* **2007**, *16*, 265–280. [CrossRef]
2. Aguilar, R.; Quesada, M.; Ashworth, L.; Herrerias-Diego, Y.; Lobo, J. Genetic consequences of habitat fragmentation in plant populations: Susceptible signals in plant traits and methodological approaches. *Mol. Ecol.* **2010**, *17*, 5177–5188. [CrossRef] [PubMed]
3. Laurance, W.F.; Sayer, J.; Cassman, K.G. Agricultural expansion and its impacts on tropical nature. *Trends Ecol. Evol.* **2014**, *29*, 107–116. [CrossRef] [PubMed]
4. Saunders, D.A.; Hobbs, R.J.; Margules, C.R. Biological consequences of ecosystem fragmentation: A review. *Conserv. Biol.* **1991**, *5*, 18–32. [CrossRef]
5. Tilman, D.; May, R.M.; Lehman, C.L.; Nowak, M.A. Habitat destruction and the extinction debt. *Nature* **1994**, *371*, 65–66. [CrossRef]
6. Fahrig, L.; Merriam, G. Conservation of fragmented populations. *Conserv. Biol.* **1994**, *8*, 50–59. [CrossRef]
7. Frankham, R. Genetics and extinction. *Biol. Conserv.* **2005**, *126*, 131–140. [CrossRef]
8. Young, A.; Boyle, T.; Brown, T. The population genetic consequences of habitat fragmentation for plants. *Trends Ecol. Evol.* **1996**, *11*, 413–418. [CrossRef]
9. Sato, Y.; Kudoh, H. Fine-scale genetic differentiation of a temperate herb: Relevance of local environments and demographic change. *AoB Plants* **2014**, *6*, plu070. [CrossRef]
10. Fontúrbel, F.E.; Jordano, P.; Medel, R.; Bartomeus, I. Scale-dependent responses of pollination and seed dispersal mutualisms in a habitat transformation scenario. *J. Ecol.* **2015**, *103*, 1334–1343. [CrossRef]
11. Wang, R.; Compton, S.G.; Chen, X.Y. Fragmentation can increase spatial genetic structure without decreasing pollen-mediated gene flow in a wind-pollinated tree. *Mol. Ecol.* **2011**, *20*, 4421–4432. [CrossRef] [PubMed]
12. Yuan, N.; Comes, H.P.; Mao, Y.R.; Qi, X.S.; Qiu, Y.X. Genetic effects of recent habitat fragmentation in the thousand-island lake region of southeast china on the distylous herb *Hedyotis chrysotricha* (Rubiaceae). *Am. J. Bot.* **2012**, *99*, 1715–1725. [CrossRef] [PubMed]
13. Riveros, G.M.; Barría, O.R.; Humaña, P.A.M. Self-compatibility in distylous *Hedyotis salzmannii* (Rubiaceae). *Plant Syst. Evol.* **1995**, *194*, 1–8. [CrossRef]
14. Maki, M.; Matsumua, S.; Yamshiro, T. Allozyme diversity and history of distribution expansion in the maritime perennial plant *Hedyotis strigulosa* (Rubiaceae), distributed over the wide latitudes in the Japanese Archipelago. *Biol. J. Linn. Soc.* **2008**, *93*, 679–688. [CrossRef]
15. Wang, Y.; Zhang, J.; Feelfy, K.J.; Jiang, P.; Ding, P. Life-history traits associated with fragmentation vulnerability of lizards in the Thousand-Island Lake, China. *Anim. Conser.* **2009**, *12*, 329–337. [CrossRef]
16. Wang, J.T.; Wang, P.X. The relationship between sea level rising and climate change in East China. *Chin. J. Geogr.* **1980**, *35*, 299–313, (In Chinese with English abstract). [CrossRef]
17. Zhou, J.L. *The Comprehensive Agriculture Programme of Zhoushan City*; Zhejiang Peoples Press: Hangzhou, China, 1987.
18. Yuan, N.; Comes, H.P.; Cao, Y.N.; Guo, R.; Zhang, Y.H.; Qiu, Y.X. A comparative study on genetic effects of artificial and natural habitat fragmentation on *Loropetalum chinense* (Hamamelidaceae) in Southeast China. *Heredity* **2015**, *114*, 544–551. [CrossRef]
19. Yu, W.; Wu, B.; Wang, X.; Yao, Z.; Li, Y.; Liu, Y. Scale-dependent effects of habitat fragmentation on the genetic diversity of *Actinidia chinensis* populations in China. *Hortic. Res.* **2020**, *7*, 172. [CrossRef]
20. Fuller, M.R.; Doyle, M.W. Gene flow simulations demonstrate resistance of long-lived species to genetic erosion from habitat fragmentation. *Conserv. Genet.* **2018**, *19*, 1439–1448. [CrossRef]
21. Chen, T.; Zhu, H.; Chen, J.R.; Charlotte, M.T.; Friedrich, E.; Henrik, L.; Michele, F.; Christian, P. Rubiaceae. In *Flora of China*; Wu, Z.Y., Raven, P.H., Eds.; Science Press: Beijing, China; Missouri Botanical Garden Press: St. Louis, MO, USA, 2011; Volume 19, p. 159.
22. Yuan, N.; Sun, Y.; Nakamura, K.; Qiu, Y.X. Development of microsatellite markers in heterostylous *Hedyotis chrysotricha* (Rubiaceae). *Am. J. Bot.* **2012**, *99*, e43–e45. [CrossRef]
23. Shibayama, Y.; Kadono, Y. Floral morph composition and pollen limitation in the seed set of *Nymphoides indica* populations. *Ecol. Res.* **2003**, *18*, 725–737. [CrossRef]
24. Kéry, M.; Matthies, D.; Schmid, B. Demographic stochasticity in population fragments of the declining distylous perennial *Primula veris* (Primulaceae). *Basic Appl. Ecol.* **2003**, *4*, 197–206. [CrossRef]
25. Washitani, I. Creeping "fruitless falls": Reproductive failure in heterostylous plants in fragmented landscapes. In *The Biology of Biodiversity*; Kato, M., Ed.; Springer: Tokyo, Japan, 2000; pp. 133–145. [CrossRef]
26. Rossum, F.V.; Sousa, S.; Triest, L. Morph-specific differences in reproductive success in the distylous *primula veris* in a context of habitat fragmentation. *Acta Oecolog.* **2006**, *30*, 426–433. [CrossRef]
27. Ågren, J.; Ericson, L. Population structure and morph-specific fitness differences in tristylous *Lythrum salicaria*. *Evolution* **1996**, *50*, 126–139. [CrossRef]
28. Jacquemyn, H.; van Rossum, F.; Brys, R.; Endes, P.; Hermy, M.; Triest, L.; de Blust, G. Effects of agricultural land use and fragmentation on genetics, demography and population persistence of the rare *Primula vulgaris*, and implications for conservation. *Belg. J. Bot.* **2003**, *136*, 5–22.

29. Van Rossum, F.; Triest, L. Fine-scale spatial genetic structure of the distylous *Primula veris* in fragmented habitats. *Plant Biol.* **2007**, *9*, 374–382. [CrossRef]
30. Cornuet, J.M.; Luikart, G. Description and power analysis of two tests for detecting recent population bottlenecks from allele frequency data. *Genetics* **1996**, *144*, 2001–2014. [CrossRef]
31. Landguth, E.L.; Fedy, B.C.; Oyler-McCance, S.J.; Garey, A.L.; Emel, S.L.; Mumma, M.; Wagner, H.H.; Fortin, M.J.; Cushman, S.A. Effects of sample size, number of markers, and allelic richness on the detection of spatial genetic pattern. *Mol. Ecol. Resour.* **2012**, *12*, 276–284. [CrossRef]
32. Hoban, S.M.; Gaggiotti, O.E.; Bertorelle, G. The number of markers and samples needed for detecting bottlenecks under realistic scenarios, with and without recovery: A simulation based study. *Mol. Ecol.* **2013**, *22*, 3444–3450. [CrossRef]
33. Savolainen, O.; Lascoux, M.; Meril, J. Ecological genomics of local adaptation. *Nat. Rev. Genet.* **2013**, *14*, 807–820. [CrossRef]
34. Leimu, R.; Vergeer, P.; Angeloni, F.; Ouborg, N.J. Habitat fragmentation, climate change, and inbreeding in plants. *Ann. N. Y. Acad. Sci.* **2010**, *1195*, 84–98. [CrossRef] [PubMed]
35. Byers, D.L.; Meagher, T.R. Mate availability in small populations of plant species with homomorphic sporophytic self-incompatibility. *Heredity* **1992**, *68*, 353–359. [CrossRef]
36. Washitani, I. Predicted genetic consequences of strong fertility selection due to pollinator loss in an isolated population of *Primula sieboldii*. *Conserv. Biol.* **1996**, *10*, 59–64. [CrossRef]
37. Schaal, B.A.; Leverich, W.J. Molecular variation in isolated plant populations. *Plant Species Biol.* **1996**, *11*, 33–40. [CrossRef]
38. Ewers, R.M.; Didham, R.K. Confounding factors in the detection of species responses to habitat fragmentation. *Biol. Rev.* **2006**, *81*, 117–142. [CrossRef]
39. Ornduff, R. Heterostyly, Population Composition, and Pollen Flow in *Hedyotis caerulea*. *Am. J. Bot.* **1980**, *67*, 95–103. [CrossRef]
40. Soomers, H.; Karssenberg, D.; Soons, M.B.; Verweij, P.A.; Verhoeven, J.T.A.; Wassen, M.J. Wind and water dispersal of wetland plants across fragmented landscapes. *Ecosystems* **2013**, *16*, 434–451. [CrossRef]
41. Catchen, J.; Hohenlohe, P.A.; Bassham, S.; Amores, A.; Cresko, W.A. Stacks: An analysis tool set for population genomics. *Mol. Ecol.* **2013**, *22*, 3124–3140. [CrossRef]
42. Hohenlohe, P.A.; Amish, S.J.; Catchen, J.M.; Allendorf, F.W.; Luikart, G. Next-generation RAD sequencing identifies thousands of SNPs for assessing hybridization between rainbow and westslope cutthroat trout. *Mol. Ecol. Resour.* **2011**, *11*, 117–122. [CrossRef]
43. Excoffier, L.; Laval, G.; Schneider, S. Arlequin (version 3.0): An integrated software package for population genetics data analysis. *Evol. Bioinform.* **2007**, *23*, 47–50. [CrossRef]
44. Pritchard, J.K.; Stephens, P.; Donnelly, P. Inference of population structure using multilocus genotype data. *Genetics* **2000**, *155*, 945–959. [CrossRef] [PubMed]
45. Evanno, G.S.; Regnaut, S.J.; Goudet, J. Detecting the number of clusters of individuals using the software STRUCTURE: A simulation study. *Mol. Ecol.* **2005**, *14*, 2611–2620. [CrossRef] [PubMed]
46. Peakall, R.; Smouse, P.E. GENALEX 6: Genetic analysis in Excel. Population genetic software for teaching and research. *Mol. Ecol. Notes* **2006**, *6*, 288–295. [CrossRef]
47. Wilson, G.A.; Rannala, B. Bayesian inference of recent migration rates using multilocus genotypes. *Genetics* **2003**, *163*, 1177–1191. [CrossRef]
48. Rambaut, A.; Drummond, A.J.; Xie, D.; Baele, G.; Suchard, M.A. Posterior summarisation in Bayesian phylogenetics using Tracer 1.7. *Syst. Biol.* **2018**, *67*, 901–904. [CrossRef]
49. Keenan, K.; McGinnity, P.; Cross, T.F.; Crozier, W.W.; Prodohl, P.A. diveRsity: An R package for the estimation of population genetics parameters and their associated errors. *Methods Ecol. Evol.* **2013**, *4*, 782–788. [CrossRef]
50. Sundqvist, L.; Keenan, K.; Zackrisson, M.; Prodöhl, P.; Kleinhans, D. Directional genetic differentiation and relative migration. *Ecol. Evol.* **2016**, *6*, 3461–3475. [CrossRef]
51. Piry, S.; Luikart, G.; Cornuet, J.M. Bottleneck: A computer program for detecting recent reductions in the effective size using allele frequency data. *J. Hered.* **1999**, *90*, 502–503. [CrossRef]

plants

MDPI

Article

Elucidating Biological Functions of 9-*cis*-Epoxycarotenoid Dioxygenase Genes Involved in Seed Dormancy in *Paeonia lactiflora*

Riwen Fei [1,2], Shixin Guan [2,3,*], Siyang Duan [2,3], Jiayuan Ge [2,3], Tianyi Sun [2,3] and Xiaomei Sun [2,3,*]

[1] College of Horticulture, Shenyang Agricultural University, Shenyang 110866, China
[2] Key Laboratory of Forest Tree Genetics Breeding and Cultivation of Liaoning Province, College of Forestry, Shenyang Agricultural University, Shenyang 110866, China
[3] College of Forestry, Shenyang Agricultural University, Shenyang 110866, China
* Correspondence: shixinguan@126.com (S.G.); sxm7280@syau.edu.cn (X.S.); Tel.: +86-152-0404-0905 (S.G.); +86-182-0241-3441 (X.S.)

Abstract: Abscisic acid (ABA) is a major phytohormone affecting seed dormancy and germination in plants. ABA is synthesized mainly through the C40 carotenoid pathway. In the ABA biosynthesis pathway, 9-*cis*-epoxycarotenoid dioxygenase (NCED) is a key rate-limiting enzyme that regulates the accumulation and content of ABA. However, the role of the *NCED* gene in perennial plants with complex seed dormancy remains largely unknown. Here, we cloned two differentially expressed paralogs of herbaceous peony *NCED* genes, named *PlNCED1* and *PlNCED2*, and further identified their involvement in seed dormancy from perennial herbaceous peony experiencing complex double seed dormancy. The deduced PlNCED amino acid sequences had high sequence homology with *NCED* sequences from other plants and contained the typical conserved RPE65 domain of the NCED family. Phylogenetic analysis showed that *PlNCED1* and *PlNCED2* have a close relationship with *PoNCED* in *Paeonia ostii* and *VvNCED6* in *Vitis vinifera*, respectively. A subcellular localization assay demonstrated that the *PlNCED1* protein resided within the nucleus, while the *PlNCED2* protein was located in the cytoplasm, indicating their different roles in the biosynthesis of ABA. Furthermore, the content of endogenous ABA in transgenic calluses showed that *PlNCEDs* were positively correlated with ABA content. Both *PlNCED* transgenic *Arabidopsis* lines and the functional complementation of *Arabidopsis NCED* mutants found that *PlNCEDs* promoted seed dormancy and delayed seed germination. These results reveal that *PlNCEDs* participate in the seed dormancy of herbaceous peony by regulating the accumulation of endogenous ABA.

Keywords: *Paeonia lactiflora* pall.; abscisic acid; *NCEDs*; seed dormancy; germination

Citation: Fei, R.; Guan, S.; Duan, S.; Ge, J.; Sun, T.; Sun, X. Elucidating Biological Functions of 9-*cis*-Epoxycarotenoid Dioxygenase Genes Involved in Seed Dormancy in *Paeonia lactiflora*. *Plants* **2023**, *12*, 710. https://doi.org/10.3390/plants12040710

Academic Editors: Aiping Song and Yu Chen

Received: 20 December 2022
Revised: 19 January 2023
Accepted: 2 February 2023
Published: 6 February 2023

1. Introduction

Phytohormones are important regulators of seed dormancy, among which ABA plays a major role [1–3]. Studies have shown that ABA-deficient mutants of *A. thaliana*, tomato, and maize undergo early dormancy breaking and enter the germination stage, whereas plants overexpressing the ABA biosynthetic enzyme show prolonged dormancy [4–7]. Studies also demonstrated that the change in endogenous ABA content significantly positively correlates with the degree of seed dormancy [8,9].

In plants, it is known that several phytohormones are involved in seed dormancy and germination [10–12]. Among them, endogenous ABA content is significantly positively correlated with herbaceous peony seed dormancy, while low ABA content promotes seed germination [13]. The final concentration of endogenous ABA in plant seeds depends on the dynamic balance of ABA synthesis and catabolism [1,14]. *NCED* and *CYP707A* are two key enzymes in ABA anabolic and catabolic pathways, respectively. Studies show that NCED has a common role in regulating ABA synthesis and seed dormancy in plants. For example,

A. thaliana contains five *NCED* family members, of which *AtNCED5* is up-regulated at the late stage of seed maturation and cooperates with *AtNCED6* and *AtNCED9* to enhance seed dormancy by controlling ABA levels [15,16]. *LeNCED1* transgenic tomato plants enhanced ABA biosynthesis and increased seed dormancy [17]. The variation trend of ABA content was consistent with that of *AhNCED1* gene transcription in peanut [18]. *PtNCED1* directly regulated orchid seed dormancy and was involved in ABA content [19]. In addition, *NCED* can also affect other physiological functions of plants by changing endogenous ABA content. Overexpressing *PvNCED1* enhanced drought tolerance by manipulating ABA levels in tobacco [4]. Silencing *AcNCED1* blocked ABA biosynthesis and delayed kiwifruit softening [20]. *PpNCED1* and *PpNCED5* can cooperatively control ABA biosynthesis and affect fruit ripening and senescence in peach fruit [21]. However, not all *NCED* family members regulate ABA synthesis. For example, a total of 23 *NCED* genes were identified in cotton. Among them, only the expression of *GhNCED5*, *GhNCED6*, and *GhNCED13* was similar to the change in ABA content, which could play a role in ABA biosynthesis [22].

Herbaceous peony (*Paeonia lactiflora* Pall.) is an herbaceous perennial flower of *Paeoniaceae*. In the long-term systematic evolutionary process, herbaceous peony seeds evolved a unique double dormancy characteristic of the epicotyl and hypocotyl [23]. In the breeding process, seed dormancy is often not released or incompletely released, which greatly reduces the germination rate and seriously affects cultivation and production [24]. Thus, understanding the mechanisms associated with herbaceous peony seed dormancy is beneficial to greatly promote the breeding of new varieties. At present, research on the seed dormancy release technology for herbaceous peony mainly focuses on mechanical scarification, low temperature, and exogenous hormone treatment [25–27]. However, the knowledge of the molecular mechanisms underlying the seed dormancy of herbaceous peony is relatively limited. Using previously published transcriptomes from herbaceous peony seeds pre- and post-double dormancy release [28], here, we were able to identify candidate genes that were associated with double dormancy in herbaceous peony. Specifically, we identified ten family members of *PlNCEDs* involved in ABA biosynthesis. Among them, *PlNCED1* (c53147_g1) and *PlNCED2* (c69372_g1) showed significantly differential expression. We subsequently cloned and performed expressional analysis, subcellular localization analysis, and functional characterization of *PlNCED1* and *PlNCED2* in *Paeonia lactiflora*. Our results demonstrate that the genes encoding NCED1 and NCED2 regulate ABA synthesis and consequentially affect the herbaceous peony seed dormancy process.

2. Materials and Methods

2.1. Plant Material and Growth Condition

Herbaceous peony hybrid seeds ('Fen Yu Nu' × 'Fen Yu Lou') were harvested in the Shenyang Agricultural University germplasm resources nursery (Shenyang, Liaoning, China) in August 2019. Filled hybrid seeds were used for variable temperature stratification using a previous method [29]. According to the observation of the seed anatomical structure [29], seeds in six key dormancy release stages were collected: stage 1 (S1: dry seed), stage 2 (S2: imbibition seed), stage 3 (S3: the radicle breaking of seed coat), stage 4 (S4: the length of the seed root is 3–4 cm), stage 5 (S5: the basal part of the seed root turns red), and stage 6 (S6: seed germ breakout) (Figure 1). The cotyledons used as explants were obtained using the conventional embryo induction method [30]. Then, the explants were transferred to an MS callus induction and proliferation medium containing 0.5 mg/L 2,4-dichlorophenoxyacetic acid, 0.5 mg/L α-naphthalene acetic acid, 0.5 mg/L thidiazuron, and 1 g/L polyvinyl pyrrolidone (PVP). The *nced5-2* (GK_328D05) and *nced9-1* (SALK_033388) genes, which are in the Col-0 background, were obtained from the Arabidopsis Biological Resource Center (ABRC, http://abrc.osu.edu). Homozygous mutants were screened and validated by PCR using the left and right genomic primers (LP and RP) and the T-DNA left border primer (LB) (Supplementary Table S1). Seeds of *A. thaliana* WT (Col-0) and mutants were grown following previously reported methods [31].

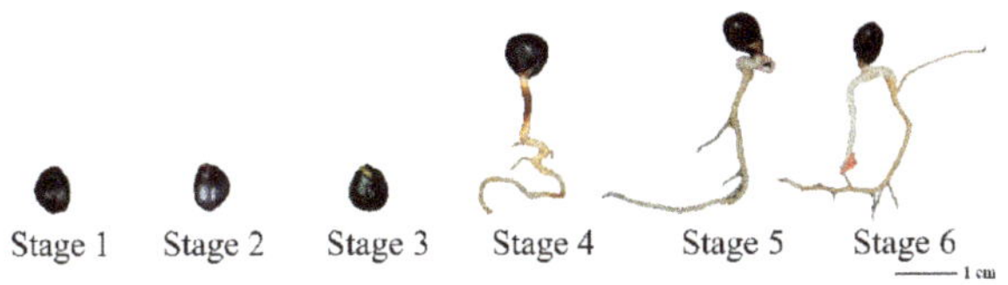

Figure 1. Seeds of herbaceous peony at six different dormancy release stages.

2.2. RNA Extraction, cDNA Synthesis, and qRT-PCR

Total RNA was extracted using the RNAprep pure Plant Kit (TianGen, Beijing, China). cDNA was synthesized using the PrimeScript™ RT Master Mix kit (Perfect Real Time) (Takara, Dalian, China). Based on the full-length coding sequences (CDSs) of *PlNCEDs* in transcriptome data, qRT-PCR primers were designed with the Primer Premier 5.0 software. *PlACTIN* (GenBank accession number JN105299.1) was used as the reference gene [32]. The primers used for qRT-PCR are listed in Supplementary Table S1. qRT-PCR was performed using TB Green® Premix Ex Taq™ II (Tli RNaseH Plus) (Takara, Dalian, China). The reactions were accomplished according to the two-step method—holding stage: 95 °C for 30 s; cycling stage: 40 cycles of 95 °C for 5 s, 60 °C for 30 s; and melt curve stage: 95 °C for 15 s, 60 °C for 1 min, 95 °C for 15 s. Each experiment was performed with three biological and three technical replicates. The relative expression levels of genes were calculated according to the $2^{-\Delta\Delta Ct}$ method, and the error bars represent the standard error from three independent experiments. The results were analyzed by GraphPad Prism 8.0 for ANOVA.

2.3. Cloning and Sequence Analysis

We obtained the CDSs of *PlNCEDs* from transcriptome data. The amino acid sequences of *PlNCEDs* were deduced using ORF Finder (https://www.ncbi.nlm.nih.gov/orffinder/) (accessed on 8 December 2022). A phylogenetic tree was constructed using MEGA 7.0 software with the neighbor-joining method, applying bootstrap analysis with 1000 replicates, and iTOL v6 (https://itol.embl.de/) (accessed on 13 January 2023) was used to optimize the trees. The conserved domains were predicted by National Center for Biotechnology Information (NCBI) online software (https://www.ncbi.nlm.nih.gov/Structure/cdd/wrpsb.cgi) (accessed on 13 January 2023). The protein physicochemical properties were analyzed using the Expasy ProtParam tool (http://web.expasy.org/protparam/) (accessed on 8 December 2022).

Using total RNA as the template, 1st-strand cDNA was synthesized using 3′ RACE adaptor primers. According to the CDSs of *PlNCEDs*, we designed the specific outer and inner primers (Supplementary Table S1) to amplify the 3′ untranslated region (UTR) sequences of *PlNCEDs* using 3′-Full RACE Core Set with PrimeScript™ RTase Kit (Takara, Dalian, China). The miRNA binding sites of 3′ UTR sequences were predicted using MiRanda software.

Genomic DNA was extracted using the Plant Genome DNA Rapid Extraction Kit (Aidlab, Beijing, China). According to the known verified intronless sequences, three specific primers were designed, namely, SP1, SP2, and SP3 (Supplementary Table S1), to amplify the 5′ end sequences of *PlNCEDs* containing 5′ UTR and promoter regions using the Genome Walking Kit (Takara, Dalian, China). *Cis*-acting elements of the promoter were analyzed using PlantCARE.

2.4. Subcellular Localization Analysis

Arabidopsis thaliana leaf protoplasts were extracted using the *Arabidopsis* Protoplast Preparation and Transformation Kit (Coolaber, Beijing, China) for subcellular localization. The CDSs of *PlNCEDs* were cloned into the 16318-hGFP vector and fused in-frame with the hGFP sequence under the control of the CaMV 35S promoter. The 16318-hGFP empty vector was used as a blank control. After 16 h of incubation in darkness, the green fluorescence protein (GFP) fluorescence was captured by an ultra-high-resolution laser scan confocal microscope (Leica TCS SP8 STED, Wetzlar, Germany).

2.5. Functional Analysis

To silence *PlNCED* expression in the herbaceous peony callus, the fragments of *PlNCEDs* (*PlNCED1*: 565 bp; *PlNCED2*: 387 bp) were each amplified and recombined into the linearized pTRV2 empty vector. The CDSs of *PlNCEDs* were separately inserted downstream of 35S in the pCAMBIA1300-35S-flag vector. The pTRV2-PlNCED and 35S::PlNCED vectors were transformed into EHA105-competent cells. *Agrobacterium* containing pTRV1 and *Agrobacterium* containing pTRV2-PlNCEDs were mixed in a 1:1 volume ratio for the preparation of the callus infection solution. The overexpression and silencing experiment in the herbaceous peony callus was performed according to our previous infection method [33]. After the resistance screening of the culture, part of the PlNCED transgenic callus was taken for qRT-PCR identification. The extraction, purification, and determination of the endogenous levels of ABA in the positive transgenic callus of *PlNCEDs* using an enzyme-linked immunosorbent assay (ELISA) were performed as described by He [34].

We obtained two *A. thaliana* homozygous T-DNA mutants (Figure S1): *nced5-2* and *nced9-1*. *NCED* mutants have interrupted 9-*cis*-epoxycarotenoid dioxygenase genes that result in plants that are deficient in the plant growth regulator abscisic acid. To make the transgenic line and functional complementation line of *A. thaliana*, the 35S::PlNCED vectors were transformed into *Agrobacterium* strain GV3101 and then used to infect inflorescences of *A. thaliana* WT (Col-0) and homozygous mutants (*nced5-2* and *nced9-1*) using the floral-dip method [35], respectively. The transgenic line and functional complementation line were screened on 1/2 MS medium plates that contained 50 mg/L kanamycin. The seed germination rate was measured in WT, homozygous mutants, functional complementation lines, and transgenic lines (stable T_3-generation genetic lines) of *A. thaliana*, which were grown at the same time under 16 h light and 8 h dark conditions at 22 °C.

3. Results

3.1. Cloning and Sequence Analysis of PlNCEDs

The full-length cDNA sequences of *PlNCEDs* were isolated and deposited in GenBank (GenBank accession numbers—*PlNCED1*: OL744236; *PlNCED2*: OL744237). *PlNCED1* and *PlNCED2* contained 1518 bp and 1326 bp open reading frames, encoding 505 and 441 amino acids, respectively (Figure S2). NCBI tblastx results displayed their homology to the *NCED* genes of other plant species. PoNCED from *Paeonia ostii* (74.70%) and JrNCED from *Juglans regia* (62.43%) had the highest identity with PlNCED1 and PlNCED2, respectively (Supplementary Tables S2 and S3). This similarity demonstrated that PlNCEDs are relatively conserved among diverse plant species. PlNCEDs had the typical conserved RPE65 domain of the NCED family (Figure S3), which is related to the degradation of carotenoids in plants. Phylogenetic analysis results indicated that PlNCED1 and PlNCED2 are closely related to the NCED proteins in *Paeonia ostii*, *Vitis riparia*, and *Vitis vinifera*, respectively (Figure 2). In addition, based on DNA sequences, we identified putative microRNA (miRNA) and transcription factor binding sites in the 3′UTR and promoter regions of *PlNCEDs* (Supplementary Tables S4–S7). Among the miRNAs that may target the 3′UTRs of *PlNCEDs* genes, six miRNAs, namely, miR837-5p, miR5640, miR319c, miR6425a/b/c/d/e-5p, miR168a, and miR5304, are related to plant development (Supplementary Tables S4 and S5). Among the *cis*-acting elements that may bind to the promoter sequences of *PlNCED* genes, six cis-acting elements, namely, TCA-element, ABRE, AuxRR-core, TGA-box, TGACG-motif, and CGTCA-motif, are related to phytohormone responses (Supplementary Tables S6 and S7). These results suggest that *PlNCEDs* may be involved in plant development mediated by ABA and other phytohormones.

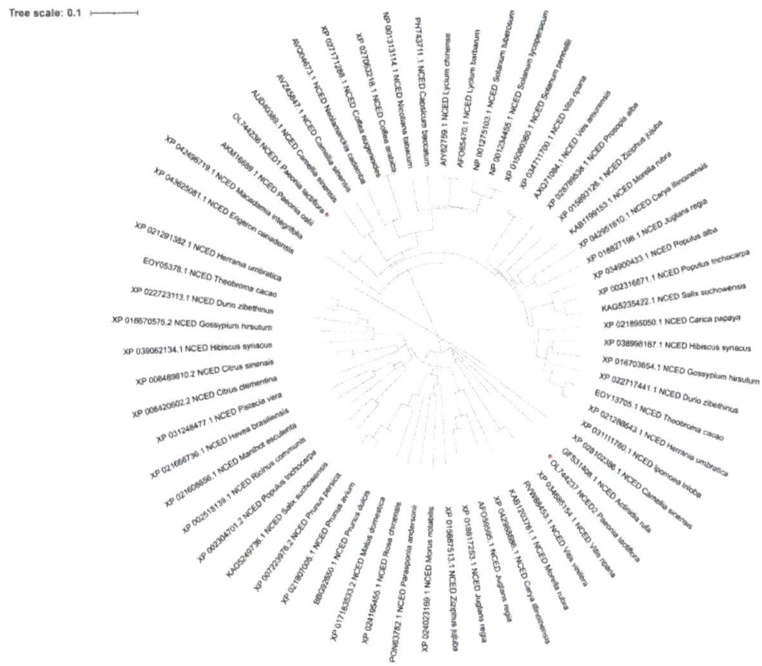

Figure 2. Neighbor-joining phylogenetic tree of *PlNCEDs*. PlNCED1 and PlNCED2 are indicated with red dots. A neighbor-joining tree was constructed based on amino acid sequences of NCED protein from different plants using MEGA 7.0 software. The bootstrap values of the branches were obtained by testing the tree 1000 times.

3.2. Expression Analysis of PlNCEDs

To explore the correlation between the expression of *PlNCED* genes and ABA accumulation in herbaceous peony seeds, we performed qRT-PCR using seeds at stages 1-6. Our results indicated that the expression levels and trends of *PlNCED1* and *PlNCED2* dynamically varied during seed dormancy release (Figure S4). The expression level of *PlNCED1* increased sharply from stage 1 to stage 2 and then decreased significantly at stage 3, where seeds experienced hypocotyl dormancy release (Figure S4). After the completion of the hypocotyl dormancy release process, the expression level of *PlNCED1* increased incrementally from stage 4 to stage 6 (Figure S4). Conversely, we observed different trajectories for the expression of the *PlNCED2* gene. Overall, *PlNCED2* maintained a high level of expression at stages 1-2, and it displayed a decreasing trend from stages 2 to 6 (Figure S4). Correspondingly, the ABA content in each of the six seed dormancy release stages largely declined from S1 to S6 (Figure S5) [36]. By comparing the dynamics of *PlNCED* expression and ABA content during the process of dormancy release, we demonstrated that only *PlNCED2* expression was positively associated with ABA accumulation.

3.3. Subcellular Localization of PlNCEDs

Protein maintains its optimal function in a specific subcellular localization. Therefore, to unravel the cellular functions of *PlNCEDs* during seed dormancy and its release process, we also carried out fluorophore tagging of the protein using green fluorescent protein to locate the presence of *PlNCED* proteins within the cell. Fluorescence microscopic analysis showed that the GFP fluorescence signal was distributed in the cell membrane, nucleus, and cytoplasm of *A. thaliana* protoplast containing the empty 16318-hGFP vector (Figure 3). In contrast, the 16318-hGFP-PlNCED1 and 16318-hGFP-PlNCED2 fusion proteins were only observed in the nucleus and cytoplasm, respectively (Figure 3). This differential cellular localization of two PlNCEDs implies that PlNCED1 may play a genetic role similar to that of transcription factors, whereas PlNCED2 may act as a functional enzyme to synthesize ABA.

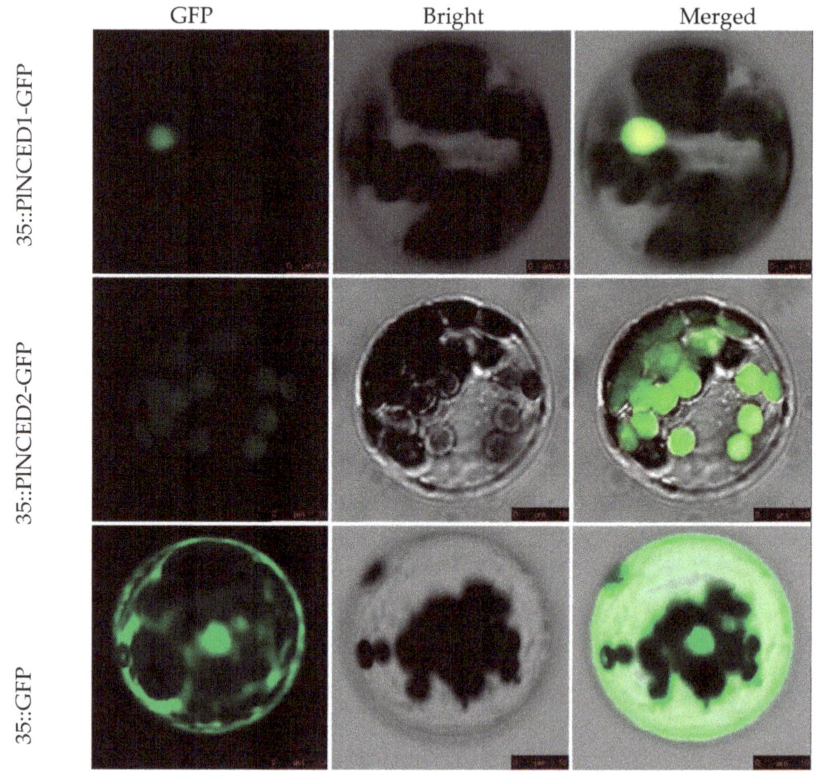

Figure 3. Subcellular localization of the 16318-hGFP-PlNCED fusion protein in protoplast.

3.4. Functional Analysis of PlNCEDs

We first used the transgenic herbaceous peony callus to identify the impact of *PlNCED1* and *PlNCED2* on in vivo ABA content. The results showed that the expression of *PlNCED1* and *PlNCED2* in transgenic herbaceous peony callus was significantly altered compared with the control groups. The expression levels of *PlNCED1* and *PlNCED2* in the overexpressed callus were about 14.5-fold higher than that in the control callus, and the expression levels in the silenced callus were about 0.35-fold lower than that in the control callus, indicating that transgenic calluses were successfully obtained (Figure S6). As shown in Figure 4, the ABA content of the overexpressed *PlNCED* transgenic callus was significantly higher than that of the wild-type callus, while that of the silenced *PlNCED* transgenic callus was much lower, indicating a positive correlation between *PlNCEDs* and endogenous ABA content. In particular, the increase in ABA content caused by *PlNCED2* overexpression was larger than that generated by *PlNCED1* overexpression. We illustrated that *PlNCED2* specifically affects endogenous ABA content by regulating its biosynthesis.

To identify the functional involvement of *PlNCEDs* in seed dormancy, we recorded the seed germination times of *A. thaliana* WT, mutant, transgenic lines, and complementation lines. The germination rates of *A. thaliana* WT and mutant seeds reached about 90% at 48 h, but the seeds of overexpression transgenic lines did not germinate at 48h (Figure 5a,b), indicating that the overexpression of *PlNCEDs* inhibited seed germination. At 68 h, the

PlNCED1 overexpression transgenic line seeds began to germinate, while the *PlNCED2* overexpression transgenic line seeds began to germinate at 78 h (Figure 5a,b), indicating that the inhibitory effects of *PlNCEDs* on seed germination were different, with *PlNCED1* having a slightly weaker impact on seed germination. Compared to *Atnced9-1* and/or *Atnced5-2* mutants, complementation lines induced seed dormancy with a delay and a lower rate of seed germination. In particular, the *PlNCED2* complementation line had a stronger effect (Figure 5c–f). Overall, we demonstrated that *PlNCEDs* inhibited seed dormancy release, and the inhibitory effect of *PlNCED2* was stronger.

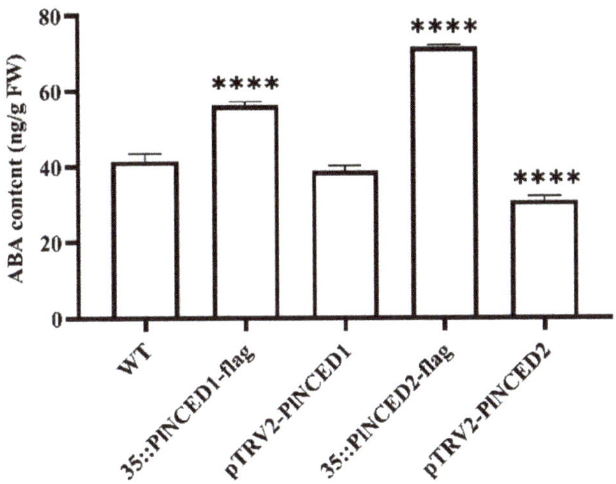

Figure 4. Changes in ABA content in the overexpressed callus of *PlNCEDs*. *PlNCED1* expression in the experimental group (35S::PlNCED1-flag/pTRV2-PlNCED1), *PlNCED2* expression in the experimental group (35S::PlNCED2-flag/pTRV2-PlNCED2), and empty control group (WT). Significant differences (**** $p \leq 0.0001$) are indicated by asterisks. One-Way ANOVA (*F*-test) analysis was performed using GraphPad Prism 8.0. WT was used as a control.

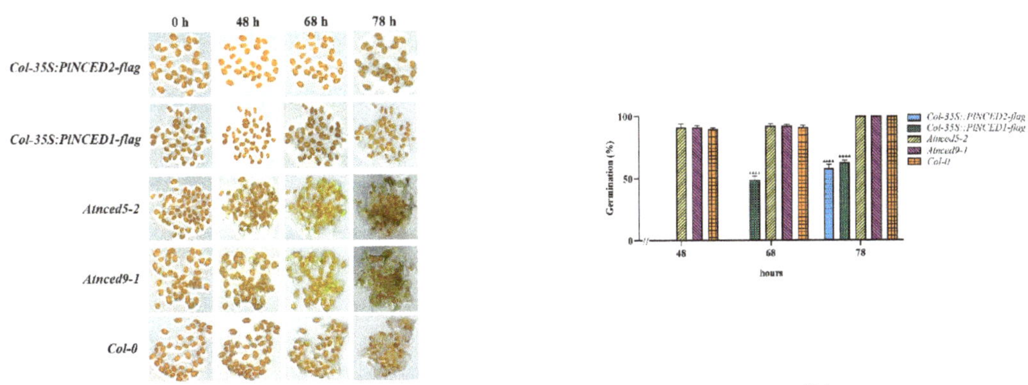

Figure 5. *Cont.*

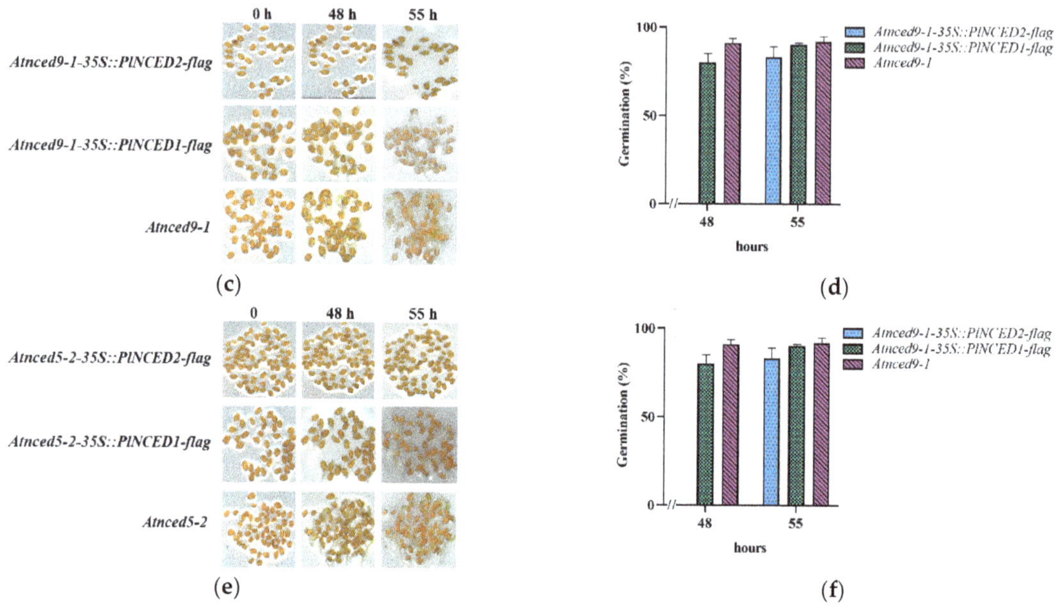

Figure 5. Observation of seed germination rate of different types of *A. thaliana*. (**a**,**b**) Seed germination rate of overexpressed *PlNCED* transgenic lines (under Col-0 background), Col-0, and mutants (*nced9-1* and *nced5-2*); (**c**,**d**) seed germination rate of overexpressed *PlNCED* transgenic lines (under *nced9-1* background) and *nced9-1*; (**e**,**f**) seed germination rate of overexpressed *PlNCED* transgenic lines (under *nced5-2* background) and *nced5-2*. Significant differences (**** $p \leq 0.0001$) are indicated by asterisks. One-Way ANOVA (*F*-test) analysis was performed using GraphPad Prism 8.0. Col-0 and *Atnced5-2-35S::PlNCED2-flag* in (**b**) and (**f**) were used as controls, respectively.

4. Discussion

Paeonia lactiflora is the most familiar herbaceous peony seen in gardens and produces some of the best cut flowers in the floral industry. Though herbaceous peony is one of the most easily grown hardy perennials, its complex double seed dormancy hinders seed germination and consequently imposes adverse effects on breeding and cultivar improvements [24]. Practically, breaking herbaceous peony seed dormancy can be handily achieved through physical (e.g., cold treatments, slitting the seed coat) and biological means (e.g., hormone treatment) [25,26]. The content and level of phytohormones, particularly ABA, are the key factors for natural seed dormancy release. The final concentration of endogenous ABA depends on the dynamic balance between ABA synthesis and catabolism. Therefore, it is critical to know the genes encoding ABA metabolic enzymes and their impacts on herbaceous peony seed dormancy and germination.

To search for genes related to ABA biosynthesis, we identified two *NCED* genes (*PlNCED1* and *PlNCED2*) with differential transcription pre- and post-germination (Figures S4 and S7) based on previously published transcriptome data. Studies have shown that *NCED* genes are the key factors that control the responses of endogenous ABA content to environmental stimuli [37]. The Conserved Domain Database (CDD) search for protein sequences of *PlNCED1* and *PlNCED2* in the NCBI database indicates that *PlNCED1* and *PlNCED2* proteins belong to the RPE65 family, a characteristic conserved domain of enzymes involved in carotenoid cleavage dioxygenase [38,39]. Furthermore, the phylogenetic analysis of *NCED2* clearly revealed its intimate genetic relationship among *P. lactiflora*, *P. ostii*, and *V. vinifera*. This conclusion suggests that *PlNCED1* and *PlNCED2* have similar functions to *PoNCED and VvNCED*, which play an important rate-limiting role in ABA biosynthesis [40,41].

Previous reports indicate that most NCED proteins are located in chloroplasts [19,42,43]. However, our data show that PlNCED1 is located in the nucleus, but it has no Nuclear Localization Signal (NLS), a short peptide acting as a signal fragment and mediating the transport of proteins from the cytoplasm into the nucleus. Previous studies have shown that not all nucleus-expressed proteins require an NLS, and multiple additional pathways can also mediate their nuclear import [44,45]. One of these pathways is that these proteins without NLSs enter the nucleus by interacting with proteins with NLSs or with other nuclear localization proteins [46]. Therefore, our experiments imply that PlNCED1 may enter the nucleus by relying on the NLSs of other proteins. Additionally, our results suggest that the expression of *PlNCEDs* may be regulated by several miRNAs located in the 3′UTR regions of *PlNCEDs*, as well as *cis*-acting elements located upstream of transcripts (Supplementary Tables S4 and S5). Given the characteristics of these miRNAs, we further demonstrated that *PlNCED1* and *PlNCED2* might play a certain role in seed development and biotic and abiotic stresses, which is consistent with the role of ABA in plant growth. These *cis*-acting elements from the promoter sequences of *PlNCED* genes were presumed to be involved in salicylic acid, ABA, auxin, and jasmonic acid (Supplementary Tables S6 and S7). By combining *PlNCED* genes and their associated transcriptional regulatory binding site predictions, our data provide a preliminary path to explore the molecular mechanisms of ABA and other phytohormones involved in seed dormancy and germination.

The ABA content is proportionally related to the process of herbaceous peony seed dormancy and germination, while ABA accumulation in seeds gradually decreases from dormancy to germination (Figure S5) [36]. By comparing the transcription levels of *PlNCED* genes with ABA contents, we clearly show that the expression dynamics of *PlNCED2* are directly associated with ABA biosynthesis and accumulation after seeds imbibe water (Figures S4 and S5) [36], suggesting that *PlNCED2* is the crucial causal factor for ABA-mediated seed dormancy release in herbaceous peony. The deviation of *PlNCED* expression from ABA content in the imbibition stage may be due to the self-regulation of ABA metabolism genes to adapt to the dynamic balance among endogenous phytohormones after seed imbibition (Figures S4 and S5) [36]. Lee et al. (2018) also found that the level of *NCED1* in orchids (*PtNCED1*) was low in the early stage of seed development but gradually increased and then declined slightly when seeds germinated, and the resulting changes in the seed's endogenous ABA content played a key role in seed germination [19]. In contrast, the expression level of *NCED* in habanero pepper was high during seed germination, but the content of ABA gradually decreased during the same time span. This indicates that *NCED* has less of an effect on ABA synthesis and a weaker impact on seed germination in habanero pepper [47]. All of these reports indicate that *NCED* has different regulatory effects on ABA content in different plant species. Lastly, our transgenic experiment in herbaceous peony has also proved that *PlNCED2* plays a major role in ABA biosynthesis and accumulation and subsequently contributes to seed dormancy maintenance.

The ultimate way to dissect the biological function of *PlNCED* genes involved in seed dormancy is to generate transgenic lines of herbaceous peony. However, it is difficult to establish a stable and efficient genetic transformation system for herbaceous peony [48]. Although the genetic transformation system of herbaceous peony has been used for the functional analysis of some genes in recent years [49–51], it takes at least 3 years to obtain transgenic seeds from transgenic seedlings due to the growth characteristics of herbaceous peony. In order to efficiently verify the effect of *PlNCED* genes on the seed germination rate, we used *A. thaliana*, a typical short-growth-cycle model plant, for subsequent gene transformation experiments in this study [52]. *Arabidopsis thaliana* mutants play an important role in revealing the growth and development of different plants. *AtNCED5* and *AtNCED9* mutants in *A. thaliana* have been found to affect ABA production in the embryo and endosperm, leading to seed dormancy [15,16]. Therefore, we built *PlNCED* transgenic lines in wild-type *A. thaliana* as well as *PlNCED* complementation lines in *A. thaliana NCED* mutants to explore the role of *PlNCED* in seed dormancy and germination. Phenotyping

PlNCED transgenic lines indeed showed that weak seed dormancy was induced to a certain extent by the overexpression of *PlNCEDs* in *A. thaliana* (Figure 5a,b). More significantly, the seed germination rate of the *PlNCED2* complementation line was significantly lower than that of the *nced5-2* mutant (Figure 5f), indicating that overexpressed *PlNCED2* rescued the partial function of *Arabidopsis NCED* genes. Collectively, the biological function of *PlNCEDs* is consistent to that of *A. thaliana NCEDs* by promoting seed dormancy and delaying germination.

5. Conclusions

In summary, we identified and cloned two genes from the *NCED* gene family, which are related to ABA synthesis in herbaceous peony. By comparing their protein sequences and phylogenetics with those of homologs in other plant species, we were able to detect their functional conserved domain. The dynamics of *PlNCED2* transcription epistatically regulated endogenous ABA biosynthesis and accumulation. Using transgenic and complementation rescue lines in *Arabidopsis*, we were able to demonstrate the phenotypic traits of *PlNCED* genes, which induce seed dormancy and hinder seed germination. Our data and analysis provide the first step to understanding the underlying molecular and genetic mechanisms of complex double seed dormancy in herbaceous peony.

Supplementary Materials: The following supporting information can be downloaded at https://www.mdpi.com/article/10.3390/plants12040710/s1. Figure S1: Identification of *nced* homozygous mutants in *A. thaliana.*; Figure S2: Nucleotide and amino acid sequences of the *PlNCED* gene coding regions; Figure S3: Conserved domain prediction for the proteins encoded by *PlNCEDs*; Figure S4: *PlNCED* expression at different stages of seed dormancy release in herbaceous peony; Figure S5: Changes in ABA content at different seed dormancy release stages in herbaceous peony; Figure S6: Identification of *PlNCED* transgenic callus; Figure S7: Other *PlNCED* family members' expression at different stages of seed dormancy release in herbaceous peony; Table S1: Primers used in this study; Table S2: The identification of PlNCED1 from different plant species (%); Table S3: The identification of PlNCED1 from different plant species (%); Table S4: Prediction of miRNA targets in 3'UTR of *PlNCED1*; Table S5: Prediction of miRNA targets in 3'UTR of *PlNCED2*; Table S6: Prediction of *cis*-acting elements in promoter sequence of *PlNCED1*; Table S7: Prediction of *cis*-acting elements in promoter sequence of *PlNCED2*.

Author Contributions: Conceptualization, X.S. and S.G.; methodology, X.S.; software, S.D.; validation, R.F., J.G., and T.S.; formal analysis, R.F.; investigation, S.G.; resources, X.S.; data curation, S.G.; writing—original draft preparation, R.F.; writing—review and editing, S.D.; visualization, R.F.; supervision, X.S.; project administration, S.G.; funding acquisition, X.S. All authors have read and agreed to the published version of the manuscript.

Funding: This research was funded by the National Natural Science Foundation of China, grant number 32071814, and the China Agriculture Research System of MOF and MARA, grant number CARS-23.

Institutional Review Board Statement: Not applicable.

Informed Consent Statement: Not applicable.

Data Availability Statement: The datasets generated and analyzed during the current study are available in the NCBI repository—*PlNCED1*: OL744236, https://www.ncbi.nlm.nih.gov/nuccore/OL744236, *PlNCED2*: OL744237, https://www.ncbi.nlm.nih.gov/nuccore/OL744237.

Conflicts of Interest: The authors declare no conflict of interest. The funders had no role in the design of the study; in the collection, analyses, or interpretation of data; in the writing of the manuscript; or in the decision to publish the results.

References

1. Eggels, S.; Avramova, V.; Schon, C.C.; Poppenberger, B.; Rozhon, W. Assay for abscisic acid 8′-hydroxylase activity of cloned plant cytochrome P450 oxidases in *Saccharomyces cerevisiae*. *Anal. Biochem.* **2018**, *553*, 24–27. [CrossRef]
2. Pan, J.J.; Wang, H.P.; Hu, Y.R.; Yu, D.Q. *Arabidopsis* VQ18 and VQ26 proteins interact with ABI5 transcription factor to negatively modulate ABA response during seed germination. *Plant J.* **2018**, *95*, 529–544. [CrossRef]
3. Shen, W.; Yao, X.; Ye, T.; Ma, S.; Liu, X.; Yin, X.; Wu, Y. *Arabidopsis* aspartic protease ASPG1 affects seed dormancy, seed longevity and seed germination. *Plant Cell Physiol.* **2018**, *59*, 1415–1431. [CrossRef]
4. Qin, X.Q.; Zeevaart, J.A.D. Overexpression of a 9-*cis*-epoxycarotenoid dioxygenase gene in *Nicotiana plumbaginifolia* increases abscisic acid and phaseic acid levels and enhances drought tolerance. *Plant Physiol.* **2002**, *128*, 544–551. [CrossRef]
5. Wu, X.L.; Gong, F.P.; Yang, L.; Hu, X.L.; Tai, F.J.; Wang, W. Proteomic analysis reveals differential accumulation of small heat shock proteins and late embryogenesis abundant proteins between ABA-deficient mutant *vp5* seeds and wild-type *Vp5* seeds in maize. *Front. Plant Sci.* **2014**, *5*, 801. [CrossRef]
6. Martin-Rodriguez, J.A.; Huertas, R.; Ho-Plagaro, T.; Ocampo, J.A.; Tureckova, V.; Tarkowska, D.; Ludwig-Muller, J.; Garcia-Garrido, J.M. Gibberellin-abscisic acid balances during arbuscular mycorrhiza formation in tomato. *Front. Plant Sci.* **2016**, *7*, 1273. [CrossRef]
7. Jia, K.P.; Mi, J.; Ali, S.; Ohyanagi, H.; Moreno, J.C.; Ablazov, A.; Balakrishna, A.; Berqdar, L.; Fiore, A.; Diretto, G.; et al. An alternative, zeaxanthin epoxidase-independent abscisic acid biosynthetic pathway in plants. *Mol. Plant* **2021**, *14*, 1–6. [CrossRef]
8. Wang, D.L.; Gao, Z.Z.; Du, P.Y.; Xiao, W.; Tan, Q.P.; Chen, X.D.; Li, L.; Gao, D.S. Expression of ABA metabolism-related genes suggests similarities and differences between seed dormancy and bud dormancy of Peach (*Prunus persica*). *Front. Plant Sci.* **2015**, *6*, 1248. [CrossRef]
9. Tognacca, R.S.; Botto, J.F. Post-transcriptional regulation of seed dormancy and germination: Current understanding and future directions. *Plant Commun.* **2021**, *2*, 100169. [CrossRef]
10. Xia, Q.; Saux, M.; Ponnaiah, M.; Gilard, F.; Perreau, F.; Huguet, S.; Balzergue, S.; Langlade, N.; Bailly, C.; Meimoun, P.; et al. One way to achieve germination: Common molecular mechanism induced by ethylene and after-ripening in sunflower seeds. *Int. J. Mol. Sci.* **2018**, *19*, 2464. [CrossRef]
11. Kim, S.Y.; Warpeha, K.M.; Huber, S.C. The brassinosteroid receptor kinase, BRI1, plays a role in seed germination and the release of dormancy by cold stratification. *J. Plant Physiol.* **2019**, *241*, 153031. [CrossRef]
12. Munguia-Rodriguez, A.G.; Lopez-Bucio, J.S.; Ruiz-Herrera, L.F.; Ortiz-Castro, R.; Guevara-Garcia, A.A.; Marsch-Martinez, N.; Carreon-Abud, Y.; Lopez-Bucio, J.; Martinez-Trujillo, M. *YUCCA4* overexpression modulates auxin biosynthesis and transport and influences plant growth and development via crosstalk with abscisic acid in *Arabidopsis thaliana*. *Genet. Mol. Biol.* **2020**, *43*, e20190221. [CrossRef]
13. Li, X.T. Functional analysis of *PlCYP707As* related to dormancy release in *Paeonia lactiflora* seed based on transcriptome analysis. Ph.D. Thesis, Shenyang Agricultural University, Shenyang, China, 2020.
14. Yan, M.; Yao, Y.D.; Mou, K.P.; Dan, Y.Y.; Li, W.T.; Wang, C.L.; Liao, W.B. The involvement of abscisic acid in hydrogen gas-enhanced drought resistance in tomato seedlings. *Sci. Hortic.* **2022**, *292*, 110631. [CrossRef]
15. Frey, A.; Effroy, D.; Lefebvre, V.; Seo, M.; Perreau, F.; Berger, A.; Sechet, J.; To, A.; North, H.M.; Marion-Poll, A. Epoxycarotenoid cleavage by NCED5 fine-tunes ABA accumulation and affects seed dormancy and drought tolerance with other NCED family members. *Plant J.* **2012**, *70*, 501–512. [CrossRef]
16. Lefebvre, V.; North, H.; Frey, A.; Sotta, B.; Seo, M.; Okamoto, M.; Nambara, E.; Marion-Poll, A. Functional analysis of *Arabidopsis NCED6* and *NCED9* genes indicates that ABA synthesized in the endosperm is involved in the induction of seed dormancy. *Plant J.* **2006**, *45*, 309–319. [CrossRef]
17. Thompson, A.J.; Jackson, A.C.; Symonds, R.C.; Mulholland, B.J.; Dadswell, A.R.; Blake, P.S.; Burbidge, A.; Taylor, I.B. Ectopic expression of a tomato 9-*cis*-epoxycarotenoid dioxygenase gene causes over-production of abscisic acid. *Plant J.* **2000**, *23*, 363–374. [CrossRef]
18. Hu, B.; Wan, X.R.; Liu, X.H.; Guo, D.L.; Li, L. Abscisic acid (ABA)-mediated inhibition of seed germination involves a positive feedback regulation of ABA biosynthesis in *Arachis hypogaea* L. *Afr. J. Biotechnol.* **2010**, *9*, 1578–1586. [CrossRef]
19. Lee, Y.I.; Chen, M.C.; Lin, L.; Chung, M.C.; Leu, W.M. Increased expression of 9-*Cis*-Epoxycarotenoid dioxygenase, *PtNCED1*, associated with inhibited seed germination in a terrestrial orchid, *Phaius tankervilliae*. *Front. Plant Sci.* **2018**, *9*, 1043. [CrossRef]
20. Gan, Z.; Shan, N.; Fei, L.; Wan, C.; Chen, J. Isolation of the 9-*cis*-epoxycarotenoid dioxygenase (*NCED*) gene from kiwifruit and its effects on postharvest softening and ripening. *Sci. Hortic.* **2020**, *261*, 109020. [CrossRef]
21. Wang, P.; Lu, S.; Zhang, X.; Hyden, B.; Qin, L.; Liu, L.; Bai, Y.; Han, Y.; Wen, Z.; Xu, J.; et al. Double NCED isozymes control ABA biosynthesis for ripening and senescent regulation in peach fruits. *Plant Sci.* **2021**, *304*, 110739. [CrossRef]
22. Li, Q.; Yu, X.; Chen, L.; Zhao, G.; Li, S.; Zhou, H.; Dai, Y.; Sun, N.; Xie, Y.; Gao, J.; et al. Genome-wide identification and expression analysis of the NCED family in cotton (*Gossypium hirsutum* L.). *PLoS ONE* **2021**, *16*, e0246021. [CrossRef] [PubMed]
23. Zhang, K.; Yao, L.; Zhang, Y.; Baskin, J.M.; Baskin, C.C.; Xiong, Z.; Tao, J. A review of the seed biology of Paeonia species (Paeoniaceae), with particular reference to dormancy and germination. *Planta* **2019**, *249*, 291–303. [CrossRef] [PubMed]
24. Li, J.Y. *Chinese Tree Peony and Herbaceous Peony*; China Forestry Publishing House: Beijing, China, 1999; pp. 153–155.

25. Ren, X.X. Mechanism of Low Temperature Regulating *Paeonia ostii* 'Fengdan' Seed Hypocotyl Germination. Master's Thesis, Chinese Academy of Agricultural Sciences, Beijing, China, 2016.
26. Sun, X.M.; Zhang, M.M.; Gao, H.D.; Yang, H.G. Study on Characteristic for Seed Coat of *Paeonia lactiflora*. *North. Hort.* **2012**, *36*, 55–57. (In Chinese) [CrossRef]
27. Jia, X.; Zhang, M.X.; Li, Y.; Xue, Y.; Zhang, Y.; Yu, J.; Wang, X.Q. Study on activities of seed germination inhibitors and methods of inhibition elimination in *Paeonia lactiflora*. *J. Chin. Med. Mater.* **2021**, *44*, 2521–2525. (In Chinese) [CrossRef]
28. Ma, Y.L.; Cui, J.Q.; Lu, X.J.; Zhang, L.J.; Chen, Z.J.; Fei, R.W.; Sun, X.M. Transcriptome Analysis of Two Different Developmental Stages of *Paeonia lactiflora* Seeds. *Int. J. Genom.* **2017**, *2017*, 8027626. [CrossRef] [PubMed]
29. Fei, R.W.; Sun, X.M.; Yang, P.P.; Chen, Z.J.; Ma, Y.L. Anatomical observation of *Paeonia lactiflora* seeds during stratification process. *J. Shenyang Agric.Univ.* **2017**, *48*, 354–359. (In Chinese) [CrossRef]
30. Liu, X.T.; Sun, X.M.; Li, Y.; Li, H.; Li, X.T.; Song, L.L. Study on initiation culture of *Paeonia lactiflora* embryo and induction of multiple shoots. *J. Shenyang Agric.Univ.* **2020**, *51*, 312–320. (In Chinese) [CrossRef]
31. Fei, R.W. Cloning of the *PlSAURs* Genes from *Paeonia lactiflora* Pall. and Transformation of *Arabidopsis thaliana*. Master dissertation, Shenyang Agricultural University, Shenyang, China, 2018.
32. Zhao, D.Q.; Tao, J.; Han, C.X.; Ge, J.T. An actin gene as the internal control for gene expression analysis in herbaceous peony (*Paeonia lactiflora* Pall.). *Afr. J. Agric. Res.* **2012**, *7*, 2153–2159. [CrossRef]
33. Duan, S.; Xin, R.; Guan, S.; Li, X.; Fei, R.; Cheng, W.; Pan, Q.; Sun, X. Optimization of callus induction and proliferation of *Paeonia lactiflora* Pall. and *Agrobacterium*-mediated genetic transformation. *Front. Plant Sci.* **2022**, *13*, 996690. [CrossRef]
34. He, Z. *A Laboratory Guide to Chemical Control Technology on Field Crop*; Beijing Agricultural University Press: Beijing, China, 1993; pp. 60–68.
35. Clough, S.J.; Bent, A.F. Floral dip: A simplified method for *Agrobacterium*-mediated transformation of *Arabidopsis thaliana*. *Plant J.* **1998**, *16*, 735–743. [CrossRef]
36. Li, X.T.; Fei, R.W.; Chen, Z.J.; Fan, C.Z.; Sun, X.M. Plant hormonal changes and differential expression profiling reveal seed dormancy removal process in double dormant plant-herbaceous peony. *PLoS ONE* **2020**, *15*, e0231117. [CrossRef]
37. Nambara, E.; Marion-Poll, A. Abscisic acid biosynthesis and catabolism. *Annu. Rev. Plant Biol.* **2005**, *56*, 165–185. [CrossRef]
38. Kiser, P.D. Retinal pigment epithelium 65 kDa protein (RPE65): An update. *Prog. Retin. Eye Res.* **2021**, *88*, 101013. [CrossRef]
39. Tian, X.; Ji, J.; Wang, G.; Jin, C.; Guan, C.; Wu, D.; Li, Z. Cloning and Expression Analysis of 9-*cis*-Epoxycarotenoid Dioxygenase Gene 1 Involved in Fruit Maturation and Abiotic Stress Response in *Lycium chinense*. *J. Plant Growth Regul.* **2015**, *34*, 465–474. [CrossRef]
40. Xue, J.Q.; Wang, S.L.; Zhang, P.; Zhu, F.Y.; Ren, X.X.; Liu, C.J.; Zhang, X.X. On the role of physiological substances, abscisic acid and its biosynthetic genes in seed maturation and dormancy of tree peony (*Paeonia ostii* 'Feng Dan'). *Sci. Hortic.* **2015**, *182*, 92–101. [CrossRef]
41. Parada, F.; Noriega, X.; Dantas, D.; Bressan-Smith, R.; Perez, F.J. Differences in respiration between dormant and non-dormant buds suggest the involvement of ABA in the development of endodormancy in grapevines. *J. Plant Physiol.* **2016**, *201*, 71–78. [CrossRef]
42. Chen, Y.; Xiang, Z.; Liu, M.; Wang, S.; Zhang, L.; Cai, D.; Huang, Y.; Mao, D.; Fu, J.; Chen, L. ABA biosynthesis gene OsNCED3 contributes to preharvest sprouting resistance and grain development in rice. *Plant Cell Environ.* **2022**. [CrossRef]
43. Jia, Y.; Liu, J.; Bai, Z.; Ding, K.; Li, H.; Liang, Z. Cloning and functional characterization of the *SmNCED3* in *Salvia miltiorrhiza*. *Acta Physiol. Plant.* **2018**, *40*, 133. [CrossRef]
44. Lu, J.; Wu, T.; Zhang, B.; Liu, S.; Song, W.; Qiao, J.; Ruan, H. Types of nuclear localization signals and mechanisms of protein import into the nucleus. *Cell Commun. Signal* **2021**, *19*, 60. [CrossRef]
45. Wagstaff, K.M.; Jans, D.A. Importins and beyond: Non-conventional nuclear transport mechanisms. *Traffic* **2009**, *10*, 1188–1198. [CrossRef]
46. Poon, I.K.; Jans, D.A. Regulation of nuclear transport: Central role in development and transformation? *Traffic* **2005**, *6*, 173–186. [CrossRef]
47. Santos, H.O.; Von Pinho, E.V.R.; Von Pinho, I.V.; Dutra, S.M.F.; Andrade, T.; Guimarães, R.M. Physiological quality and gene expression during the development of habanero pepper (*Capsicum chinense* Jacquin) seeds. *Gen. Mol. Res.* **2015**, *14*, 5085–5098. [CrossRef] [PubMed]
48. Shen, M.M.; Tang, Z.J.; da Silva, J.A.T.; Yu, X.N. Induction and proliferation of axillary shoots from in vitro culture of *Paeonia lactiflora* Pall. mature zygotic embryos. *N. Z. J. Crop Hortic. Sci.* **2015**, *43*, 42–52. [CrossRef]
49. Tang, Y.; Lu, L.; Sheng, Z.; Zhao, D.; Tao, J. An R2R3-MYB Network Modulates Stem Strength by Regulating Lignin Biosynthesis and Secondary Cell Wall Thickening in Herbaceous Peony. *Plant J.* **2023**. [CrossRef]
50. Zhao, D.; Luan, Y.; Shi, W.; Tang, Y.; Huang, X.; Tao, J. Melatonin enhances stem strength by increasing lignin content and secondary cell wall thickness in herbaceous peony. *J. Exp. Bot.* **2022**, *73*, 5974–5991. [CrossRef]

51. Zhang, T.; Tang, Y.; Luan, Y.; Cheng, Z.; Wang, X.; Tao, J.; Zhao, D. Herbaceous peony AP2/ERF transcription factor binds the promoter of the tryptophan decarboxylase gene to enhance high-temperature stress tolerance. *Plant Cell Environ* **2022**, *45*, 2729–2743. [CrossRef] [PubMed]
52. Gepstein, S.; Horwitz, B.A. The impact of *Arabidopsis* research on plant biotechnology. *Biotechnol. Adv.* **1995**, *13*, 403–414. [CrossRef] [PubMed]

plants

MDPI

Article

IlAP2, an AP2/ERF Superfamily Gene, Mediates Cadmium Tolerance by Interacting with *IlMT2a* in *Iris lactea* var. *chinensis*

Zhiquan Wang [1,2], Longjie Ni [3], Liangqin Liu [1,2], Haiyan Yuan [1,2] and Chunsun Gu [1,2,3,*]

1 Institute of Botany, Jiangsu Province and Chinese Academy of Sciences (Nanjing Botanical Garden Memorial Sun Yat-Sen), Nanjing 210014, China
2 Jiangsu Key Laboratory for the Research and Utilization of Plant Resources, Nanjing 210014, China
3 College of Forest Sciences, Nanjing Forestry University, Nanjing 210037, China
* Correspondence: chunsungu@cnbg.net; Tel.: +86-25-84347051

Abstract: Cadmium (Cd) stress has a major impact on ecosystems, so it is important to find suitable Cd-tolerant plants while elucidating the responsible molecular mechanism for phytoremediation to manage Cd soil contamination. *Iris lactea* var. *chinensis* is an ornamental perennial groundcover plant with strong tolerance to Cd. Previous studies found that *IlAP2*, an AP2/ERF superfamily gene, may be an interacting partner of the metallothionein gene *IlMT2a*, which plays a key role in Cd tolerance. To study the role of *IlAP2* in regulating Cd tolerance in *I. lactea*, we analyzed its regulation function and mechanism based on a yeast two-hybrid assay, a bimolecular fluorescence complementation test, quantitative real-time PCR, transgenics and transcriptome sequencing. The results showed that IlAP2 interacts with IlMT2a and may cooperate with other transcription factors to regulate genes involved in signal transduction and plant hormones, leading to reduced Cd toxicity by hindering Cd transport. These findings provide insights into the mechanism of *IlAP2*-mediated stress responses to Cd and important gene resources for improving plant stress tolerance in phytoremediation.

Keywords: cadmium; *Iris lactea* var. chinensis; *IlAP2*; *IlMT2a*

Citation: Wang, Z.; Ni, L.; Liu, L.; Yuan, H.; Gu, C. *IlAP2*, an AP2/ERF Superfamily Gene, Mediates Cadmium Tolerance by Interacting with *IlMT2a* in *Iris lactea* var. *chinensis*. *Plants* **2023**, *12*, 823. https://doi.org/10.3390/plants12040823

Academic Editor: Andrzej Bajguz

Received: 25 December 2022
Revised: 10 February 2023
Accepted: 10 February 2023
Published: 12 February 2023

1. Introduction

Human activity, natural pollution such as volcanic ash, climate change and the use of recycled water, among other reasons, have resulted in an increase in the area of cadmium (Cd)-contaminated soil [1–3]. The presence of Cd in soil is a serious environmental hazard, as the metal is not only toxic to plants but also transmissible through the food chain, threatening ecosystems and human health [4,5]. Compared with physicochemical treatment, phytoremediation is an economical, natural and effective method for managing Cd soil contamination [6,7]. Therefore, it is important to find suitable Cd-tolerant plants and elucidate their relevant molecular mechanisms.

Cd toxicity can inhibit carbon fixation and reduce chlorophyll content and photosynthetic activity [8]. When plants are exposed to Cd-contaminated soil, osmotic stress occurs, reducing the relative moisture content, stomatal conductance and transpiration [9]. Meanwhile, an excessive occurrence of reactive oxygen species (ROS) can damage plant membranes and destroy cell biomolecules and organelles [10]. Cd also interferes with the transport and uptake of mineral elements [9]. To avoid the harm caused by Cd stress, plants have undergone a series of morphological, physiological and biochemical evolutionary changes [11]. One major mechanism is external exclusion, which prevents plant cells from absorbing excessive Cd and hinders Cd transport within the plant [12,13]. Another mechanism is tolerance to Cd accumulation, whereby Cd exists in a non-biologically active conjugated form within the plant through chelation and compartmentalization [14]. Studies have reported that when Cd is absorbed by a plant, proteins such as metallothioneins (MTs) can form stable chelates with the Cd ion, thus reducing its toxicity [4,15]. The regulation of transcription factors is a vital mechanism in plant cells that protects them from heavy metal exposure [16]. For example, previous studies showed that metal-responsive transcription factor 1 (MTF1) is involved in cellular protection against Cd stress signals [17]. Under Cd stress conditions, MTF1 can translocate into the nucleus and bind to metal-responsive elements (MREs), mediating the transcription of a series of downstream genes such as MTs, ZIP10, ferroportin 1 (FPN 1), selenoprotein H (Sel H) and selenoprotein W (Sel W) [16].

Iris lactea var. *chinensis*, an ornamental perennial groundcover plant that grows fast and forms a large biomass, is resistant to several stresses [18,19]. In particular, it has a very strong tolerance to Cd and the ability to accumulate Cd, which makes it suitable for improving Cd-contaminated soil [20]. We preliminarily analyzed the regulatory mechanism of *I. lactea* in response to Cd stress through transcriptome sequencing and mined an MT gene, *IlMT2a*, that may play a key role in Cd tolerance [21]. When constitutively expressed in *Arabidopsis thaliana*, *IlMT2a* leads to greater root length under Cd stress compared to the wild type (WT) [22]. In turn, we identified several proteins interacting with IlMT2a using a yeast two-hybrid assay (Y2H) with a yeast library constructed from cadmium-treated *I. lactea* seedlings and found that IlAP2, an apetala2/ethylene responsive factor, may be an interacting partner of IlMT2a [3].

Many transcription factors can be regulated by Cd stress and induce MTs to confer Cd tolerance [23–25], and *AP2* genes have been specifically induced by Cd in many plants, including *A. thaliana*, rice and kenaf [26–28]. Many reports have documented that AP2s are important regulators involved in plant growth and development, hormonal regulation, plant metabolite biosynthesis and the conferring of stress tolerance to plants [29,30]. Recently, potato *StAP2/ERF*s were found to be indispensable for Cd uptake and tolerance and may be useful in designing gene-modified plants with improved Cd tolerance [31]. To study the role of *IlAP2* in regulating the Cd tolerance of *I. lactea*, we cloned the open reading frame of *IlAP2* and confirmed the interaction between IlMT2a and IlAP2 using Y2H and a bimolecular fluorescence complementation test (BiFC). Following that, the regulation function and mechanism were explored through genetic transformation and transcriptome sequencing. This study provides information on the molecular mechanism of Cd tolerance and a gene resource for improving plant tolerance to Cd.

2. Results

2.1. Cloning and Analysis of the I1AP2 Sequence

Based on transcriptome data, primers were designed to obtain the full-length ORF sequence of IlAP2. The ORF of *IlAP2* was 1296 bp and encoded 432 amino acids. Several proteins with high similarity to IlAP2 in other plant species were selected using Blast on the NCBI website, and their amino acid sequences were analyzed using ClustalX software (Figure 1a). The alignment results showed that IlAP2 contained two conserved AP2 domains (Figure 1a). A phylogenetic tree was constructed using the neighbor-joining method to further analyze the evolutionary relationship between the *IlAP2* and *AP2* genes in other plants. The results showed that *IlAP2* was most closely related to the genes of *Phoenix dactylifera* (Figure 1b).

To further understand the cytological function of the IlAP2 protein, we inserted *IlAP2* into a pCAMBIA expression vector with an eGFP tag and transgenically transformed *A. thaliana* protoplasts to observe the subcellular distribution of IlAP2 proteins. The *IlAP2*–GFP fusion protein was only expressed in the nucleus, while the empty GFP vector was expressed in all parts of the cell, which indicates that *IlAP2* is a nuclear-localized transcription factor (Figure 2).

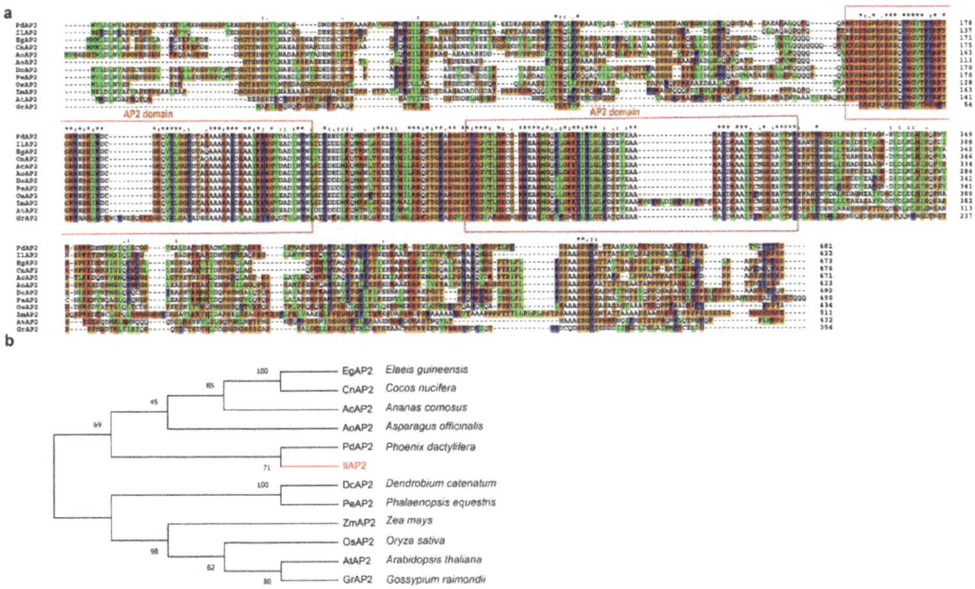

Figure 1. Analysis of the AP2 sequence: (**a**) alignment result of IlAP2 amino acid sequence (the red boxes represent functional domains, '*' indicates positions which have a single, fully conserved residue); and (**b**) phylogenetic tree of AP2 proteins from *I. lactea* (IlAP2) and other species.

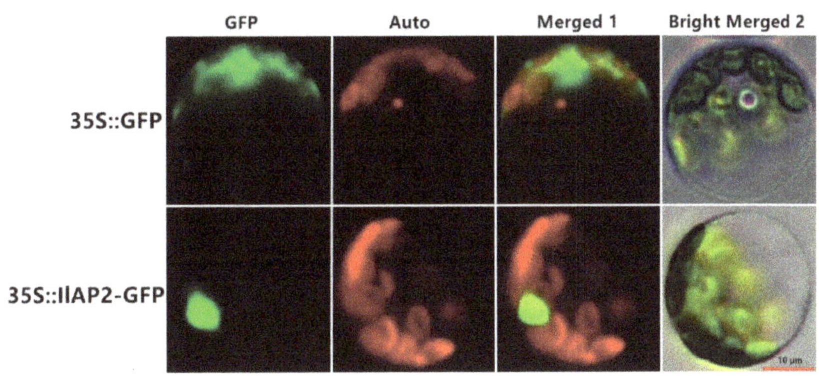

Figure 2. Subcellular localization of IlAP2 in *A. thaliana* leaf cells, with GFP as control. Green fluorescence protein (GFP), chlorophyll autofluorescence (Auto) and merged images (Merged 1 and Bright Merged 2) are shown. Scale bar: 10 μm.

2.2. Confirmation of Interaction between IlAP2 and IlMT2a

We re-inserted the complete *IlAP2* sequence into a pGADT7 vector and verified the interaction between IlAP2 and IlMT2a by Y2H based on the screening results using SD-Trp-Leu-Ade-His (QDO plates) [3,32]. The positive control (BD-53 + AD-T), negative control (BD-Lam + AD-T) and yeast cells co-transformed with PGADT7–*IlAP2* and PGBKT7–*IlMT2a* all grew normally in the SD-Trp-Leu medium, while only the positive plasmid and yeast cells co-transformed with PGADT7–*IlAP2* and PGBKT7–*IlMT2a* turned blue in the SD-Trp-Leu/X-α-Gal medium, indicating that IlAP2 and IlMT2a interact in yeast (Figure 3a). In addition, we carried out a BiFC experiment. *A. tumefaciens* containing *IlAP2*–NYFP and *IlMT2a*–CYFP plasmids were co-injected into tobacco leaves. The obvious interaction between IlAP2 and IlMT2a was verified again. The BiFC fluorescence was located in the nuclear region (Figure 3b).

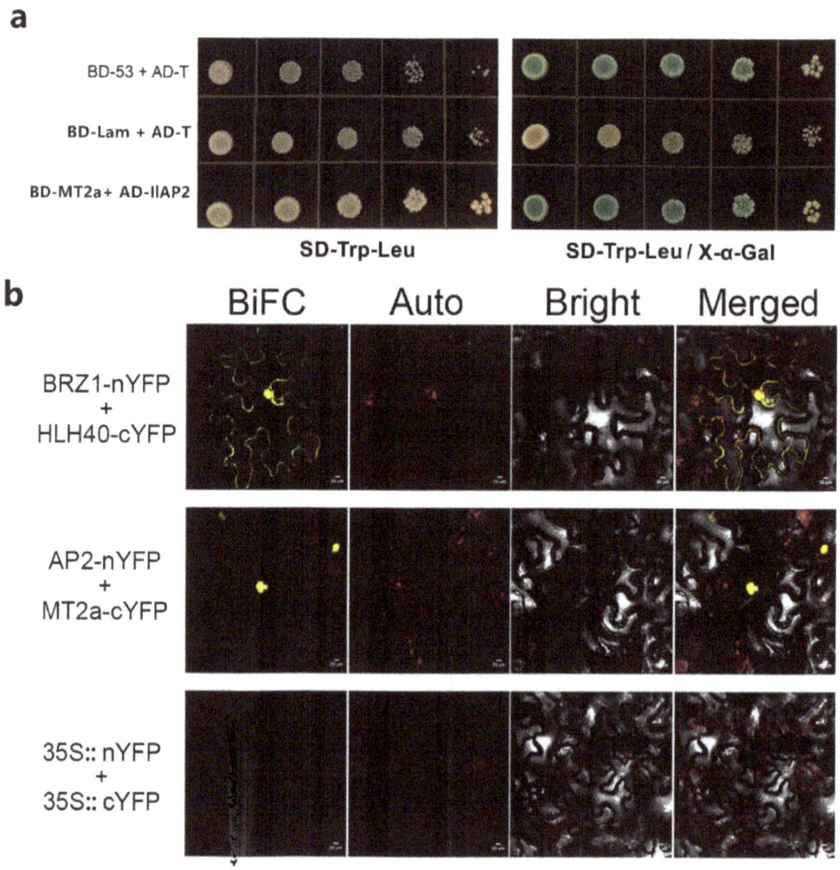

Figure 3. Interaction between IlAP2 and IlMT2a: (**a**) yeast two-hybrid analysis of IlAP2 and IlMT2a, with BD-53 + AD-T as positive control and BD-Lam + AD-T as negative control; and (**b**) BiFC analysis of IlAP2 and IlMT2a, with BRZ1-nTFP + HLH40-cYFP as positive control and 35S::nYFP + 35S::cYFP as negative control. Ruler: 20 μm.

2.3. *IlAP2 Expression Pattern Analysis*

To explore the expression pattern of *IlAP2* in response to Cd stress, semi-quantitative PCR and RT-qPCR were used to analyze the expression of *IlAP2* in *I. lactea* roots under 50 μM $CdCl_2$ stress at different times. Semi-quantitative PCR showed up-regulated *IlAP2* expression after stress treatment, which was further demonstrated by RT-qPCR (Figure 4). There was no significant difference in *IlAP2* expression level among the different treatment times (Figure 4).

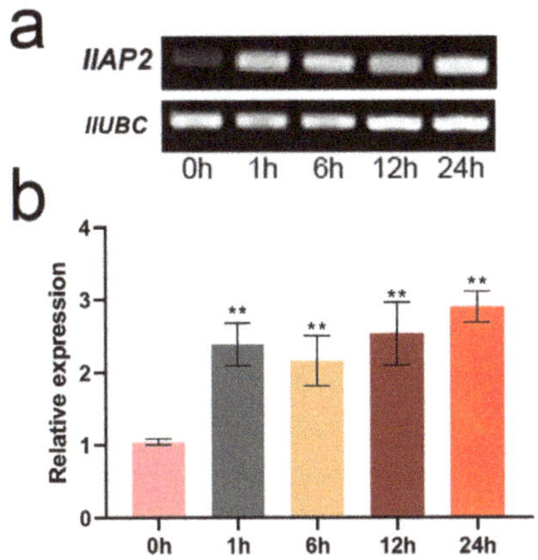

Figure 4. Analysis of *IlAP2* expression pattern in *I. lactea* roots at different times of Cd stress: (**a**) semi-quantitative PCR and (**b**) RT-qPCR. '**' shows the significance level of $p < 0.01$.

2.4. Functional Verification of *I1AP2* in *A. thaliana*

To investigate whether *IlAP2* can improve Cd tolerance, we constructed *I1AP2*-overexpressing *A. thaliana* and obtained eight homozygous T3-generation transgenic lines. Three independent transgenic lines were selected for functional verification: OE2, OE4 and OE5. After 10 days of growth (3-day-old seedlings were exposed to cadmium for 7 days), there was no significant difference in root length and growth between the WT and transgenic plants grown on normal half-MS medium. Under a 7-day Cd treatment, although the root length and growth of the WT and transgenic plants decreased continuously, the degree of decline for the transgenic plants was slower than that of the WT plants (Figure 5), suggesting that *IlAP2* improved Cd tolerance in *A. thaliana*.

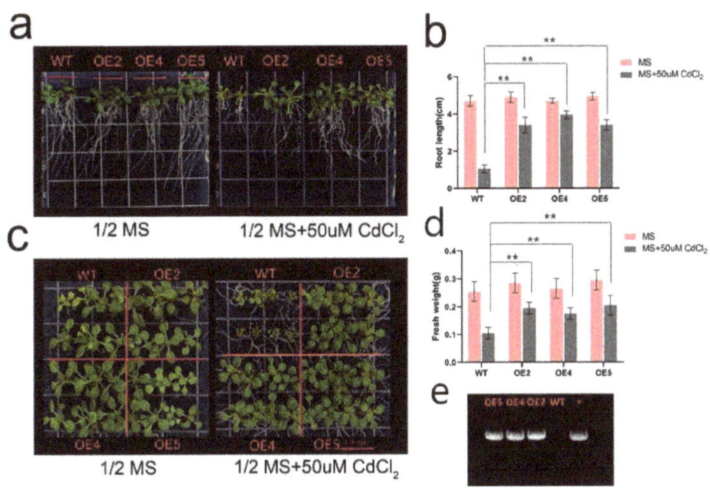

Figure 5. Functional verification analysis of *I1AP2* in *A. thaliana*. (**a**) Root length of wild-type (WT) and transgenic *A. thaliana* under $CdCl_2$ stress. (**b**) Root length statistics of WT and transgenic *A. thaliana* under $CdCl_2$ stress. (**c**) Comparison of growth of WT and transgenic *A. thaliana* under $CdCl_2$ stress. (**d**) Fresh weight of WT and transgenic *A. thaliana* under $CdCl_2$ stress. (**e**) Agarose gel electrophoresis profile for the expression of *I1AP2* in WT and transgenic *A. thaliana* (OE2, OE4 and OE5). The + band was amplified with *I. lactea* cDNA and ORF primers of *I1AP2*, serving as positive control. '**' shows the significance level of $p < 0.01$.

2.5. Analysis of the Potential Mechanism of *IlAP2* under Cd Stress

To further explore the mechanism by which *IlAP2* regulates Cd tolerance, RNA-seq analysis was performed on transgenic *A. thaliana* seedlings. The comparison between the transgenic *A. thaliana* and the WT identified 2210 genes as DEGs: 1340 up-regulated and 870 down-regulated (Figure 6a). The GO enrichment analysis showed that the enrichment degree of transferase activity was the highest in molecular function, indicating that ion signal transduction played an important role in *I. lactea*'s response after 24 h of Cd stress (Figure 6b). The KEGG enrichment analysis showed that plant hormone signal transduction was significantly enriched (Figure 6c), suggesting that plant hormones also played an important role in response to Cd stress. In addition, we found that bHLH, WRKY, NAC, MYB, GRAS, AP2/ERF and other transcription factors were all affected by *IlAP2* after 24 h of Cd stress (Figure 6d). We randomly chose six differentially expressed transcription factors and analyzed the relative changes in expression using qRT-PCR (Figure S1). The results showed that the expression patterns of most of the transcription factors tested with qRT-PCR were similar to those obtained from the sequencing data. Additionally, it was proven again that the transcription factors were affected by *IlAP2*.

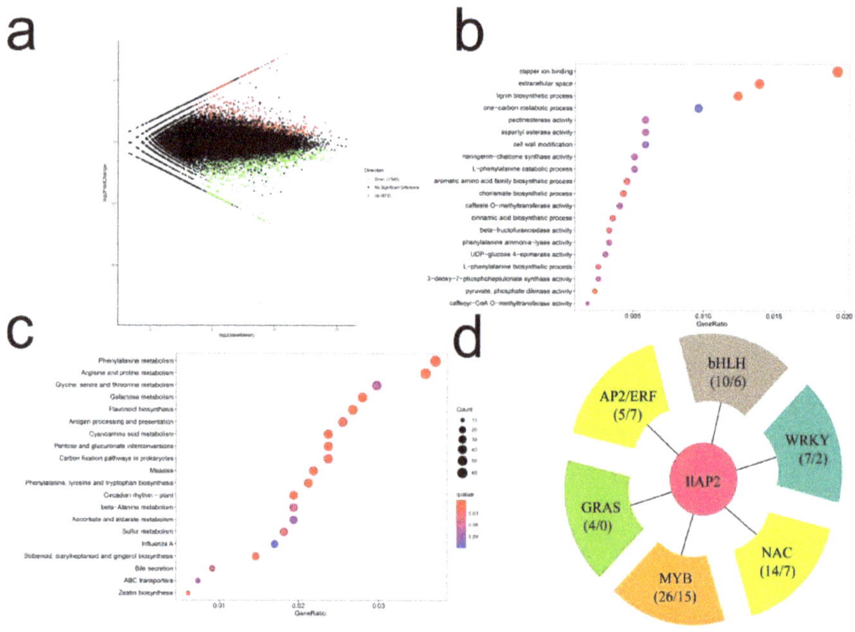

Figure 6. RNA-seq analysis of WT and transgenic *A. thaliana*. (**a**) DEG analysis results, (**b**) GO enrichment results, (**c**) KEGG enrichment results and (**d**) gene families possibly affected by *IlAP2*.

3. Discussion

Heavy metal pollution of soil and water is a serious environmental problem. Among heavy metal pollutants, Cd is considered to be one of the most toxic ones [24]. Cd can be quickly absorbed by plants due to its high solubility in water, which is the main way it enters the food chain [33]. Plant roots can absorb and transport Cd to nutritional and reproductive organs even at low Cd concentrations, which has a negative impact on the regulation of the plant's nutrition and its mineral balance during growth [34].

Several genes in plants, such as those encoding an ATP-binding transporter, ZRT/IRT-like proteins and glutathione, have been reported to participate in reducing Cd ion toxicity by inhibiting Cd ion absorption and transport, improving Cd ion chelation ability and other processes [35,36]. Gene *BcMT2* could enhance Cd tolerance and reduce the production of reactive oxygen species (ROS) in *A. thaliana* [37]. Similarly, we previously discovered that *IlMT2a* in *I. lactea* increased plant tolerance to Cd by potentially reducing the accumulation of H_2O_2 and $O_2^{\bullet -}$ [22]. However, *IlMT3*, another MT, was identified as an interacting protein of *IlMT2a* but did not provide protection against Cd toxicity in transgenic *A. thaliana* [3]. In this study, an AP2/ERF superfamily gene that encodes an interacting protein of IlMT2a, further validated its role in *I. lactea*'s response to Cd stress.

The transformation of transcriptomics under Cd stress has been studied in several plant species, which has helped to identify many Cd-responding transcription factors [24]. Different transcription factors, such as ERF (ethylene-responsive factor), WRKY, bZIP (basic leucine zipper) and MYB (myeloblastosis protein), control the expression of specific stress-related genes under Cd stress and play important roles [24]. Based on the ORF sequence amplified by PCR, the gene was named *IlAP2* following the phylogenetic tree and amino acid sequence analyses. To further understand the cytological function of the IlAP2 protein, we studied its expression and found that it was localized in the nucleus. In addition, the BiFC assays showed that the site of interaction between *IlAP2* and *IlMT2a* was also in the nucleus. The expression pattern analysis indicated that *IlAP2* was a stress-responsive transcription factor that could be regulated by Cd stress and maintained a steady high level of expression under Cd stress treatment. These findings indicate that *IlAP2* may be involved in the response to Cd stress and has a positive effect on Cd-stress tolerance, similar to some *StAP2/ERF* genes in kenaf [23].

Plant-specific AP2/ERF transcription factors can regulate plant responses to environmental stimuli or plant growth and development, depending on the presence of a highly conserved 60-amino-acid AP2 domain in the protein [38,39]. For example, rice *OsEREBP1* attenuates disease caused by *Xanthomonas* and confers drought and submergence tolerance by activating the jasmonate and abscisic acid signaling pathways, thereby priming the rice plants for enhanced survival under abiotic or biotic stress conditions [40]. *A. thaliana RAP2.6* participates in abiotic stress, including abscisic acid (ABA), salt and osmotic stress, possibly through the ABA-dependent pathway [41]. A previous study found that *StAP2/ERFs* are indispensable in Cd uptake and tolerance and may be useful in designing gene-modified potato plants with improved Cd tolerance [31]. To further demonstrate that *IlAP2* can improve plant Cd tolerance, we performed phenotype analysis on overexpressing *A. thaliana* plants. Under normal conditions, the phenotypic difference between the transgenic and WT *A. thaliana* plants was not significant. After Cd treatment, the growth of all *A. thaliana* plants was inhibited, but the root length and growth of the WT were inhibited to a greater extent, indicating that *IlAP2* influences the response of *A. thaliana* to Cd stress.

The overexpression of *TdSHN1*, a wheat ERF transcription factor, in transgenic tobacco conferred Cd tolerance, produced less ROS under excess Cd compared to WT plants and had higher activities of ROS-scavenging enzymes, which might contribute to Cd tolerance [42]. IlAP2 was confirmed to interact with IlMT2a and improve Cd tolerance. Although there is no report on the overexpression of *AtAP2* in *Arabidopsis*, the expression of potato *StAP2/ERF075/077/126* has been shown to increase tolerance to Cd [31]. Therefore, *AP2* may play an important role in the regulation of cadmium tolerance. Based on these results, we used transgenic *Arabidopsis* to preliminarily explore the regulatory mechanism through transcriptome sequencing, laying the foundation for subsequent research. A total of 1,340 and 870 genes were identified as up- and down-regulated, respectively. Enrichment analysis showed that *IlAP2* could alter ion signal transduction, plant hormones and other transcription factors at the transcriptional level, playing an important role in the plant's response to Cd stress. Previous studies have preliminarily analyzed the regulatory mechanism of *I. lactea* in response to Cd stress through proteome data and found that Cd regulatory proteins were mainly involved in signal transduction, ion transport, redox reactions, ion binding and other functions [43]. These findings indicate that *IlAP2* interacts with *IlMT2a* in the nucleus and may cooperate with other transcription factors to regulate genes related to signal transduction and plant hormones, thereby reducing Cd toxicity.

4. Materials and Methods

4.1. Plant Materials

The plant materials were preserved in Nanjing Botanical Garden Memorial Sun Yat-Sen. Selected seeds of *I. lactea* were cultivated under a photoperiod of 16/8 h (day/night) in a greenhouse incubator for 6 weeks. Seedlings about 10 cm high were washed and transferred to 50 mL centrifuge tubes containing 1/2 Hoagland solution (Table S1) for 1 week. Following that, the seedlings were transferred again to centrifuge tubes containing 1/2 Hoagland solution with 50 μM $CdCl_2$, as described in a previous study [21]. The roots were sampled at time points of 0, 1, 3, 6, 12 and 24 h, with three biological replicates for each time point. Samples at 0 h served as controls. All sampled plant materials were immediately frozen in liquid nitrogen and stored at −80 °C.

Tobacco (*Nicotiana benthamiana*) and *A. thaliana* (ecotype Columbia) seeds were selected and sown in plastic pots containing mixed culture medium (peat, vermiculite and perlite, 1:1:1 by volume) after soaking in sterile water for 24 h. The leaves obtained were used for research on subcellular localization and BiFC testing. Transgenic and WT *A. thaliana* seeds were prepared and sterilized with 10% NaClO for 10 min, then sown in 1/2 Murashige and Skoog (1/2 MS) medium supplemented with 3% (*w*/*v*) sucrose and 0.8% agar, incubated in darkness at 4 °C for 24 h and grown in a sterile environment.

4.2. Sequence Alignment and Subcellular Localization

The open reading frame (ORF) was amplified using Takara LA Taq (Takara, Dalian, China) based on the sequence obtained from transcriptome data (Table 1), and the PCR product was purified and cloned into the pMD19-T vector (Takara) for sequencing verification. Protein sequence was used for alignment using ClustalW (v. 2.1) software, and a phylogenetic tree was constructed using MEGA 7.0 with the neighbor-joining method following 1000 bootstrap replications.

Table 1. Primer sequences of the genes for full-length open reading frames (ORFs) and semi-quantitative PCR and quantitative real-time PCR.

Primer	Forward PCR Primer	Reverse PCR Primer
IlAP2-ORF	ATGTCGTTCGACCTGAACTTC	TCAGCTCCTGGAGTGGTAATG
IlAP2-qRT-PCR (semi-quantitative PCR)	GAATTTGAGGATTTCTCAACC	TACTTGATGATGCTCAGGAAC

A ClonExpress II One Step Cloning Kit (Vazyme, Nanjing, China) was selected to construct the recombinant plasmid vector containing the target sequence without the stop codon and the pBI121–GFP vector. The recombinant plasmids were transformed into *A. thaliana* protoplasts, which were isolated with enzymatic hydrolysis [44,45] using the PEG transformation method. Confocal laser scanning microscopy (Zeiss LSM 710 META, Jena, Germany) was used to determine the localization of gene expression through observing GFP fluorescence [46].

4.3. Semi-Quantitative PCR and Quantitative Real-Time PCR (qRT-PCR)

Total RNA of *I. lactea* roots under Cd stress was extracted using Plant RNeasy Mini Kit (Vazyme) and cDNA was generated with PrimeScript® RT kit (Takara, Dalian, China). Oligo 6.0 was used to design primers for semi-quantitative PCR and qRT-PCR (Table 1). Semi-quantitative PCR and qRT-PCR were performed using *IlUBC* as the reference gene [20]. Each reaction mix of semi-quantitative PCR contained 1 μL Taq enzyme, 2 μL 10 × PCR Buffer (Mg^{2+} Free), 1.5 μL Mg^{2+} (25 mmol·L^{-1}), 1.3 μL dNTP mixture (2.5 mmol·L^{-1} each), 6 pmol of each primer and 1 μL cDNA (about 1000 ng·μL^{-1}). PCR was performed under the following cycling conditions: 94 °C for 4 min, 25 cycles of 94 °C for 30 s, 55°C for 30 s, 72°C for 30 s and 72 °C for 10 min. PCR products were detected by 1% agarose gel electrophoresis. qRT-PCR was performed using AceQ qPCR SYBR Green Master Mix

(Vazyme, Nanjing, China) and StepOnePlus real-time PCR system (Applied Biosystems, Foster City, CA, USA) according to the manufacturer's instructions. The reaction mixture consisted of 2 μL diluted cDNA, 10 μL 2 × SYBR Green Master Mix (Bimake, Houston, TX, USA), 0.4 μL ROX (Dye I), 1 μL of each forward and reverse primer and ddH_2O to a total volume of 20 μL. The PCR procedure included an initial denaturation step set at 95 °C for 10 min, followed by 40 cycles of 95°C for 15 s, 60 °C for 30 s and 72 °C for 30 s. Melting curve analyses of the amplified products were conducted at 60–95 °C. Three biological replicates were performed for each treatment and three technical replicates for each reaction. Expression levels were analyzed with the $2^{-\Delta\Delta CT}$ calculation method.

4.4. Y2H and BiFC

The full-length ORF sequence of *IlAP2* was inserted into the PGADT7 vector using a ClonExpress II One Step Cloning Kit (Vazyme) to form the recombinant vector PGADT7–*IlAP2*. According to the Yeast Protocols Handbook (PT3024-1, Clontech, USA), the negative control (PGBKT7-LAM + PGADT7-T), positive control (PGBKT7-53 + PGADT7-T) and recombinant vector (PGADT7–*IlAP2* + PGBKT7–*IlMT2a*) were transformed into YH2GOLD yeast strain using the Yeastmaker™ Yeast Transformation System 2 (Takara). After growing in SD-Trp-Leu and SD-Trp-Leu/x-α-Gal medium (Takara) for 3 days, the monoclones were inoculated into YPDA liquid medium for 24 h of dark culture. Ten-fold serial dilutions (1, 1:10, 1:10^2, 1:10^3 and 1:10^4) were spotted onto SD-Trp-Leu and SD-Trp-Leu /x-α-Gal solid media for 3 days to observe the phenotype.

The full-length ORF sequence of *IlAP2* was inserted into the PSM35S–NYFP vector using the ClonExpress II One Step Cloning Kit (Vazyme). The full-length *IlMT2a* ORF sequence was inserted into the PSM35S–CyFP vector to form the recombinant vector PSM35S–*IlMT2a*–CyFP. The constructed vector plasmids were transferred into *Agrobacterium tumefaciens GV3101* using electro-transformation method and cultured for 2 days at 30 °C. Subsequently, the constructed vector plasmids were mixed in a 1:1 ratio and injected into the lower epidermis of tobacco leaves. The results were observed with confocal laser scanning microscopy (Zeiss LSM 710 META) after 2 days under low light conditions with BRZ1-nTFP + HLH40-cYFP as positive control [47–49].

4.5. Construction and Stress Treatment of Transgenic A. thaliana

The *IlAP2* sequence was inserted into the pCAMBIA1305 vector with 35S promoter using the ClonExpress II One Step Cloning Kit (Vazyme), then the 35S:*IlAP2* plasmid was transformed into *A. tumefaciens GV3101*. *A. thaliana* Columbia was used for transformation by tidbit infection, and homozygous T3 transgenic *A. thaliana* lines were identified with 1/2 MS medium containing selective antibiotic hygromycin (50 mg/L). The expression of *I1AP2* in WT and transgenic *A. thaliana* was confirmed by agarose gel electrophoresis with PCR products amplified using IlAP2-ORF primers and cDNA from WT and transgenic *A. thaliana*. Total RNA from WT and transgenic *A. thaliana* under Cd stress was extracted using Plant RNeasy Mini Kit (Vazyme), and cDNA was generated using PrimeScript® RT kit (Takara). Three-day-old WT and transgenic *A. thaliana* seedlings were selected and transplanted to 1/2 MS medium containing 50 μM $CdCl_2$. The root length and fresh weight of seedlings were measured after 7 days of growth.

4.6. RNA-seq and Analysis

Twenty-day-old transgenic and WT *A. thaliana* seedlings were treated with 50 μM $CdCl_2$ for 24 h and sampled for RNA extraction [50]. RNA-seq libraries were created following the manufacturer's instructions for the NEBNext® UltraTM RNA Library Prep Kit for Illumina® (NEB, Ipswich, MA, USA). The cDNA fragments, ranging in length from 250 to 300 bp, were screened using AMPure XP system (Beckman Coulter, Beverly, CA, USA). Illumina Hiseq 2000 (Illumina, San Diego, CA, USA) was utilized for library sequencing and paired-end read production.

Genes were assembled using the full-length transcript sequence as a reference [51]. The FPKM value obtained from read counts was converted to a TPM value to obtain the expression level of each isoform. Based on analysis using differential expression analysis software DESeq2, genes with padj < 0.05 and $|\log_2\text{foldchange}| > 1$ were identified as differentially expressed genes (DEGs). GO enrichment analysis was performed on DEGs using Wallenius' noncentral hypergeometric distribution of the GOseq R package. The KEGG enrichment analysis was performed using KOBAS software.

5. Conclusions

In conclusion, the AP2/ERF superfamily gene *IlAP2*, confirmed to interact with *IlMT2a* in this study, was analyzed using quantitative real-time PCR, transgenics and transcriptome sequencing to understand its regulation function and mechanism in improving plant tolerance to cadmium stress. The results showed that *IlAP2* could interact with IlMT2a in the nucleus and may cooperate with other transcription factors to regulate genes related to signal transduction and plant hormones, leading to reduced cadmium toxicity. These findings provide insights into the mechanism of *IlAP2*-mediated stress responses to cadmium and highlight *IlAP2* as an important gene resource for improving plant stress tolerance in phytoremediation to manage cadmium soil contamination.

Supplementary Materials: The following supporting information can be downloaded at: https://www.mdpi.com/article/10.3390/plants12040823/s1, Figure S1: Relative changes in expression of genes analyzed using qRT-PCR; Table S1: 1/2 Hoagland's solution composition.

Author Contributions: Conceptualization, Z.W. and C.G.; formal analysis, L.N. and Z.W.; investigation, Z.W., L.N., L.L. and H.Y.; writing—original draft preparation, Z.W. and L.N.; writing—review and editing, Z.W. and C.G. All authors have read and agreed to the published version of the manuscript.

Funding: This research was funded by the National Natural Science Foundation of China (31772351) and the 333 Talents Project of Jiangsu Province (BRA2017498).

Data Availability Statement: No new data were created or analyzed in this study. Data sharing is not applicable to this article.

Conflicts of Interest: The authors declare no conflict of interest.

References

1. Gill, S.S.; Tuteja, N. Cadmium stress tolerance in crop plants: Probing the role of sulfur. *Plant Signal. Behav.* **2011**, *6*, 215–222. [CrossRef] [PubMed]
2. Chen, H.; Li, Y.; Ma, X.; Guo, L.; He, Y.; Ren, Z.; Kuang, Z.; Zhang, X.; Zhang, Z. Analysis of potential strategies for cadmium stress tolerance revealed by transcriptome analysis of upland cotton. *Sci. Rep.* **2019**, *9*, 86. [CrossRef]
3. Wang, Z.; Ni, L.; Liu, L.; Yuan, H.; Huang, S.; Gu, C. Screening and identification of host proteins interacting with *Iris lactea* var. Chinensis metallothionein IlMT2a by yeast two-hybrid assay. *Genes* **2021**, *12*, 554.
4. Verbruggen, N.; Hermans, C.; Schat, H. Mechanisms to cope with arsenic or cadmium excess in plants. *Curr. Opin. Plant Biol.* **2009**, *12*, 364–372. [CrossRef]
5. Nawrot, T.; Plusquin, M.; Hogervorst, J.; Roels, H.A.; Celis, H.; Thijs, L.; Vangronsveld, J.; Van Hecke, E.; Staessen, J.A. Environmental exposure to cadmium and risk of cancer: A prospective population-based study. *Lancet Oncol.* **2006**, *7*, 119–126. [CrossRef] [PubMed]
6. Kanamarlapudi, S.; Chintalpudi, V.K.; Muddada, S. Application of biosorption for removal of heavy metals from wastewater. *Biosorption* **2018**, *18*, 70–116.
7. Monteiro, C.M.; Castro, P.; Malcata, F.X. Metal uptake by microalgae: Underlying mechanisms and practical applications. *Biotechnol. Prog.* **2012**, *28*, 299–311. [CrossRef]
8. Gallego, S.; Pena, L.; Barcia, R.; Azpilicueta, C.; Iannone, M.; Rosales, E.; Benavides, M. Unravelling cadmium toxicity and tolerance in plants: Insight into regulatory mechanisms. *Environ. Exp. Bot.* **2012**, *83*, 33–46. [CrossRef]
9. Haider, F.; Liqun, C.; Coulter, J.; Cheema, S.; Wu, J.; Zhang, R.; Wenjun, M.; Farooq, M. Cadmium toxicity in plants: Impacts and remediation strategies. *Ecotoxicol. Environ. Saf.* **2021**, *211*, 111887. [PubMed]
10. Abbas, T.; Rizwan, M.; Ali, S.; Adrees, M.; Zia-ur-Rehman, M.; Qayyum, M.F.; Ok, Y.S.; Murtaza, G. Effect of biochar on alleviation of cadmium toxicity in wheat (*Triticum aestivum* L.) grown on Cd-contaminated saline soil. *Environ. Sci. Pollut. Res.* **2018**, *25*, 25668–25680. [CrossRef]

11. Liu, S.; Li, L.; Deng, Y.; Bai, Y.; Sun, C.; Huang, S.; Zhou, J.; Shi, L.; Yang, X.; Li, L. *BRPNAC895* and *BRPABI449* coregulate the transcription of the afflux-type cadmium transporter *BRPHMA2* in *Brassica parachinensis*. *Hortic. Res.* **2022**, *9*, uhac044. [CrossRef]
12. Tong, Y.-P.; Kneer, R.; Zhu, Y.-G. Vacuolar compartmentalization: A second-generation approach to engineering plants for phytoremediation. *Trends Plant Sci.* **2004**, *9*, 7–9. [CrossRef] [PubMed]
13. Chen, G.; Liu, Y.; Wang, R.; Zhang, J.; Owens, G. Cadmium adsorption by willow root: The role of cell walls and their subfractions. *Environ. Sci. Pollut. Res.* **2013**, *20*, 5665–5672. [CrossRef]
14. Deng, D.; Shu, W.; Zhang, J.; Zou, H.; Lin, Z.; Ye, Z.; Wong, M.H. Zinc and cadmium accumulation and tolerance in populations of sedum alfredii. *Environ. Pollut.* **2007**, *147*, 381–386. [CrossRef] [PubMed]
15. Weiti, C.; Huiping, C.; Kaikai, Z.; Qijiang, J.; Yanjie, X.; Jin, C.; Yan, X.; Jing, Z.; Wenbiao, S.; Ji-Hong, L. Cadmium-induced hydrogen sulfide synthesis is involved in cadmium tolerance in medicago sativa by reestablishment of reduced (homo)glutathione and reactive oxygen species homeostases. *PLoS ONE* **2014**, *9*, e109669.
16. Talukder, M.; Bi, S.; Jin, H.T.; Ge, J.; Zhang, C.; Lv, M.; Li, J. Cadmium induced cerebral toxicity via modulating MTF1-MTs regulatory axis. *Environ. Pollut.* **2021**, *285*, 117083. [CrossRef] [PubMed]
17. Saydam, N.; Georgiev, O.; Nakano, M.; Greber, U.; Schaffner, W. Nucleo-cytoplasmic trafficking of metal-regulatory transcription factor 1 is regulated by diverse stress signals. *J. Biol. Chem.* **2001**, *276*, 25487–25495. [CrossRef]
18. Tang, J.; Liu, Q.; Yuan, H.; Zhang, Y.; Huang, S. Molecular analysis of a novel alkaline metal salt (nacl)-responsive WRKY transcription factor gene IlWRKY1 from the halophyte *Iris lactea* var. *Chinensis*. *Int. Biodeterior. Biodegrad.* **2018**, *127*, 139–145. [CrossRef]
19. Han, Y.-L.; Huang, S.-Z.; Gu, J.-G.; Qiu, S.; Chen, J.-M. Tolerance and accumulation of lead by species of *Iris* L. *Ecotoxicology* **2008**, *17*, 853–859. [CrossRef]
20. Gu, C.; Xu, S.; Wang, Z.; Liu, L.; Zhang, Y.; Deng, Y.; Huang, S. De novo sequencing, assembly, and analysis of *Iris lactea* var. Chinensis roots' transcriptome in response to salt stress. *Plant Physiol. Biochem.* **2018**, *125*, 1–12.
21. Gu, C.-S.; Liu, L.-Q.; Deng, Y.-M.; Zhang, Y.-X.; Wang, Z.-Q.; Yuan, H.-Y.; Huang, S.-Z. De novo characterization of the *Iris lactea* var. Chinensis transcriptome and an analysis of genes under cadmium or lead exposure. *Ecotoxicol. Environ. Saf.* **2017**, *144*, 507–513.
22. Gu, C.-S.; Liu, L.-q.; Zhao, Y.-H.; Deng, Y.-m.; Zhu, X.-d.; Huang, S.-Z. Overexpression of *Iris. Lactea* var. Chinensis metallothionein IlMT2a enhances cadmium tolerance in Arabidopsis thaliana. *Ecotoxicol. Environ. Saf.* **2014**, *105*, 22–28. [PubMed]
23. Chen, P.; Li, Z.; Luo, D.; Jia, R.; Lu, H.; Tang, M.; Hu, Y.; Yue, J.; Huang, Z. Comparative transcriptomic analysis reveals key genes and pathways in two different cadmium tolerance kenaf (*Hibiscus cannabinus* L.) cultivars. *Chemosphere* **2021**, *263*, 128211. [CrossRef]
24. DalCorso, G.; Farinati, S.; Furini, A. Regulatory networks of cadmium stress in plants. *Plant Signal. Behav.* **2010**, *5*, 663–667. [CrossRef] [PubMed]
25. Chen, J.; Yang, L.; Yan, X.; Liu, Y.; Wang, R.; Fan, T.; Ren, Y.; Tang, X.; Xiao, F.; Liu, Y. Zinc-finger transcription factor *ZAT6* positively regulates cadmium tolerance through the glutathione-dependent pathway in *Arabidopsis*. *Plant Physiol.* **2016**, *171*, 707–719. [CrossRef] [PubMed]
26. Weber, M.; Trampczynska, A.; Clemens, S. Comparative transcriptome analysis of toxic metal responses in *Arabidopsis thaliana* and the cd^{2+}-hypertolerant facultative metallophyte Arabidopsis halleri. *Plant Cell Environ.* **2006**, *29*, 950–963. [CrossRef]
27. He, F.; Liu, Q.; Zheng, L.; Cui, Y.; Shen, Z.; Zheng, L. RNA-seq analysis of rice roots reveals the involvement of post-transcriptional regulation in response to cadmium stress. *Front. Plant Sci.* **2015**, *6*, 1136. [CrossRef]
28. Ogawa, I.; Nakanishi, H.; Mori, S.; Nishizawa, N.K. Time course analysis of gene regulation under cadmium stress in rice. *Plant Soil* **2009**, *325*, 97–108. [CrossRef]
29. Feng, K.; Hou, X.; Xing, G.; Liu, J.; Duan, A.; Xu, Z.; Li, M.; Zhuang, J.; Xiong, A. Advances in AP2/ERF super-family transcription factors in plant. *Crit. Rev. Biotechnol.* **2020**, *40*, 750–776. [CrossRef]
30. Sun, F.; Ding, L.; Feng, W.; Cao, Y.; Lu, F.; Yang, Q.; Li, W.; Li, Y.; Shabek, N.; Fu, F.; et al. Maize transcription factor ZmBES1/BZR1-5 positively regulates kernel size. *J. Exp. Bot.* **2021**, *72*, 1714–1726. [CrossRef] [PubMed]
31. Tian, W.; Huang, Y.; Li, D.; Meng, L.; He, T.; He, G. Identification of *STAP2/ERF* genes of potato (*Solanum tuberosum*) and their multiple functions in detoxification and accumulation of cadmium in yest: Implication for genetic-based phytoremediation. *Sci. Total Environ.* **2022**, *810*, 152322. [CrossRef] [PubMed]
32. Zhu, H.; Zhou, Y.; Zhai, H.; He, S.; Zhao, N.; Liu, Q. A novel sweetpotato WRKY transcription factor, *IbWRKY2*, positively regulates drought and salt tolerance in transgenic *Arabidopsis*. *Biomolecules* **2020**, *10*, 506. [CrossRef]
33. Buchet, J.; Lauwerys, R.; Roels, H.; Bernard, A.; Bruaux, P.; Claeys, F.; Ducoffre, G.; Plaen, P.; Staessen, J.; Amery, A.; et al. Renal effects of cadmium body burden of the general population. *Lancet* **1990**, *336*, 699–702. [CrossRef] [PubMed]
34. Farinati, S.; DalCorso, G.; Varotto, S.; Furini, A. The *Brassica juncea BjCdR15*, an ortholog of Arabidopsis TGA3, is a regulator of cadmium uptake, transport and accumulation in shoots and confers cadmium tolerance in transgenic plants. *New Phytol.* **2010**, *185*, 964–978. [CrossRef]
35. Lin, Y.F.; Aarts, M.G.M. The molecular mechanism of zinc and cadmium stress response in plants. *Cell. Mol. Life Sci.* **2012**, *69*, 3187–3206.
36. Narendrula-Kotha, R.; Theriault, G.; Mehes-Smith, M.; Kalubi, K.; Nkongolo, K. Metal toxicity and resistance in plants and microorganisms in terrestrial ecosystems. *Rev. Environ. Contam. Toxicol.* **2019**, *249*, 1–27.

37. Lv, Y.; Deng, X.; Quan, L.; Xia, Y.; Shen, Z. Metallothioneins *BcMT1* and *BcMT2* from *Brassica campestris* enhance tolerance to cadmium and copper and decrease production of reactive oxygen species in *Arabidopsis thaliana*. *Plant Soil* **2013**, *367*, 507–519. [CrossRef]
38. Nakano, T.; Suzuki, K.; Fujimura, T.; Shinshi, H. Genome-wide analysis of the ERF gene family in *Arabidopsis* and rice. *Plant Physiol.* **2006**, *140*, 411–432. [CrossRef] [PubMed]
39. Sharoni, A.M.; Nuruzzaman, M.; Satoh, K.; Shimizu, T.; Kondoh, H.; Sasaya, T.; Choi, I.-R.; Omura, T.; Kikuchi, S. Gene structures, classification and expression models of the AP2/EREBP transcription factor family in rice. *Plant Cell Physiol.* **2011**, *52*, 344–360. [CrossRef] [PubMed]
40. Jisha, V.; Dampanaboina, L.; Vadassery, J.; Mithöfer, A.; Kappara, S.; Ramanan, R. Overexpression of an AP2/ERF type transcription factor oserebp1 confers biotic and abiotic stress tolerance in rice. *PLoS ONE* **2015**, *10*, e0127831. [CrossRef]
41. Zhu, Q.; Zhang, J.; Gao, X.; Tong, J.; Xiao, L.; Li, W.; Zhang, H. The Arabidopsis AP2/ERF transcription factor RAP2. 6 participates in ABA, salt and osmotic stress responses. *Gene* **2010**, *457*, 1–12. [CrossRef] [PubMed]
42. Djemal, R.; Khoudi, H. The ethylene-responsive transcription factor of durum wheat, TDSHN1, confers cadmium, copper, and zinc tolerance to yeast and transgenic tobacco plants. *Protoplasma* **2022**, *259*, 19–31. [CrossRef] [PubMed]
43. Liu, Q.; Zhang, Y.; Wang, Y.; Wang, W.; Gu, C.; Huang, S.; Yuan, H.; Dhankher, O.P. Quantitative proteomic analysis reveals complex regulatory and metabolic response of *Iris lactea* pall. Var. Chinensis to cadmium toxicity. *J. Hazard. Mater.* **2020**, *400*, 123165. [CrossRef]
44. Bossche, R.V.; Demedts, B.; Vanderhaeghen, R.; Goossens, A. Transient expression assays in tobacco protoplasts. In *Jasmonate Signaling*; Springer: Totowa, NJ, USA, 2013; pp. 227–239.
45. Yoo, S.D.; Cho, Y.H.; Sheen, J. Arabidopsis mesophyll protoplasts: A versatile cell system for transient gene expression analysis. *Nat. Protoc.* **2007**, *2*, 1565–1572. [CrossRef]
46. Wei, F.; Tang, D.; Li, Z.; Kashif, M.H.; Khan, A.; Lu, H.; Jia, R.; Chen, P. Molecular cloning and subcellular localization of six HDACs and their roles in response to salt and drought stress in kenaf (*Hibiscus cannabinus* L.). *Biol. Res.* **2019**, *52*, 20. [CrossRef] [PubMed]
47. Chen, W.C.; Wang, Q.; Cao, T.J.; Lu, S. UBC19 is a new interacting protein of ORANGE for its nuclear localization in *Arabidopsis thaliana*. *Plant Signal. Behav.* **2021**, *16*, 1964847. [CrossRef]
48. Fang, Z.; Ji, Y.; Hu, J.; Guo, R.; Sun, S.; Wang, X. Strigolactones and brassinosteroids antagonistically regulate the stability of D53-OsBZR1 complex to determine FC1 expression in rice tillering. *Mol. Plant* **2020**, *13*, 586–597. [CrossRef]
49. Kong, L.; Duan, Y.; Ye, Y.; Cai, Z.; Wang, F.; Qu, X.; Qiu, R.; Wu, C.; Wu, W. Screening and analysis of proteinsinteracting with OsMADS16 in rice (*Oryza sativa* L.). *PLoS ONE* **2019**, *14*, e0221473. [CrossRef]
50. Ni, L.; Wang, Z.; Liu, X.; Wu, S.T.; Hua, J.; Liu, L.; Yin, Y.; Li, H.; Gu, C. Genome-wide study of the GRAS gene family in *Hibiscus hamabo* Sieb. et Zucc and analysis of *HhGRAS14*-induced drought and salt stress tolerance in *Arabidopsis*. *Plant Sci.* **2022**, *319*, 111260. [CrossRef]
51. Ni, L.; Wang, Z.; Guo, J.; Pei, X.; Liu, L.; Li, H.; Yuan, H.; Gu, C. Full-length transcriptome sequencing and comparative transcriptome analysis to evaluate drought and salt stress in *Iris lactea* var. Chinensis. *Genes* **2021**, *12*, 434. [CrossRef]

plants

MDPI

Article

Transcriptomic Analysis Reveals the Regulatory Mechanism of Color Diversity in *Rhododendron pulchrum* Sweet (Ericaceae)

Nanyan Zhu [1] and Chunhua Zhou [1,2,*]

1 College of Animal Science and Technology, Yangzhou University, 30 Wenhui East Rd., Yangzhou 225009, China; dx120200148@stu.yzu.edu.cn

2 College of Horticulture and Landscape Architecture, Yangzhou University, 30 Wenhui East Rd., Yangzhou 225009, China

* Correspondence: chzhou@yzu.edu.cn

Abstract: *Rhododendron pulchrum* Sweet is a traditional ornamental plant cultivated in China and presents a great variation in petal coloration. However, few studies have been performed to reveal the genes involved and the regulatory mechanism of flower color formation in this plant. In this study, to explore the underlying genetic basis of flower color formation, transcriptome analysis was performed by high-throughput sequencing techniques on four petal samples of different colors: purple, pink, light pink, and white. Results show that a total of 35.55 to 40.56 million high-quality clean reads were obtained, of which 28.56 to 32.65 million reads were mapped to the reference genome. For their annotation, 28,273, 18,054, 24,301, 19,099, and 11,507 genes were allocated to Nr, Swiss-Prot, Pfam, GO, and KEGG databases, correspondingly. There were differentially expressed genes among the four different petal samples, including signal-transduction-related genes, anthocyanin biosynthesis genes, and transcription factors. We found that the higher expressed levels of genes associated with flavonol synthase (FLS) might be the key to white formation, and the formation of red color may be related to the higher expression of flavanone 4-reductase (DFR) families. Overall, our study provides some valuable information for exploring and understanding the flower color intensity variation in *R. pulchrum*.

Keywords: *Rhododendron pulchrum*; transcriptome analysis; qRT-PCR; anthocyanin; gene expression

Citation: Zhu, N.; Zhou, C. Transcriptomic Analysis Reveals the Regulatory Mechanism of Color Diversity in *Rhododendron pulchrum* Sweet (Ericaceae). *Plants* **2023**, *12*, 2656. https://doi.org/10.3390/plants12142656

Academic Editors: Aiping Song and Yu Chen

Received: 12 April 2023
Revised: 12 July 2023
Accepted: 13 July 2023
Published: 15 July 2023

1. Introduction

Rhododendron is the largest genus belonging to the family Ericaceae, which is known for colorful flowers [1]. In recent years, varieties of *Rhododendron* have been grown as ornamental plants, making them become the most popular evergreen shrubs all over the world [2]. As one of the important characteristics of *Rhododendron*, color has long been a concern of breeders and consumers. The author (N.Z.) and her team found that there were a few white and pink flowers in the purple *Rhododendron pulchrum* Sweet community. Petals are an extremely essential part of numerous ornamental plants, and differences in petal color, tones, and intensity directly influence the ornamental value of a plant [3]. Previous studies have shown that flower color is attributed to specific pigments in petal cells, including flavonoids, carotenoids, and alkaloids [4,5]. In addition, flavonoids are major pigments in flowers that are responsible for the coloration of plant petals [6,7]. Flavonoids are naturally occurring polyphenols in plants. According to their structural characteristics, flavonoids are usually classified into anthocyanins and flavonoid alcohols [8,9]. Among them, anthocyanins are the main water-soluble pigments in flowers that constitute the color of plant petals. They are mainly accumulated in the vacuoles of petal epidermal cells and impart to the petals a colorful appearance from light pink to purple [10]. Color modification of flowers can be achieved by enhancing the accumulation of anthocyanins [11]. Flavonoids, on the other hand, have auxiliary effects on anthocyanins. The petals of plants with a higher concentration of flavonoids usually show bright colors [12]. In general, anthocyanins are synthesized through the secondary metabolic pathway, which occurs in a wide range of plants, such as *Gerbera hybrida* and *Triticum aestivum* [13,14]. However, the molecular mechanisms regulating anthocyanin synthesis have not been elucidated in *R. pulchrum* due to the structural diversity of anthocyanins among the different plant species [15].

Anthocyanins are derived from the branches of the flavonoid biosynthesis pathway, which also leads to the production of isoflavonoids and flavonols. The anthocyanin biosynthesis pathway has been extensively studied in a range of plant species and involves several enzymes encoded by different structural genes [16,17]. First, phenylalanine ammonia-Lyase (PAL) catalyzes the deamination of phenylalanine to cinnamic acid in the initial step of the flavonoid pathway [18], while chalcone synthase (CHS) catalyzes the synthesis of naringenin chalcone with 4-coumaroyl CoA and malonyl CoA as substrates in the first committed step of flavonoid biosynthesis [19]. Subsequently, chalcone isomerase (CHI) catalyzes the conversion of naringenin chalcone to naringenin [20,21]. The naringenin is then catalyzed to dihydrokaempferol (DHK) via flavanone-3-hydroxylase (F3H). Moreover, DHK can be further hydroxylated by flavonoid 3′-hydroxylase (F3′H) to produce dihydroquercetin (DHQ). In the third stage, all obtained DHQ are reduced to leucocyanidin by dihydroflavonol 4-reductase (DFR) and then are further converted into anthocyanidins by anthocyanidin synthase (ANS). Of the main key enzymes discovered, CHS catalyzes the first step of flavonoid biosynthesis, and DFR is the first committed enzyme of anthocyanin biosynthesis [22].

It is known that the MBW complexes (MYB-*bHLH*-WD40) consist of the MYB transcription factors (TFs), *bHLH* TFs, and one WD-40 repeat factor (WDR) [23], which play important roles in regulation of the expression of the genes involved in the anthocyanin biosynthetic pathway at the transcriptional level [24]. The MYB regulator is a large family of proteins with diverse functions, and most of the MYB genes in plants belong to *R2R3-MYB* TFs. Several studies have shown that *R2R3-MYB* TFs control the transcriptional regulation of anthocyanin structural genes [25]. Another crucial TF regulating anthocyanin biosynthesis is the *bHLH* protein, which is critical for the activity of *R2R3-MYB*. For example, the *bHLH* TF in *Arabidopsis* interacts with the *R2R3-MYB* protein to regulate DFR gene expression in anthocyanin biosynthesis [26]. It was considered that the WDR proteins could interact with different *R2R3-MYB* and *bHLH* to form transcription complexes, which play vital roles in anthocyanin accumulation in vegetative tissues [27].

Currently, to our best knowledge, research on flower coloration in *R. pulchrum* is very limited, and the molecular mechanism regulating the flower coloration remains unknown. It would help to clarify the mechanisms controlling anthocyanin accumulation through the analysis of the expression pattern of the key genes related to color formation in *Rhododendron*. In recent years, with the development of next-generation sequencing, transcriptome sequencing (RNA-seq) has been widely used to identify differentially expressed genes (DEGs) in many plants [28,29]. Therefore, transcriptome sequencing of *R. pulchrum* flowers will provide meaningful knowledge to unravel the molecular mechanism of color formation. In this current study, four varieties with different petal colors were used as the experimental materials to determine the genotypic difference at the transcriptional level using RNA-seq technology. Our findings will provide a useful resource for further analyzing the molecular mechanism of color formation and intensity of *R. pulchrum*.

2. Results

2.1. The Contents of Anthocyanins Components and Total Flavonoids

The petals of *R. pulchrum* have deep-color blotches in the center. The components and contents of anthocyanins both in petals and blotches were analyzed by HPLC and compared with the standard. The results show that the peak value of pigments appeared in samples A, B, and C around 7.32, 9.07, 10.55, 12.30, and 15.6 min, which is consistent with the detection time of five pigment standards (delphinidin, 6.95; cyanidin, 8.91; pelargonidin, 10.82; peonidin, 11.95; malvidin, 14.80 min) (Figure 1). In petal samples, peonidin was the main anthocyanin in sample A. The content of peonidin in B and C decreased significantly, and pelargonidin pigment became the main anthocyanin that determined the color of samples B and C (Table S1). The main anthocyanins in the deep-color blotches were peonidin. Anthocyanin components were not detected in the petals of D, and only a small amount of anthocyanins were detected in the blotches. The analysis of flavonoid content in sample petals significantly showed that the lighter the color of the petals, the higher the flavonoid content, and there was no significant difference in the flavonoid content in the dark spots of the petals (Table 1).

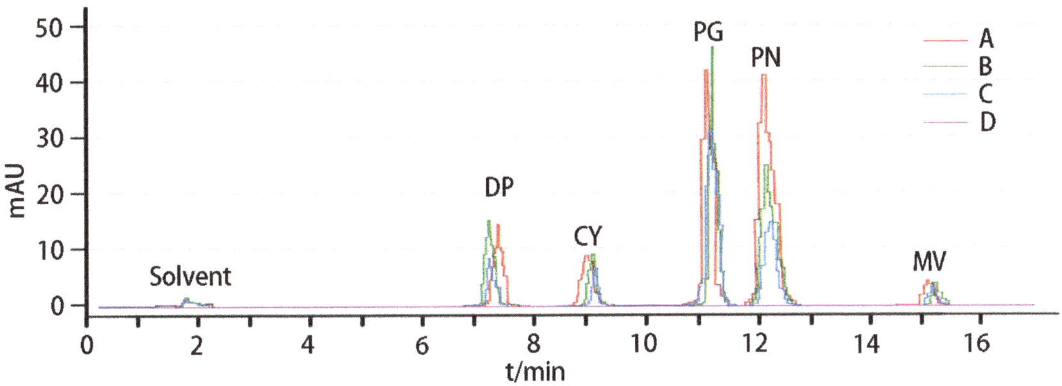

Figure 1. High-performance liquid chromatograms (HPLC) of mixed anthocyanins in petals. The *y*-axis represents the peak area. DP, delphinidin; CY, cyanidin; PG, pelargonidin; PN, peonidin; MV, malvidin.

Table 1. The content of total flavonoids.

Sample	Total Flavonoids (QE mg/g DW)	
	Petal	Spot
A	1.14 ± 0.12 [d]	1.03 ± 0.07 [b]
B	1.71 ± 0.06 [c]	1.03 ± 0.03 [b]
C	1.97 ± 0.06 [b]	1.44 ± 0.10 [a]
D	2.24 ± 0.06 [a]	1.57 ± 0.08 [a]

Means are followed by ± standard deviations. Different lowercase letters indicate significant difference between treatments based on one-way ANOVA ($p < 0.05$). QE mg/g DW, mg quercetin equivalent per g of dry weight.

2.2. Overview of RNA-Seq Data and Sequence Assembly

To further research the molecular mechanism of *R. pulchrum* petal coloration, transcriptome analysis was performed via RNA-seq. Libraries were prepared from three biological replicates of each variety, and twelve libraries were established. The cDNA libraries were then submitted for transcriptome sequencing analysis on the Illumina HiSeq 2000 platform. A total of 38.67 to 43.72 million raw sequencing reads were generated for each variety (Table 2). After removing low-quality reads, 35.55 to 40.56 million high-quality clean reads were obtained, which accounted for over 90% of the raw reads. The percentage of bases with Q30 (high sequencing quality) was not less than 89.80%. The clean reads were then aligned to the corresponding reference genome using Hisat2 (version 2.1.0) software. The number of mapped reads ranged from 28.56 to 32.65 million, and the mapping ratio of each sample ranged between 79.32 and 80.66%. In addition, the percentage of reads that were only aligned to one position (uniquely mapped reads) was greater than 94.70% (Table 2).

Table 2. Summary of the transcriptome sequencing dataset.

Sample	Raw Reads	Clean Reads	Q30 (%)	Total Mapped	Unique Mapped
A1	42,949,574	38,765,084 (90.25%)	91.04	31,266,555 (80.66%)	29,622,723 (94.74%)
A2	39,308,136	36,123,834 (91.89%)	90.85	29,098,513 (80.55%)	27,596,151 (94.84%)
A3	38,671,858	35,545,248 (91.91%)	90.84	28,562,447 (80.36%)	27,047,544 (94.70%)
B1	40,011,542	36,950,600 (92.34%)	91.08	29,530,300 (79.92%)	28,227,592 (95.59%)
B2	42,082,072	38,694,198 (91.94%)	89.80	30,691,892 (79.32%)	29,280,727 (95.40%)
B3	43,723,232	40,563,542 (92.77%)	91.24	32,648,377 (80.49%)	31,192,551 (95.54%)
C1	42,767,546	39,166,056 (91.57%)	91.52	31,415,529 (80.21%)	29,824,376 (94.94%)
C2	43,187,530	39,626,174 (91.75%)	91.14	31,744,384 (80.11%)	30,137,984 (94.94%)
C3	41,563,920	38,298,794 (92.14%)	90.59	30,513,325 (79.67%)	28,991,127 (95.01%)
D1	42,163,278	38,866,774 (92.18%)	91.29	31,243,679 (80.39%)	29,832,406 (95.48%)
D2	41,325,128	38,027,950 (92.02%)	91.07	30,559,996 (80.36%)	29,179,418 (95.48%)
D3	40,312,100	37,029,356 (91.85%)	91.05	29,625,922 (80.01%)	28,282,098 (95.46%)

Sample: petal sample's name; Raw Reads: the total number of original reads before filtering; Clean Reads: the remaining reads after filtering, where the percentage in parentheses represents clean reads relative to raw reads; Q30 (%): the proportion of bases with a base recognition error rate of 0.1% or less; Total Mapped: the total number of reads mapped to the reference genome, where the proportion in parentheses is total mapped/clean reads; Uniquely Mapped: the reads which can be mapped to the reference genome at only one site, where the proportion in parentheses is uniquely mapped/total mapped.

The abundance of each gene was calculated based on the fragments per kb per million reads (FPKM) method. Moreover, Pearson's correlation test was applied to calculate the relation of expression patterns among the different samples. The gene expression levels showed similar patterns within sample groups and differences between groups (Figure 2a), indicating that the analysis results are reliable. Principal component analysis (PCA) was used to further explore the differences and similarities, which showed a clear separation among the four varieties with different colors. Meanwhile, the biological replicates of each sample were clustered together, indicating a high degree of transcriptional similarity (Figure 2b). Collectively, these results show that the RNA sequencing quality was suitable for further analysis.

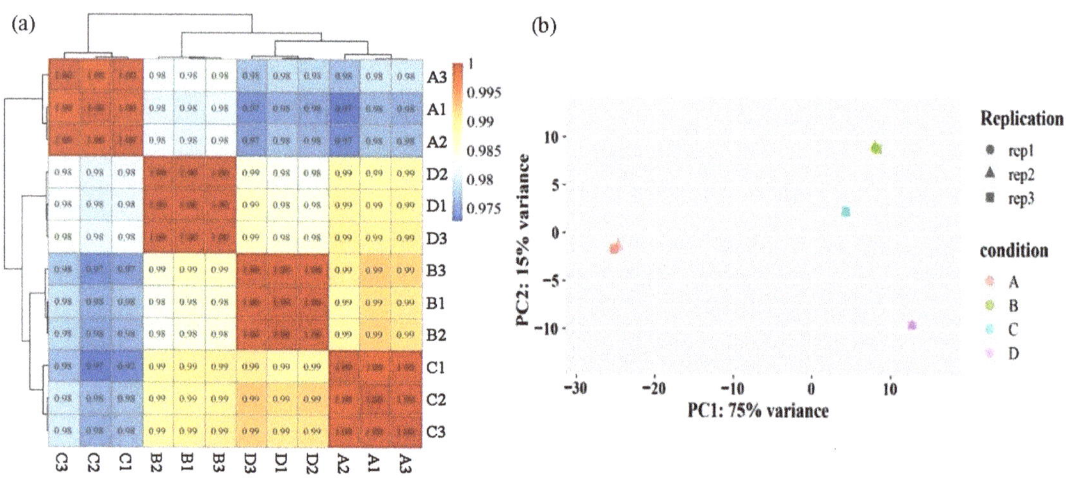

Figure 2. Different samples analysis. (**a**) Pearson correlation coefficient of biological replicates of different samples. The correlation coefficient between two samples was calculated based on the FPKM values of those samples. The left and upper sides of the figure show sample clustering, and the right and lower sides show sample names. (**b**) Principal component analysis (PCA) of the similarities and differences between the four samples used for RNA-seq.

2.3. Annotation of *R. pulchrum* Transcriptome

Five public databases were used for annotation of the unigenes in *R. pulchrum*. The results show that a total of 32,999 unigenes were annotated according to the BLASTx results, of which 28,273, 18,054, 24,301, 19,099, and 11,507 genes could be annotated using the Nr database (85.68%), Swiss-Prot database (54.71%), Pfam database (73.64%), GO database (57.88%), and KEGG database (34.87%), respectively (Table 3).

Table 3. Summary of functional annotation of transcripts in the five public databases searched.

Annotated Databases	Gene Number	Matching Proportion (%)
Nr	28,273	85.68
Swiss-Prot	18,054	54.71
Pfam	24,301	73.64
GO	19,099	57.88
KEGG	11,507	34.87

2.4. Differentially Expressed Genes Analysis

In the current study, differentially expressed genes (DEGs) were selected using a cut-off value of $|\log_2 \text{FoldChange}| \geq 1$ and an adjusted p-value < 0.05. Thousands of DEGs were identified through pairwise comparisons (Figure 3a). Among them, the B vs. A group had the largest number of DEGs, followed by the D vs. A and C vs. A groups. It is worth noting that the number of up-regulated genes was significantly higher than that of down-regulated genes, with only one exception: the C vs. B comparison. Interestingly, the largest difference between the numbers of DEGs was observed in the D vs. A comparison, reaching a total of 1003 genes. Furthermore, the common DEGs among the D vs. A, C vs. A, and B vs. A groups were selected using the Venn Diagram Plotter tool (http://omics.pnl.gov/software/venn-diagram-plotter (accessed on 22 January 2022)). The results indicate that there were 1684 genes found to be differentially expressed in the three-way comparisons (Figure 3b), which suggests that these DEGs might have different functions, resulting in different petal color pattern forms.

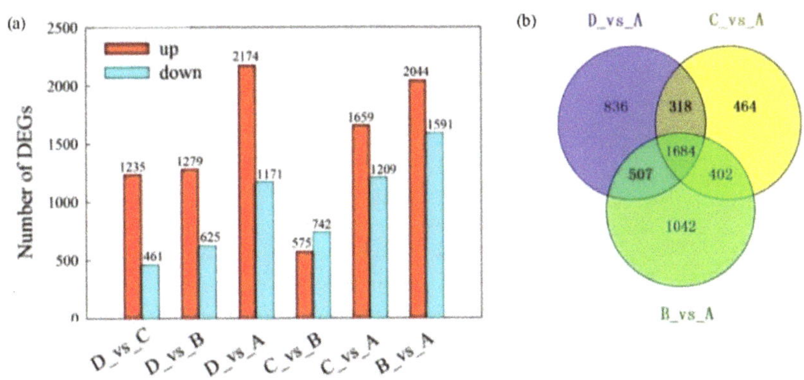

Figure 3. Number of differentially expressed genes (DEGs) in pairwise comparisons of the four varieties. (**a**) The total number of DEGs is up-regulated and down-regulated. The red and cyan colors indicate the up-regulated and down-regulated genes, respectively. (**b**) Venn diagram of DEGs for making a comparison among the three groups.

To explore the differences in gene expression patterns, we performed a hierarchical clustering analysis in a heatmap. Overall, three main clusters of samples were identified (Figure 4). The first cluster mainly consisted of A samples, and cluster two mainly consisted of D samples, whereas cluster three comprised B and C samples. The smallest number of up- and down- regulated DEGs was identified in the C vs. B comparison (Figure 3a), which implied transcriptional similarity between B and C samples. However, the contents of anthocyanins between B and C were distinctly different (Table S1). Therefore, we may assume that the 167 unique up-regulated genes in B samples might be related to the anthocyanin biosynthetic pathway. Meanwhile, we found that the expression patterns of most DEGs were opposite between A and D samples.

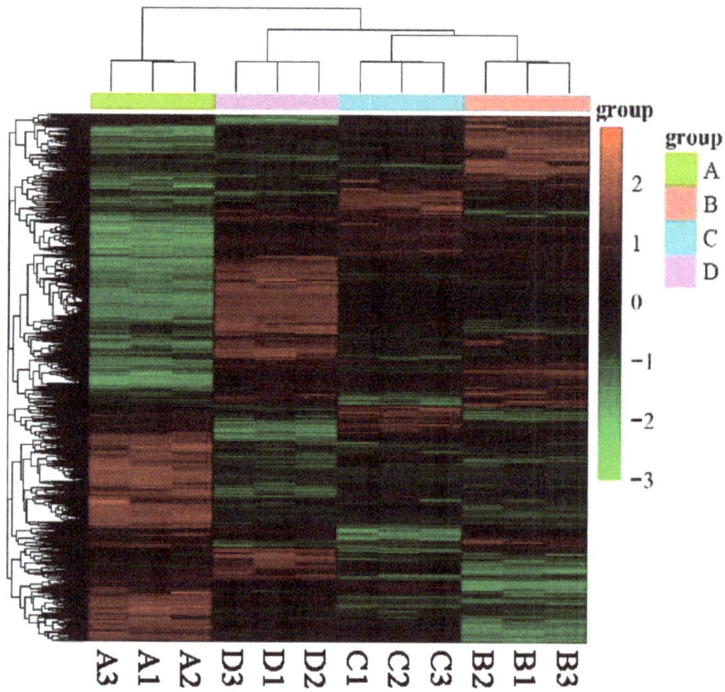

Figure 4. Heatmap showing the relative expression profiles of DEGs by Euclidean distance among the four samples. A selection of genes is shown in rows, and each column represents an individual sample. The upper dendrogram showed gene clustering by complete linkage, while the left showed clustering of gene expression quantity. The filled colors indicate the gene expression levels (from low—green to high—red).

To verify the reliability of the RNA-seq analysis, we randomly selected several genes for validation using quantitative RT-PCR (qRT-PCR). All the primers used in this study are provided in Table S2, and these genes are responsible for L-ascorbate oxidase, pectin esterase, and beta-glucuronosyltransferase. The results show that the coefficient of determination (R^2) between the qRT-PCR data and RNA-seq data was 0.82 (Figure 5), which indicates the high reliability of the RNA-seq results in our study.

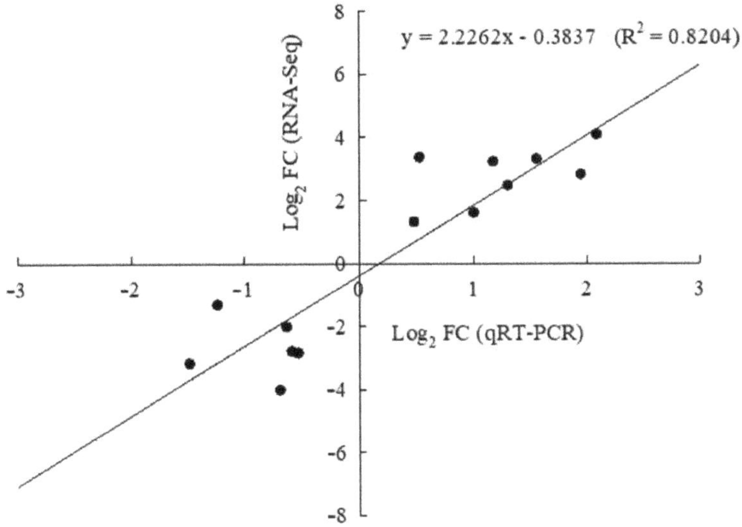

Figure 5. Regressions of gene expression ratio between RNA-seq and qRT-PCR with GAPDH as a reference. The *x*-axis represents relative gene expression fold change obtained by qRT-PCR, and the *y*-axis represents relative gene expression fold change obtained by RNA-seq. Each point represents the average of the three biological replicates.

2.5. Classification of GO and KEGG Terms

With gene ontology (GO) annotation, DEGs were classified into three major functional categories: biological process (BP), cellular component (CC), and molecular function (MF). The three types of comparisons presented similar distribution patterns, while the numbers and types of enriched pathways were different (Figures S1 and S2). For the category of BP, the pathway "localization" (GO: 0051179)–"oligosaccharide transport" (GO: 0015772)–"sucrose transport" (GO: 0015770) had a higher number of genes than others (Figure S2a–c). Among the CC functions, the pathway "cell periphery" (GO: 0071944)–"plasma membrane" (GO: 0005886)–"obsolete intrinsic component of plasma membrane" (GO: 0031226) presents the higher number of DEGs (Figure S2d–f). In the MF category, most of the DEGs were mapped to "nutrient reservoir activity" (GO: 0045735) and "catalytic activity" (GO: 003824) (Figure S2g–i).

GO terms with a corrected p-value < 0.05 were considered to be significantly enriched. For simplicity, only the top twenty most significant GO terms were selectively presented in each pairwise comparison. GO analysis revealed that most enriched GO terms were assigned to molecular function, followed by biological process (Figure S3 and Table S3). In the D vs. A comparison, these DEGs were associated with six BP terms, five CC terms, and nine MF terms. Comparison of the C sample with the A sample revealed that there were six, five, and nine functional terms assigned to the main categories of three ontologies. Among the 20 GO terms, 2, 3 and 15 GO terms belonged to these 3 functional terms in the B vs. A comparison. Notably, we found that most of the significantly enriched GO MF terms contain genes involved in "sucrose transmembrane transporter activity" (GO: 008515) and "disaccharide transmembrane transporter activity" (GO: 0015154).

The KEGG enrichment analysis was performed to further explore the various metabolic and biosynthesis pathways of the DEGs obtained in different comparisons. The top 20 most significantly enriched pathways that were associated with the DEGs are listed in Figure 6. For example, "Plant hormone signal transduction", "Phenylpropanoid biosynthesis", and "Anthocyanin biosynthesis" were significantly enriched in the D vs. A comparison. "Biosynthesis of secondary metabolites" and "Anthocyanin biosynthesis" were most highly enriched in the C vs. A comparison. "Glutathione metabolism", "Anthocyanin biosynthesis", and "Phenylpropanoid biosynthesis" were detected in the B vs. A comparison. In particular, pathways related to "phenylpropanoid biosynthesis" (ko00940), and "Anthocyanin biosynthesis" (ko00942) were all significantly enriched in the three pairwise comparisons.

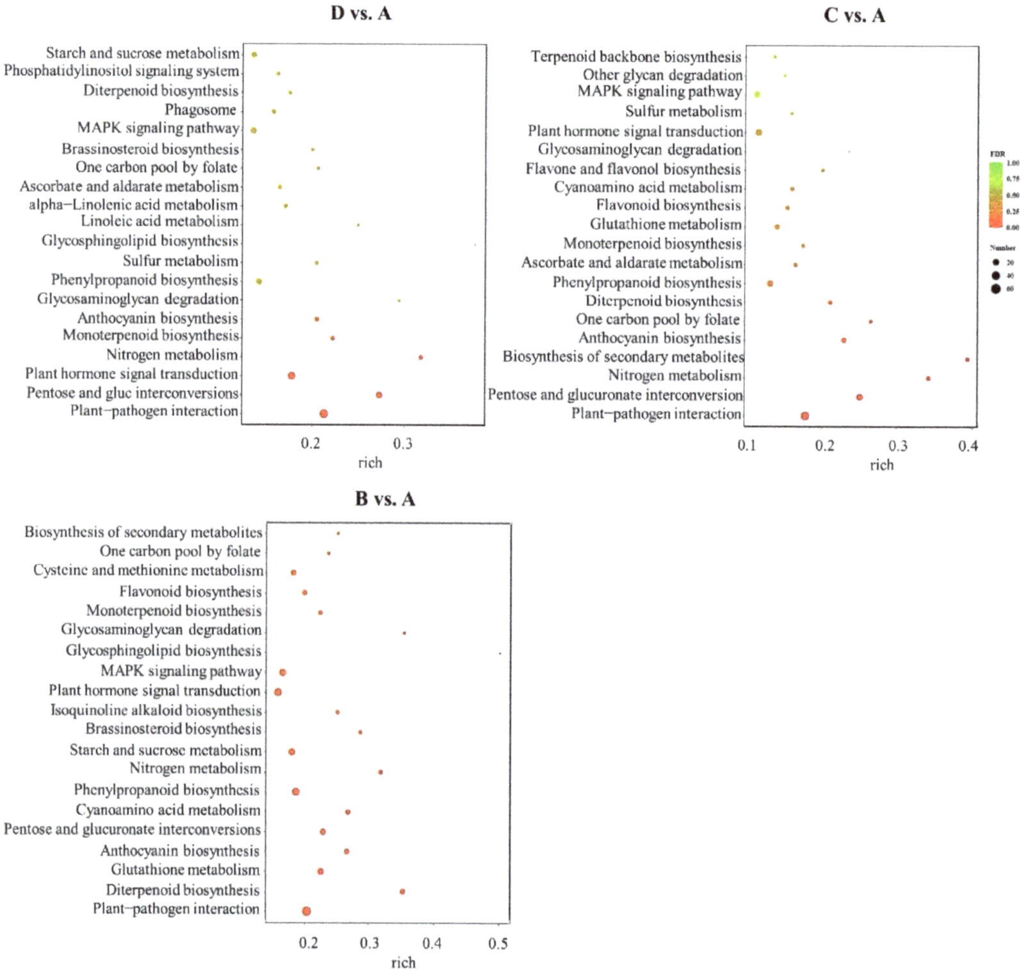

Figure 6. KEGG pathway enrichment analysis of the identified DEGs among the three comparison groups. The x-axis represents the rich factor, the y-axis indicates the name of the KEGG pathway. The size of the dot indicates the number of DEGs in each pathway, while dot color represents the q-value of the enrichment.

2.6. The Key DEGs Involved in Anthocyanin Biosynthesis Pathway

Since the contents of anthocyanin components in four peals were correlated with color intensity (Table S1), to investigate the differences in anthocyanin biosynthesis in the petals of the four varieties, DEGs in the anthocyanin biosynthesis pathway were detected, including the phenylpropanoid biosynthesis, flavonoid biosynthesis, and anthocyanin biosynthesis pathways (Table 4). Phenylalanine is the major amino acid precursor for phenylpropanoids. In our study, the gene (*Rhsim04G0145800*) encoding phenylalanine ammonia-Lyase (PAL) was down-regulated in the A sample relative to the B and D samples.

Similarly, one gene (*Rhsim01G0211600*) linked to 4-coumarateCoA ligase (4CL) in converting 4-coumaric acid to 4-coumaroyl-CoA was down-regulated in the A sample, as compared to the C and D samples. Additionally, several enzymes were identified, such as flavonoid 3′,5′-hydroxylase (F3′5′H), flavonol synthase (FLS), flavanone 4-reductase (DFR), and anthocyanidin synthase (ANS), which were associated with two, six, three, and one DEGs, respectively. It has been noticed that six genes (*RhsimUnG0095500, RhsimUnG0143300, RhsimUnG0134400, Rhsim04G0219900, RhsimUnG0007900, RhsimUnG0105700*) encoding FLS were highly expressed in the B, C, and D samples, especially in the D sample. These results suggest that the genes encoding FLS might be the key to absence of pigments in white petals, leading to more dihydroflavonol entering the flavonols branch. In contrast, three genes in the A sample associated with DFR (*Rhsim06G0030600, Rhsim06G0030500,* and *Rhsim06G0030400*) were more highly expressed than in the B, C, or D samples (Table 4). These results indicate that the genes encoding DFR might be critical for color formation, due to causing more dihydroflavonols to enter into the anthocyanin pathway. Interestingly, DFR in the B sample compared to the A sample showed a greater degree of up-regulation than in the C sample, and the high expression of ANS (*Rhsim07G0096600)* in the B sample may explain the similarity of anthocyanins content between A and B samples.

Table 4. Gene expression levels of anthocyanin biosynthesis in pairwise comparisons.

Id	$\log_2$(Fold Change)			Description
	D vs. A	C vs. A	B vs. A	
Rhsim04G0145800	1.02		1.00	PAL
Rhsim01G0211600	1.44	1.68		4CL
Rhsim07G0135800			−1.08	4CL
Rhsim13G0208200		−5.79		F3′5′H
Rhsim04G0208200		−2.05		F3′5′H
RhsimUnG0095500	2.12	1.75	1.52	FLS
RhsimUnG0143300	3.35	2.01	3.29	FLS
RhsimUnG0134400	2.97		1.68	FLS
Rhsim04G0219900		3.64		FLS
RhsimUnG0007900	1.18		2.15	FLS
RhsimUnG0105700			3.54	FLS
Rhsim12G0144700	2.70	2.94	3.48	Leucoanthocyanidin reductase
Rhsim06G0030600	−4.36	−4.22	−6.99	DFR
Rhsim06G0030500	−1.22	−2.45	−2.55	DFR
Rhsim06G0030400		−1.15	−1.30	DFR
Rhsim07G0096600			1.33	ANS

Note: Blank present in the table indicates that there is no significant difference in gene expression.

As we know, anthocyanins and flavonoids are mainly pigments in flowers. In this current study, DEGs involved in the anthocyanin biosynthesis pathway were detected according to KEGG pathway analysis. In total, 17 genes were detected in the D vs. A comparison. Compared with the D sample, 12 genes were up-regulated and 5 genes were down-regulated in the A sample. Correspondingly, 19 DEGs were detected between C and A samples, including 14 up-regulated genes and 5 down-regulated genes in the A sample. Additionally, 22 genes were identified in the B vs. A comparison, of which 15 genes were up-regulated and 7 genes down-regulated in the A sample (Table S4). The overlaps among the three pairwise comparisons were calculated, and 11 DEGs were found in the overlapping regions. In addition, eight genes were up-regulated in the A sample relative to the B, C, and D samples, including one, two, and five DEGs encoding UGAT, 5AT, and UGT75C1, respectively (Table S4).

2.7. Identification of Transcription Factors Regulating Petal Color Formation

Transcription factors (TFs) are important regulators that play key roles in regulating the expression of the genes during anthocyanin biosynthesis in flowering plants [30]. According to the annotation in Plant TF Database v4.0, a total of 302 TFs from 36 families were identified in the *R. pulchrum*, of which 162, 139, and 179 TFs were found in the comparisons of D vs. A, C vs. A, and B vs. A (Figure 7a and Table S5). Most of the genes encoding TFs were down-regulated in the A sample relative to the B, C, and D samples. In the D vs. A comparison, 101 TFs were up-regulated and 61TFs were down-regulated. In the C vs. A comparison, 95 TFs were up-regulated and 44 TFs were down-regulated. In the B vs. A comparison, 96 TFs were up-regulated and 83 TFs were down-regulated. Moreover, 11 TF families containing more than 10 DEGs were identified: ERF (42), MYB (29), *bHLH* (24), WRKY (20), MYB related (16), FAR1 (15), NAC (14), LBD (12), C2H2 (12), bZIP (12), and TCP (11) (Figure 7b). Notably, the ERF, MYB, *bHLH*, and WRKY families were relatively large TF families in the three pairwise groups (Table S5).

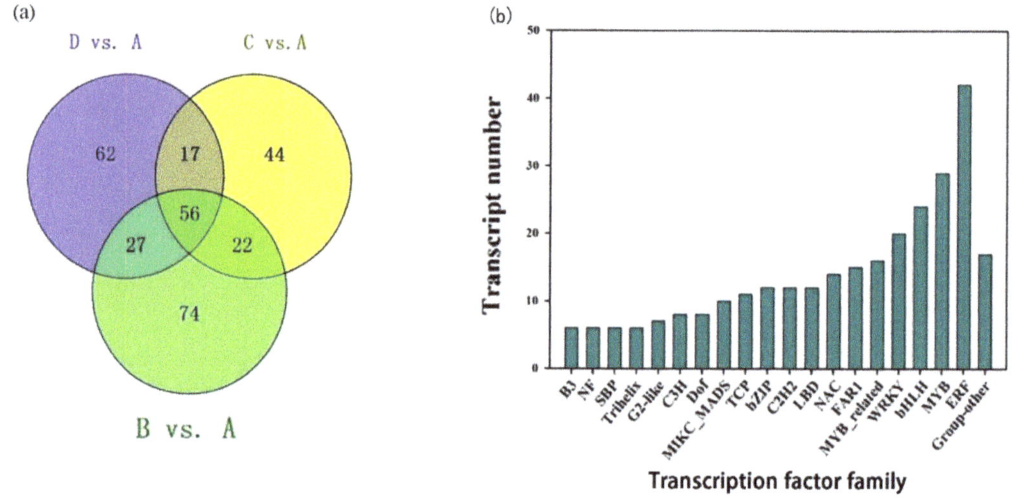

Figure 7. Differentially expressed TFs in the three pairwise comparisons. (**a**) Distribution of the TF families that include more than 10 DEGs. (**b**) Venn diagram of differentially expressed TFs in D vs. A, C vs. A, B vs. A comparisons.

Furthermore, we investigated ERF, MYB, and *bHLH* TFs to enhance our understanding of their involvement in regulating color formation in *R. pulchrum*. The overlaps among the three pairwise comparisons were calculated, and 56 TFs were found in the overlapping regions (Table S5). In our present study, five DEGs (*Rhsim02G0040100, Rhsim03G0176600, Rhsim07G0072500, Rhsim08G0214100, Rhsim12G0192100*) encoding ERF and three DEGs (*Rhsim03G0180500, Rhsim06G0163100, Rhsim09G0042000*) encoding MYB were up-regulated in the D vs. A, C vs. A, and B vs. A comparisons. Moreover, the expression level of one gene (*Rhsim08G0230000*) linked to *bHLH* was up-regulated in the A sample, as compared to the B, C, and D samples (Table S5).

2.8. DEGs Related to Hormone Signaling

Plant hormones (IAA, abscisic acid, ethylene, brassinosteroids, etc.) play a direct role in the biosynthesis of anthocyanins [31,32]. KEGG analysis showed that there were differences in phytohormone signaling transduction in the three pairwise comparisons. To thoroughly analyze the molecular mechanism of anthocyanin transformation, we investigated the gene expression differences in hormone signaling in the three pairwise groups. The results show that most of the genes involved in phytohormone metabolic pathways were down-regulated in the A sample relative to the B, C, and D samples. In the D vs. A comparison, 42 hormone-signaling-related DEGs were up-regulated and 17 DEGs were down-regulated. Meanwhile, 38 DEGs were screened in the C vs. A comparison, with 22 up-regulated and 16 down-regulated. For the comparison of B vs. A, 52 DEGs revealed differences, including 30 up-regulated and 22 down-regulated genes. (Table S6).

The overlapping DEGs among the three pairwise comparisons were screened using a Venn diagram (Figure S4). The results indicate that 21 genes were differentially expressed, and we found that the genes involved in auxin (IAA) or brassinosteroid (BR) signaling were the most enriched (Table S6). Three DEGs (*RhsimUnG0255100*, *Rhsim06G0190800*, *Rhsim13G0033000*) involved in IAA signaling were identified, and the abundance of these transcripts was significantly decreased in the A sample relative to the B, C, and D samples (Table S6). Moreover, one (*Rhsim07G022810*) and eight DEGs (*Rhsim09G0181000*, *Rhsim07G0028200*, *RhsimUnG0200100*, etc.) involved in ABA and BR signaling, respectively, were found in the three comparison groups. Interestingly, these genes were more lowly expressed in the A sample relative to the B, C, and D samples.

3. Discussion

Anthocyanins and flavonoids are the key pigments that determine flower colors. In general, cyanidin, pelargonidin, and paeoniflorin are beneficial to the formation of red color; delphinium and malvidin are conducive to the formation of blue color [9,33]. The combined effect of these pigments makes the flowers of *Rhododendron* present white, pink, red, purple, and other colorful colors. The results of HPLC show that the main anthocyanins in different samples were the same, but the content of each pigment was significantly different. It is inferred that the color depth of *R. pulchrum* is related to the content of main anthocyanins in the petals. The high content of peonidin in sample A may be an important factor in the formation of deep color. Peng et al. also put forward a similar conclusion that the anthocyanin biosynthesis pathway is the main metabolic pathway of flower color formation in their study on *Hydrangea macrophylla* cv. 'Forever Summer' [34]. The biosynthesis and floral color regulation mechanism of flavonoids in *R. pulchrum* was elaborated in Xia et al.'s research [12]. In the four samples of this experiment, flavonoid content was negatively correlated with colors. This shows that the color depth of *R. pulchrum* is produced by the combined action of anthocyanins and flavonoids, which is consistent with the research results of Liu et al. [35].

The intensity of petal color change was correlated with the concentration of anthocyanin in plants [36]. Xue [37] reported that the anthocyanin content increased as the flower color deepened, and the expression of key structural genes was significantly increased in red-flowered strawberry in comparison with white-flowered strawberry. Previous results suggested that the competition expression of FLS and DFR genes led to the redirection of flavonol and anthocyanin accumulation [38,39]. In this study, the difference in gene expression was consistent with the accumulation of anthocyanins in different samples. This discovery is like that of blueberry fruit development, in which a partial correlation between anthocyanin-related gene expression and anthocyanin concentration was found [40]. Meanwhile, transcriptome analysis revealed that the FLS and DFR genes were differentially expressed in the flower color of the four different varieties. Six genes encoding FLS were highly expressed, especially in the D sample, and three genes in the A sample associated with DFR were more highly expressed than in the B, C, or D samples. These results indicate that the expression levels of genes encoding DFR were higher in deeper color flowers, which significantly promoted the conversion of dihydroflavonols to anthocyanins. On the contrary, the increased expression of FLS in the white flowers promoted the conversion of dihydroflavonols into flavonols (Table 4). There may be substrate competition between FLS and DFR, controlling the metabolic flux via the branching of the flavonoid biosynthetic pathway, resulting in changes in flower color intensity in *R. pulchrum*.

Transcription factors perform important roles in all plants. In the current study, most of the TFs were enriched in ERF, MYB, *bHLH*, WRKY, MYB related, FAR1, and NAC families during anthocyanin transformation (Figure 7b). ERF plays a particularly important role in the color expression of light petals, especially pink petals, by affecting the upstream pathway of anthocyanin biosynthesis [41]. The class of TFs was previously involved in the regulation of petal color formation [42]. The *bHLH* proteins, forming one of the largest TF families, play vital roles in various metabolic, physiological, and developmental processes in plants [43]. In our study, one gene (*Rhsim08G0230000*) encoding *bHLH* was more highly expressed in the A sample, indicating that this *bHLH* was a positive regulator of anthocyanins biosynthesis in *R. pulchrum.* It has been reported that MYB is closely related to flavonoid metabolic pathways [44]. However, some studies showed that there were also inhibitory factors in the MYB family that inhibit anthocyanin biosynthesis [45]. In our study, one gene (*Rhsim09G0042000*) encoding the MYB family members was significantly down-regulated in the D vs. A, C vs. A, and B vs. A comparisons (Table S5), which suggested it inhibited the anthocyanin accumulation.

Phytohormones are also critical internal factors affecting anthocyanin biosynthesis in many plant species. Wang et al. [46] reported that auxin inhibits anthocyanin accumulation and reduces the expression of genes linked to anthocyanin biosynthesis in apples. Based on our transcriptome data (Table S6), six genes associated with auxin-mediated signaling pathways were detected in the three groups, four of which were down-regulated in the A sample. In addition, the transcript level of two genes (*Rhsim09G0181000* and *Rhsim07G0028200*) encoding brassinosteroid was increased in the A sample. Similarly, Peng et al. [47] found that BR affects anthocyanin accumulation by regulating the anthocyanin biosynthesis genes in *Arabidopsis* seedlings. In addition, the response proteins in signal transduction of IAA and ABA were prominently regulated during anthocyanin transformation, and the functions of these hormone-responsive proteins deserve further investigation.

The results for anthocyanin composition in petals and deep-color blotches obtained by HPLC show that the accumulation of delphinidin and peonidin may be the pigment basis for the formation of deep-color blotches (Table S1). In the four samples, the down-regulation of DFR genes related to anthocyanin synthesis (Table 4) was inconsistent with the decreasing trend in anthocyanin contents. This is possibly due to the defect that petals and deep-color blotches were not sequenced separately for RNA-seq. The presence of deep-color blotches resulted in high expression of DFR genes in B and C samples. Studies of tree peony (*Paeonia suffruticosa*) have already shown that petal coloration and spot coloration are differentially regulated by MYB transcription factor [25,48]. Separating petals and deep-color blotches for independent analysis, revealing the transcriptional and metabolic regulatory mechanisms of petal and deep-color blotches formation, will be the focus of our subsequent research.

4. Materials and Methods

4.1. Plant Materials

Rhododendron pulchrum Sweet used in this study was cultivated at Yangzhou University (32°23′ N, 119°25′ E), which located in Jiangsu Province, China. The purple (purple petals, purple blotches) flower variety of *R. pulchrum* is "Zihe", the pink (pink petals, purple blotches) and light pink (light pink petals, purple blotches) flower varieties are "Fenhe", and the white (white petals, no or very few blotches) flower variety is "Baihe". To explore the petal color regulation mechanism of *R. pulchrum*, we selected well-grown plants that had been planted for more than three years and collected full-bloom flowers having four different petal colors. The samples of purple, pink, light pink and white petals are labeled as A, B, C, and D, respectively (Figure 8). Fresh petals were collected on 24 April 2021, and flowers of the same color gradient from three individual plants were pooled as one sample. They were then frozen and ground in liquid nitrogen and subsequently stored in ultralow-temperature freezer under −80 °C for subsequent analysis.

Figure 8. Color comparison among four phenotypes of *Rhododendron pulchrum* Sweet. The samples of purple, pink, light pink and white petals are labeled as (**A**–**D**).

4.2. Determination of Contents of Anthocyanins Components and Total Flavonoids

The frozen petal tissue (0.5 g) was added to 1.5 mL of 2 M HCl in methanol solution for grinding and homogenization, and low-temperature ultrasound for 15 min was followed by centrifugation [49]. The supernatant was used for anthocyanin extract. The contents and components of anthocyanins in the four samples were determined by high-performance liquid chromatography (HPLC), and chromatograms were mapped by chromatographic workstation software (LC-WS100) [50]. The total flavonoid content in each sample was measured by UV spectrophotometry [51]. The results are expressed as mg quercetin equivalent per g of dry weight (QE mg/g DW).

4.3. RNA Extraction and Transcriptome Sequencing

For the RNA-seq study, the petals of each variety were completely cut and mixed (including the deep-color blotches in the centers of the petals), and three biological replicates were conducted. Total RNA extraction was completed using the RNA simple Total RNA Kit (Tiangen, Shanghai, China) from petals following the manufacturer's instructions. Extracted RNA purity was determined in a Nanodrop spectrophotometer (Thermo Scientific, Waltham, MA, USA), and RNA integrity was assessed using the Agilent 2100 Bioanalyzer (Agilent Technologies, Santa Clara, CA, USA). RNA-seq libraries were generated using the TruSeq RNA Sample Preparation Kit (Illumina, San Diego, CA, USA). After preparation, twelve libraries were sequenced on the Illumina HiSeq platform [52]. The RNA-seq analysis was performed by Panomix Biomedical Tech Co., Ltd. (Suzhou, China).

4.4. RNA-Seq Data Analysis

After removing adapter sequences, ploy-N-containing reads, and low-quality sequences from the raw data, the remaining clean reads were mapped to the reference genome via Hisat2 (version 2.1.0) software [53]. The expression levels of each gene were calculated and normalized as fragments per kilobase million (FPKM) measurements. Functional annotations of the identified transcripts were then aligned against various databases, including NR [54], Swiss-Prot [55], protein family (Pfam) [56], Gene Ontology (GO) [57], and Kyoto Encyclopedia of Genes and Genomes (KEGG) [58]. Differential gene expression analysis across varieties was performed using the DESeq2 package (version 1.16.1) [59], and genes with an adjusted p-value < 0.05 and absolute $\log_2$ fold changes (FC) $\geq$ 1 were set as the screening criteria for significantly differential expression. Based on the identified DEGs, three types of GO categories were obtained, including biological processes (BP), cellular components (CC), and molecular function (MF). Additionally, the functional enrichment analysis of GO and KEGG was implemented to identify which genes were significantly enriched. GO terms and KEGG pathway with a threshold of false discovery rate (FDR) < 0.05 were considered as the significance level.

4.5. Verification of RNA-Seq by qRT-PCR

To confirm the reliability of RNA-seq data, we randomly selected several genes and tested their expression profiles using qRT-PCR. Total RNA was isolated from petals as already described above. The first strand of complementary DNA (cDNA) was synthesized from total RNA using the cDNA synthesis kit (Vazyme Biotech, Nanjing, China). Finally, qRT-PCR was performed on the CFX96 real-time PCR system (Bio-Rad, Hercules, CA, USA) using the ChamQ Universal SYBR qPCR Master Mix (Vazyme Biotech, Nanjing, China) according to the manufacture recommended protocol. The amplification reaction volume was 20 μL, containing 10 μL of 2 × SYBR qPCR Master Mix and 10 μM of forward and reverse primers. All reactions were run in triplicate assays using the following cycling conditions: initial denaturation at 95 °C for 30 s followed by 40 cycles of PCR consisting of denaturation at 95 °C for 10 s and annealing at 60 °C for 30 s. All gene-specific primers for qRT-PCR were designed using the online primer design software (https://www.ncbi.nlm.nih.gov/tools/primer-blast/ (accessed on 22 January 2022)), and all the primers used in this study are presented in Table S2. The relative gene expression levels were calculated using the $2^{-\Delta\Delta CT}$ method [60].

4.6. Statistical Analyses and Bioinformatics

Data on the contents of anthocyanins and flavonoids were subjected to analysis of variance (one-way ANOVA) using SPSS 20.0 software (IBM Corp, Armonk, NY, USA), while graphs were produced using Microsoft Excel (v2013, Microsoft Corp., USA) and Origin 8.0 (Origin Lab Corporation, Northampton, MA, USA) [61]. Treatment means were presented as the mean ± standard deviations of the mean using the three replicates [62]. Bioinformatics data were analyzed using R statistical software package in R Studio version 0.99.446 (RStudio, Inc. 2015) [63].

5. Conclusions

In summary, we used transcriptome approaches to study the molecular mechanism of petal coloration of four varieties of *R. pulchrum*. Our results show that variation in the flower color was related to the expression level of anthocyanin biosynthesis genes. The higher expression levels of genes associated with FLS might be the key to white formation, leading to more dihydroflavonol entering the flavonols branch. The formation of red color may be related to the higher expression of DFR genes, which promoted the accumulation of anthocyanin. The structural genes and their regulators (TFs) in this research provided valuable molecular information on the flower color intensity variation in *R. pulchrum*.

Supplementary Materials: The following are available online at https://www.mdpi.com/article/10.3390/plants12142656/s1. Figure S1: The distribution and number of DEGs in three main categories in the three pairwise comparisons; Figure S2: Enrichment of top 10 GO pathways; Figure S3: GO enrichment analysis of DEGs in the three pairwise comparison groups; Figure S4: Distribution of the genes involved in phytohormone metabolic pathways in the three pairwise comparisons. Table S1: The contents of anthocyanin components; Table S2: Primer sequences of genes used in qRT-PCR validation; Table S3: Significantly enriched GO terms in the three pairwise comparison groups; Table S4: Genes involved in the anthocyanin biosynthesis pathway showing genotypic difference expression; Table S5: List of differentially expressed TFs identified in the three pairwise comparisons; Table S6: List of phytohormone signal-transduction- and biosynthesis-related DEGs.

Author Contributions: Conceptualization, N.Z. and C.Z.; methodology, C.Z.; software, N.Z.; validation, C.Z.; formal analysis, N.Z.; writing—original draft preparation, N.Z.; writing—review and editing, C.Z.; project administration, C.Z. All authors have read and agreed to the published version of the manuscript.

Funding: This research was funded by Jiangsu Modern agricultural industrial technology system construction project, grant number JATS [2022] 489.

Data Availability Statement: The data is contained within the article and Supplementary Materials.

Conflicts of Interest: The authors declare no conflict of interest.

References

1. Chang, Y.H.; Yao, G.; Neilsen, J.; Liu, D.T.; Zhang, L.; Ma, Y.P. *Rhododendron kuomeianum* (Ericaceae), a new species from northeastern Yunnan (China), based on morphological and genomic data. *Plant Divers.* **2021**, *43*, 292–298. [CrossRef] [PubMed]
2. Shuzhen, W.; Zhiliang, L.; Weibin, J.; Yuanping, F.; Qiaofeng, Y.; Jun, X. Transcriptome analysis and identification of genes associated with flower development in *Rhododendron pulchrum* Sweet (Ericaceae). *Gene* **2018**, *679*, 108–118.
3. Alappat, B.; Alappat, J. Anthocyanin Pigments: Beyond Aesthetics. *Molecules* **2020**, *25*, 5500. [CrossRef] [PubMed]
4. Grotewold, E. The genetics and biochemistry of floral pigments. *Annu. Rev. Plant Biol.* **2006**, *57*, 761. [CrossRef]
5. Noman, A.; Aqeel, M.; Deng, J.; Khalid, N.; Sanaullah, T.; Shuilin, H. Biotechnological advancements for improving floral attributes in ornamental plants. *Front. Plant Sci.* **2017**, *8*, 530. [CrossRef]
6. Yang, F.-S.; Nie, S.; Liu, H.; Shi, T.-L.; Tian, X.-C.; Zhou, S.-S.; Bao, Y.-T.; Jia, K.-H.; Guo, J.-F.; Zhao, W.; et al. Chromosome-level genome assembly of a parent species of widely cultivated azaleas. *Nat. Commun.* **2020**, *11*, 5269. [CrossRef]
7. Dong, X.M.; Zhang, W.; Zhang, S.B. Selection and validation of reference genes for quantitative real-time PCR analysis of development and tissue-dependent flower color formation in *Cymbidium lowianum*. *Int. J. Mol. Sci.* **2022**, *23*, 738. [CrossRef]
8. Liu, Q.; Liaquat, F.; He, Y.; Munis, M.F.H.; Zhang, C. Functional annotation of a full-length transcriptome and identification of genes associated with flower development in *Rhododendron simsii* (Ericaceae). *Plants* **2021**, *10*, 649. [CrossRef]
9. Du, H.; Lai, L.; Wang, F.; Sun, W.; Zhang, L.; Li, X.; Wang, L.; Jiang, L.; Zheng, Y. Characterisation of flower colouration in 30 *Rhododendron* species via anthocyanin and flavonol identification and quantitative traits. *Plant Biol.* **2018**, *20*, 121–129. [CrossRef]
10. Fleschhut, J.; Kratzer, F.; Rechkemmer, G.; Kulling, S.E. Stability and biotransformation of various dietary anthocyanins in vitro. *Eur. J. Nutr.* **2006**, *45*, 7–18. [CrossRef]
11. Naing, A.H.; Kang, H.H.; Jeong, H.Y.; Soe, M.T.; Kim, C.K. Overexpression of the *Raphanus sativus RsMYB1* using the flower-specific promoter (InMYB1) enhances anthocyanin accumulation in flowers of transgenic Petunia and their hybrids. *Mol. Breed.* **2020**, *40*, 97. [CrossRef]
12. Xia, X.; Gong, R.; Zhang, C.Y. Integrative analysis of transcriptome and metabolome reveals flavonoid biosynthesis regulation in *Rhododendron pulchrum* petals. *BMC Plant Biol.* **2022**, *22*, 401. [CrossRef]
13. Deng, X.; Bashandy, H.; Ainasoja, M.; Kontturi, J.; Pietiäinen, M.; Laitinen, R.A.E.; Albert, V.A.; Valkonen, J.P.T.; Elomaa, P.; Teeri, T.H. Functional diversification of duplicated chalcone synthase genes in anthocyanin biosynthesis of *Gerbera hybrida*. *New Phytol.* **2014**, *201*, 1469–1483. [CrossRef] [PubMed]
14. Jiang, W.; Liu, T.; Nan, W.; Jeewani, D.C.; Niu, Y.; Li, C.; Wang, Y.; Shi, X.; Wang, C.; Wang, J.; et al. Two transcription factors TaPpm1 and TaPpb1 co-regulate anthocyanin biosynthesis in purple pericarps of wheat. *J. Exp. Bot.* **2018**, *69*, 2555–2567. [CrossRef]
15. Fujino, N.; Tenma, N.; Waki, T.; Ito, K.; Komatsuzaki, Y.; Sugiyama, K.; Yamazaki, T.; Yoshida, S.; Hatayama, M.; Yamashita, S.; et al. Physical interactions among flavonoid enzymes in snapdragon and torenia reveal the diversity in the flavonoid metabolon organization of different plant species. *Plant J.* **2018**, *94*, 372–392. [CrossRef] [PubMed]
16. Springob, K.; Nakajima, J.; Yamazaki, M.; Saito, K. Recent advances in the biosynthesis and accumulation of anthocyanins. *Nat. Prod. Rep.* **2003**, *20*, 288–303. [CrossRef]
17. Cazzonelli, C.; Pogson, B.J. Source to sink: Regulation of carotenoid biosynthesis in plants. *Trends Plant Sci.* **2010**, *15*, 266. [CrossRef] [PubMed]
18. Vogt, T. Phenylpropanoid biosynthesis. *Mol. Plant* **2010**, *3*, 2–20. [CrossRef]
19. Dao, T.T.; Linthorst, H.J.; Verpoorte, R. Chalcone synthase and its functions in plant resistance. *Phytochem. Rev.* **2011**, *10*, 397–412. [CrossRef]
20. Jaakola, L. New insights into the regulation of anthocyanin biosynthesis in fruits. *Trends Plant Sci.* **2013**, *18*, 477–483. [CrossRef]
21. Sun, Y.; Zhang, D.; Zheng, H.; Wu, Y.; Mei, J.; Ke, L.; Yu, D.; Sun, Y. Biochemical and expression analyses revealed the involvement of proanthocyanidins and/or their derivatives in fiber pigmentation of *Gossypium stocksii*. *Int. J. Mol. Sci.* **2022**, *23*, 1008. [CrossRef] [PubMed]
22. Tanaka, Y.; Sasaki, N.; Ohmiya, A. Biosynthesis of plant pigments: Anthocyanins, betalains and carotenoids. *Plant J.* **2008**, *54*, 733–749. [CrossRef] [PubMed]
23. Xu, W.; Dubos, C.; Lepiniec, L. Transcriptional control of flavonoid biosynthesis by MYB-bHLH-WDR complexes. *Trends Plant Sci.* **2015**, *20*, 176–185. [CrossRef] [PubMed]
24. Lloyd, A.; Brockman, A.; Aguirre, L.; Campbell, A.; Bean, A.; Cantero, A.; Gonzalez, A. Advances in the MYB-bHLH-WD repeat (MBW) pigment regulatory model: Addition of a WRKY factor and co-option of an anthocyanin MYB for betalain regulation. *Plant Cell Physiol.* **2017**, *58*, 1431–1441. [CrossRef]
25. Yuting, L.; Yuhan, T.; Xin, W.; Cong, X.; Jun, T.; Daqiu, Z. Tree peony R2R3-MYB transcription factor PsMYB30 promotes petal blotch formation by activating the transcription of the anthocyanin synthase gene. *Plant Cell Physiol.* **2022**, *63*, 1101–1116.
26. Zimmermann, I.M.; Heim, M.A.; Weisshaar, B.; Uhrig, J.F. Comprehensive identification of *Arabidopsis thaliana* MYB transcription factors interacting with R/B-like BHLH proteins. *Plant J.* **2004**, *40*, 22–34. [CrossRef]
27. Koes, R.; Verweij, W.; Quattrocchio, F. Flavonoids: A colorful model for the regulation and evolution of biochemical pathways. *Trends Plant Sci.* **2005**, *10*, 236–242. [CrossRef]

28. He, C.; Liu, X.; Da Silva, J.A.T.; Liu, N.; Zhang, M.; Duan, J. Transcriptome sequencing and metabolite profiling analyses provide comprehensive insight into molecular mechanisms of flower development in *Dendrobium officinale* (Orchidaceae). *Plant Mol. Biol.* **2020**, *104*, 529–548. [CrossRef]
29. Mei, J.; Wu, Y.; Niu, Q.; Miao, M.; Zhang, D.; Zhao, Y.; Cai, F.; Yu, D.; Ke, L.; Feng, H.; et al. Integrative analysis of expression profiles of mRNA and microRNA provides insights of cotton response to *Verticillium dahliae*. *Int. J. Mol. Sci.* **2022**, *23*, 4702. [CrossRef]
30. Livak, K.J.; Schmittgen, T.D. Analysis of relative gene expression data using real-time quantitative PCR and the 2(-Delta Delta C (T)) Method. *Methods* **2001**, *25*, 402–408. [CrossRef]
31. Hichri, I.; Barrieu, F.; Bogs, J.; Kappel, C.; Delrot, S.; Lauvergeat, V. Recent advances in the transcriptional regulation of the flavonoid biosynthetic pathway. *J. Exp. Bot.* **2011**, *62*, 2465–2483. [CrossRef] [PubMed]
32. Holton, T.A.; Cornish, E.C. Genetics and biochemistry of anthocyanin biosynthesis. *Plant Cell* **1995**, *7*, 1071–1083. [CrossRef] [PubMed]
33. Katja, K.; Declan, J.L.; Nick, W.A.; Nelli, M.; Tony, M.; Andrew, C.A.; Bilal, M.A.; Hely, H.; Richard, V.E.; Laura, J. MYBA and MYBPA transcription factors co-regulate anthocyanin biosynthesis in blue-coloured berries. *New Phytol.* **2021**, *232*, 1350–1367.
34. Peng, J.Q.; Dong, X.J.; Xue, C.; Liu, Z.M.; Cao, F.X. Exploring the molecular mechanism of blue flower color formation in *Hydrangea macrophylla* cv. "Forever Summer". *Front. Plant Sci.* **2021**, *12*, 585665. [CrossRef] [PubMed]
35. Liu, L.; Zhang, L.Y.; Wang, S.L.; Niu, X.Y. Analysis of anthocyanins and flavonols in petals of 10 *Rhododendron* species from the Sygera Mountains in Southeast Tibet. *Plant Physiol. Bioch.* **2016**, *104*, 250–256. [CrossRef]
36. Gichuki, D.K.; Li, Q.; Hou, Y.; Liu, Y.; Ma, M.; Zhou, H.; Xu, C.; Zhu, Z.; Wang, L.; Musila, F.M.; et al. Characterization of flavonoids and transcripts involved in their biosynthesis in different organs of *Cissus rotundifolia* Lam. *Metabolites* **2021**, *11*, 741. [CrossRef]
37. Peng, Q.Z.; Zhu, Y.; Liu, Z.; Du, C.; Li, K.G.; Xie, D.Y. An integrated approach to demonstrating the ANR pathway of proanthocyanidin biosynthesis in plants. *Planta* **2012**, *236*, 901–918. [CrossRef]
38. Petroni, K.; Tonelli, C. Recent advances on the regulation of anthocyanin synthesis in reproductive organs. *Plant Sci.* **2011**, *181*, 219–229. [CrossRef]
39. Davies, K.M.; Schwinn, K.E.; Deroles, S.C.; Manson, D.G.; Lewis, D.H.; Bloor, S.J. Enhancing anthocyanin production by altering competition for substrate between flavonol synthase and dihydroflavonol 4-reductase. *Euphytica* **2003**, *131*, 259–268. [CrossRef]
40. Jaakola, L.; Määttä, K.; Pirttilä, A.M.; Törrönen, R.; Kärenlampi, S.; Hohtola, A. Expression of genes involved in anthocyanin biosynthesis in relation to anthocyanin, proanthocyanidin, and flavonol levels during bilberry fruit development. *Plant Physiol.* **2002**, *130*, 729–739. [CrossRef]
41. Xia, Y.; Chen, W.; Xiang, W.; Wang, D.; Xue, B.; Liu, X.; Xing, L.; Wu, D.; Wang, S.; Guo, Q.; et al. Integrated metabolic profiling and transcriptome analysis of pigment accumulation in *Lonicera japonica* flower petals during colour-transition. *BMC Plant Biol.* **2021**, *21*, 98. [CrossRef]
42. Lu, J.; Zhang, Q.; Lang, L.; Jiang, C.; Wang, X.; Sun, H. Integrated metabolome and transcriptome analyses reveal the molecular mechanism of a color mutation in *Miniature Roses*. *BMC Plant Biol.* **2019**, *20*, 611.
43. Pires, N.; Dolan, L. Early evolution of bHLH proteins in plants. *Plant Signal Behav.* **2010**, *5*, 911–912. [CrossRef]
44. Bogs, J.; Jaffe, F.W.; Takos, A.M.; Walker, A.R.; Robinson, S.P. The grapevine transcription factor VvMYBPA1 regulates proanthocyanidin synthesis during fruit development. *Plant Physiol.* **2007**, *143*, 1347–1361. [CrossRef]
45. Yoshida, K.; Ma, D.; Constabel, C.P. The MYB182 protein down-regulates proanthocyanidin and anthocyanin biosynthesis in poplar by repressing both structural and regulatory flavonoid genes. *Plant Physiol.* **2015**, *3*, 693–710. [CrossRef]
46. Wang, C.K.; Han, P.L.; Zhao, Y.W.; Ji, X.L.; Yu, J.Q.; You, C.X.; Hu, D.G.; Hao, Y.J. Auxin regulates anthocyanin biosynthesis through the auxin repressor protein MdIAA26. *Biochem. Biophys. Res. Commun.* **2020**, *533*, 717–722. [CrossRef]
47. Peng, Z.; Han, C.; Yuan, L.; Zhang, K.; Huang, H.; Ren, C. Brassinosteroid enhances jasmonate-induced anthocyanin accumulation in *Arabidopsis* seedlings. *J. Integr. Plant Biol.* **2011**, *53*, 632–640. [CrossRef]
48. Gu, Z.Y.; Zhu, J.; Hao, Q.; Yuan, Y.W.; Duan, Y.W.; Men, S.Q.; Wang, Q.Y.; Hou, Q.Z.; Liu, Z.A.; Shu, Q.Y.; et al. A novel R2R3-MYB transcription factor contributes to petal blotch formation by regulating organ-specific expression of *PsCHS* in tree peony (*Paeonia suffruticosa*). *Plant Cell Physiol.* **2019**, *60*, 599–611. [CrossRef] [PubMed]
49. Daria, N.; Robert, B.; Thomas, D.; Rares, C.L.; Heidi, H. First genome edited poinsettias: Targeted mutagenesis of flavonoid 3′-hydroxylase using CRISPR/Cas9 results in a colour shift. *Plant Cell* **2021**, *147*, 49–60.
50. Tang, Y.C.; Liu, Y.J.; He, G.R.; Cao, Y.W.; Bi, M.M.; Song, M.; Yang, P.P.; Xu, L.F.; Ming, J. Comprehensive analysis of secondary metabolites in the extracts from different lily bulbs and their antioxidant ability. *Antioxidants* **2021**, *10*, 1634. [CrossRef] [PubMed]
51. Wan, H.; He, Y.; He, Z.; He, F. Determination of quercetin, plumbagin and total flavonoids in *Drosera peltata Smith var. glabrata* Y.Z.Ruan. *Phcog Mag.* **2012**, *8*, 263. [CrossRef] [PubMed]
52. Qi, S.; Yang, L.; Wen, X.; Hong, Y.; Song, X.; Zhang, M.; Dai, S. Reference gene selection for RT-qPCR analysis of flower development in *Chrysanthemum morifolium* and *Chrysanthemum lavandulifolium*. *Front. Plant Sci.* **2016**, *7*, 287. [CrossRef]
53. Wang, M.; Ren, T.; Marowa, P.; Du, H.; Xu, Z. Identification and selection of reference genes for gene expression analysis by quantitative real-time PCR in *Suaeda glauca*'s response to salinity. *Sci. Rep.* **2021**, *11*, 8569. [CrossRef] [PubMed]
54. Kim, D.; Langmead, B.; Salzberg, S.L. HISAT: A fast spliced aligner with low memory requirements. *Nat. Methods* **2015**, *12*, 357–360. [CrossRef] [PubMed]

55. Yu, K.; Zhang, T. Construction of customized sub-databases from NCBI-nr database for rapid annotation of huge metagenomic datasets using a combined BLAST and MEGAN approach. *PLoS ONE* **2013**, *8*, e59831. [CrossRef] [PubMed]
56. Gasteiger, E.; Jung, E.; Bairoch, A. SWISS-PROT: Connecting biomolecular knowledge via a protein database. *Curr. Issues Mol. Biol.* **2001**, *3*, 47–55. [PubMed]
57. Mistry, J.; Chuguransky, S.; Williams, L.; Qureshi, M.; Salazar, G.A. Pfam: The protein families database in 2021. *Nucleic Acids Res.* **2021**, *49*, 412–419. [CrossRef] [PubMed]
58. The Gene Ontology Consortium. Expansion of the gene ontology knowledgebase and resources. *Nucleic Acids Res.* **2017**, *45*, 331–338. [CrossRef] [PubMed]
59. Kanehisa, M.; Sato, Y.; Kawashima, M.; Furumichi, M.; Tanabe, M. KEGG as a reference resource for gene and protein annotation. *Genet. Epidemiol.* **2016**, *44*, 457–462. [CrossRef]
60. Liu, Y.; Lv, J.; Liu, Z.; Wang, J.; Yang, B.; Chen, W.; Ou, L.; Dai, X.; Zhang, Z.; Zou, X. Integrative analysis of metabolome and transcriptome reveals the mechanism of color formation in pepper fruit (*Capsicum Annuum* L.). *Food Chem.* **2020**, *306*, 125629. [CrossRef]
61. Wu, B.S.; Mansoori, M.; Trumpler, K.; Addo, P.W.; MacPherson, S.; Lefsrud, M. Effect of amber (595 nm) light supplemented with narrow blue (430 nm) light on tomato biomass. *Plants* **2023**, *12*, 2457. [CrossRef] [PubMed]
62. Villagra, J.; Sancho, L.G.; Alors, D. Macrolichen communities depend on phorophyte in Conguillío national park, Chile. *Plants* **2023**, *12*, 2452. [CrossRef] [PubMed]
63. Papadopoulos, E.I.; Petraki, C.; Gregorakis, A.; Fragoulis, E.G.; Scorilas, A. Clinical evaluation of microRNA-145 expression in renal cell carcinoma: A promising molecular marker for discriminating and staging the clear cell histological subtype. *Biol. Chem.* **2016**, *397*, 529–539. [CrossRef] [PubMed]

plants

Review

The Role of Transcription Factors in the Regulation of Plant Shoot Branching

Lingling Zhang, Weimin Fang, Fadi Chen * and Aiping Song *

State Key Laboratory of Crop Genetics and Germplasm Enhancement, Key Laboratory of Landscaping, Ministry of Agriculture and Rural Affairs, College of Horticulture, Nanjing Agricultural University, Nanjing 210095, China; 2021204041@stu.njau.edu.cn (L.Z.); fangwm@njau.edu.cn (W.F.)
* Correspondence: chenfd@njau.edu.cn (F.C.); aiping_song@njau.edu.cn (A.S.)

Abstract: Transcription factors, also known as trans-acting factors, balance development and stress responses in plants. Branching plays an important role in plant morphogenesis and is closely related to plant biomass and crop yield. The apical meristem produced during plant embryonic development repeatedly produces the body of the plant, and the final aerial structure is regulated by the branching mode generated by axillary meristem (AM) activities. These branching patterns are regulated by two processes: AM formation and axillary bud growth. In recent years, transcription factors involved in regulating these processes have been identified. In addition, these transcription factors play an important role in various plant hormone pathways and photoresponses regulating plant branching. In this review, we start from the formation and growth of axillary meristems, including the regulation of hormones, light and other internal and external factors, and focus on the transcription factors involved in regulating plant branching and development to provide candidate genes for improving crop architecture through gene editing or directed breeding.

Keywords: transcription factors; branching; axillary meristem; development

Citation: Zhang, L.; Fang, W.; Chen, F.; Song, A. The Role of Transcription Factors in the Regulation of Plant Shoot Branching. *Plants* **2022**, *11*, 1997. https://doi.org/10.3390/plants11151997

Academic Editor: Giovanna Frugis

Received: 7 July 2022
Accepted: 29 July 2022
Published: 31 July 2022

1. Introduction

Transcription factors (TFs), also known as trans-acting factors, are proteins with special structures that regulate plant growth and development. Transcription factors bind to specific DNA sequences (cis-acting elements) in the upstream promoter region of target genes through their DNA-binding domain (DBD), thereby regulating the specific expression of target genes in different cell types of plants or under different environmental conditions [1]. In plant morphogenesis, selective expression of genes leads to the differentiation of phenotypes, and transcription factors play an important regulatory role in these processes. TFs are divided into different gene families, such as the bHLH, TCP, MADS, bZIP, KNOX, WOX, AP2/ERF, NAC, GATA and ARF families, according to differences in DNA-binding domains and conserved motifs.

Shoot branching is a common phenomenon in plant growth and plays a very important role in plant morphogenesis. Branching also affects plant competitiveness against weeds or pests [2,3]. Therefore, research on branching mechanisms has become a popular topic worldwide. Studies have shown that axillary meristems initiate from cell groups detached from the primary SAM that retain their meristematic identity (Figure 1). Alternatively, axillary meristems may originate de novo later in development from partially or fully differentiated cells. Development of the lateral branch involves two important processes, axillary meristem formation and axillary bud growth [4,5].

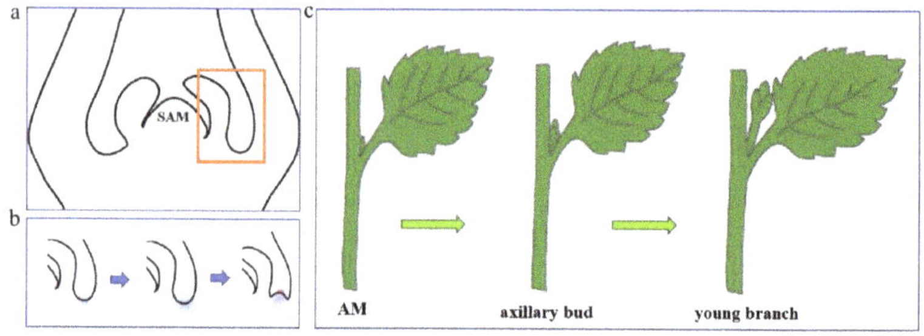

Figure 1. Steps of plant shoot branching. (**a**) indicates the axillary meristem at the leaf primordium axils. (**b**) indicates the formation of axillary meristems (area shown in yellow box in (**a**)). (**c**) represents the development of plant axillary meristem into young branches.

Through a study on the regulatory mechanism of plant branching, a series of branching-related transcription factors (Table 1) have been isolated from *Oryza sativa*, *Arabidopsis thaliana*, *Lycopersicon esculentum*, *Zea mays* and other plants. However, the nomenclature of homologous genes in different species is confusing, and related studies on transcription factors involved in the regulation of branching development are lacking in systematic elaboration. Branch formation is regulated at two developmental stages: axillary bud meristem formation and axillary bud emergence. The latter is induced by bud dormancy release and regulated by the synergistic effect of plant hormones such as auxin (IAA), strigolactones (SLs), cytokinins (CKs), abscisic acid (ABA) and brassinosteroids (BRs) [6,7]. Of course, many endogenous and developmental signals can be integrated to determine the fate of buds and the number and location of new buds growing on plants. This regulation is also strongly dependent on environmental factors, and plants adjust their branching ability according to their environmental conditions [8].

Table 1. Transcription factors involved in regulating plant branching.

Name	Homologs in Other Species	Family	Function
AtAGL6 (AGAMOUS-LIKE6)		MADS	Facilitates the formation of axillary meristems
OsMADS34			Coordinates with LAX1 to regulate the number of primary branches
OsMADS57			Is expressed predominantly in the SAM and axillary buds and is involved in SL signaling to enhance axillary bud growth and subsequent tillering
AtFUL (FRUITFULL)			Is involved in development of the axillary meristem, the expression of which is controlled by auxin
AtCUC1-3 (CUP SHAPED COTYLEDON1-3)		NAC	Is negatively regulated by BRs and involved in AM initiation
AtSTM (SHOOT MERISTEMLESS)	OsOSH1 (*O. sativa* homeobox1)	HB-KNOX	Is involved in initiation or maintenance of undifferentiated cell fate in very early stages of AM formation
AtLOF1 (LATERAL ORGAN FUSION1)			Is involved in lateral organ separation and axillary meristem formation
AtLOF2 (LATERAL ORGAN FUSION2)		MYB	
AtAS1 (ASYMMETRIC LEAVES1)			Inhibits branching and downregulates STM when cells start to differentiate

Table 1. *Cont.*

Name	Homologs in Other Species	Family	Function
AtRAX1 (REGULATOR OF AXILLARY MERISTEMS 1)	Bl (Blind), *S. lycopersicum*		Is involved in the early steps of AM initiation and development
AtRAX2-3			
MsMYB112			Inhibits collateral growth
AtMYB2			Inhibits branching and reduces cytokinin concentrations by inhibiting expression of IPTs in *Arabidopsis*
AtWUS (WUSCHEL)	OsTAB1 (TILLERS ABSENT1); OsMOC3 (MONOCULM 3)		Promotes branching and is involved in maintenance of meristematic stem cell function and regulation of cell division
OsWOX4		WOX	Is involved in AM initiation
AtWOX4			Regulates *Arabidopsis* secondary growth by SL signaling
MdWUS2 (WUSCHEL 2)			Regulates branching by inhibiting the activity of MdTCP12 (BRC2 homolog)
AtLAS (LATERAL SUPPRESSOR)	Ls, *S. lycopersicum*; ERA (ERAMOSA, *A. majus*); OsMOC1; HaLSL (LATERAL SUPPRESSOR LIKE)	GRAS	Is necessary for maintenance of the meristematic potential of the cells in the axils of leaf primordia
HaROXL (REGULATOR OF AXILLARY MERISTEM FORMATION LIKE)	ZmBA1 (BARREN STALK1); OsLAX1 (Lax Panicle 1); AtROX (REGULATOR OF AXILLARY MERISTEM FORMATION)	bHLH	Is involved in development of the SAM and lateral young leaf primordia
OsLAX2 (LAX PANICLE2)			Is involved in development of the SAM and lateral young leaf primordia
AtPIF4/5 (PHYTOCHROME INTERACTING FACTORs 4/5)			Inhibits the branching caused by phyB dysfunction and low R:FR
OsFZP (FRIZZLE PANICLE)	ZmBD1; COM2, *H. vulgare*		Represses axillary meristem formation
AtEBE (ERF BUD ENHANCER)		AP2/ERF-ERF	Is involved in cell proliferation and axillary bud growth
AtDRN (DORNRÖSCHEN) AtDRNL (DORNRÖSCHEN-LIKE)			Regulates *STM* expression and AM initiation
AtERF053			Is involved in cytokinin control of stem branching
OsRFL (RICE FLORICULA/LEAFY)	OsAPO2 (PANICLE ORGANIZATION 2)		Promotes AM specificity through its action on LAX1 and CUC genes
ZmBAD1 (BRANCH ANGLE DEFECTIVE1)		TCP	Promotes the formation of lateral meristems (e.g., branches) and axillary organs (e.g., leaf pillows) in wild-type maize
AtBRC1 (BRANCHED1)	OsTB1 (TEOSINTE BRANCHED 1); OsFC1 (FINECULM1); VvBRC		Negatively regulates axillary bud growth
AtBRC2 (BRANCHED2)	MdTCP12		Has a redundant role with BRC1 in regulation of axillary bud growth
AtAS2 (ASYMMETRIC LEAVES1)		LOB	Inhibits branching and downregulates STM when cells start to differentiate

Table 1. *Cont.*

Name	Homologs in Other Species	Family	Function
AtLOB1 (LATERAL ORGAN BOUNDARIES 1)			Is negatively regulated by BRs to reduce cell division and expansion in the border zone
AtWRKY71/EXB1		WRKY	Is expressed in tissues surrounding the AM start site
WRKY72			Positively regulates bud branching
AtREV (REVOLUTA)		HD-ZIP	Upregulates *STM* expression and promotes AM initiation
HB21 (Homeobox21) HB40 (Homeobox40) HB53 (Homeobox53)			Inhibits branching, directly downstream of BRC1
AtSPL13 (SQUAMOSA PROMOTER BINDING PROTEIN-LIKE 13)		SBP	Inhibits the growth of axillary buds
AtIPA1 (IDEAL PLANT ARCHITECTUREL1)	OsSPL14		Acts with D53 to mediate SL-regulated tiller development in rice
AtARR1 (ARABIDOPSIS RESPONSE REGULATOR 1)		GARP-ARR-B	Promotes branching, acts Downstream of cytokinins and promotes LAS expression by binding to their promoters
AtBES1 (BRI1-EMS-SUPPRESSOR1)		BES1	Inhibits branching and negatively regulates cambium activity in the SL signaling pathway in *Arabidopsis*

Note: The transcription factor prefix indicates the species to which it belongs. At—Arabidopsis thaliana; Os—Oryza sativa; Ha—Helianthus annuus; Cr—Ceratopteris richardii; Zm—Zea mays; Sl—Solanum lycopersicum; Md—Malus pumila; Vv—Vitis vinifera.

In this review, we begin with the formation and growth of axillary meristems to elaborate on the research progress of transcription factors involved in regulating plant branching development to provide target genes for manipulating plant branching.

2. Axillary Meristem Formation

The first step of branching is the development of axillary meristems in leaves. In recent years, a series of transcription factors that regulate the initiation of leaf axil meristems have been found in *Arabidopsis*, rice, maize and tomato. The AM is formed in the center of the frontal boundary zone of the leaf base. This region is not only a boundary but also plays an important role in the maintenance of meristem and organ development [9].

The origin of AMs is a controversial topic. There is a major view that AMs originate from meristem cell groups that become detached from the shoot apical meristem (SAM) as leaves form and never lose their meristem identity [10]. Early AM development depends on the maintenance of the specificity and meristem ability of axillary cells. In *Arabidopsis* (Figure 2), *WUSCHEL* (*WUS*) is a homologous domain transcription factor that is expressed in the center of the SAM and specifies the fate of meristem cells in this region. The *wus* mutation leads to an inability to maintain stem cell division ability [11]. ARABIDOPSIS RESPONSE REGULATOR 1 (ARR1) is a transcription factor downstream of CK that promotes *LAS* expression by binding to its promoter, promoting AM initiation [12,13]. Cytokinins also activate *WUS* expression through ARR1, enabling stem cell differentiation and axillary bud formation [14].

MERISTEMLESS (*STM*) is another important factor for maintaining branch organization. *STM*, a KNOX gene, is expressed in the whole SAM but is excluded from the organ primordium that maintains the function of undifferentiated cells in the SAM [15]. The molecular markers of the AM include concentrated and strong expression of *STM* in the center of the boundary region [16]. Once cells begin to differentiate, *STM* is downregulated by the MYB family transcription factor ASYMMETRIC LEAVES1 (AS1) and the LBD family transcription factor AS2 [17]. This indicates that the cells in the border region maintain the ability to recover to the meristem stage within a limited period of time. During the developmental stage, AMs began to form [18].

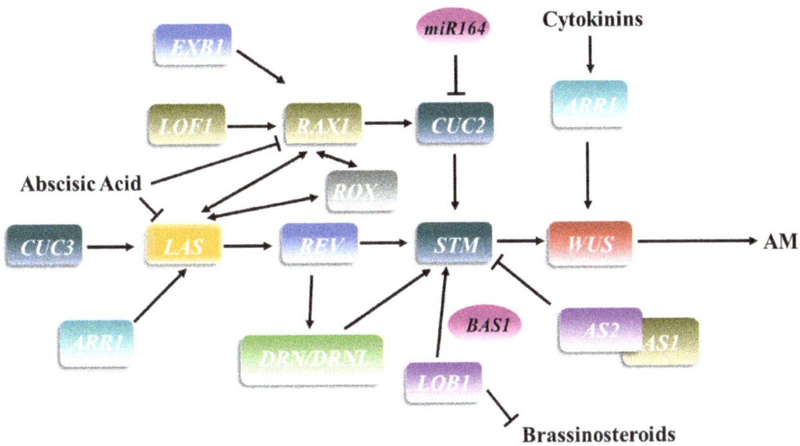

Figure 2. The pattern of transcription factors involved in regulating axillary meristem formation in *Arabidopsis thaliana*. The rectangular box represents a transcription factor, the same color represents the same family, and ellipses represent genes other than transcription factors. The arrow and rough line represent positive and negative regulation, respectively.

Before axillary bud formation, REVOLUTA (REV) upregulates *STM* expression and promotes AM initiation. Subsequently, CK reactivates *WUS* expression to establish the AM [1]. Preliminary evidence shows that REV acts upstream of *STM* and *WUS* and that Ls/LAS acts upstream of *STM* to activate expression [19,20]. However, the upstream region of *REV* is regulated by LAS. As an HD-ZIP transcription factor, REV itself is necessary for all lateral meristem formation. In addition to RNA accumulation in other modes, *REV* is expressed in the near-axis position of the developmental leaf primordium in a region similar to *RAX1*, which can produce the position signal of *RAX1* expression and control the radial mode [21,22]. The AP2 family transcription factor DORNRÖSCHEN (DRN) also plays a role in embryonic meristem and lateral organ development. Although AM initiation is affected in the *drn-1* mutant, the *drn-1 drnl-1* double mutant shows more serious defects in axillary bud formation than the single mutant, indicating that *drn* and *drnl* have important redundant functions in AM initiation [1]. Further studies have shown that DRN and DRNL preferentially affect the initiation pathway of the AM at the early nutritional stage rather than at late and reproductive stages. DRN, DRNL and REV can directly activate *STM* expression by binding to the same promoter region. In summary, DRN and DRNL redundantly promote AM initiation at the vegetative growth stage, and DRN/DRNL and REV synergistically upregulate *STM* transcription in mature leaf axils [23].

REGULATOR OF AXILLARY MERISTEM FORMATION (ROX) encodes direct homologs of bHLH transcription factors, namely, LAX1 and BA1. In these mutants, axillary bud formation during vegetative bud development is damaged, and their combination with REGULATOR OF AXILLARY MERISTEMS1 (*rax1*) and *las* mutations enhance these branching defects [24], indicating that ROX regulates AM formation by cooperating with RAX1 and LAS [18]. In the nutritional and reproductive development process, orthologous bHLH transcription factors seem to be involved in the formation of boundary regions in eudicotyledons and Gramineae plants, as delayed growth is needed. In subsequent studies, ABA was found to significantly inhibit the expression of the *RAX1* and *LAS* genes, thereby affecting the growth of axillary buds. RAX1 is a member of the largest MYB TF R2R3-MYB family in *Arabidopsis*. *RAX2* and *RAX3* genes function in the early stages of AM initiation and development [22,24,25]. In addition, RAX1 promotes the early stage of AM formation; it also negatively regulates the gibberellin level in shoot tips to modulate AM formation and affect the timing of development phase transition [26]. Studies have shown that RAX1 is involved in the determination of axillary meristems by generating a tissue environment conducive to the establishment of meristems to control the spatial pattern of AM development [22]. In sunflower, the rax-like gene R2R3-MYB2 was also found to play a key role in AM formation to establish or maintain the leaf axillary stem cell niche [24]. The three genes are expressed at the regional boundary between the shoot apical meristem and leaf primordium prior to the establishment of axillary meristem [22]. Further evidence shows that LAS and RAX1 can replace ROX to some extent and regulate axillary meristem formation [27].

LATERAL ORGAN FUSION1 (LOF1) and LOF2 encode MYB transcription factors that play roles in lateral organ separation and axillary meristem formation, partly through interaction with CUC2, CUC3 and STM [16,28]. *LOF1* is expressed at organ boundaries and acts upstream of *RAX1, LAS* and *CUC* [16,29,30]. The *lof1* mutant exhibits defects in organ separation, which is the result of abnormal cell division and amplification during early boundary formation. In addition, low concentrations of BRs in the border area promote specific expression of the *CUC* gene and initiation of the AM [21].

The NAC domain proteins CUP SHAPED COTYLEDON1 (CUC1), CUC2 and CUC3 are involved in the initiation of the *Arabidopsis* SAM via *STM* expression [16,22,27,28]. The number of axillary buds in its mutants is significantly decreased [31]. Among the three CUC family members, CUC3 plays a major role in determining the formation and location of axillary meristems [30–32]. Further studies have shown that CUC1 and CUC2 control the development of the axillary meristem by regulating LAS and that CUC3 may play a role independent of LAS [30]. *CUC2* is also regulated by RAX1 in early AM development to establish or maintain the stem cell niche formed by the AM [22,31]. In particular, transcription of *CUC2* is continuously downregulated in the *raxl* mutant, indicating that RAX1 affects AM initiation by regulating the expression of *CUC2* [21]. Thus, RICE FLORICULA/LEAFY (RFL) promote AM specificity through an effect on *LAX1* and *CUC* [33]. *RFL* is expressed in the vegetative axillary meristem and very young tillering buds, and its expression pattern is similar to that of STM, which may be related to the maintenance of the meristem cell zone [34].

Guo et al. [35] identified the *EXCESSIVE BRANCHES1* (*EXB1*) gene, which encodes a WRKY transcription factor previously known as WRKY71 that is mainly expressed in tissues around the AM initiation site. The functional *exb1-D* mutant displays an obvious increase in branching, which is due to the combined effect of excessive AM priming and increased bud activity. Quantitative data show that EXB1 controls the initiation of the AM by positively regulating the transcription of *RAX1, RAX2* and *RAX3*. Subsequent data indicate that *EXB1* may be located upstream of the *RAX* gene and regulate AM formation [21,36]. EXB1 also regulates branching in *Arabidopsis* through negative regulation of auxin signaling [37]. Auxin is a well-known bud growth inhibitor, and AM initiation requires minimal auxin [28,38]. In the presence of apical buds exist, auxin is transported from top to bottom in the axilla and inhibits the growth of axillary buds. This phenomenon is called apical dominance [39]. WRKY proteins in the EXB1 clade regulate auxin pathways. Similarly, overexpression of rice *WRKY72* in *Arabidopsis* also increases bud branching, suggesting that the role of EXB1 in bud branching may be evolutionarily conserved between monocots and dicots [40]. Furthermore, a key factor in the establishment of the AM boundary region in *Arabidopsis* is the transcription factor *LATERAL ORGAN BOUNDARYES1* (*LOB1*), which induces *PHYB ACTIVATION TAGGED SUPPRESSOR1* (*BAS1*) expression and encodes a protein that brassinosteroid-inactivation capacity. In leaf axils, BR accumulation is negatively regulated by LOB1, an important boundary-specific transcription factor [18]. LOB1 directly upregulates *BAS1* to produce low BR concentrations to reduce cell division and expansion in border areas [21]. Unlike all other known regulators, AGAMOUS-LIKE 6 (AGL6) specifically promotes stem branching only at the leaf axils of stem leaves in *Arabidopsis* [41].

Similarly, in rice, *TILLERS ABSENT1* (*TAB1*) and *WUSCHEL-RELATED HOMEOBOX4* (*WOX4*) are two *WUS* genes that are necessary for AM initiation [31]. *TAB1* is normally expressed only in the anterior portion of the meristem and is not usually expressed in the SAM or mature AM [42]. Interestingly, *wus* mutation does not affect axillary meristem development in *Arabidopsis*, but rice TAB1 seems to control AM formation through mechanisms different from those of WUS in *Arabidopsis* [43]. The other WUS family transcription factor, WOX4 (the close paralogous homolog of TAB1), plays a role in the development of the AM by alternating with TAB1 [44]. However, unlike *TAB1*, *WOX4* is not expressed in the premeristem but contributes to the maintenance of the AM after almost complete AM establishment. TAB1 forms an AM by enhancing the expression of *O. sativa homeobox 1* (*OSH1*) and *WOX4* [31]. During rice AM formation, *OSH1* is preferentially expressed in the AM, and a significant decrease in its expression and a decrease in tillering in its mutant indicate that the gene is necessary for the initiation or maintenance of the fate of undifferentiated cells at the very early stages of AM formation [45]. *MONOCULM 3* (*MOC3*) is a direct homolog of WUS in rice and is necessary for the formation of tillering buds and interacts with key components of the CK pathway controlling rice tillering [46]. CKs antagonize auxin at the top. Even in the presence of auxin provided by the growing stem apex or the apex, CK applied to buds is sufficient to initiate growth [47].

LAX PANICLE1 (LAX1) and MONOCULM1 (MOC1) encode bHLH family and GRAS family transcription regulators, respectively, which are necessary for the initiation and maintenance of the AM in rice panicles [24,48]. *LAX* is expressed on the boundaries of apical meristem- and neomeristem-forming regions, specifically controlling the initiation or maintenance of new meristems [49]. Studies have shown a significant reduction in the number of spikelets in the *lax1* mutant, and AMs cannot be formed during vegetative development. Similarly, the maize *barren stalk1* (*ba1*) mutant cannot initiate AMs at any stage of the life cycle [24,50]. These two genes accumulate at the proximal axis boundary of the axillary meristems formed during nutrition and reproductive development [27]. *MADS34* in rice encodes a MADS-box transcription factor that coregulates the number of primary and secondary branches with LAX1 [51]. *MOC1* (a direct homolog of *LATERAL SUPPRESSOR, LS* in tomato and *LATERAL SUPPRESSOR, LAS* in *Arabidopsis*) was the first key transcription factor identified [52,53] controlling rice tillering; it is mainly expressed in leaf axils and axillary buds during AM development, positively regulating rice tillering [43,54]. As expression of *MOC1* and *LAX1* is not altered in the *tab 1* mutant, *TAB1* plays a role in an independent pathway or downstream of MOC1 and LAX1. In addition, LAX2, together with LAX1 and MOC1, plays a role in different AM maintenance pathways to control branching at vegetative and reproductive stages [45]. Similarly, disruption of *LAS* in *Arabidopsis* leads to AM loss during vegetative development, causing loss of branching or tillering and indicating that these genes are highly conserved [5]. In conclusion, both *LAS* and *MOC1* are specifically expressed in the initiation region of the AM [21]. However, in the *moc1* mutant, which lacks axillary buds and tillers, *OSH1* expression completely disappears in leaf axils but is not affected in the SAM. Moreover, *MOC1* is expressed earlier than *LAX1*, *LAX2* and *TAB1* during AM formation, indicating that the sequence or independent role of these genes is the cause of axillary bud formation and that multiple pathways contribute to the development of the AM [31].

Similarly, *REGULATOR OF AXILLARY MERISTEM FORMATION LIKE* (*ROXL*) isolated from sunflower is a homologous gene of *ROX/LAX1*. In situ results of a cross-section show accumulation of *ROXL* transcripts at specific points in the boundary region between the apical meristem and the lateral leaf primordium, displaying a similar pattern in *Arabidopsis* [27]. Based on in situ hybridization, *Ha-ROXL* exhibits clear-boundary transcription in vegetative branches, though the expression pattern of *LATERAL SUPPRESSOR LIKE* (*LSL*) is confined to the boundary region because signals can also be detected in other cellular domains of vegetative and reproductive branches. Transcription of *LSL* is also expanded at the early stage of lateral primordium development, indicating that this gene is involved in the early development of the lateral primordium and in the initiation of the AM [24,55].

BLIND (*Bl*) is the homologous gene of the *RAX1* gene in tomato and *Arabidopsis* [21,25]. It encodes an R2R3 Myb transcription factor and regulates the early steps of AM initiation. There are fewer axillary buds due to defects in AM initiation caused by the *bl* mutant [31]. Double-mutation analysis has shown that different members of the Bl-related subgroup of the *R2R3 Myb* gene regulate axillary meristem formation in a partially redundant manner [22]. Double mutants of *ls* and *bl* in tomato display an additive phenotype, suggesting that at least two pathways are involved in AM initiation [20,22,27]. It is noteworthy that these genes are expressed in leaf axils during AM development. The homologous compounds of BLIND and RAX do not play a role in AM formation in rice, indicating that the BLIND/RAX pathway is unique to true dicotyledonous plants [31]. It should also be noted that in rice, axillary meristems usually develop into tillers from the basal nodes of plants, forming a typical cluster plant structure [34]. *FRIZZLE PANICLE* (*FZP*) is a very important gene in rice tillering development. Its overexpression inhibits *RFL/PANICLE ORGANIZATION 2* (*APO2*), which seriously suppresses the formation of axillary meristems at the vegetative stage and leads to a significant decrease in tiller number [48]. In maize, *BAD1* transcripts are detected mainly in the AM boundary region and between lateral branches, and its functional deficiency results in organ fusion via a reduction in the number and angle of branches [56]. Nevertheless, homologs of this gene, *COMPOSITUM 1* (*COM1*) and *RETARDED PALEA1* (*REP1*), in barley and rice have no similar functions. In addition, *BELL1-like homeobox 12* (*BLH12*) and *BLH14* play an important role in maintaining axillary meristems and possess the redundant functions necessary for axillary bud development throughout nutritional and reproductive development [19].

In addition to common plants such as *Arabidopsis*, rice and maize, other species are predicted to have transcription factors involved in branching or tillering. For example, it was found that the formation of the AM in *Antirrhinum majus* requires *ERAMOSA* (*ERA*), a gene encoding a GRAS transcription factor (orthologous to *LAS* in *Arabidopsis*). The basic role of *ERA* in AM formation is consistent with that of *LAS/MOC1/ERA* in preventing cell differentiation in the boundary area and in stimulating AM formation [57].

3. Axillary Meristem Outgrowth

The structure of mature plants is determined by the occurrence of axillary meristems, control of bud growth and subsequent dynamics of branching growth. Changes in these parameters result in the high morphological diversity observed in different plant species and even among individuals of a particular species [18]. After the formation of the AM, its growth as a branch repeats the development pattern of primary branches and endows plants with a branching structure [33]. Although a series of genetic studies have revealed the molecular mechanism and genes involved in SAM formation and maintenance, little is known about the generation and growth control of axillary buds [20].

The transcription factor *BRANCHED1* (*BRC1*) in *Arabidopsis* is homologous to *TEOSINTE BRANCHED 1* (*TB1*) in maize and plays a core inhibitory role in regulating axillary bud growth (Figure 3). *BRC2* is a paralog of *BRC1*, which is also expressed in axillary buds and plays a redundant role in the regulation of axillary bud growth [43]. Similarly, *VvBRC* plays a key and negative role in branching in grape [58]. LAS and REV act upstream of *BRC1*. As expression of *BRC1* is significantly downregulated in the *max2* mutant, the MAX-mediated pathway seems to control the activity of BRC1 [59]. *Homeobox 21* (*HB21*), *HB40* and *HB53* act directly downstream of BRC1 to regulate branching [60]. Moreover, BRC1 was identified as an inhibitor of downstream branches of SL signal transduction [61,62]. SL, a plant hormone synthesized by carotenoid catabolism, moves from root to stem, inhibiting stem branching by preventing the growth of axillary buds [63]. In *Arabidopsis*, another transcription factor involved in the SL pathway, namely, BRI1-EMS-SUPPRESSOR1 (BES1), negatively regulates cambium activity in the SL signaling pathway. Further studies have shown that SL signaling regulates the expression level of *WOX4* through BES1, controlling secondary growth [64]. CK is another class of hormones that has an important role in regulating apical dominance and axillary bud outgrowth. At the later stage of development, MYB2 reduces the concentration of CK by inhibiting the expression of *IPTs*, suppressing the growth of axillary buds [47]. In functional exploration of the MADS domain factor *FRUITFULL* (*FUL*) in *Arabidopsis thaliana*, it was found that the combination of auxin and BR strongly induced the growth regulator *SMALL AUXIN UPREGULATED RNA 10* (*SAUR10*) to be directly modulated by FUL, thus participating in branching angle regulation [65]. It should be noted that FUL also responds to a decrease in R:FR light by regulating the SAUR10 pathway and affecting *Arabidopsis* branching [65]. The auxin-regulated branching pathway in *Arabidopsis* involves a class of photochrome-targeted transcription factors, *PHYTOCHROME INTERACTING FACTOR 4* (*PIF4*)/*PIF5* [66]. When these transcription factors participate in photoreactions, they inhibit the branching caused by phyB dysfunction and low R: FR. Hence, R:FR plays an important role in branching.

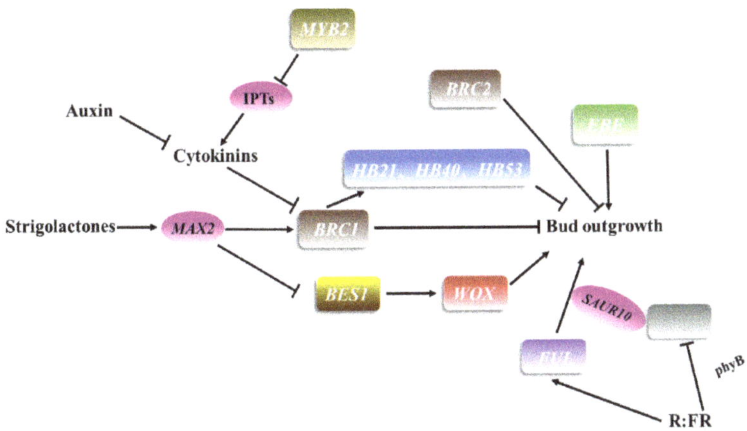

Figure 3. Patterns of transcription factors involved in regulating axillary meristem growth in *Arabidopsis thaliana*. The rectangular box represents transcription factors, the same color represents the same family, and ellipses represent genes other than transcription factors. The arrow and rough line represent positive and negative regulation, respectively.

Overall, the interaction between ABA and auxin may mediate the effect of PIF4/PIF5 on branching [67]. Mohammad et al. [68] reported another transcription factor, ERF BUD ENHANCER (EBE), that affects cell proliferation, axillary bud growth and branching in *Arabidopsis*. The gene encodes the AP2/ERF transcription factor and is strongly expressed in proliferating cells. Moreover, overexpression of *EBE* promotes cell proliferation, shortens the cell cycle in growing calli and stimulates the formation and growth of axillary buds [68].

TB1, also known as *FINECULM1* (*FC1*), is a TCP family transcription factor that negatively regulates rice tillering and inhibits the subsequent growth of axillary buds [69]. This gene encodes a TCP TF that is expressed at the base of these buds and SAMs. Its overexpression leads to a significant reduction in tiller number, whereas the *tb1* mutant exhibits an increase in tiller number [31]. In addition, TB1 may be a common target for the CK and SL pathways and act downstream of SL [70,71]. *MADS57* is a transcription factor of the MADS domain family that participates in the regulation of axillary bud growth in rice. RNA in situ hybridization analysis has shown that MADS57 is mainly expressed in meristems and axillary buds. Additionally, its expression is higher in the tillering and stem elongation stages than in other stages of rice growth [70]. Interaction of TB1 with MADS57 reduces the negative regulatory activity of MADS57 on *D14* expression, allowing MADS57 to affect tillering [71].

In a study of branching-related genes in chrysanthemum, *CmERF053* was found to be rapidly upregulated in axillary buds when apical dominance was relieved, which may be related to the growth of lateral branches. The gene belongs to the AP2/ERF family and is mainly expressed in stem and root organs. Further transcriptome analysis showed that CK-mediated control of stem branching may be related to the transcription level of *CmERF053* [26]. *IDEAL PLANT ARCHITECTUREL1* (*IPA1*), also known as *SQUAMOSA PROMOTER BINDING PROTEIN-LIKE 14* (*SPL14*), is another key regulator that determines plant structure but not meristem activity, and upregulation of *IPA1* expression leads to fewer tillers in rice [72]. Furthermore, *IPA1* overexpression lines display a reduced tillering phenotype, whereas tillering in the *ipa1* mutant increases [73]. This gene acts as a direct downstream component of D53 in regulating tiller number and SL-induced gene expression. In fact, *D53* is one of the only known transcription targets of SL, and D53 inhibits *IPA1* upregulation [74]. IPA1 directly binds to the *D53* promoter and plays a key role in the feedback regulation of SL-induced *D53* expression. In summary, *IPA1* may constitute a long-term transcription factor that acts with D53 to mediate SL-regulated rice tillering development [75]. In later vegetative axillary bud growth, OsMADS57 enhances axillary bud growth and subsequent tillering through SL signal transduction via direct inhibition of expression. OsMADS57 directly inhibits *D14* transcription to regulate tillering during organogenesis [70].

The SQUAMOSA PROMOTER BINDING PROTEIN-LIKE 13 (SPL13) gene encodes an SBP transcription factor that is mainly expressed in meristems and is essential for regulating branching and vegetative growth in *Medicago* [63]. Overexpression of *SPL13* inhibits axillary bud growth, thereby reducing lateral branches. In this process, *MYB112* is targeted and downregulated by SPL13. Compared with WT, *MsMYB112* RNAi plants show more branches, which confirms that MYB112 itself also inhibits the lateral branch growth of alfalfa [63,76]. Furthermore, MdWUS2 in apple can regulate branches by inhibiting the expression of *MdTCP12* [77].

Related to ABI3 and Viviparous 1 (*RAV1*) is a circadian rhythm gene in *Castanea mollissima* that is homologous to the *TEM* gene in *Arabidopsis*. It reaches its peak expression at noon under vegetative growth conditions and is highly expressed during winter dormancy and in response to low temperature [78]. When the CsRAV1 protein was overexpressed in hybrid poplar, high sylleptic branching was induced during the same growing season as lateral bud formation.

4. Application of Branching-Regulating TFs

Genetic control of branches is the main determinant of yield, seed number regulation and harvestability. For example, interference mutations in *FZP* and *IPA1* in rice, *TB1* in maize, and *BRC* in grape lead to increased branching and increasing yield [20,48,75,79]. Overexpression of *Bl* in tomato increases the number of branches, also increasing yield. At the same time, the change in plant type caused by branching provides an opportunity to explore the ornamental value of plants. For example, overexpression of *CmERF053* significantly increases the number of branches in chrysanthemum. In summary, regulation of these key transcription factors can significantly increase the number of branches, thereby increasing crop yield or quality.

In addition, fine regulation of branching has become an important strategy for plants to morphologically adapt to various environments. Chestnut CsRAV1, a circadian rhythm response factor, participates in the winter dormancy and low-temperature response of poplar and increases branching. Thus, manipulating this gene may lead to the possibility of producing trees with greater biomass. In actual cultivation, plants usually inhibit axillary bud growth in response to reductions in the ratio of red light to far-red light (R:FR) caused by the presence of competitive neighbors. Overexpression of *PIF4* and *PIF5* significantly inhibit the branching caused by this shade-avoidance syndrome, providing opportunities for practical cultivation.

5. Environmental Pathways Involved in the Control of Shoot Branching

It is well known that plant types have remarkable plasticity. Branch development is affected by many external factors, such as light, temperature and soil nutrients. Light is a powerful environmental factor that affects the branching of plants [80]. For example, low-intensity light reduces tillering in *Triticum aestivum* [81], whereas high-intensity light increases branching in hybrid roses [82]. Low R:FR and a decrease in blue light intensity trigger SAS, which leads to a decrease in axillary bud growth ability, such as in *Rhododendron* and *Hordeum vulgare* [83]. However, in *Lilium*, FR light strongly inhibits bud outgrowth [84], and blue light can increase or decrease the length of branches and internodes [85]. In general, UV radiation exposure reduces the length of branches [86], and studies have shown that photoperiod is one of the environmental factors involved in regulating branching, altering the branching pattern [87]. In summary, light is of great significance for the regulation of branching. In addition to light, temperature, moisture, carbon dioxide and other environmental factors affect the branching of plants. High temperature can inhibit branching, and CO_2 reduces the negative impact of high temperature on branch growth [88]. Of course, water and nutrients (such as nitrogen and phosphorus) are decisive factors in regulating plant shoot branching [89]. Many TFs responding to plant stress responses have been reported. However, research on TFs involved in branch response stress is scarce. Therefore, further exploration of TFs involved in branching response stress is worthy of attention.

6. Perspectives

Branching determines plant architecture and crop yield and plays an important role in plant morphogenesis. Therefore, research on branching regulation mechanisms is a popular topic worldwide. The regulation of plant branching by different transcription factors through mutual connections is one of the main directions of the current study of branching development patterns. Previous research and discussion on a single transcription factor have been the premise and foundation for studying the transcription regulatory network.

Notably, the functions of some transcription factors are not conserved. For example, *RAV1* in chestnut is a circadian rhythm gene that is homologous to the *TEM* gene in *Arabidopsis*. However, the two genes may lead to different phenotypes in woody and herbaceous plants. In view of the transcription factors related to branching discovered to date, we found exploration of the new functions of known transcription factors to be innovative, even though conserved transcription factors appear to provide key targets for the branching regulation mechanism.

Author Contributions: Conceptualization, A.S. and L.Z.; writing—original draft preparation, L.Z. and A.S.; writing—review and editing, W.F. and F.C.; funding acquisition, A.S., W.F. and F.C. All authors have read and agreed to the published version of the manuscript.

Funding: This research was funded by the National Natural Science Foundation of China (31872149, 32172609, 31870694 and 32072609), the Natural Science Fund of Jiangsu Province (BK20190076) and a project funded by the Priority Academic Program Development of Jiangsu Higher Education Institution.

Conflicts of Interest: The authors declare that the research was conducted in the absence of any commercial or financial relationships that could be construed as a potential conflict of interest.

References

1. Zhang, C.; Wang, J.; Wenkel, S.; Chandler, J.W.; Werr, W.; Jiao, Y. Spatiotemporal control of axillary meristem formation by interacting transcriptional regulators. *Development* **2018**, *145*, dev158352. [CrossRef]
2. Acker, R.; Weise, S.F.; Swanton, C.J. Influence of interference from a mixed weed species stand on soybean (*Glycine max* (l.) merr.) growth. *Can. J. Plant Sci.* **1993**, *73*, 1293–1304. [CrossRef]
3. Costes, E.; Lauri, P.E.; Simon, S.; Andrieu, B. Plant architecture, its diversity and manipulation in agronomic conditions, in relation with pest and pathogen attacks. *Eur. J. Plant Pathol.* **2013**, *135*, 455–470. [CrossRef]
4. Wang, Y.; Jiao, Y. Axillary meristem initiation-a way to branch out. *Curr. Opin. Plant Biol.* **2018**, *41*, 61–66. [CrossRef] [PubMed]

5. Greb, T.; Clarenz, O.; Schafer, E.; Muller, D.; Herrero, R.; Schmitz, G.; Theres, K. Molecular analysis of the LATERAL SUPPRESSOR gene in *Arabidopsis* reveals a conserved control mechanism for axillary meristem formation. *Genes Dev.* **2003**, *17*, 1175–1187. [CrossRef]
6. Long, J.; Barton, M.K. Initiation of axillary and floral meristems in *Arabidopsis*. *Dev. Biol.* **2000**, *218*, 341–353. [CrossRef] [PubMed]
7. Yang, M.; Jiao, Y. Regulation of Axillary Meristem Initiation by Transcription Factors and Plant Hormones. *Front. Plant Sci.* **2016**, *7*, 183. [CrossRef]
8. Rameau, C.; Bertheloot, J.; Leduc, N.; Andrieu, B.; Foucher, F.; Sakr, S. Multiple pathways regulate shoot branching. *Front. Plant Sci.* **2014**, *5*, 741. [CrossRef]
9. Yan, K.; Li, C.C.; Wang, Y.; Wang, X.Q.; Wang, Z.M.; Wei, D.Y.; Tang, Q.L. AGL18-1 delays flowering time through affecting expression of flowering-related genes in Brassica juncea. *Plant Biotechnol.* **2018**, *35*, 357–363. [CrossRef]
10. McSteen, P.; Leyser, O. Shoot branching. *Annu. Rev. Plant Biol.* **2005**, *56*, 353–374. [CrossRef]
11. Groover, A.T.; Mansfield, S.D.; DiFazio, S.P.; Dupper, G.; Fontana, J.R.; Millar, R.; Wang, Y. The Populus homeobox gene ARBORKNOX1 reveals overlapping mechanisms regulating the shoot apical meristem and the vascular cambium. *Plant Mol. Biol.* **2006**, *61*, 917–932. [CrossRef] [PubMed]
12. Jia, T.; Zhang, K.; Li, F.; Huang, Y.; Fan, M.; Huang, T. The AtMYB2 inhibits the formation of axillary meristem in *Arabidopsis* by repressing RAX1 gene under environmental stresses. *Plant Cell Rep.* **2020**, *39*, 1755–1765. [CrossRef] [PubMed]
13. Lee, D.Y.; An, G. Two AP2 family genes, supernumerary bract (SNB) and Osindeterminate spikelet 1 (OsIDS1), synergistically control inflorescence architecture and floral meristem establishment in rice. *Plant J.* **2012**, *69*, 445–461. [CrossRef] [PubMed]
14. Tian, C.; Zhang, X.; He, J.; Yu, H.; Wang, Y.; Shi, B.; Han, Y.; Wang, G.; Feng, X.; Zhang, C.; et al. An organ boundary-enriched gene regulatory network uncovers regulatory hierarchies underlying axillary meristem initiation. *Mol. Syst. Biol.* **2014**, *10*, 755. [CrossRef]
15. Ding, L.; Yan, S.; Jiang, L.; Liu, M.; Zhang, J.; Zhao, J.; Zhao, W.; Han, Y.Y.; Wang, Q.; Zhang, X. HANABA TARANU regulates the shoot apical meristem and leaf development in cucumber (*Cucumis sativus* L.). *J. Exp. Bot.* **2015**, *66*, 7075–7087. [CrossRef]
16. Lee, D.K.; Geisler, M.; Springer, P.S. LATERAL ORGAN FUSION1 and LATERAL ORGAN FUSION2 function in lateral organ separation and axillary meristem formation in *Arabidopsis*. *Development* **2009**, *136*, 2423–2432. [CrossRef]
17. Ikezaki, M.; Kojima, M.; Sakakibara, H.; Kojima, S.; Ueno, Y.; Machida, C.; Machida, Y. Genetic networks regulated by ASYMMETRIC LEAVES1 (AS1) and AS2 in leaf development in *Arabidopsis thaliana*: KNOX genes control five morphological events. *Plant J.* **2010**, *61*, 70–82. [CrossRef]
18. Teichmann, T.; Muhr, M. Shaping plant architecture. *Front. Plant Sci.* **2015**, *6*, 233. [CrossRef]
19. Tsuda, K.; Abraham-Juarez, M.J.; Maeno, A.; Dong, Z.; Aromdee, D.; Meeley, R.; Shiroishi, T.; Nonomura, K.I.; Hake, S. KNOTTED1 Cofactors, BLH12 and BLH14, Regulate Internode Patterning and Vein Anastomosis in Maize. *Plant Cell* **2017**, *29*, 1105–1118. [CrossRef]
20. Schmitz, G.; Tillmann, E.; Carriero, F.; Fiore, C.; Cellini, F.; Theres, K. The tomato Blind gene encodes a MYB transcription factor that controls the formation of lateral meristems. *Proc. Natl. Acad. Sci. USA* **2002**, *99*, 1064–1069. [CrossRef]
21. Guo, D.; Zhang, J.; Wang, X.; Han, X.; Wei, B.; Wang, J.; Li, B.; Yu, H.; Huang, Q.; Gu, H.; et al. The WRKY Transcription Factor WRKY71/EXB1 Controls Shoot Branching by Transcriptionally Regulating RAX Genes in *Arabidopsis*. *Plant Cell* **2015**, *27*, 3112–3127. [CrossRef] [PubMed]
22. Muller, D.; Schmitz, G.; Theres, K. Blind homologous R2R3 Myb genes control the pattern of lateral meristem initiation in *Arabidopsis*. *Plant Cell* **2006**, *18*, 586–597. [CrossRef] [PubMed]
23. Chandler, J.W.; Cole, M.; Flier, A.; Grewe, B.; Werr, W. The AP2 transcription factors DORNROSCHEN and DORNROSCHEN-LIKE redundantly control *Arabidopsis* embryo patterning via interaction with PHAVOLUTA. *Development* **2007**, *134*, 1653–1662. [CrossRef] [PubMed]
24. Fambrini, M.; Salvini, M.; Pugliesi, C. Molecular cloning, phylogenetic analysis, and expression patterns of LATERAL SUPPRESSOR-LIKE and REGULATOR OF AXILLARY MERISTEM FORMATION-LIKE genes in sunflower (*Helianthus annuus* L.). *Dev. Genes Evol.* **2017**, *227*, 159–170. [CrossRef] [PubMed]
25. Keller, T.; Abbott, J.; Moritz, T.; Doerner, P. Arabidopsis REGULATOR OF AXILLARY MERISTEMS1 controls a leaf axil stem cell niche and modulates vegetative development. *Plant Cell* **2006**, *18*, 598–611. [CrossRef]
26. Nie, J.; Wen, C.; Xi, L.; Lv, S.; Zhao, Q.; Kou, Y.; Ma, N.; Zhao, L.; Zhou, X. The AP2/ERF transcription factor CmERF053 of chrysanthemum positively regulates shoot branching, lateral root, and drought tolerance. *Plant Cell Rep.* **2018**, *37*, 1049–1060. [CrossRef]
27. Yang, F.; Wang, Q.; Schmitz, G.; Muller, D.; Theres, K. The bHLH protein ROX acts in concert with RAX1 and LAS to modulate axillary meristem formation in *Arabidopsis*. *Plant J.* **2012**, *71*, 61–70. [CrossRef]
28. Li, Y.; Xia, T.; Gao, F.; Li, Y. Control of Plant Branching by the CUC2/CUC3-DA1-UBP15 Regulatory Module. *Plant Cell* **2020**, *32*, 1919–1932. [CrossRef]
29. Hibara, K.; Karim, M.R.; Takada, S.; Taoka, K.; Furutani, M.; Aida, M.; Tasaka, M. Arabidopsis CUP-SHAPED COTYLEDON3 regulates postembryonic shoot meristem and organ boundary formation. *Plant Cell* **2006**, *18*, 2946–2957. [CrossRef]
30. Raman, S.; Greb, T.; Peaucelle, A.; Blein, T.; Laufs, P.; Theres, K. Interplay of miR164, CUP-SHAPED COTYLEDON genes and LATERAL SUPPRESSOR controls axillary meristem formation in *Arabidopsis thaliana*. *Plant J.* **2008**, *55*, 65–76. [CrossRef]
31. Wai, A.H.; An, G. Axillary meristem initiation and bud growth in rice. *J. Plant Biol.* **2017**, *60*, 440–451. [CrossRef]

32. Katayama, N.; Tanaka, R.; Fujinami, R.; Imaichi, R. Expression pattern of CUC3 ortholog in Zeylanidium tailichenoides (Podostemaceae) infers organization of a unique distichous shoot in Podostemoideae. *J. Plant Res.* **2019**, *132*, 521–529. [CrossRef] [PubMed]
33. Deshpande, G.M.; Ramakrishna, K.; Chongloi, G.L.; Vijayraghavan, U. Functions for rice RFL in vegetative axillary meristem specification and outgrowth. *J. Exp. Bot.* **2015**, *66*, 2773–2784. [CrossRef]
34. Rao, N.N.; Prasad, K.; Kumar, P.R.; Vijayraghavan, U. Distinct regulatory role for RFL, the rice LFY homolog, in determining flowering time and plant architecture. *Proc. Natl. Acad. Sci. USA* **2008**, *105*, 3646–3651. [CrossRef]
35. Guo, D.; Qin, G. EXB1/WRKY71 transcription factor regulates both shoot branching and responses to abiotic stresses. *Plant Signal Behav.* **2016**, *11*, e1150404. [CrossRef] [PubMed]
36. Eulgem, T.; Rushton, P.J.; Robatzek, S.; Somssich, I.E. The WRKY superfamily of plant transcription factors. *Trends Plant Sci.* **2000**, *5*, 199–206. [CrossRef]
37. Jiang, Y.; Liang, G.; Yang, S.; Yu, D. *Arabidopsis* WRKY57 functions as a node of convergence for *jasmonic acid*- and auxin-mediated signaling in jasmonic acid-induced leaf senescence. *Plant Cell* **2014**, *26*, 230–245. [CrossRef]
38. Muller, D.; Leyser, O. Auxin, cytokinin and the control of shoot branching. *Ann. Bot.* **2011**, *107*, 1203–1212. [CrossRef]
39. Domagalska, M.A.; Leyser, O. Signal integration in the control of shoot branching. *Nat. Rev. Mol. Cell Biol.* **2011**, *12*, 211–221. [CrossRef]
40. Yu, S.; Ligang, C.; Liping, Z.; Diqiu, Y. Overexpression of OsWRKY72 gene interferes in the abscisic acid signal and auxin transport pathway of *Arabidopsis*. *J. Biosci.* **2010**, *35*, 459–471. [PubMed]
41. Huang, X.; Effgen, S.; Meyer, R.C.; Theres, K.; Koornneef, M. Epistatic natural allelic variation reveals a function of AGAMOUS-LIKE6 in axillary bud formation in *Arabidopsis*. *Plant Cell* **2012**, *24*, 2364–2379. [CrossRef] [PubMed]
42. Nardmann, J.; Werr, W. The shoot stem cell niche in angiosperms: Expression patterns of WUS orthologues in rice and maize imply major modifications in the course of mono- and dicot evolution. *Mol. Biol. Evol.* **2006**, *23*, 2492–2504. [CrossRef] [PubMed]
43. Tanaka, W.; Ohmori, Y.; Ushijima, T.; Matsusaka, H.; Matsushita, T.; Kumamaru, T.; Kawano, S.; Hirano, H.Y. Axillary Meristem Formation in Rice Requires the WUSCHEL Ortholog TILLERS ABSENT1. *Plant Cell* **2015**, *27*, 1173–1184. [CrossRef] [PubMed]
44. Ohmori, Y.; Tanaka, W.; Kojima, M.; Sakakibara, H.; Hirano, H.Y. WUSCHEL-RELATED HOMEOBOX4 is involved in meristem maintenance and is negatively regulated by the CLE gene FCP1 in rice. *Plant Cell* **2013**, *25*, 229–241. [CrossRef]
45. Tabuchi, H.; Zhang, Y.; Hattori, S.; Omae, M.; Shimizu-Sato, S.; Oikawa, T.; Qian, Q.; Nishimura, M.; Kitano, H.; Xie, H.; et al. LAX PANICLE2 of rice encodes a novel nuclear protein and regulates the formation of axillary meristems. *Plant Cell* **2011**, *23*, 3276–3287. [CrossRef]
46. Lu, Z.; Shao, G.; Xiong, J.; Jiao, Y.; Wang, J.; Liu, G.; Meng, X.; Liang, Y.; Xiong, G.; Wang, Y.; et al. MONOCULM 3, an ortholog of WUSCHEL in rice, is required for tiller bud formation. *J. Genet. Genom.* **2015**, *42*, 71–78. [CrossRef]
47. Guo, Y.; Gan, S. AtMYB2 regulates whole plant senescence by inhibiting cytokinin-mediated branching at late stages of development in *Arabidopsis*. *Plant Physiol.* **2011**, *156*, 1612–1619. [CrossRef]
48. Bai, X.; Huang, Y.; Mao, D.; Wen, M.; Zhang, L.; Xing, Y. Regulatory role of FZP in the determination of panicle branching and spikelet formation in rice. *Sci. Rep.* **2016**, *6*, 19022. [CrossRef]
49. Komatsu, K.; Maekawa, M.; Ujiie, S.; Satake, Y.; Furutani, I.; Okamoto, H.; Shimamoto, K.; Kyozuka, J. LAX and SPA: Major regulators of shoot branching in rice. *Proc. Natl. Acad. Sci. USA* **2003**, *100*, 11765–11770. [CrossRef]
50. Gallavotti, A.; Zhao, Q.; Kyozuka, J.; Meeley, R.; Ritter, M.; Doebley, J.; Pe, M.; Schmidt, R. Haemangioblast commitment is initiated in the primitive streak of the mouse embryo. *Nature* **2004**, *432*, 630–635. [CrossRef]
51. Meng, Q.; Li, X.; Zhu, W.; Yang, L.; Liang, W.; Dreni, L.; Zhang, D. Regulatory network and genetic interactions established by OsMADS34 in rice inflorescence and spikelet morphogenesis. *J. Integr. Plant Biol.* **2017**, *59*, 693–707. [CrossRef] [PubMed]
52. Li, C. Toward understanding the stem-cell origin and molecular regulation of rice tillering. *J. Genet. Genom.* **2015**, *42*, 47–48. [CrossRef] [PubMed]
53. Xu, C.; Wang, Y.; Yu, Y.; Duan, J.; Liao, Z.; Xiong, G.; Meng, X.; Liu, G.; Qian, Q.; Li, J. Degradation of MONOCULM 1 by APC/C(TAD1) regulates rice tillering. *Nat. Commun.* **2012**, *3*, 750. [CrossRef]
54. Li, X.; Qian, Q.; Fu, Z.; Wang, Y.; Xiong, G.; Zeng, D.; Wang, X.; Liu, X.; Teng, S.; Hiroshi, F.; et al. Control of tillering in rice. *Nature* **2003**, *422*, 618–621. [CrossRef] [PubMed]
55. Basile, A.; Fambrini, M.; Tani, C.; Shukla, V.; Licausi, F.; Pugliesi, C. The Ha-ROXL gene is required for initiation of axillary and floral meristems in sunflower. *Genesis* **2019**, *57*, e23307. [CrossRef] [PubMed]
56. Lewis, M.W.; Bolduc, N.; Hake, K.; Htike, Y.; Hay, A.; Candela, H.; Hake, S. Gene regulatory interactions at lateral organ boundaries in maize. *Development* **2014**, *141*, 4590–4597. [CrossRef] [PubMed]
57. Mizzotti, C.; Galliani, B.M.; Dreni, L.; Sommer, H.; Bombarely, A.; Masiero, S. ERAMOSA controls lateral branching in snapdragon. *Sci. Rep.* **2017**, *7*, 41319. [CrossRef] [PubMed]
58. Min, Z.; Chen, L.; Zhang, Y.; Li, Z.Y.; Liu, M.; Li, W.P.; Ju, Y.L.; Fang, Y.L. VvBRC inhibits shoot branching in grapevine. *Sci. Hortic.* **2021**, *289*, 110370. [CrossRef]
59. Yang, Y.; Nicolas, M.; Zhang, J.; Yu, H.; Guo, D.; Yuan, R.; Zhang, T.; Yang, J.; Cubas, P.; Qin, G. The TIE1 transcriptional repressor controls shoot branching by directly repressing BRANCHED1 in *Arabidopsis*. *PLoS Genet.* **2018**, *14*, e1007296. [CrossRef]
60. Wang, B.; Smith, S.M.; Li, J. Genetic Regulation of Shoot Architecture. *Annu. Rev. Plant Biol.* **2018**, *69*, 437–468. [CrossRef]

61. Tsuji, H.; Tachibana, C.; Tamaki, S.; Taoka, K.; Kyozuka, J.; Shimamoto, K. Hd3a promotes lateral branching in rice. *Plant J.* **2015**, *82*, 256–266. [CrossRef] [PubMed]
62. Mashiguchi, K.; Sasaki, E.; Shimada, Y.; Nagae, M.; Ueno, K.; Nakano, T.; Yoneyama, K.; Suzuki, Y.; Asami, T. Feedback-regulation of strigolactone biosynthetic genes and strigolactone-regulated genes in *Arabidopsis*. *Biosci. Biotechnol. Biochem.* **2009**, *73*, 2460–2465. [CrossRef] [PubMed]
63. Gao, R.; Gruber, M.Y.; Amyot, L.; Hannoufa, A. SPL13 regulates shoot branching and flowering time in Medicago sativa. *Plant Mol. Biol.* **2018**, *96*, 119–133. [CrossRef] [PubMed]
64. Hu, J.; Hu, X.; Yang, Y.; He, C.; Hu, J.; Wang, X. Strigolactone signaling regulates cambial activity through repression of WOX4 by transcription factor BES1. *Plant Physiol.* **2022**, *188*, 255–267. [CrossRef]
65. Bemer, M.; van Mourik, H.; Muino, J.M.; Ferrandiz, C.; Kaufmann, K.; Angenent, G.C. FRUITFULL controls SAUR10 expression and regulates *Arabidopsis* growth and architecture. *J. Exp. Bot.* **2017**, *68*, 3391–3403. [CrossRef] [PubMed]
66. Li, L.; Ljung, K.; Breton, G.; Schmitz, R.J.; Pruneda-Paz, J.; Cowing-Zitron, C.; Cole, B.J.; Ivans, L.J.; Pedmale, U.V.; Jung, H.S.; et al. Linking photoreceptor excitation to changes in plant architecture. *Genes Dev.* **2012**, *26*, 785–790. [CrossRef] [PubMed]
67. Holalu, S.V.; Reddy, S.K.; Blackman, B.K.; Finlayson, S.A. Phytochrome interacting factors 4 and 5 regulate axillary branching via bud abscisic acid and stem auxin signalling. *Plant Cell Environ.* **2020**, *43*, 2224–2238. [CrossRef]
68. Mehrnia, M.; Balazadeh, S.; Zanor, M.I.; Mueller-Roeber, B. EBE, an AP2/ERF transcription factor highly expressed in proliferating cells, affects shoot architecture in *Arabidopsis*. *Plant Physiol.* **2013**, *162*, 842–857. [CrossRef]
69. Takeda, T.; Suwa, Y.; Suzuki, M.; Kitano, H.; Ueguchi-Tanaka, M.; Ashikari, M.; Matsuoka, M.; Ueguchi, C. The OsTB1 gene negatively regulates lateral branching in rice. *Plant J.* **2003**, *33*, 513–520. [CrossRef]
70. Guo, S.; Xu, Y.; Liu, H.; Mao, Z.; Zhang, C.; Ma, Y.; Zhang, Q.; Meng, Z.; Chong, K. The interaction between OsMADS57 and OsTB1 modulates rice tillering via DWARF14. *Nat. Commun.* **2013**, *4*, 1566. [CrossRef] [PubMed]
71. Minakuchi, K.; Kameoka, H.; Yasuno, N.; Umehara, M.; Luo, L.; Kobayashi, K.; Hanada, A.; Ueno, K.; Asami, T.; Yamaguchi, S.; et al. FINE CULM1 (FC1) works downstream of strigolactones to inhibit the outgrowth of axillary buds in rice. *Plant Cell Physiol.* **2010**, *51*, 1127–1135. [CrossRef] [PubMed]
72. Luo, L.; Li, W.; Miura, K.; Ashikari, M.; Kyozuka, J. Control of tiller growth of rice by OsSPL14 and Strigolactones, which work in two independent pathways. *Plant Cell Physiol.* **2012**, *53*, 1793–1801. [CrossRef] [PubMed]
73. Cao, J.; Liu, K.; Song, W.; Zhang, J.; Yao, Y.; Xin, M.; Hu, Z.; Peng, H.; Ni, Z.; Sun, Q.; et al. Pleiotropic function of the SQUAMOSA PROMOTER-BINDING PROTEIN-LIKE gene TaSPL14 in wheat plant architecture. *Planta* **2021**, *253*, 44. [CrossRef] [PubMed]
74. Kerr, S.C.; Beveridge, C.A. IPA1: A direct target of SL signaling. *Cell Res.* **2017**, *27*, 1191–1192. [CrossRef] [PubMed]
75. Song, X.; Lu, Z.; Yu, H.; Shao, G.; Xiong, J.; Meng, X.; Jing, Y.; Liu, G.; Xiong, G.; Duan, J.; et al. IPA1 functions as a downstream transcription factor repressed by D53 in strigolactone signaling in rice. *Cell Res.* **2017**, *27*, 1128–1141. [CrossRef]
76. Stracke, R.; Werber, M.; Weisshaar, B. The R2R3-MYB gene family in *Arabidopsis thaliana*. *Curr. Opin. Plant Biol.* **2001**, *4*, 447–456. [CrossRef]
77. Li, G.; Tan, M.; Ma, J.; Cheng, F.; Li, K.; Liu, X.; Zhao, C.; Zhang, D.; Xing, L.; Ren, X.; et al. Molecular mechanism of MdWUS2-MdTCP12 interaction in mediating cytokinin signaling to control axillary bud outgrowth. *J. Exp. Bot.* **2021**, *72*, 4822–4838. [CrossRef]
78. Moreno-Cortes, A.; Hernandez-Verdeja, T.; Sanchez-Jimenez, P.; Gonzalez-Melendi, P.; Aragoncillo, C.; Allona, I. CsRAV1 induces sylleptic branching in hybrid poplar. *New Phytol.* **2012**, *194*, 83–90. [CrossRef]
79. Lewis, J.M.; Mackintosh, C.A.; Shin, S.; Gilding, E.; Kravchenko, S.; Baldridge, G.; Zeyen, R.; Muehlbauer, G.J. Overexpression of the maize Teosinte Branched1 gene in wheat suppresses tiller development. *Plant Cell Rep.* **2008**, *27*, 1217–1225. [CrossRef]
80. Leduc, N.; Roman, H.; Barbier, F.; Péron, T.; Huché-Thélier, L.; Lothier, J.; Demotes-Mainard, S.; Sakr, S. Light Signaling in Bud Outgrowth and Branching in Plants. *Plants* **2014**, *23*, 223–250. [CrossRef]
81. Evers, J.B.; Vos, J.; Andrieu, B.; Struik, P.C. Cessation of tillering in spring wheat in relation to light interception and red: Far-red ratio. *Ann. Bot.* **2006**, *97*, 649–658. [CrossRef] [PubMed]
82. Girault, T.; Bergougnoux, V.; Combes, D.; Viemont, J.-D.; Leduc, N. Light controls shoot meristem organogenic activity and leaf primordia growth during bud burst in *Rosa sp. Plant Cell Environ.* **2008**, *31*, 1534–1544. [CrossRef] [PubMed]
83. Franklin, K.A. Shade avoidance. *New Phytol.* **2008**, *179*, 930–944. [CrossRef] [PubMed]
84. Ballaré, C.L.; Casal, J.J. Light signals perceived by crop and weed plants. *Field Crops Res.* **2000**, *67*, 149–160.83. [CrossRef]
85. Abidi, F.; Girault, T.; Douillet, O.; Guillemain, G.; Sintes, G.; Laffaire, M.; Ahmed, H.B.; Smiti, S.; Huché-Thélier, L.; Leduc, N. Blue light effects on rose photosynthesis and photomorphogenesis. *Plant Biol. (Stuttg.)* **2013**, *15*, 67–74. [CrossRef]
86. Kuhlmann, F.; Muller, C. MüllerDevelopment-dependent effects of UV radiation exposure on broccoli plants and interactions with herbivorous insects Environ. *Exp. Bot.* **2009**, *66*, 61–68. [CrossRef]
87. Fustec, J.; Beaujard, F. Effect of photoperiod and nitrogen supply on basal shoots development in *Rhododendron Catawbiense*. *Biol. Plant* **2000**, *43*, 511–515. [CrossRef]
88. Martin, C.A.; Stutz, J.C.; Kimball, B.A.; Idso, S.B.; Akey, D.H. Growth and topological changes of citrus limon (l.) burm. f. 'eureka' in response to high temperatures and elevated atmospheric carbon dioxide. *Am. Soc. Hortic. Sci.* **1995**, *120*, 1025–1031. [CrossRef]
89. Yan, Y.Y.; Zhao, N.; Tang, H.M.; Gong, B.; Shi, Q.H. Shoot branching regulation and signaling. *Plant Growth Regul.* **2020**, *92*, 131–140. [CrossRef]

plants

Article

Genome-Wide Identification and Expression Analysis of *XTH* Gene Family during Flower-Opening Stages in *Osmanthus fragrans*

Yang Yang [1,2,3], Yunfeng Miao [1,3], Shiwei Zhong [1,3], Qiu Fang [1,3], Yiguang Wang [1,3], Bin Dong [1,2,3,*] and Hongbo Zhao [1,2,3,*]

1 Zhejiang Provincial Key Laboratory of Germplasm Innovation and Utilization for Garden Plants, Hangzhou 311300, China; 2019105051027@stu.zafu.edu.cn (Y.Y.); 2019105052024@stu.zafu.edu.cn (Y.M.); zsw1105@zafu.edu.cn (S.Z.); fangqiu@zafu.edu.cn (Q.F.); wangyiguang1990@zafu.edu.cn (Y.W.)
2 Key Laboratory of National Forestry and Grassland Administration on Germplasm Innovation and Utilization for Southern Garden Plants, Hangzhou 311300, China
3 School of Landscape Architecture, Zhejiang Agriculture and Forestry University, Hangzhou 311300, China
* Correspondence: dongbin@zafu.edu.cn (B.D.); zhaohb@zafu.edu.cn (H.Z.)

Abstract: *Osmanthus fragrans* is an aromatic plant which is widely used in landscaping and garden greening in China. However, the process of flower opening is significantly affected by ambient temperature changes. Cell expansion in petals is the primary factor responsible for flower opening. Xyloglucan endoglycolase/hydrolase (XTH) is a cell-wall-loosening protein involved in cell expansion or cell-wall weakening. Through whole-genome analysis, 38 *OfXTH* genes were identified in *O. fragrans* which belong to the four main phylogenetic groups. The gene structure, chromosomal location, synteny relationship, and cis-acting elements prediction and expression patterns were analyzed on a genome-wide scale. The expression patterns showed that most *OfXTHs* were closely associated with the flower-opening period of *O. fragrans*. At the early flower-opening stage (S1 and S2), transcriptome and qRT-PCR analysis revealed the expression of *OfXTH24, 27, 32, 35* and *36* significantly increased under low ambient temperature (19 °C). It is speculated that the five genes might be involved in the regulation of flower opening by responding to ambient temperature changes. Our results provide solid foundation for the functional analysis of *OfXTH* genes and help to explore the mechanism of flower opening responding to ambient temperature in *O. fragrans*.

Keywords: *Osmanthus fragrans*; xyloglucan endoglycolase/hydrolase; flower opening; ambient temperature

Citation: Yang, Y.; Miao, Y.; Zhong, S.; Fang, Q.; Wang, Y.; Dong, B.; Zhao, H. Genome-Wide Identification and Expression Analysis of *XTH* Gene Family during Flower-Opening Stages in *Osmanthus fragrans*. *Plants* **2022**, *11*, 1015. https://doi.org/10.3390/plants11081015

Academic Editors: Aiping Song and Yu Chen

Received: 18 March 2022
Accepted: 2 April 2022
Published: 8 April 2022

1. Introduction

Xyloglucan endotransglucosylase/hydrolase (XTH), a cell-wall-relaxation factor, modifies the cellulose–xylan complex structure through catalyzing the breaking and rejoining of xyloglucan molecules in the plant cell wall [1]. XTH belongs to the Glycoside Hydrolase Family 16 (GH16) of the Carbohydrate Active Enzyme Family Database (CAZy). XTH proteins contain a Glyco_hydro_16 protein domain and a XET_C domain [2,3]. Accumulating evidence has demonstrated that XTH proteins were related to the two kinks enzyme activity of xyloglucan transglycosylation (XET) and xyloglucan hydrolysis (XEH) [4]. In *Arabidopsis thaliana*, *XTH* gene family can be divided into four groups, and the different groups have different functions or enzyme activity. For instance, Group I/II and Group IIIB genes show XET activity, while Group IIIA genes show XEH activity [2,5].

Increasing reports showed that XTH family genes play important roles in many processes of plant growth and development. During root development stages, high expression of *SlXTH1* can be increased by XET activity in the hypocotyl elongation zone and affect the hypocotyl growth and cell wall extensibility in tomato [6]. Similarly, *AtXTH21* gene played a principal role in the growth of the primary roots by altering the deposition of cellulose and the elongation of the cell wall [7]. In addition, *AtXTH9* was positively correlated with cell elongation of the inflorescence stem [8]. In flowering plants, flower opening is dependent on petal growth, which is closely related to the cell expansion. In *Ipomoea nil*, *InXTH2*, *InXTH3*, and *InXTH4* were upregulated by dark period and their expression was positively correlated with the rate of flower opening [9]. *EgXTH1* from *Eustoma grandiflorum* was involved in cell expansion, and the expression was increased with the flower-opening process compared with the control [10]. In *Rosa hybrida*, *RhXTH1* were relevant to the typical growth in different part of petals, and also increased markedly during petal growth [11]. Conversely, if *XTH* genes are not expressed or inhibited, the petal growth will not continue, thereby effectively inhibiting flower opening [12]. Therefore, this evidence suggests that *XTH* genes are an important regulator, and significantly influence petal cell expansion during the flower-opening process of most species.

Osmanthus fragrans is one of the Chinese top ten traditional flowers, known for its strong fragrance, and is widely cultivated as a garden tree in most Asian countries due to its high ornamental value and aromatic scent. However, the changes in ambient temperature significantly affected the opening and longevity of the flower [13]. In our previous research, we found that the flower opening is significantly promoted by low ambient temperature 19 °C in *O. fragrans*, and the flower-opening process is accompanied by petal cells significant expansion [14,15]. To clarify whether *OfXTH* genes regulate flower opening in response to ambient temperature, the expression profiles of *OfXTH* genes were analyzed during flower-opening process under different temperature treatment. Our results could deepen understanding of the *OfXTH* gene family and help to explore the mechanism of *O. fragrans* petal expansion responding to ambient temperature changes.

2. Results

2.1. Identification of XTH Gene Family in O. fragrans

Through the HMM search with the XTH protein domains (PF00722 and PF06955) and BLASTp with 33 AtXTH proteins, a total of 38 *OfXTH* genes were identified from the *O. fragrans* genome. Then, the characteristics of OfXTH proteins were investigated, including protein length, molecular weight (MW), and theoretical isoelectric point (pI), by the online software Expasy. The signal peptide (SP) and subcellular localization were predicted by the online program SignalP 4.1 Service and Plant-mPLoc. The results showed that the length of amino acids was between 234 and 355, and the molecular weight of the proteins was between 26.8 and 40.4 kD (Table 1), corresponding with the protein length. The theoretical isoelectric point varied from 4.83 to 9.47; this is possibly due to the different polarities of the amino acids. In addition, a total of 29 OfXTH proteins were predicted to contain a signal peptide, while the other nine proteins had no signal peptide. Additionally, the subcellular localization prediction revealed that 32 OfXTH proteins were located in the cell wall, and six proteins were predicted to locate in the cell wall or cytoplasm (Table 1).

Table 1. The characteristic of *OfXTH* genes in *O. fragrans*.

Gene	Length	MW	PI	SP	Subcellular Localization
OfXTH1	298	34,863.67	4.83	28	cell wall
OfXTH2	298	34,958.87	5.03	28	cell wall
OfXTH3	298	33,595.75	5.57	29	cell wall
OfXTH4	289	32,550.53	8.85	32	cell wall
OfXTH5	305	34,591.01	6.51	26	cell wall
OfXTH6	338	39,335.44	9.05	—	cell wall
OfXTH7	255	29,221.92	9.43	20	cell wall
OfXTH8	260	29,772.14	8.24	—	cell wall
OfXTH9	355	40,429.32	8.84	20	cell wall
OfXTH10	294	34,293.78	8.08	23	cell wall

Table 1. *Cont.*

Gene	Length	MW	PI	SP	Subcellular Localization
OfXTH11	294	34,213.84	8.84	25	cell wall/cytoplasm
OfXTH12	293	33,385.14	5.74	21	cell wall
OfXTH13	294	34,263.86	8.83	25	cell wall
OfXTH14	297	33,314.48	5.21	17	cell wall
OfXTH15	308	35,606.44	9.47	33	cell wall
OfXTH16	324	37,516.58	9.03	—	cell wall
OfXTH17	295	33,485.54	6.89	32	cell wall/cytoplasm
OfXTH18	270	30,312.99	5.23	23	cell wall
OfXTH19	324	37,516.58	9.03	—	cell wall
OfXTH20	295	33,463.49	6.43	32	cell wall/cytoplasm
OfXTH21	325	37,273.13	8.93	24	cell wall/cytoplasm
OfXTH22	282	31,918.7	8.69	—	cell wall
OfXTH23	234	26,856.03	8.79	—	cell wall
OfXTH24	295	34,651.93	5.14	26	cell wall
OfXTH25	321	37,347.8	8.96	26	cell wall
OfXTH26	261	30,242.91	5.98	—	cell wall
OfXTH27	250	28,403.93	9.1	—	cell wall
OfXTH28	302	35,166.08	5.7	20	cell wall
OfXTH29	295	33,937.19	6.75	18	cell wall
OfXTH30	339	38,895.06	6.4	31	cell wall
OfXTH31	298	34,746.73	5.02	28	cell wall
OfXTH32	289	32,124.53	5.1	23	cell wall/cytoplasm
OfXTH33	287	32,580.54	8.78	25	cell wall
OfXTH34	298	33,611.81	5.57	29	cell wall
OfXTH35	304	34,732.12	9.08	—	cell wall
OfXTH36	288	32,847.03	9.22	21	cell wall/cytoplasm
OfXTH37	293	33,556.83	6.31	30	cell wall
OfXTH38	294	34,368.01	8.69	25	cell wall

2.2. Phylogenetic Analysis of XTH Proteins

The phylogenetic tree was constructed using 38 OfXTHs from *O. fragrans*, 33 AtXTHs from *A. thaliana*, and 29 OsXTHs from *Oryza sativa* to analyze the evolutionary relationship (Figure 1). According to phylogenetic analysis, thirty-eight OfXTH proteins were classified into four groups in *O. fragrans*. Twenty-six OfXTHs belong to Group I/II. Group IIIA contained five OfXTH proteins. Group IIIB contained six OfXTH proteins. Only one protein, *OfXTH28*, belonged to Ancestral Group.

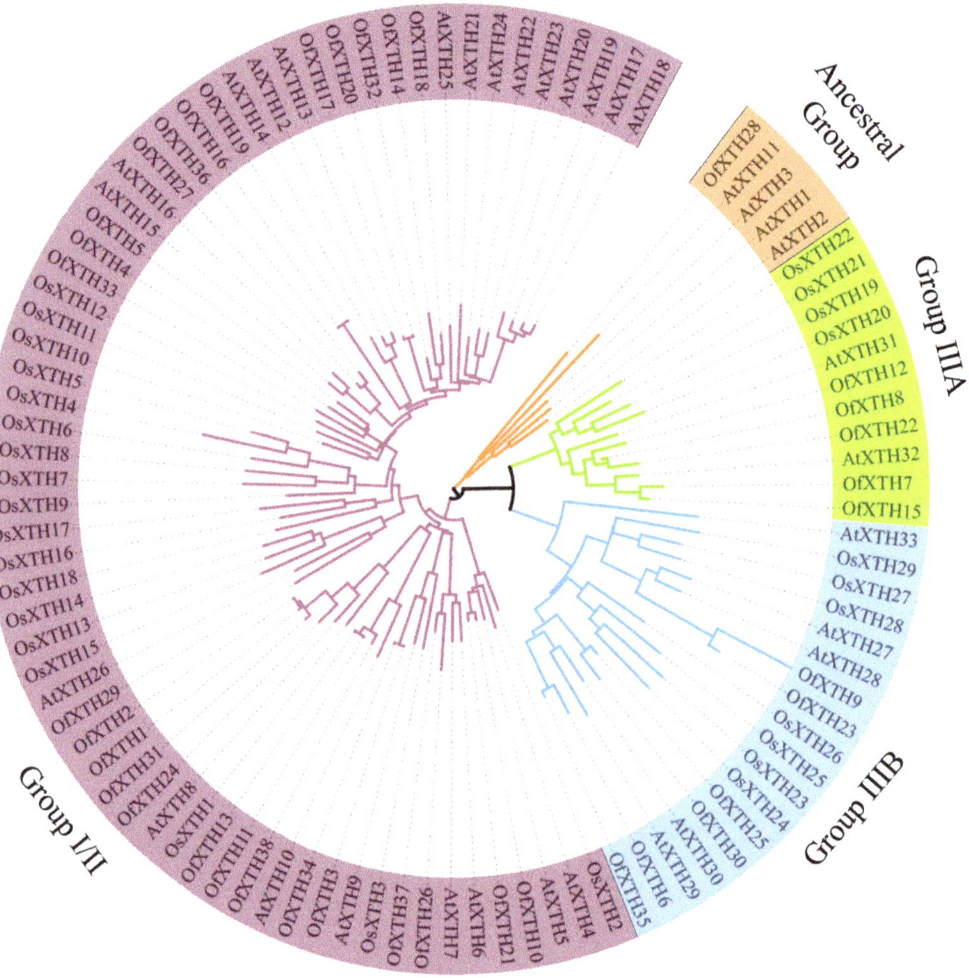

Figure 1. Phylogenetic analysis of XTH proteins from *O. fragrans*, *A. thaliana*, and *O. sativa*. Each subgroup is distinguished by a different color.

2.3. Gene Structure and Motif Analyses

Gene Structure Display Server (GSDS) was used to investigate the *OfXTHs* gene structure. The number of exons in thirty-seven *OfXTHs* ranged widely within 3–5 (Figure 2). Only *OfXTH23* contained two exons. The conservative motif, ExDxE, was present in every OfXTH protein. Motifs 3, 5, 8, and 9 were highly conserved in all proteins. Group I/II genes contained common motifs 2–9, while OfXTH16 lacked motifs 4, 6, 7, and OfXTH34 lacked motif 7. In addition, motif 10 was only present in proteins of Group I/II. Motifs 1–5 and 7–9 were found in Group IIIA, and motifs 1–3, 5, 8 and 9 presented in Group IIIB. In Ancestral Group, OfXTH28 shared the same motif with other OfXTHs, except for motif 2.

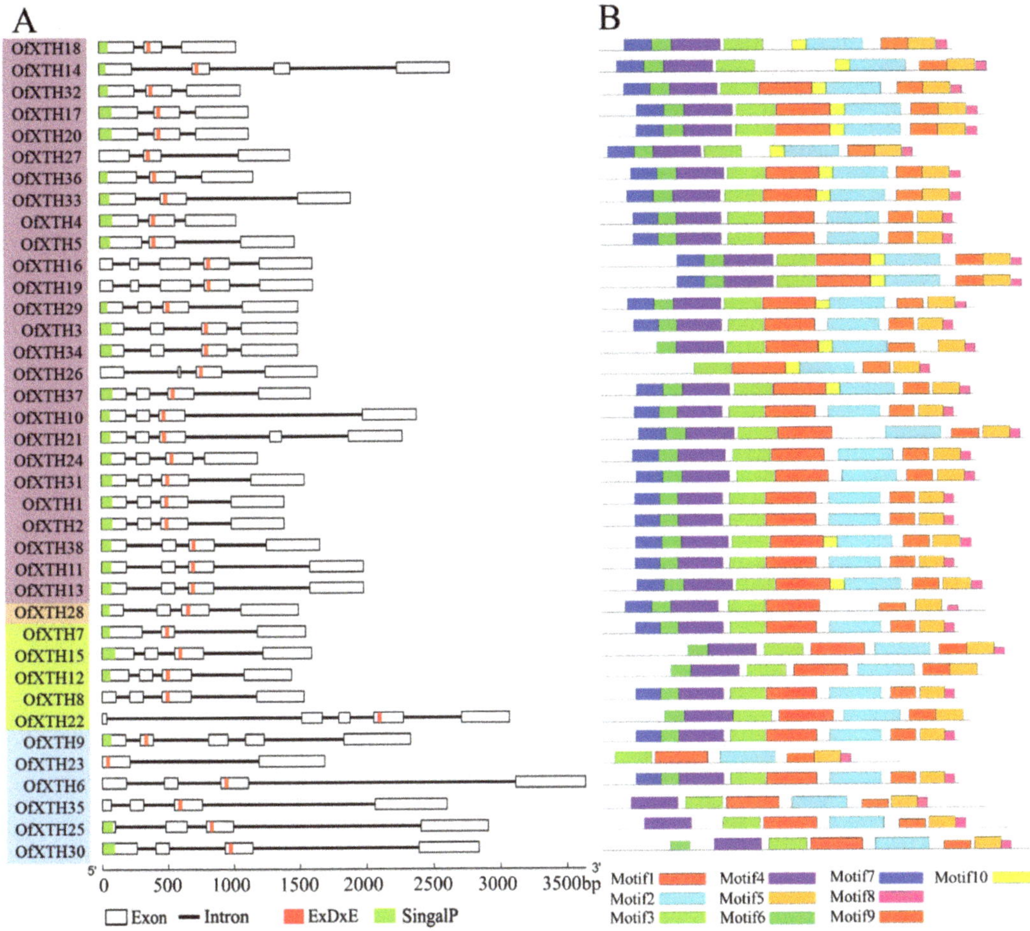

Figure 2. Gene structure and motif distribution of OfXTH proteins. (**A**) Gene structure, putative signal peptide sequences, and conservative motif sequence. Each subgroup is distinguished by a different color. White boxes and black lines represent exons and introns, respectively. The green and red strips indicate putative signal peptide sequences and conservative motif sequence, respectively. (**B**) The motif distribution. Motifs 1–10 in the OfXTH proteins are represented by different colored boxes.

2.4. Chromosomal Location and Synteny Analysis

As shown in Figure 3, 38 *OfXTHs* were distributed on the 17 chromosomes. Notably, seven genes, *OfXTH14–20*, were densely located on the Chr08, which was the chromosome distributing with the largest number of *OfXTH* genes. No *OfXTH* gene located on Chr05, Chr07, Chr14, Chr17, and Chr21. In addition, the other chromosomes contained 1–4 *OfXTH* genes. The locations of *OfXTH* genes on the chromosomes are described in detail in the mentary Materials (Table S1). Duplication events are related to plant evolutionary patterns. In this study, thirty segmentally duplicated genes were identified, and no tandem duplicated gene was found (Figure 4). All the synteny gene pairs were in the same cluster group. The substitution rates of Ka (non-synonymous) and Ks (synonymous) were calculated to estimate the selection pressure and determine the divergence time of the duplicated events. A total of 20 segmentally duplicated gene pairs were identified from

38 *OfXTHs* (Table 2). Interestingly, the Ka/Ks ratios of these 20 *OfXTHs* segmentally duplicated gene pairs were less than 1, suggesting that all segmentally duplicated gene pairs were purified selected instead of positively selected. The divergence time of *OfXTH* genes suggested that the duplicated events dated back to 59.93 million years ago (Mya) and lasted up to 0.31 Mya, and most of the duplicated events occurred between 5 and 35 Mya (Table 2).

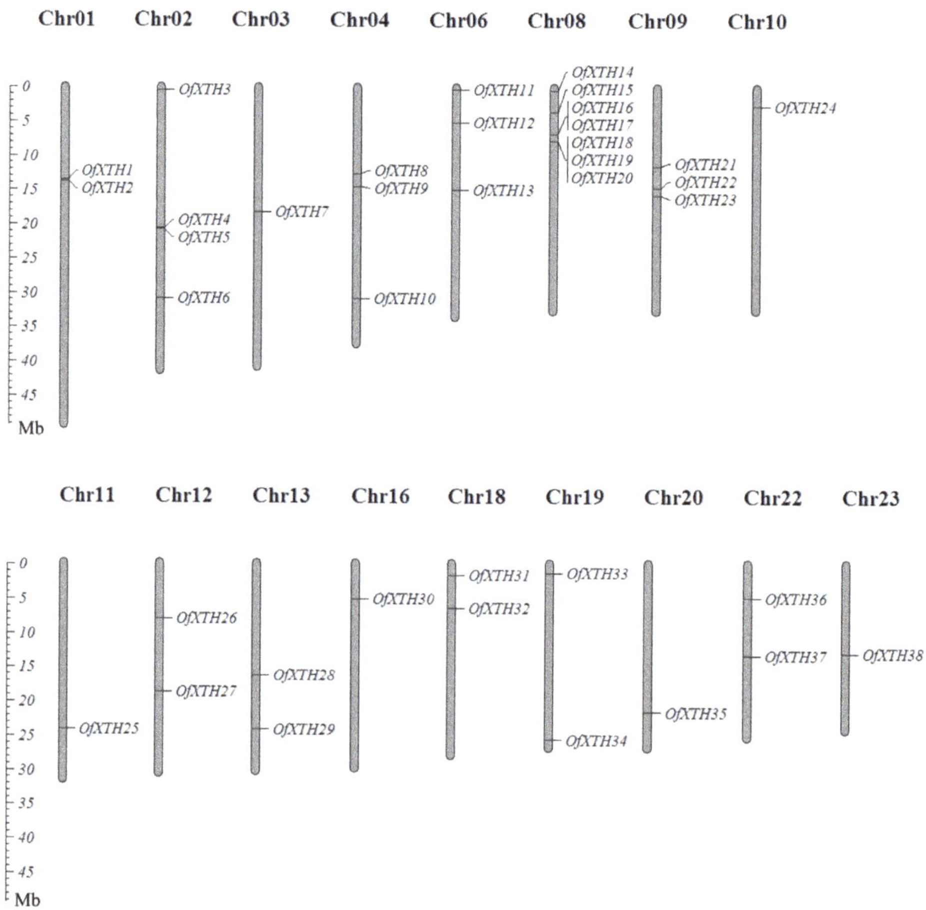

Figure 3. Chromosomal distribution of *OfXTH* genes.

Table 2. Ka/Ks analysis and estimated divergence time of *OfXTHs*.

Duplicated Gene Pairs	Ka	Ks	Ka/Ks	Type of Selection	Divergence Time (Mya)
OfXTH1-OfXTH24	0.118	0.987	0.119	Purifying	32.91
OfXTH1-OfXTH31	0.017	0.346	0.050	Purifying	11.55
OfXTH2-OfXTH31	0.023	0.355	0.065	Purifying	11.82
OfXTH4-OfXTH33	0.221	1.798	0.123	Purifying	59.93
OfXTH6-OfXTH35	0.083	0.394	0.212	Purifying	13.14

Table 2. *Cont.*

Duplicated Gene Pairs	Ka	Ks	Ka/Ks	Type of Selection	Divergence Time (Mya)
OfXTH7-OfXTH15	0.059	0.371	0.158	Purifying	12.36
OfXTH8-OfXTH12	0.121	0.666	0.181	Purifying	22.21
OfXTH8-OfXTH22	0.051	0.264	0.194	Purifying	8.82
OfXTH9-OfXTH23	0.262	0.568	0.461	Purifying	18.93
OfXTH10-OfXTH21	0.027	0.274	0.100	Purifying	9.13
OfXTH11-OfXTH13	0.003	0.011	0.267	Purifying	0.36
OfXTH11-OfXTH38	0.034	0.205	0.166	Purifying	6.84
OfXTH12-OfXTH22	0.130	0.806	0.161	Purifying	26.87
OfXTH13-OfXTH38	0.038	0.194	0.194	Purifying	6.48
OfXTH16-OfXTH19	0.000	0.009	0.000	Purifying	0.31
OfXTH18-OfXTH32	0.143	0.958	0.150	Purifying	31.92
OfXTH24-OfXTH31	0.123	1.094	0.112	Purifying	36.48
OfXTH25-OfXTH30	0.078	0.345	0.226	Purifying	11.49
OfXTH26-OfXTH37	0.171	0.526	0.326	Purifying	17.54
OfXTH27-OfXTH36	0.038	0.222	0.169	Purifying	7.40

Figure 4. Synteny analysis of 38 *OfXTH* genes in *O. fragrans*. Orange rectangles represent chromosomes of *O. fragrans*, and blue lines represent segment duplicate gene pairs.

2.5. Analysis of Cis-Acting Regulatory Elements in the Promoters of OfXTH Genes

For further understanding of the cis-acting regulatory elements in the promoter of *OfXTHs*, 2 kb upstream sequences of the transcription initiation site were subjected to PlantCARE. A large number of cis-acting elements in response to hormones were found in *OfXTH* promoter sequences (Figure 5), including ABA (abscisic acid), MeJA (methyl jasmonate),

GA (gibberellin), SA (salicylic acid), and auxin-responsive elements. Except for *OfXTH15* and *OfXTH32*, at least one ABA-responsive element was found in other *OfXTH* promoter sequences. Cis-acting regulatory elements involved in MeJA were presented in 29 *OfXTHs*. In addition, other cis-acting regulatory elements, including anaerobic induction, drought inducibility, low temperature, etc., were found in 33, 26, and 11 *OfXTH* promoters, respectively.

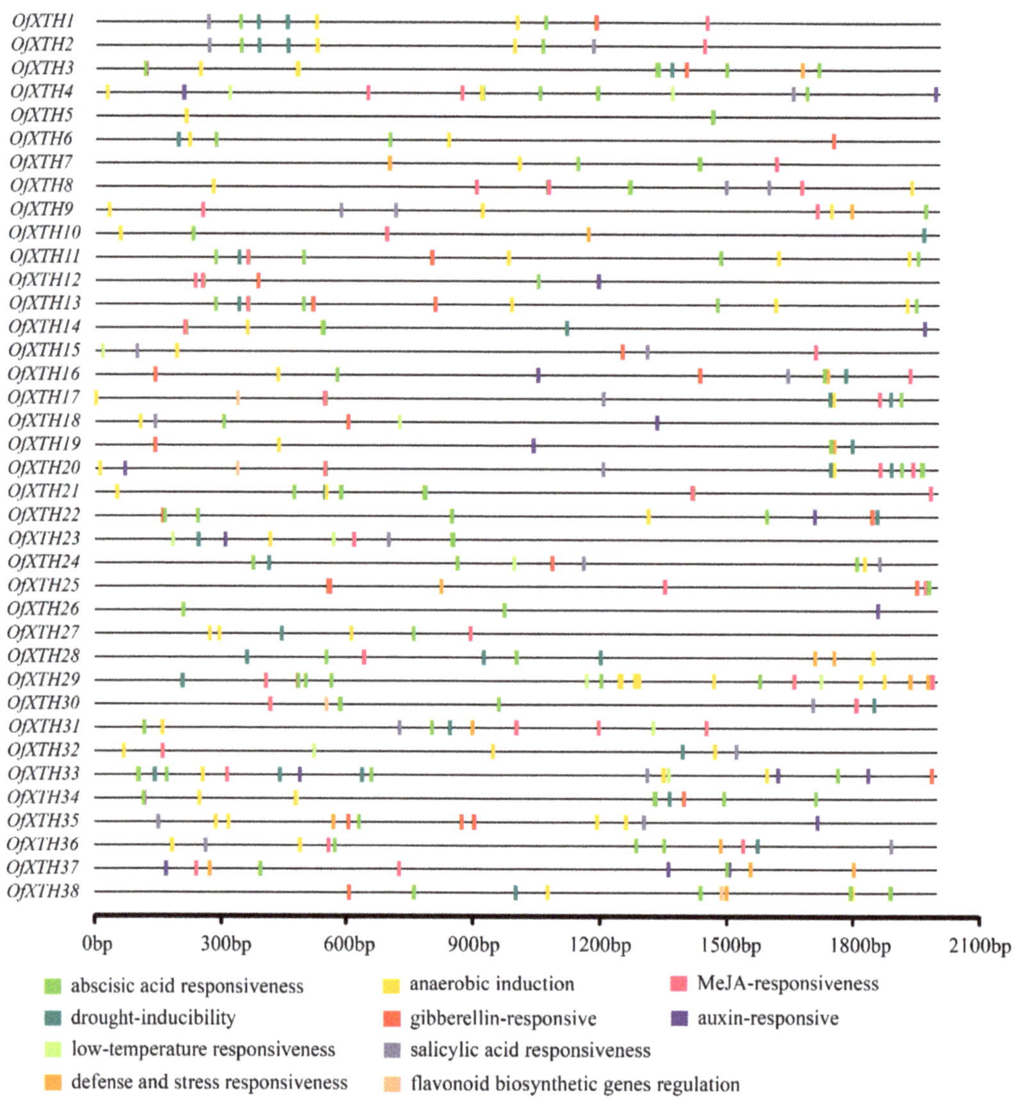

Figure 5. Prediction of cis-responsive elements in the *OfXTH* promoters. Different cis-responsive elements are represented by different colored boxes.

2.6. OfXTH Genes Expression Analysis during Flower-Opening Stages

To understand which *OfXTH* genes are involved in the regulation of the flowering processes, the quantitative real-time PCR was performed to detect *OfXTH* expression patterns during different flowering stages in *O. fragrans*. Flower bud phenotype at different stages are shown in Figure S1. According to the expression profiles, *OfXTHs* can be classified into three major groups (Figure 6). The first group, including four genes (*OfXTH8, 12, 26* and *37*), exhibited that these *OfXTH* genes were highly expressed in the S0 period and then decreased at other flowering stages (Figure 6A). Twenty-eight genes belong to the second group, including *OfXTH1–7, 9–15, 17, 21–24, 27–32, 34–36* and *38*. The expression pattern analysis showed that these *OfXTH* genes significantly upregulated in the S1 and S2 period, or only expressed in the S2 period (Figure 6B). Finally, six *OfXTH* genes were categorized into the third group, including *OfXTH16, 18–20, 25* and *33*. This group of genes have a high expression level in flowering periods S4 and S5 (Figure 6C).

2.7. Transcriptome Analysis of OfXTH Genes under Different Ambient Temperature Treatment

From our previous investigated results, ambient temperature of 19 °C significantly promoted the flower-opening process in *O. fragrans*, while ambient temperature of 23 °C inhibited the flowering process [15]. In our study, the flower buds were also treated under 19 °C and 23 °C conditions. In the 19 °C treatment for 2 and 5 days, the traits of flower buds were reached at S1 and S2 stage, respectively. However, the flower buds still remained in the S1 stage when treated at 23 °C for either 2 days or 5 days. Therefore, the flower buds were collected at the second and fifth day under 19 °C and 23 °C treatment for transcriptome sequencing, and then we explored which *OfXTHs* are involved in the regulation of flower-opening processes responding to ambient temperature changes. Based on log2(ratio of abundance) ≤ -1 or ≥ 1 with FDR < 0.001, a total of 5702 and 8082 differentially expressed genes (DEGs) were identified in 2 d, 19 °C vs. 2 d, 23 °C and 5 d, 19 °C vs. 5 d, 23 °C, respectively (Table S2). Among these obtained DEGs, five and sixteen differentially expressed *OfXTH* genes were identified at the second and fifth day under 19 °C treatment relative to the control 23 °C, respectively (Table 3). The differentially expressed *OfXTH* genes including *OfXTH24, 27, 32, 35,* and *36* were all induced by 19 °C at the second day. Compared with the control of 23 °C at the fifth day, fifteen upregulated *OfXTH* genes (*OfXTH4–7, 12, 14, 23–25, 27, 28, 31–33, 35* and *36*) were identified under 19 °C treatment. The expression of five *OfXTH* genes, including *OfXTH24, 27, 32, 35,* and *36,* were significantly increased at the second and fifth day. Then, qRT-PCR was employed to validate the differentially expressed *OfXTH* genes. Consistent with the transcriptome profile, the expression of *OfXTH24, 27, 32, 35* and *36* were strongly induced by 19 °C (Figure 7).

Table 3. DEGs of *OfXTHs* treated at 19 °C and 23 °C for 2 and 5 days.

Gene	2 d, 19 °C FPKM	2 d, 23 °C FPKM	5 d, 19 °C FPKM	5 d, 23 °C FPKM	Log2 (Ratio of Abundance)			
					2 d, 19 °C vs. 2 d, 23 °C	FDR	5 d, 19 °C vs. 5 d, 23 °C	FDR
OfXTH4	3.97	0.38	18.99	0.05	3.37	9.4×10^{-2}	8.69	1.6×10^{-15}
OfXTH5	1.36	2.00	15.91	1.92	−0.56	2.8×10^{-1}	3.05	5.9×10^{-21}
OfXTH6	0.02	0.05	2.75	0.10	−1.64	1.0×10^{0}	4.79	2.5×10^{-8}
OfXTH7	1.90	1.47	3.10	0.18	0.38	3.8×10^{-5}	4.09	4.9×10^{-8}
OfXTH12	1.75	3.00	0.64	4.07	−0.78	1.5×10^{-2}	−2.68	7.8×10^{-7}
OfXTH14	1.98	0.85	88.96	1.97	1.22	7.9×10^{-1}	5.50	3.6×10^{-16}
OfXTH23	5.13	7.06	9.33	3.72	−0.46	2.4×10^{-1}	1.33	7.6×10^{-6}
OfXTH24	14.69	6.25	38.07	5.38	1.23	3.8×10^{-9}	2.82	2.4×10^{-58}
OfXTH25	3.22	3.90	19.84	3.93	−0.28	9.6×10^{-2}	2.33	2.0×10^{-26}
OfXTH27	58.21	6.08	87.85	5.47	3.26	3.1×10^{-34}	4.01	3.1×10^{-109}
OfXTH28	0.39	0.43	104.94	0.54	−0.15	8.0×10^{-1}	7.60	4.0×10^{-47}
OfXTH31	0.29	0.30	1.53	0.13	−0.06	8.9×10^{-1}	3.55	1.7×10^{-4}
OfXTH32	12.50	3.46	10.53	0.66	1.85	2.1×10^{-30}	3.99	1.8×10^{-4}
OfXTH33	0.31	0.04	1.71	0.02	2.83	4.5×10^{-1}	6.21	1.8×10^{-5}
OfXTH35	1.97	0.94	3.23	0.46	1.06	4.7×10^{-4}	2.81	4.2×10^{-7}
OfXTH36	23.03	2.76	70.54	2.32	3.06	3.1×10^{-19}	4.93	2.3×10^{-134}

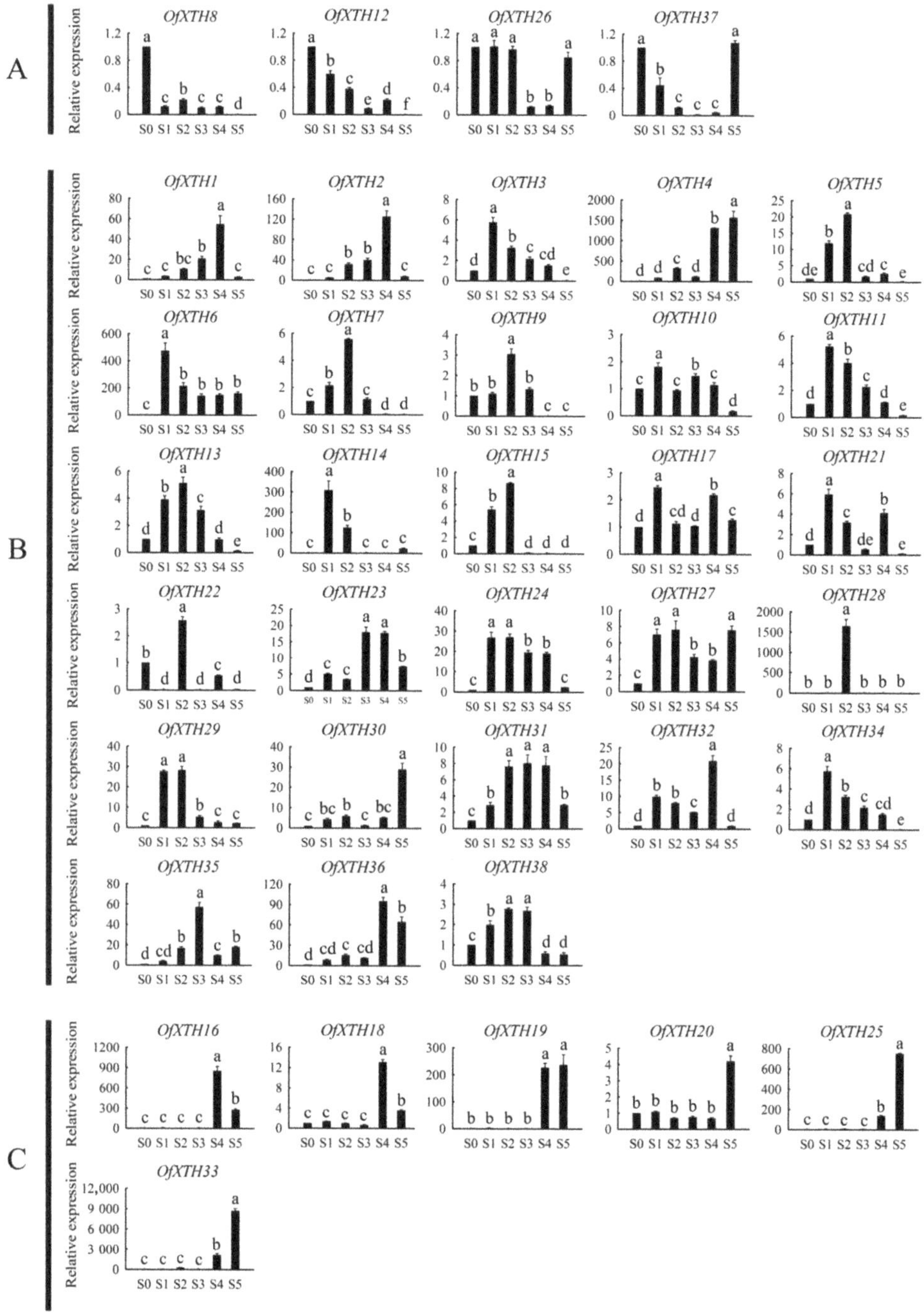

Figure 6. Expression profiles of *OfXTHs* at different flowering stages. Significant differences are identified by SPSS (version 22) with Duncan's test ($p < 0.05$) and are represented by different letters above the error bars. (**A**) *OfXTH* genes highly expressed in the S0 period. (**B**) *OfXTH* genes upregulated in the S1 and S2 period. (**C**) *OfXTH* genes upregulated in the S4 and S5 period.

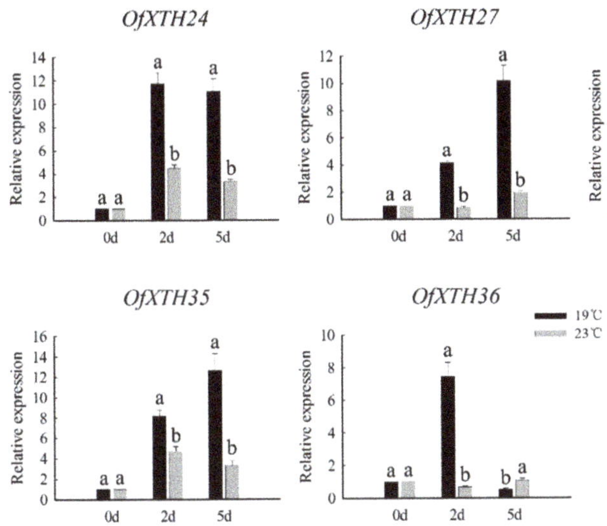

Figure 7. The qRT-PCR analysis of five differentially expressed *OfXTH* genes treated at 19 °C and 23 °C. Significant differences are identified by SPSS (version 22) with Duncan's test ($p < 0.05$) and are represented by different letters above the error bars.

3. Discussion

3.1. The Characteristics of XTH Gene Family in O. fragrans

In this study, a total of 38 *OfXTH* genes were identified, and the corresponding proteins were analyzed for their physical–chemical properties in *O. fragrans* genome (Table 1). All of the OfXTH proteins were predicted to be located in the cell wall, which is consistent with the function where XTH proteins are involved in cell-wall reconstruction [5,16]. A signal peptide was shown in most OfXTH proteins. When the signaling peptide of *DkXTH6* was absent, the subcellular localization of *DkXTH6* protein was transformed from the cell wall to the whole cell [17]. It implied that a signal peptide played an important role in the transmembrane transport of XTH protein and guided the protein to the plant cell wall. All of the *OfXTH* genes were categorized into four subfamilies by polygenetic analysis (Figure 1). The number of 26, 5, and 6 XTHs were contained in Group I/II, IIIA, and IIIB, respectively, whereas only one XTH was contained in Ancestry Group. Compared with *A. thaliana*, the numbers of XTH proteins in Group I/II, Group IIIA, and Group IIIB were increased, while Ancestral Group was decreased in *O. fragrans* (Figure 1). This might be due to the gene duplication or loss during the evolution of the *XTH* gene family [18]. Most *OfXTHs* are involved in multiple segmental duplications, which might contribute to the expansion and evolution of the OfXTH gene family [19]. It might be the reason why the number of the *XTH* family members in *O. fragrans* exceeds other species, such as *A. thaliana* (33) [20], *O. sativa* (29) [21], and *Hordeum vulgare* (24) [22]. The Ka/Ks values of all duplicated *OfXTH* gene pairs were <1.00 (Table 2), suggesting that the evolution of duplicated *OfXTH* gene pairs occurred through purifying selection and evolved slowly in *O. fragrans*.

Cis-acting elements are specific DNA sequences connected in tandem with structural genes and are binding sites for transcription factors [23]. Through analysis of *OfXTH* promoters, we found that ABA and MeJA responsiveness elements presented in most *OfXTH* promoters (Figure 5), which suggested that *OfXTH* genes might be regulated by abscisic acid and jasmonic acid, similar to the *XTH* genes in sweet cherry [24]. Recent research showed that the expression of *PavXTH15* was elevated by the phytohormone ABA and MeJA in sweet cherry [24]. Cis-acting elements involved in SA responsiveness were found in *OfXTH* promoters, indicating that the *OfXTH* gene might be the target gene of the SA signaling pathway. Similar findings reported previously showed that the accumulation of SA suppressed cell division and elongation through reducing *AtXTH8* and *AtXTH31* expression [25]. Besides responding to plant hormones, the promoters of *OfXTHs* also contain cis-acting regulatory elements involved in drought inducibility, low-temperature responsiveness, defense, and stress responsiveness. The result suggested that *OfXTH* may respond to environmental factors and be related to biotic and abiotic stress resistance. Similar results have been reported before. Overexpression of *DkXTH1* enhances tolerance to abiotic stress salt and drought stresses in transgenic *Arabidopsis* plants with respect to root and leaf growth, and survival [26]. The *XTH19* mutant showed reduced freezing tolerance after both cold and sub-zero acclimation [27].

3.2. OfXTHs Are Closely Associated with the Flower-Opening Period of O. fragrans

The flower-opening process is accompanied by the rapid expansion of petal cells. The size of the adaxial petal epidermal cells and the abaxial petal epidermal cells increased by 37.5% and 6.5% from S1 to S2, respectively [15]. Previous works have reported that XTHs participated in flower-opening processes in many species. *DcXTH2* and *DcXTH3* transcripts were markedly accumulated in *Dianthus caryophyllus* petals of opening flowers and showed high XET activity in petal claw, which is the main part of carnation petal elongation [28]. *LhXTH1* transcript levels in the petals markedly increased during *Lilium* flower opening and were higher in adaxial epidermal cells contributing to petal expansion [29]. In addition, it is reported that *XTH* plays an important role in the rapid petal growth period of *E. grandiflorum* [10]. In our study, we focused on the *OfXTH* expression changes in the S1 and S2 periods. A total of 28 *OfXTH* genes were found and significantly upregulated in S1 and S2 periods relative to the S0 period (Figure 6B), implying that these genes might be involved in petal cell expansion and contribute to the regulation of flower opening in *O. fragrans*. In addition, it has been reported that the expression of *RbXTH1* and *RbXTH2* in *Rosa bourboniana* leads to petal abscission [30]. Overexpression of persimmon *DkXTH8* causes cells to be easily destroyed and accelerates leaf senescence in transgenic plants [31]. Six *OfXTH* genes have high expression level in the flowering periods of S4 and S5 (Figure 6C); we hypothesized that these *OfXTH* genes may contribute to the cellular senescence.

3.3. OfXTHs Response to Ambient Temperature in Regulation of Flower Opening

Flowering in many species is influenced by ambient temperature. In *Arabidopsis*, the transition to flowering is delayed in low ambient temperatures [32]. Phased constant temperature treatment and a lower temperature could shorten the flower differentiation process and promote the flowering of *Crocus sativus* [33]. Similarly, the change in ambient temperature affected flower-opening processes significantly in *O. fragrans*. The flower opening is significantly promoted by low ambient temperature (19 °C) in *O. fragrans*, and the flower buds could not open until ambient temperature was below 23 °C [15]. However, there are fewer reports on the role of *XTHs* in responding to low ambient temperature; most studies focused on the high temperature regulating cell growth. For example, the expression of *AtXTH9* and *AtXTH11* was strongly increased by high temperature in stem and root, respectively, to accelerate cell elongation [8,34]. Through transcriptome analysis, we found that five and sixteen *OfXTH* genes were differentially expressed at the second and fifth day under 19 °C treatment compared with the control 23 °C, respectively (Table 3). Specifically, *OfXTH24*, *27*, *32*, *35*, and *36* were significantly upregulated at both the second and fifth day after 19 °C treatment, which indicated that *OfXTH24*, *27*, *32*, *35*, and *36* might respond to ambient temperature changes and contribute to the process of low ambient temperature accelerated flower opening of *O. fragrans*.

4. Conclusions

In this study, thirty-eight *OfXTH* genes were identified from the *O. fragrans* genome database. All of the OfXTH proteins were predicted to locate in the cell wall and were divided into four groups. Multiple segment duplication happened in the evolution of the *OfXTH* genes and contributed to the expansion of *OfXTH* gene family. The expression profiles showed that most *OfXTHs* are closely associated with the flower-opening processes in *O. fragrans*. Using transcriptome and qRT-PCR analysis, five *OfXTH* genes (*OfXTH24, 27, 32, 35* and *36*) were identified and significantly upregulated at the S1 and S2 stages under 19 °C treatment, indicating that these *OfXTH* genes might be key regulators responding to ambient temperature to modulate petal cell expansion during flower opening. In conclusion, the present research increased our knowledge of the role of the XTH gene family in flower-opening processes in *O. fragrans*.

5. Materials and Methods

5.1. Identification of XTH Gene Family in O. fragrans

The whole genome of *O. fragrans* was obtained from the website http://117.78.20.255/ [35] (accessed on 17 March 2022). The protein sequences of *AtXTH* (*A. thaliana*) and *OsXTH* (*O. sativa*) were downloaded from the Ensembl Plants database (http://plants.ensembl.org (accessed on 17 March 2022)). Two programs of TBtools1.6 [36], Simple HMM Search and BLASTp, were used to identify *OfXTH* family genes. To be specific, the Simple HMM Search program was used to search the two XTH protein domains, PF00722 (ID: Glyco_hydro_16) and PF06955 (ID: XET_C), from the database [37]. Thirty-three AtXTH proteins were used as queries to blast the *O. fragrans* genome database and identify XTH homologs in *O. fragrans*, with the parameters E-value $< 10^{-15}$ and identity > 50%. Only protein sequences present in both outcomes were regarded as OfXTH proteins and used for further analysis. SMART (http://smart.embl-heidelberg.de/ (accessed on 17 March 2022)) was used to check the conserved XTH domains Glyco_hydro_16 and XET_C with default parameters. The ProtParam tool of online software Expasy (http://web.expasy.org/protparam/ (accessed on 17 March 2022)) was used to determine the physicochemical parameters of OfXTHs protein sequence. SignalP 4.1 Service (http://www.cbs.dtu.dk/services/SignalP-4.1/ (accessed on 17 March 2022)) was used to predict the signal peptide. Plant-mPLoc (http://www.csbio.sjtu.edu.cn/bioinf/plant-multi/ (accessed on 17 March 2022)) was used to predict the sub-cellular localizations.

5.2. Phylogenetic Analysis of XTH Proteins

The Clutsalx-v1.83 program was used to align the XTH sequences from *O. fragrans*, *A. thaliana*, and *O. sativa*. MEGA6.0 was then used for phylogenetic analysis via the neighbor-joining (NJ) method with 1000 bootstrap replicates. The online software ITOL (https://itol.embl.de/ (accessed on 17 March 2022)) was used to annotate the phylogenetic tree.

5.3. Motif Identification and Gene Structure Analysis

MEME Suite 5.3.0 (http://meme-suite.org/tools/meme (accessed on 17 March 2022)) was used to identify the motifs of OfXTH proteins, at a maximum motif number of 10, with a minimum and maximum width of six and 50, respectively. The online software GSDS (http://gsds.gao-lab.org/ (accessed on 17 March 2022)) was used to predict the gene structure of *OfXTHs*.

5.4. Chromosomal Location and Synteny Analysis

MapChart (version 2.32) was used to predict the distribution of OfXTH genes on chromosomes. Quick Run MCScanX Wrapper program in TBtools1.6 was used to assess gene duplication events of *OfXTHs*. The synonymous (Ka) and non-synonymous (Ks) substitution rate were calculated by the Simple Ka/Ks Calculator program of TBtools1.6. The divergence was calculated as follows: $T = Ks/2\lambda$, where $\lambda = 1.5 \times 10^{-8}$ s for dicots [38]. The selection mode was identified using Ka/Ks value (Ka/Ks ratio >1, <1 and =1 represented positive selection, negative selection, and neutral selection, respectively) [38].

5.5. Analysis of Cis-Acting Regulatory Elements in Promoter

The upstream sequences (2 kb) of the OfXTH-coding sequences were obtained from the genome database of *O. fragrans* [35]. PlantCARE (http://bioinformatics.psb.ugent.be/webtools/plantcare/html/ (accessed on 17 March 2022)) was used to identify cis-acting regulatory elements in *OfXTH* genes promoter.

5.6. Plant Materials and Ambient Temperature Treatment

O. fragrans plants, cultivar "Yanhonggui", were potted and grown in the Zhejiang Agriculture and Forestry University Osmanthus Resource Nursery. The flower buds at different stages were collected between 9.00–10.00 a.m. every day, including S0 (the outer bud scales unfurled and the inner bud scales still furled); S1 (the bud became globular-shaped and the inside bracts covering the inflorescence was visible); S2 (the inflorescence burst through bracts and the florets closely crowded); S3 (the florets are bud-shaped and the pedicels elongate); S4 (initial flowering stage); S5 (full flowering stage). The flower bud phenotype of *O. fragrans* at different stages are shown in Figure S1. In addition, the plants in S0 stage with similar size were selected and treated with different temperatures (19 °C and 23 °C) in a growth chamber. When treated at 19 °C for 2 days and 5 days, the flower buds reached S1 and S2 stages, respectively, while flower buds still remained in S1 stage when treated at 23 °C for 2 days and 5 days. The phenotype of flower buds in 19 °C and 23 °C are shown in Figure S2. The relative humidity was about 70% and the photoperiod was a 12 h light/12 h dark regime with illumination of 10,000 lux. Each treatment included three independent biological replications. Samples of each stage were collected and stored at −80 °C before total RNA extraction.

5.7. Total RNA Extraction and Transcriptome Sequencing

RNA was isolated from 300 mg of flower buds at S1 and S2 period under 19 °C and 23 °C treatment using the EasyPure Plant RNA Kit (Tiangen, Beijing, China). The quality of total RNA satisfied the standards $1.8 \leq OD_{260/280} \leq 2.2$ and $OD_{260/230} \geq 1.8$. Five μg of total RNA was employed for cDNA libraries preparation and RNA deep sequencing (6 G × 250 bp) by Novogene Biological Information Technology Company (Beijing, China). Three biological replications were performed for the transcriptome analysis. The FPKM (fragments per kilo base of exon per million fragments mapped) was used to estimate gene expression levels and identify differentially expressed genes among 2 d, 19 °C vs. 2 d, 23 °C, and 5 d, 19 °C vs. 5 d, 23 °C. The differentially expressed genes were analyzed according to the log2(ratio of abundance) threshold, the value ≥1.0 was defined as upregulation, and ≤−1.0 was defined as downregulation, along with FDR <0.001.

5.8. Real-Time Quantitative PCR

PrimeScript RT Reagent Kit (Takara, Dalian, China) was used for first-strand cDNA synthesis following the manufacturer's instructions. LightCycler480II Real-Time PCR System was used for qRT-PCR analysis. The reaction mixture (10 μL total volume) consisted of 5 μL SYBR Premix Ex Taq (Takara, Dalian, China), 20 ng cDNA template, forward and reverse primers (10 microns/L) of 0.4 μL, was and filled with ddH_2O to 10 μL. Three biological replications were performed for each treatment. The relative expression of the target gene was calculated using the $2^{-\Delta\Delta CT}$ method. OfACT was used as a reference gene [39]. Primers of *OfXTH* genes used in RT-qPCR are shown in Table S3.

Supplementary Materials: The following supporting information can be downloaded at: https://www.mdpi.com/article/10.3390/plants11081015/s1, Figure S1: Flower bud phenotype of *O. fragrans* at different stages; Figure S2: Flower bud phenotype under temperature treatment; Table S1: Chromosome location of OfXTH genes; Table S2: DEGs of 2d-19 °C vs 2d-23 °C; Table S3: Primers of OfXTH genes used in RT-qPCR; Table S4: The FPKM value of OfXTH genes in temperature treatment transcriptome; Table S5: The RT-qPCR raw data of OfXTH genes at different flowering stages; Table S6: The RT-qPCR raw data of five differential expressed OfXTH genes treated at 19 °C and 23 °C.

Author Contributions: Conceptualization, S.Z.; methodology, Y.M.; software, Y.Y.; validation, Q.F. and Y.W.; investigation, Y.Y.; data curation, Y.Y.; writing—original draft preparation, Y.Y.; writing—review and editing, Y.M., S.Z. and B.D.; visualization, Y.Y.; supervision, B.D.; project administration, H.Z. All authors have read and agreed to the published version of the manuscript.

Funding: This research was funded by the National Natural Science Foundation of China (Grant No. 32072615 and 31902057), Zhejiang Provincial Natural Science Foundation of China (Grant No. LQ19C160012), and the key research and development program of Zhejiang Province (2019C02036).

Data Availability Statement: Data are contained within the article.

Conflicts of Interest: The authors declare no conflict of interest.

Abbreviations

XTH	Xyloglucan endotransglucosylase/hydrolase
XET	Xyloglucan endotransglucosylase
XEH	Xyloglucan hydrolase
MW	Molecular weight
pI	Theoretical isoelectric point
SP	Signal peptide
Mya	Million years ago
ABA	Abscisic acid
MeJA	Methyl jasmonate
GA	Gibberellin
SA	Salicylic acid
DEGs	Differentially expressed genes

References

1. Cosgrove, D.J. Growth of the plant cell wall. *Nat. Rev. Mol. Cell Biol.* **2005**, *6*, 850–861. [CrossRef] [PubMed]
2. Eklöf, J.M.; Brumer, H. The XTH Gene Family: An Update on Enzyme Structure, Function, and Phylogeny in Xyloglucan Remodeling. *Plant Physiol.* **2010**, *153*, 456–466. [CrossRef] [PubMed]
3. Behar, H.; Graham, S.W.; Brumer, H. Comprehensive cross-genome survey and phylogeny of glycoside hydrolase family 16 members reveals the evolutionary origin of EG16 and XTH proteins in plant lineages. *Plant J.* **2018**, *95*, 1114–1128. [CrossRef] [PubMed]
4. Rose, J.K.C.; Braam, J.; Fry, S.C.; Nishitani, K. The XTH Family of Enzymes Involved in Xyloglucan Endotransglucosylation and Endohydrolysis: Current Perspectives and a New Unifying Nomenclature. *Plant Cell Physiol.* **2002**, *43*, 1421–1435. [CrossRef] [PubMed]
5. Baumann, M.J.; Eklof, J.M.; Michel, G.; Kallas, A.M.; Teeri, T.T.; Czjzek, M.; Brumer, H., III. Structural Evidence for the Evolution of Xyloglucanase Activity from Xyloglucan Endo-Transglycosylases: Biological Implications for Cell Wall Metabolism. *Plant Cell* **2007**, *19*, 1947–1963. [CrossRef]
6. Miedes, E.; Zarra, I.; Hoson, T.; Herbers, K.; Sonnewald, U.; Lorences, E.P. Xyloglucan endotransglucosylase and cell wall extensibility. *J. Plant Physiol.* **2011**, *168*, 196–203. [CrossRef]
7. Liu, Y.-B.; Lu, S.-M.; Zhang, J.-F.; Liu, S.; Lu, Y.-T. A xyloglucan endotransglucosylase/hydrolase involves in growth of primary root and alters the deposition of cellulose in Arabidopsis. *Planta* **2007**, *226*, 1547–1560. [CrossRef]
8. Hyodo, H.; Yamakawa, S.; Takeda, Y.; Tsuduki, M.; Yokota, A.; Nishitani, K.; Kohchi, T. Active gene expression of a xyloglucan endotransglucosylase/hydrolase gene, XTH9, in inflorescence apices is related to cell elongation in Arabidopsis thaliana. *Plant Mol. Biol.* **2003**, *52*, 473–482. [CrossRef]
9. Shinozaki, Y.; Tanaka, R.; Ono, H.; Ogiwara, I.; Kanekatsu, M.; Van Doorn, W.G.; Yamada, T. Length of the dark period affects flower opening and the expression of circadian-clock associated genes as well as xyloglucan endotransglucosylase/hydrolase genes in petals of morning glory (Ipomoea nil). *Plant Cell Rep.* **2014**, *33*, 1121–1131. [CrossRef]
10. Ochiai, M.; Matsumoto, S.; Maesaka, M.; Yamada, K. Expression of mRNAs and Proteins Associated with Cell-wall-loosening during Eustoma Flower Opening. *J. Jpn. Soc. Hortic. Sci.* **2013**, *82*, 154–160. [CrossRef]
11. Yamada, K.; Takahashi, R.; Fujitani, C.; Mishima, K.; Yoshida, M.; Joyce, D.C.; Yamaki, S. Cell Wall Extensibility and Effect of Cell-Wall-Loosening Proteins during Rose Flower Opening. *J. Jpn. Soc. Hortic. Sci.* **2009**, *78*, 242–251. [CrossRef]
12. Luo, J.; Ma, N.; Pei, H.; Chen, J.; Li, J.; Gao, J. A DELLA gene, RhGAI1, is a direct target of EIN3 and mediates ethylene-regulated rose petal cell expansion via repressing the expression of RhCesA2. *J. Exp. Bot.* **2013**, *64*, 5075–5084. [CrossRef] [PubMed]

13. Van Doorn, W.G.; van Meeteren, U. Flower opening and closure: A review. *J. Exp. Bot.* **2003**, *54*, 1801–1812. [CrossRef] [PubMed]
14. Wang, Y.; Zhang, C.; Fu, J.; Zhao, H. Progresses on flower bud differentiation and flower opening in *Osmanthus fragrans*. *J. Zhejiang A F Univ.* **2016**, *33*, 340–347.
15. Fu, J.; Zhang, C.; Liu, Y.; Pang, T.; Dong, B.; Gao, X.; Zhu, Y.; Zhao, H. Transcriptomic analysis of flower opening response to relatively low temperatures in Osmanthus fragrans. *BMC Plant Biol.* **2020**, *20*, 337. [CrossRef]
16. Strohmeier, M.; Hrmova, M.; Fischer, M.; Harvey, A.J.; Fincher, G.B.; Pleiss, J. Molecular modeling of family GH16 glycoside hydrolases: Potential roles for xyloglucan transglucosylases/hydrolases in cell wall modification in the poaceae. *Protein Sci.* **2004**, *13*, 3200–3213. [CrossRef]
17. Han, Y.; Ban, Q.; Hou, Y.; Meng, K.; Suo, J.; Rao, J. Isolation and Characterization of Two Persimmon Xyloglucan Endotransglycosylase/Hydrolase (XTH) Genes That Have Divergent Functions in Cell Wall Modification and Fruit Postharvest Softening. *Front. Plant Sci.* **2016**, *7*, 624. [CrossRef]
18. Li, Y.; Li, L.; Ding, W.; Li, H.; Shi, T.; Yang, X.; Wang, L.; Yue, Y. Genome-wide identification of Osmanthus fragrans bHLH transcription factors and their expression analysis in response to abiotic stress. *Environ. Exp. Bot.* **2020**, *172*, 103990. [CrossRef]
19. Cui, L.; Wall, P.K.; Leebens-Mack, J.H.; Lindsay, B.G.; Soltis, D.E.; Doyle, J.J.; Soltis, P.S.; Carlson, J.E.; Arumuganathan, K.; Barakat, A.; et al. Widespread genome duplications throughout the history of flowering plants. *Genome Res.* **2006**, *16*, 738–749. [CrossRef]
20. Yokoyama, R.; Nishitani, K. A Comprehensive Expression Analysis of all Members of a Gene Family Encoding Cell-Wall Enzymes Allowed us to Predict cis-Regulatory Regions Involved in Cell-Wall Construction in Specific Organs of Arabidopsis. *Plant Cell Physiol.* **2001**, *42*, 1025–1033. [CrossRef]
21. Yokoyama, R.; Nishitani, K. Genomic Basis for Cell-Wall Diversity in Plants. A Comparative Approach to Gene Families in Rice and Arabidopsis. *Plant Cell Physiol.* **2004**, *45*, 1111–1121. [CrossRef] [PubMed]
22. Fu, M.-M.; Liu, C.; Wu, F. Genome-Wide Identification, Characterization and Expression Analysis of Xyloglucan Endotransglucosylase/Hydrolase Genes Family in Barley (*Hordeum vulgare*). *Molecules* **2019**, *24*, 1935. [CrossRef] [PubMed]
23. Wittkopp, P.J.; Kalay, G. Cis-regulatory elements: Molecular mechanisms and evolutionary processes underlying divergence. *Nat. Rev. Genet.* **2012**, *13*, 59–69. [CrossRef] [PubMed]
24. Zhai, Z.; Feng, C.; Wang, Y.; Sun, Y.; Peng, X.; Xiao, Y.; Zhang, X.; Zhou, X.; Jiao, J.; Wang, W.; et al. Genome-Wide Identification of the Xyloglucan endotransglucosylase/Hydrolase (XTH) and Polygalacturonase (PG) Genes and Characterization of Their Role in Fruit Softening of Sweet Cherry. *Int. J. Mol. Sci.* **2021**, *22*, 12331. [CrossRef]
25. Miura, K.; Lee, J.; Miura, T.; Hasegawa, P.M. *SIZ1* Controls Cell Growth and Plant Development in Arabidopsis Through Salicylic Acid. *Plant Cell Physiol.* **2009**, *51*, 103–113. [CrossRef]
26. Han, Y.; Han, S.; Ban, Q.; He, Y.; Jin, M.; Rao, J. Overexpression of persimmon *DkXTH1* enhanced tolerance to abiotic stress and delayed fruit softening in transgenic plants. *Plant Cell Rep.* **2017**, *36*, 583–596. [CrossRef]
27. Takahashi, D.; Johnson, K.L.; Hao, P.; Tuong, T.; Erban, A.; Sampathkumar, A.; Bacic, A.; Livingston, D.P.; Kopka, J.; Kuroha, T.; et al. Cell wall modification by the xyloglucan endotransglucosylase/hydrolase XTH19 influences freezing tolerance after cold and sub-zero acclimation. *Plant Cell Environ.* **2020**, *44*, 915–930. [CrossRef]
28. Harada, T.; Torii, Y.; Morita, S.; Onodera, R.; Hara, Y.; Yokoyama, R.; Nishitani, K.; Satoh, S. Cloning, characterization, and expression of xyloglucan endotransglucosylase/hydrolase and expansin genes associated with petal growth and development during carnation flower opening. *J. Exp. Bot.* **2010**, *62*, 815–823. [CrossRef]
29. Watanabe, Y.; Niki, T.; Norikoshi, R.; Nakano, M.; Ichimura, K. Soluble carbohydrate concentration and expression of expansin and xyloglucan endotransglucosylase/hydrolase genes in epidermal and parenchyma cells during lily flower opening. *J. Plant Physiol.* **2022**, *270*, 153615. [CrossRef]
30. Singh, A.P.; Tripathi, S.K.; Nath, P.; Sane, A.P. Petal abscission in rose is associated with the differential expression of two ethylene-responsive xyloglucan endotransglucosylase/hydrolase genes, *RbXTH1* and *RbXTH2*. *J. Exp. Bot.* **2011**, *62*, 5091–5103. [CrossRef]
31. Han, Y.; Ban, Q.; Li, H.; Hou, Y.; Jin, M.; Han, S.; Rao, J. DkXTH8, a novel xyloglucan endotransglucosylase/hydrolase in persimmon, alters cell wall structure and promotes leaf senescence and fruit postharvest softening. *Sci. Rep.* **2016**, *6*, 39155. [CrossRef] [PubMed]
32. Samach, A.; Wigge, P.A. Ambient temperature perception in plants. *Curr. Opin. Plant Biol.* **2005**, *8*, 483–486. [CrossRef] [PubMed]
33. Wang, Z.; Li, X.; Xu, J.; Yang, Z.; Zhang, Y. Effects of ambient temperature on flower initiation and flowering in saffron (*Crocus sativus* L.). *Sci. Hortic.* **2021**, *279*, 109859. [CrossRef]
34. De Caroli, M.; Manno, E.; Piro, G.; Lenucci, M.S. Ride to cell wall: Arabidopsis *XTH11*, *XTH29* and *XTH33* exhibit different secretion pathways and responses to heat and drought stress. *Plant J.* **2021**, *107*, 448–466. [CrossRef] [PubMed]
35. Yang, X.; Yue, Y.; Li, H.; Ding, W.; Chen, G.; Shi, T.; Chen, J.; Park, M.S.; Chen, F.; Wang, L. The chromosome-level quality genome provides insights into the evolution of the biosynthesis genes for aroma compounds of Osmanthus fragrans. *Hortic. Res.* **2018**, *5*, 72. [CrossRef]
36. Chen, C.; Chen, H.; Zhang, Y.; Thomas, H.R.; Frank, M.H.; He, Y.; Xia, R. TBtools: An Integrative Toolkit Developed for Interactive Analyses of Big Biological Data. *Mol. Plant* **2020**, *13*, 1194–1202. [CrossRef] [PubMed]
37. Wang, M.; Xu, Z.; Ding, A.; Kong, Y. Genome-Wide Identification and Expression Profiling Analysis of the Xyloglucan Endotransglucosylase/Hydrolase Gene Family in Tobacco (Nicotiana tabacum L.). *Genes* **2018**, *9*, 273. [CrossRef]

38. Blanc, G.; Wolfe, K.H. Widespread Paleopolyploidy in Model Plant Species Inferred from Age Distributions of Duplicate Genes. *Plant Cell* **2004**, *16*, 1667–1678. [CrossRef]
39. Zhang, C.; Fu, J.; Wang, Y.; Bao, Z.; Zhao, H. Identification of Suitable Reference Genes for Gene Expression Normalization in the Quantitative Real-Time PCR Analysis of Sweet Osmanthus (*Osmanthus fragrans* Lour.). *PLoS ONE* **2015**, *10*, e0136355. [CrossRef]

plants

Article

Genome-Wide Identification and Expression Analysis of Dof Transcription Factors in Lotus (*Nelumbo nucifera* Gaertn.)

Xiaohan Cao [1,†], Wenyang Wan [1,†], Huimin Mao [1], Dandan Yin [1], Xianhui Deng [1], Huan Yan [1] and Liping Ren [1,2,*]

[1] Key Laboratory of Horticultural Plant Biology, Biological and Food Engineering School, Fuyang Normal University, Fuyang 236037, China
[2] Fuyang Academy of Agricultural Sciences, Fuyang 236065, China
* Correspondence: renliping@fynu.edu.cn; Tel.: +86-130855827250
† These authors contributed equally to this work.

Abstract: Lotus (*Nelumbo nucifera* Gaertn.) is a traditional Chinese aquatic flower with high ornamental and economic value, but water salinity seriously affects lotus cultivation and distribution. The Dof transcription factors (TFs) play a crucial function in the regulatory network of growth and defense in plants. However, no systematic investigations of the Dof TFs in lotus have been performed. In this study, comprehensive searches of the lotus genome yielded 29 potential *NnDofs*. We carried out a series of standardized analyses, which include physical properties, multiple sequence alignment, phylogenetic analysis, gene structure, motif composition, cis-acting element prediction, chromosome distribution, and synteny analysis. The results showed that segment duplication probably caused the *NnDofs* gene family expansion. The potential functions of *NnDofs* in lotus development and stress conditions are speculated by promoter analysis. Furthermore, a complete expression investigation of *NnDofs* utilizing an RNA-seq atlas and quantitative real-time polymerase chain reaction (qRT-PCR) was performed. The majority of the *NnDofs* exhibit tissue-specific expression patterns, and many genes have been identified as being extremely sensitive to salt stressors. Overall, this study is the first to report a genome-wide assessment of the Dof family in lotus, and the findings offer vital insights for prospective functional studies on lotus salinity stress.

Keywords: *Nelumbo nucifera*; Dof transcription factor; salinity stress

Citation: Cao, X.; Wan, W.; Mao, H.; Yin, D.; Deng, X.; Yan, H.; Ren, L. Genome-Wide Identification and Expression Analysis of Dof Transcription Factors in Lotus (*Nelumbo nucifera* Gaertn.). *Plants* **2022**, *11*, 2057. https://doi.org/10.3390/plants11152057

Academic Editors: Aiping Song and Yu Chen

Received: 14 June 2022
Accepted: 3 August 2022
Published: 6 August 2022

1. Introduction

Soil salinization is a serious and growing global problem [1], and salinized areas are growing at a rate of 10% per year. Soil salinity is a severe abiotic stress that affects plant seed germination, growth and development, and reproductive development by causing oxidative stress, ionic toxicity, osmotic stress, and metabolic disturbances in plants [2,3]. Several investigations on several plant species have highlighted the complicated and crucial role of transcription factors in abiotic stress reduction [4]. The dynamic coordination of salt stress-responsive transcription factors in the interaction pathway, as well as their unique integration into the stress adaption cellular network, will serve as a stepping stone for plant tolerance to environmental stresses [5].

Transcription factors (TFs), also called trans-acting factors, interact with cis-acting elements in a particular genetic promoter region to regulate gene transcription and ensure target gene expression at a specific time, place, and intensity [6,7]. Typical transcription factors contain functional regions such as the DNA binding domain, nuclear localization signal region, oligomerization site, and transcription activation domain [8,9]. Transcription factors widely regulate plant growth and development and deeply engage in biotic and abiotic stress responses.

Studies have shown that Dof (DNA binding with one finger) proteins appear to be unique to plants. The first Dof transcription factor identified was *ZmDof* from maize [10].

With the development of bioinformatics technology, more Dof proteins have been found in the genomes of different plant species, such as *Solanum lycopersicum* [11], *Manihot esculenta* [12], *Prunus persica* [13], and *Capsicum annuum* L. [14]. Dof TFs usually consist of 200–400 amino acids with highly conserved DNA-binding domains at their N-termini, transcriptional regulatory domains at their C-termini, and nuclear localization signals [10,15]. The highly conserved domain consisting of 50–52 amino acids at the N-terminus, containing a C2-C2 zinc finger domain is composed of CX2CX21CX2C. Different transcriptional regulatory domains at the C-terminus indicate the diversity of Dof protein functions [16]. The *Dof* family was divided into seven sub-populations by Yanagisawa [17]. Several researchers collected Arabidopsis and rice *Dof* genes and classified them into four subfamilies: Aa, Bb, Cc, and Dd [18]. The 116 *Dof* genes from seven species were more comprehensively classified by Moreno-Risueno into seven subgroups, A–G [19]. *Dof* proteins play multiple roles in different biological processes, including growth and development [20,21], flowering regulation [22,23], carbon and nitrogen metabolism [24], hormone response [25], and abiotic stress [26,27] in various plant species. Overexpression of *OBP4* in Arabidopsis promotes cell proliferation in the differentiation zone and induces callus formation [28]. *ZmDof3* controls starch accumulation and aleurone development in maize endosperm by binding to the *Dof* core element promoters of *Du1* and *Su2* [29]. In addition, *ZmDof36* is important in regulating starch synthesis. Its overexpression can increase starch content and reduce soluble sugars and reduce sugars [30]. Cycling Dof factor 2 (*CDF2*) leads to photoperiod-insensitivity and delayed flowering in Arabidopsis by reducing CO mRNA levels [31].

RNA-Seq data showed that most *TaDof* genes respond to heat and PEG-induced drought stress in wheat [26]. Overexpression of *GhDof1* could notably enhance tolerance to salt and cold stresses by increasing proline content during the seedling stage [32]. Some *ClDof* genes showed significantly different expressions under salt stress, suggesting that they may contribute to salt stress adaptation in watermelon [33]. *OsDof15* coordinates the regulation of salt and ethylene, inhibiting primary root growth by affecting cell proliferation in the root apical meristem [34]. *SlDof22* can be combined with the promoter of the *SlSOS1* gene, and inhibiting *SlDof22* by significantly downregulating the *SlSOS1* gene leads to reduced tolerance to salt stress [35].

Sacred lotus (*Nelumbo nucifera* Gaertn.), which has been grown in the Far East for 5000–7000 years, is a large aquatic plant with significant ecological, scenic, and economic value [36,37]. Lotus cultivars are categorized depending on their usage and morphological characteristics: rhizome lotus, seed lotus, and ornamental lotus [38]. Lotus has high ornamental value throughout the growing period, with large flowers, various petal types, gorgeous color, green leaves, tall and straight habit, and remains attractive even during the dry leaf period. Because of its ornamental value, lotus is considered a theme plant in waterscape garden layouts. In addition, lotus has high economic value and medicinal value [39]. Lotus tea is traditionally used to clear away "summer heat", that is, relieve symptoms of heat injury, and lotus seed is rich in phospholipids, alkaloids, and flavonoids, which in Chinese traditional medicine are used to clear the heart, nourish the mind, and tonify the spleen and kidney [40].

Although salinity stress causes certain harm to the growth and development of lotus, there are few reports on the salt tolerance of ornamental lotus. Furthermore, Dof TFs have been found to be resistant to salt stress in many different plant species, but the *Dof* gene has not been identified in lotus. In this study, we identified and characterized 29 *Dof* family genes in lotus. They were unevenly distributed on the seven chromosomes and divided into six groups based on phylogenetic analysis. Its physicochemical properties, gene structure, conserved motifs, and cis-acting elements upstream of the gene were also analyzed. Tissue-specific expression of *NnDofs* and gene response to salt treatment were investigated using RNA-seq data and qRT-PCR. We predicted possible interacting proteins and regulatory networks of *NnDofs* related to these stress responses. The results provide a reference for further functional research of lotus *Dof TFs*, and they can be used as genetic resources to make lotus and other crops more tolerant of salt through molecular genetic breeding.

2. Results

2.1. Identification and Physiochemical Characteristics of NnDofs

Through the search of the lotus genome by HMM with the Dof domain (PF02701), 29 *NnDof* family members were recognized after further validation in the CDD and Pfam databases. According to the distribution of *Dof* genes on chromosomes, *NnDofs* are named *NnDof1* to *NnDof29*. We investigated the physical and chemical properties of the *NnDof* members. Analysis of NnDof proteins showed amino acid lengths ranging from 105 to 508 aa, molecular weights ranging from 12.1 to 55.4 kDa, and isoelectric points ranging from 5.67 (NnDof13) to 9.96 (NnDof29). The analysis results of the instability index showed that the *NnDof5* value was lower than 40, which indicated that the rest of the *NnDofs* were probably unstable proteins. The aliphatic indexes of NnDof proteins were all between 41.81 and 73.51. GRAVY values for all NnDof proteins ranged from −1.161 to −0.341, indicating that all NnDof members are hydrophilic proteins. The subcellular localization prediction findings indicated 17 NnDof proteins in the nucleus and the remainder are extracellular. Detailed results are listed in Table S1.

2.2. Phylogenetic Analysis and Classification of NnDofs

To fully comprehend the evolutionary relationship, a phylogenetic tree was constructed with the 29 identified *NnDofs* and the 36 reported *AtDofs* of Arabidopsis (Figure 1). The results indicated that 65 Dof proteins were grouped into seven major groups, with 29 NnDof proteins in each of the following categories: II, III, IV, V, VI, and VII. Group VI had nine NnDof proteins, accounting for 31.03 percent of the total number of *NnDofs*, while group V had only two, accounting for 6.90 percent. Groups II, III, VI, and VII each have five NnDof proteins.

Multiple sequence alignment analysis of the amino acids of the Dof structural domain of 29 lotus flowers was performed to investigate the sequence characteristics of the NnDof proteins. It was found that the NnDof proteins' structural domain sequences were highly conserved and all contained the typical CX2CX21CX2C motif, where one Zn^{2+} can covalently combine with four Cys residues (Figure 2).

2.3. Gene Structure and Conserved Motifs of NnDofs

We investigated the conserved motifs of the genes using MEME to evaluate the diversity and conservation among all 29 *NnDofs* genes. Ten different motifs were predicted and they ranged in size from 16 to 134 amino acids (Figure 3B). The conserved structural domain of Dof consists of Motif 1, and Motif 1 corresponds to the CX2CX21CX2C single zinc finger structure in the Dof structural domain, which is a highly homologous core region of the *Dof* family. NnDofs proteins associated on evolutionary branches of the phylogenetic tree have identical or comparable motif structures. Some motifs, such as Motif 2, Motif 3, Motif 5, and Motif 8, are only found in certain subgroups and may be connected to different functions.

Introns and exons constitute genes, and their numbers and distribution patterns serve as an evolutionary indicator for a gene family. Thus, we completed a comparison of the intron-exon structures of each *NnDofs* (Figure 3C). In general, the number of introns per *NnDof* gene varied very little, ranging from 0 to 1. A total of 11 *NnDof* gene members had no introns (37.9%) and 18 members had only one intron (62.1%). The exon-intron structure pattern of the *NnDofs* gene is similar to that of Arabidopsis, rice, and tomato. Most introns are usually located upstream of the *Dof* structural domain, and only seven genes (*NnDof2*, *NnDof3*, *NnDof8*, *NnDof16*, *NnDof20*, *NnDof26*, and *NnDof28*) have introns located downstream of the *Dof* structural domain. Most of the genes in the same groups showed similar exon and intron patterns. For example, most *Dof* genes in group III do not contain introns. In contrast, groups II, V, VI, and VII almost all contain one intron.

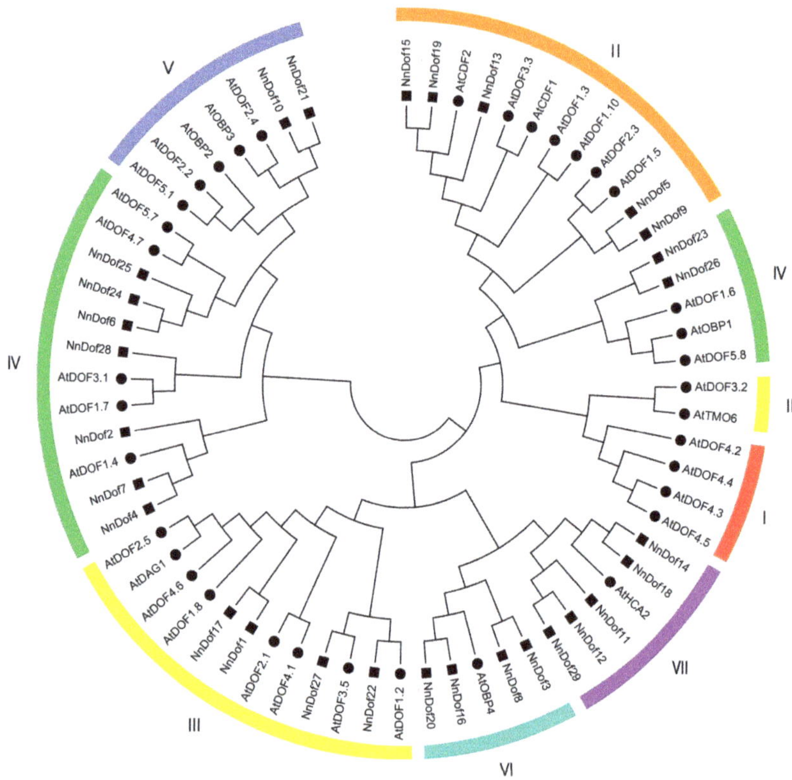

Figure 1. Phylogenetic tree of Dof transcription factors in *N. nucifera* and *A. thaliana*. The phylogenetic tree was created using MEGA 6.0, with the neighbor-joining method and bootstrap value set at 1000. Black circles represent the 36 AtDof proteins, and the 29 NnDof proteins are marked with black squares. The resulting phylogenetic tree was clustered into seven main groups, labeled I, II, III, IV, V, VI, and VII.

2.4. Promoter Analysis of NnDof Genes

The prediction of cis-regulatory elements in *NnDof* gene promoters can help us learn more about how gene family members regulate transcription (Figure 4). All 29 *NnDofs* have TATA-box and CAAT-box. It was discovered that light-responsive elements (ATTAAT, CACGTG, and GGTTAA) and meristem expression-related elements (GCCACT) are extensively distributed in the promoter area of the *NnDofs*. There were various elements implicated in abiotic stress responsiveness, including low-temperature responsive elements (CCGAAA), salicylic acid responsive elements (CCATCTTTTT and TCAGAAGAGG), anaerobic induction response elements (AAACCA), and other defense and stress responsive elements (ATTCTCTAAC). Abscisic acid-responsive elements (ABREs) such as (ACGTG, CACGTG, TACGGTC, and TACGTG), MeJA-responsive motifs such as (TGACG and CGTCA), gibberellin-responsive motifs such as (TCTGTTG and CCTTTTG), and auxin-responsive elements such as (GGTCCAT and AACGAC) were widely found. These results reveal that *NnDofs* are not only regulated by stress responses, but may also be involved in plant growth and development.

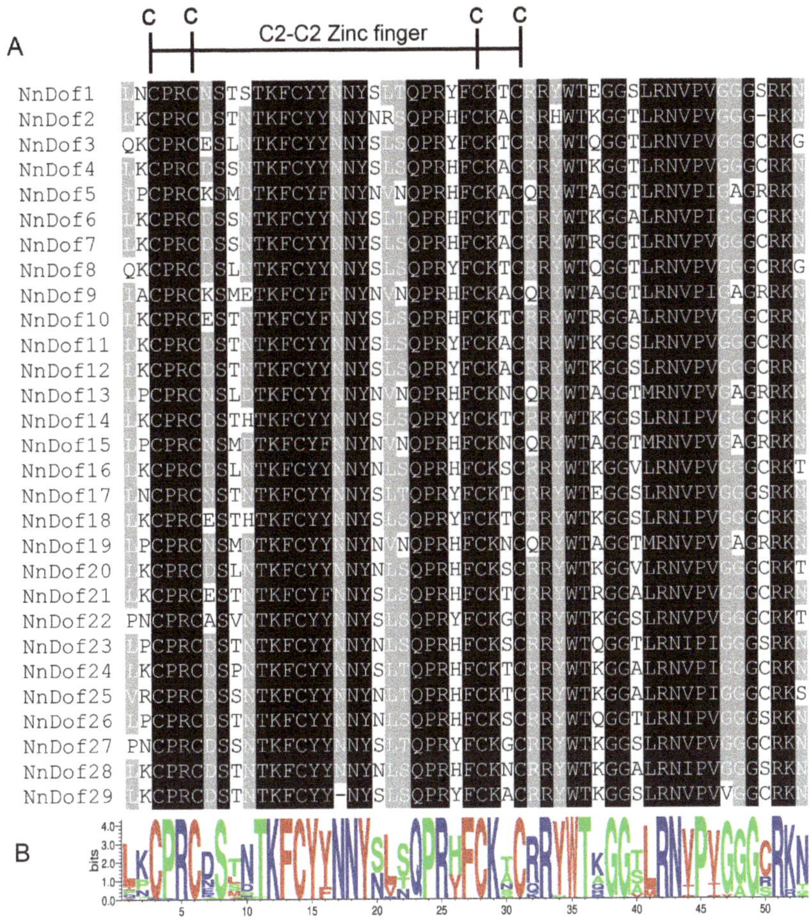

Figure 2. Dof domain sequence alignment of NnDof proteins. (**A**) Extraction of conserved domains of Dof proteins (black region—amino acid homology 100%; grey area—amino acid homology 75%); (**B**) the conserved domains shown by WebLogo.

2.5. Promoter Analysis of NnDof Genes

The prediction of cis-regulatory elements in *NnDof* gene promoters can help us learn more about how gene family members regulate transcription (Figure 4). All 29 *NnDofs* have TATA-box and CAAT-box. It was discovered that light-responsive elements (ATTAAT, CACGTG, and GGTTAA) and meristem expression-related elements (GCCACT) are extensively distributed in the promoter area of the *NnDofs*. There were various elements implicated in abiotic stress responsiveness, including low-temperature responsive elements (CCGAAA), salicylic acid responsive elements (CCATCTTTTT and TCAGAAGAGG), anaerobic induction response elements (AAACCA), and other defense and stress responsive elements (ATTCTCTAAC). Abscisic acid-responsive elements (ABREs) such as (ACGTG, CACGTG, TACGGTC, and TACGTG), MeJA-responsive motifs such as (TGACG and CGTCA), gibberellin-responsive motifs such as (TCTGTTG and CCTTTTG), and auxin-responsive elements such as (GGTCCAT and AACGAC) were widely found. These results reveal that *NnDofs* are not only regulated by stress responses, but may also be involved in plant growth and development.

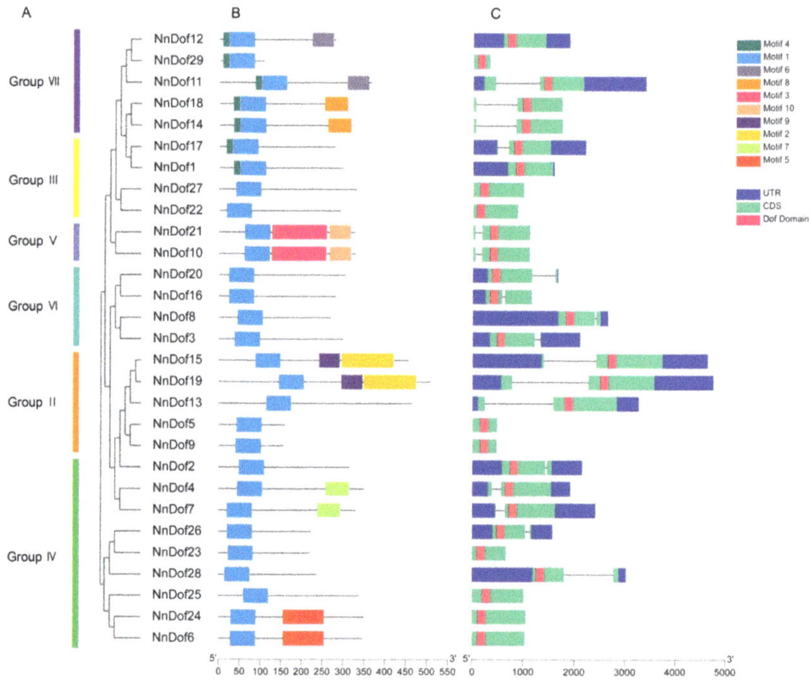

Figure 3. The phylogenetic relationships, exon–intron structures, and motif compositions of lotus Dof proteins. (**A**) The phylogenetic tree was constructed based on *NnDofs*, constructed using MEGA7 with a bootstrap of 1000 by the neighbor-joining method. (**B**) Untranslated regions (UTRs), exons, introns, and Dof domain are represented by blue, green, black lines, and pink, respectively. (**C**) A total of 10 motifs are denoted by different colored boxes.

2.6. Chromosome Distribution, Gene Duplication and Synteny Analysis of NnDofs

We investigated the evolution and expansion mechanisms of *NnDofs* genes by assessing their chromosomal location and gene duplication events. The number distribution of *NnDofs* on chromosomes was heterogeneous, with 29 *NnDofs* located on seven of the eight chromosomes and nine *NnDofs* found on Chr 1, but no *NnDof* genes located on chr 7 (Figure 6). *NnDofs* are localized in the front end of chr3 and chr4 but are separated in chr1, chr2, and chr5. Both tandem and segmental duplications contribute to gene family expansion. We evaluated duplication events in *NnDofs* and highlighted 10 segmental duplications of *NnDofs* gene pairs with red lines. Based on the research published by Holub, two pairs of *NnDofs*, NnDof3-NnDof4 and NnDof25-NnDof26, were identified as tandem duplications.

We also conducted a synteny analysis of *Dof* genes across lotus, tomato, Arabidopsis, and rice (Figure 7). There are 26 orthologous pairs between lotus and Arabidopsis, and 53 orthologous pairs between lotus and tomato. Six *NnDofs* in lotus (*NnDof11*, *NnDof12*, *NnDof15*, *NnDof17*, *NnDof19*, and *NnDof23*) have two pairs of homologous genes in Arabidopsis, and two *NnDofs* (*NnDof1* and *NnDof10*) have three pairs of homologous genes in Arabidopsis thaliana. Analysis of the number of collinear homologous genes between lotus and tomato revealed that four *NnDofs* (*NnDof12*, *NnDof14*, *NnDof18*, and *NnDof28*) have two homologous genes in tomato, three *NnDofs* (*NnDof1* and *NnDof17*) have four homologous genes in tomato genes, and three *NnDofs* (*NnDof10*, *NnDof15*, and *NnDof19*) have five homologous genes in tomato. A total of 11 *SlDofs* genes showed syntenic relationships with those in rice.

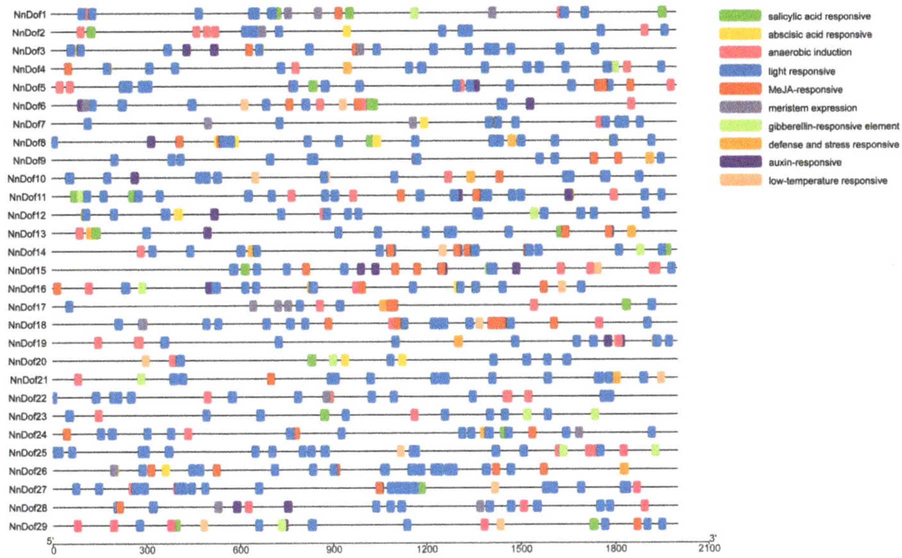

Figure 4. The prediction of a cis-acting element in the 2000 bp promoter upstream of the *NnDofs*. The main cis-acting elements are marked in the upper right corner.

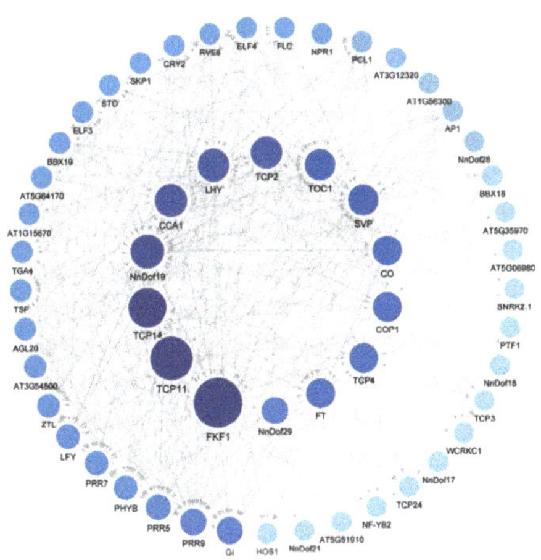

Figure 5. Protein interaction network of NnDofs protein with other Arabidopsis proteins predicted using the STRING-DB. The minimum required interaction score is a moderate confidence level of 0.400. Organism set: Arabidopsis. These lines reflect the interaction of Nndofs homologs with other proteins. The darkness of the circle color indicates the strength of the data support.

2.7. Expression Patterns of NnDofs in Lotus

RNA-seq data were used to discover how *NnDof* was expressed in different tissues of the lotus (Figure 8). We investigated the expression patterns of *NnDofs* since their distinct physiological roles are interrelated with their expression patterns in different tissues and at different developmental stages. The expression patterns of *Dof* genes in different tissues and organs may be grouped into three groups based on clustering results, and members of the same group have comparable expression profiles. In group 1, *NnDofs* expression was low in most tissues; in group 2, certain *NnDofs* were dominantly expressed in some tissues; and in group 3, *NnDofs* expression was high in most tissues. In general, genes have similar expression patterns. For example, some genes, such as NnDof19 and *NnDof28*, are highly expressed in all tissues, while *NnDof24* and *NnDof27* are lowly expressed in all tissues, and *NnDof2*, *NnDof16*, *NnDof22*, and *NnDof28* are predominantly expressed in the seed coat. Furthermore, 26 (90%) *NnDofs* were expressed in at least one tissue and developmental period ($\log_2^{(FPKM)} > 0$); six (21%) *NnDofs* were expressed in all samples tested, which suggests that these *NnDofs* act at multiple developmental stages; and only one (3%) *NnDof* was lowly expressed in all samples tested and may be specifically expressed in other tissues and developmental periods.

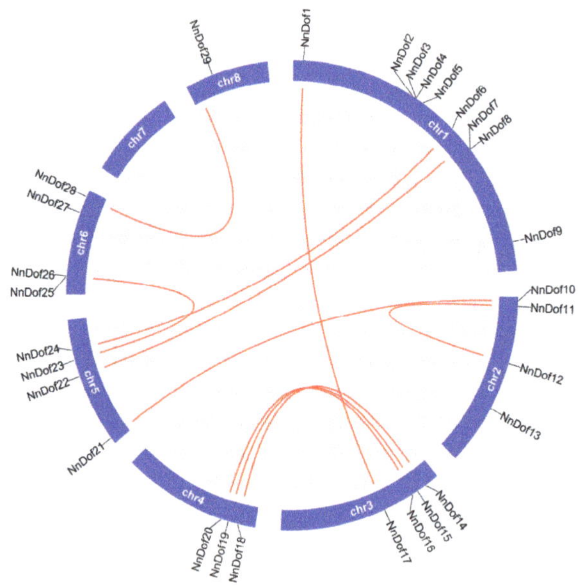

Figure 6. Chromosomal locations and segmental duplication events of NnDofs. Red lines represent segmental duplication events.

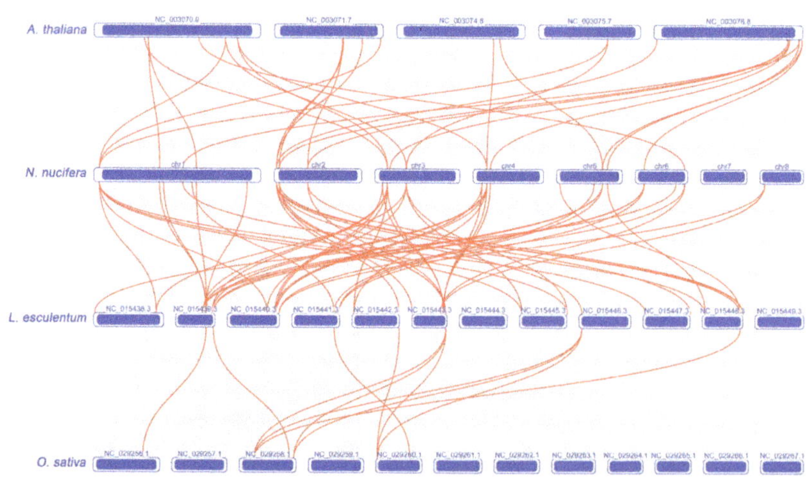

Figure 7. Synthetic analysis of *NnDofs* in lotus, rice, tomato, and Arabidopsis genomes. The red lines represent homologous gene pairs between two adjacent species.

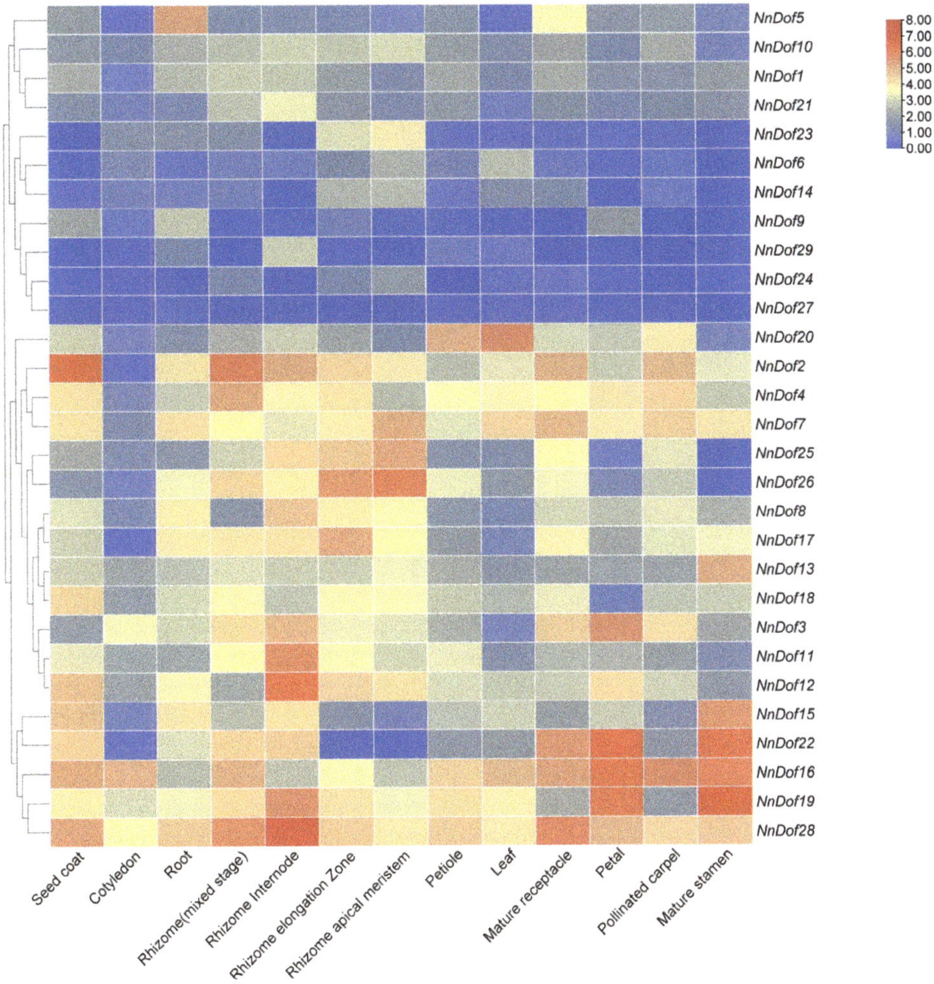

Figure 8. Expression pattern of *NnDofs* in different tissues of lotus. The expression pattern was generated based on FPKM plus 1 after log2 transformation and analyzed by heatmap hierarchical clustering.

Whole plant leaves were used when extracting total RNA to prepare for the qRT-PCR analysis, which was validated to discover how the NnDofs genes were expressed under salt stress (Figure 9). Under salt stress, the expression of *NnDofs* was rapidly upregulated at 1 h, except for *NnDof15*, *NnDof16*, *NnDof19*, and *NnDof20*. Some *NnDofs* genes were downregulated at 4 h, 12 h, and 24 h, but *NnDof3*, *NnDof11*, *NnDof12*, *NnDof17*, *NnDof18*, *NnDof22*, *NnDof25*, and *NnDof26* were upregulated at 24 h.

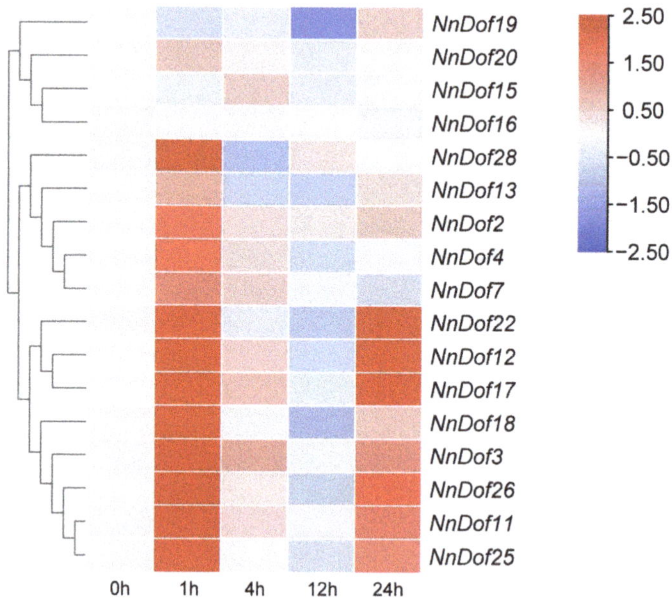

Figure 9. Expression profiles of *NnDofs* genes responding to salt stress. Blue indicates decreased expression levels and red indicates increased expression levels. Heatmaps were generated with TBtools software.

3. Discussion

The ornamental lotus is a traditional Chinese flower and an important aquatic plant, which is loved and widely cultivated for its brilliant color, rich flower shape, and fragrant smell. It has been reported that common lotus species are sensitive to salt stress, and increasing soil salt concentration can significantly affect their growth and development processes, further affecting lotus cultivation and production. Therefore, mining and analyzing the genes associated with salt stress resistance in lotus is important to improve the resistance of lotus. *Dofs* have been reported to play an important role in salt stress resistance in rice and tomato. In this study, we performed a genome-wide search for *NnDofs* with a focus on salt stress resistance. This will help us learn more about how *Dof* genes work and how to improve the genetics of lotus.

3.1. Identification, Structural and Phylogenetic Analysis, and Evolutionary Characterization of NnDofs

Assessing the structure and function of transcription factors facilitates the analysis of species-specific gene regulatory networks [41]. *Dof* genes are plant-specific transcription factors with important roles in plant growth and development [17]. Other functions of most *Dof* genes remain to be explored. *Dof* genes have been identified in Arabidopsis (36), rice (30), maize (46), tomato (34), and chrysanthemum (20) species. A total of 29 *Dof* genes were identified in the lotus genome, which is close to the number of *Dof* genes in rice and tomato. Systematic classification has important implications for the analysis of *NnDofs*. Systematic classification has important implications for the analysis of *NnDofs*. The Arabidopsis Dof proteins were incorporated with NnDof proteins to construct a phylogenetic tree, and *NnDofs* were classified into six of seven subfamilies. Interestingly, none of the *NnDofs* were classified into group I, while *AtDOF4.2* and its close homolog *AtDOF4.4*, which are classified in group I, are associated with regulation related to the regulation of branching meristem seed coat formation in Arabidopsis [42]. In addition, overexpression and RNAi-mediated silencing of *AtDOF4.2* have opposite effects on the expression levels of flavonoid biosynthesis-related genes and flavonoid accumulation [43].

Gene structure and motif distribution can be used as supporting evidence for evolutionary relationships between species or genes [44,45], while generally, members of the same subfamily have similar exon/intron structure and number distribution [19]. However, the distribution of the number of introns in the taxonomic subgroups of lotus did not follow this pattern exactly. Multiple sequence alignment was used to compare the amino acid sequences in the *Dof* structural domain of 29 different species of lotus. It was found that the *NnDofs* structural domain sequences were very similar, that they all had the typical CX2CX21CX2C pattern, and that they all had nuclear localization signals (NLS).

Gene duplication is an essential mechanism for generating new evolutionary templates in eukaryotes [46]. In the evolution of angiosperm genomes, genomic duplications have occurred extensively, including whole-genome duplications (WGDs) and segmental duplications [47,48]. Analysis of gene duplications can help us better understand how genes and organisms evolved over time [49]. Ten pairs of segmental duplications and two sets of tandem duplicated *NnDofs* were identified in the lotus genome. Most of the segmental duplicated *NnDofs* had high sequence similarity, while the Ka/Ks of the duplicated gene pairs were less than one. Indicating that all duplicated pairs of *NnDofs* were negatively selected during the evolutionary process, suggesting that to some extent, segmental duplications may be the main amplification mode of the *NnDof* gene family. To some extent, genome duplication may be the main amplification mode of the *NnDof* gene family. In a homozygous analysis of *NnDofs* in lotus and three representative plants, it was found that some *NnDof* genes, such as *NnDof1* and *NnDof10*, were linked to at least three homozygous gene pairs. It is thought that these *NnDofs* may be vital to the generation of the *NnDof* gene family.

3.2. Expression Profiling of NnDofs

NnDofs exhibit tissue-specific expression patterns, confirming previous research on *Dof* genes in other species. For instance, *NnDof2*, *NnDof5*, *NnDof13*, and *NnDof20* were highly expressed in the seed coat and rhizome (mixed stage), root and mature receptacle, mature stamen, and petiole and leaf, respectively. In addition, except for *NnDof16*, *NnDof19*, and *NnDof28*, all other *NnDofs* showed lower expression levels in the cotyledon. *Dof5.6/HCA2* has been reported to positively regulate the formation of interfascicular cambium during vascular tissue development in Arabidopsis [50]. *AtDof2.4/AtDof5.8* may be essential in the primary but distinct processes of vasculature formation [51]. Arabidopsis root hair development is ABA-dependent inhibited, attributed to *OBP4*-mediated transcriptional regulation of *RSL2* [52]. Dof genes are definitely widely involved in plant tissue differentiation and development, which explains several *NnDofs*' differential expression in rhizome (mixed stage), rhizome internode, rhizome elongation zone, and other tissues. In addition, Dof has distinct expression patterns in different tissues and developmental processes in plants. In pepper, *CaDof18* was preferentially expressed in the early stage of flower [14]. Ethylene induces specific increased expression of several *MaDofs* in bananas during fruit ripening [53]. More than half of the A and B1 Dof group members in *Brassica napus* were more highly expressed in the stems and young roots, respectively [54]. *CsDof33* was highly expressed in the terminal buds of tea plants, whereas the expression in young leaves was reversed [55].

Cis-acting elements are essential in gene expression [56], and gene promoter investigation is crucial to understanding the general control of gene expression in plants [57]. In this study, a large number of elements related to light response and meristematic tissue expression were found in the promoter region of *NnDofs*, suggesting that these *NnDofs* may be involved in light signaling pathways or meristematic tissue development. Various hormone responsive elements (ABA, GA, and MeJA response elements) and abiotic stress responsive elements are predicted.

Various investigations into the response of lotus to abiotic stresses have been published. Isolation of bZIP TFs from salt-tolerant lotus root tips enhances the adaptation of transgenic tobacco plants under salt stress [58]. The *NnCIPK6* gene was highly expressed under NaCl treatment in lotus resistant cultivars and was successfully cloned [59]. *NuSTP5*, a monosaccharide transporter family gene, produces stress responses to NaCl, drought, and cold stress [60]. All *NnWRKYs* responded to at least one of SA, JA, and submergence treatments, suggesting that they are extensively involved in abiotic stress [61]. However, studies on the lotus Dof TFs have not been reported yet. According to qRT-PCR analysis, the expression pattern of *NnDofs* was dramatically changed after salt treatment. Under salt stress, the expression of most *NnDofs* was significantly upregulated within 1 h but then progressively decreased after 4 h. It is proposed that several specific *NnDofs* may be engaged in response to salt stress in lotus. Published studies have shown that *Dof* TFs are involved in salt stress resistance of plants through various physiological pathways. Overexpression of *GhDof1* in cotton resulted in a substantial improvement throughout salinity tolerance in wild-type plants, according to prior research [32]. Several *CaDof* genes in pepper were determined as being particularly sensitive to salt stress [14]. The expression of the *ZmDof* gene in maize seedlings was dramatically elevated in those that had been exposed to salt [62]. Salt stress inhibits the transcription of *OsDof15*, which regulates ethylene generation and limits primary root growth in rice by direct contact with the *OsACS1* promoter [34]. *SlDof22* inhibition significantly reduces the expression of the *SlSOS1* gene in tomato, resulting in lower tolerance to salt stress [35]. Under salt conditions, tomato plants that overexpressed the *CDF3* gene maintained growth and boosted yield [22].

4. Conclusions

Twenty-nine *NnDofs* were identified in lotus species divided into six subfamilies, the physicochemical properties of which vary and all of which contain conserved zinc finger structures. Segmental duplications may be the primary mode of amplification for the NnDofs gene family. In addition, some *NnDofs* have distinct expression specificities in different tissues and developmental stages. Most of the *NnDofs* were significantly regulated by salt stress treatments. Therefore, *NnDofs* may be engaged in multiple cross-regulatory networks related to lotus development and salt stress responses, and their interactions would help to explain the dynamics of co-regulatory functions of signaling differently in various biological processes. This study provides a foundation for further investigation into the functional properties of the *Dof* gene family in lotus, particularly its role in salt stress resistance.

5. Materials and Methods

5.1. Identification and Physical Properties Analysis of NnDofs

The lotus genome data used in this study are available under the Nelumbo Genome Database (NGD) [63]. All protein sequences from the lotus genome were scanned by HMMER 3.0 using a hidden Markov model (HMM) of the Dof domain (PF02701), downloaded from the Pfam website (http://pfam.xfam.org/, accessed on 3 May 2021), with an E value of 1×10^{-5}. The presence of the conserved domain of Dof in the predicted protein was confirmed by NCBI Conserved Domain Database (CDD), Pfam, and SMART. Subcellular locations were predicted using ProtComp 9.0 from Softberry (http://www.softberry.com/, accessed on 3 May 2021). Protein molecular weight (Mw) and theoretical isoelectric point (pI), instability index, aliphatic index, and grand average of hydropathicity (GRAVY) were calculated by Expasy (https://web.expasy.org/protparam/, accessed on 3 May 2021).

5.2. Conserved Domain, Conserved Structure, and Phylogenetic Analysis of NnDofs

Multiple protein sequence alignments of Dof domains were constructed using ClustalW, and the results were submitted to GeneDoc software for optimization. After analyzing the full-length sequence, the gene structure pattern distribution of NnDofs was drawn using Tbtools software. The conserved motifs of NnDof sequences in lotus were scanned using

MEME (https://meme-suite.org/meme/tools/meme/, accessed on 3 May 2021), with a maximum number of motifs of 10 and the remaining parameters set to default values. The Dof-deduced sequences in Arabidopsis and lotus were aligned using MUSCLE with default settings. Phylogenetic trees were constructed using MEGA 7.0 (http://www.megasoftware.net/, accessed on 5 May 2021) with the neighbor-joining (NJ) method and 1000 bootstraps.

5.3. Gene Duplication and Syntenic Analysis of NnDofs

All NnDofs genes were mapped to lotus chromosomes based on physical location information from the database of lotus genome using Circos. The multiple collinear scanning toolkit (MCScanX, http://chibba.pgml.uga.edu/mcscan2/, accessed on 7 May 2021) was used to analyze gene duplication events in lotus, and for Arabidopsis, tomato, soybean, rice, maize, and lotus protein sequences, a collinearity analysis was performed between them. Finally, gene duplication and collinearity were shown using Circos software and Tbtool software with default parameters [64].

5.4. Cis-Acting Element Analysis and Protein Interaction Network Prediction of NnDofs

Genomic DNA sequences 2000 bp upstream of each NnDofs transcription start site were extracted and submitted to the PlantCARE database (http://bioinformatics.psb.ugent.be/webtools/plantcare/html/, accessed on 9 May 2021) to predict the cis-acting elements in the NnDofs promoter region.

All putative NnDof protein sequences were uploaded to the webserver STRING (version 11.5; http://string-db.org, accessed on 11 May 2021) to build an interaction network, and the NnDof protein interaction network was mapped using Cytoscape.

5.5. Expression Patterns Analysis of NnDofs by RNA-Seq Data

A study by Zhang describes information from transcriptomic data from various tissues of the lotus [63]. To analyze the expression patterns of NnDofs, we generated 20 groups of tissue-specific transcript abundances at different developmental stages based on the data obtained. The 20 groups of tissues are from the seed coat, cotyledon, root, rhizome (mixed), leaf, petiole, receptacle, stamen, petal, carpel, rhizome internode, rhizome elongation zone, and rhizome apical meristem. The seed coat data were taken at 18 days after pollination, and the cotyledon data were taken at 15 days after pollination. Triplicate data were averaged, data were transformed by $\log_2^{(FPKM+1)}$, and heatmaps were created using a plugin within the Tbtools software. All the detailed data are shown in Table S2.

5.6. Stress Treatments and Quantitative Real-Time PCR Analysis

Nelumbo nucifera cv. 'Taikonghongqi', a typical cultivated variety, was used throughout the study. The variety was cultivated in a greenhouse at Fuyang Normal University, Fuyang City, Anhui Province, China, from June 2021 to July 2021. 'Taikonghongqi' seeds were surface sterilized in 75% ethanol for 2 min before being peeled open and fully soaked in water. They were then provided an artificial environment with 16 h of light and 8 h of darkness at a temperature of 24 °C for four weeks. Lotus seedlings with similar development status were selected and randomly divided into two groups and cultured under the same culture conditions for 72 h. For the salt-stress treatment group, the roots of the lotus seedlings were exposed to high salt stress (200 mM NaCl) for 24 h. For all treatments, plant materials from three biological replicates were harvested immediately, frozen in liquid nitrogen, and then stored at −80 °C until RNA isolation.

Total RNA was extracted with a plant RNA extraction kit (Huayueyang, Beijing, China) and treated with RNase-free DNase I to remove potential genomic DNA contamination. Qualified RNA was selected as the template to generate first-strand cDNA by gel electrophoresis and A260/A280 ratio determination. Complementary cDNA was generated with reverse transcriptase (TransGen Biotech, Beijing, China). Specific NnDof gene primers were designed using Primer Premier 5, and the expression levels of different sampling cycles were normalized with NnActin gene as a reference gene. Supplementary Table S3 provides all primer information. qRT-PCR was performed using 2×SYBR Green qPCR Mix (with ROX) (Sparkjade, Qingdao, China) and amplified using 96-well plates and a CFX96 TouchTM RT-PCR system (Biorad, Los Angeles, CA, USA). Each reaction was performed in biological triplicate. Data from qRT-PCR amplifications were analyzed using the $2^{-\Delta\Delta Ct}$ method.

Supplementary Materials: The following supporting information can be downloaded at: https://www.mdpi.com/article/10.3390/plants11152057/s1, Table S1: List of the 29 NnDof genes identified in this study; Table S2: RNA-seq data of 29 Nndof genes that were used in this study; Table S3: Primers of NnDof genes used in RT-qPCR; Table S4: The lotus Dof gene sequence identified in this study.

Author Contributions: L.R., W.W. and X.C. conceived and designed the experiments; D.Y. and H.M. performed the experiments; W.W. analyzed the data; X.D., H.Y. and X.C. contributed reagents/materials/analysis tools; X.C. and W.W. wrote the paper. All authors have read and agreed to the published version of the manuscript.

Funding: This work was supported by a grant from the Natural Science Key Foundations of the Anhui Bureau of Education (KJ2021A0677), the research funds for postdoctoral researchers in Anhui Province (2020B434), the science and technology special project of Fuyang Normal University-Fuyang City Municipal School Cooperation (SXHZ202105) and Fuyang Normal University's major scientific and technological achievements incubator fund project (kjfh201703).

Institutional Review Board Statement: Not applicable.

Informed Consent Statement: Not applicable.

Data Availability Statement: Not applicable.

Conflicts of Interest: The authors declare no conflict of interest.

References

1. Li, J.; Pu, L.; Han, M.; Zhu, M.; Zhang, R.; Xiang, Y. Soil Salinization Research in China: Advances and Prospects. *J. Geogr. Sci.* **2014**, *24*, 943–960. [CrossRef]
2. Shrivastava, P.; Kumar, R. Soil Salinity: A Serious Environmental Issue and Plant Growth Promoting Bacteria as One of the Tools for Its Alleviation. *Saudi J. Biol. Sci.* **2015**, *22*, 123–131. [CrossRef] [PubMed]
3. Ahmad, R.; Hussain, S.; Anjum, M.A.; Khalid, M.F.; Saqib, M.; Zakir, I.; Hassan, A.; Fahad, S.; Ahmad, S. Oxidative Stress and Antioxidant Defense Mechanisms in Plants under Salt Stress. In *Plant Abiotic Stress Tolerance*; Springer: Berlin/Heidelberg, Germany, 2019; pp. 191–205.
4. Shah, W.H.; Rasool, A.; Saleem, S.; Mushtaq, N.U.; Tahir, I.; Hakeem, K.R.; Rehman, R.U. Understanding the Integrated Pathways and Mechanisms of Transporters, Protein Kinases, and Transcription Factors in Plants under Salt Stress. *Int. J. Genom.* **2021**, *2021*, 5578727. [CrossRef]
5. Golldack, D.; Lüking, I.; Yang, O. Plant Tolerance to Drought and Salinity: Stress Regulating Transcription Factors and Their Functional Significance in the Cellular Transcriptional Network. *Plant Cell Rep.* **2011**, *30*, 1383–1391. [CrossRef]
6. Boggon, T.J.; Shan, W.-S.; Santagata, S.; Myers, S.C.; Shapiro, L. Implication of Tubby Proteins as Transcription Factors by Structure-Based Functional Analysis. *Science* **1999**, *286*, 2119–2125. [CrossRef] [PubMed]
7. Gupta, S.; Malviya, N.; Kushwaha, H.; Nasim, J.; Bisht, N.C.; Singh, V.K.; Yadav, D. Insights into Structural and Functional Diversity of Dof (DNA Binding with One Finger) Transcription Factor. *Planta* **2015**, *241*, 549–562. [CrossRef]
8. Kim, J.; Kim, H.-Y. Functional Analysis of a Calcium-Binding Transcription Factor Involved in Plant Salt Stress Signaling. *FEBS Lett.* **2006**, *580*, 5251–5256. [CrossRef]
9. Nakano, T.; Suzuki, K.; Ohtsuki, N.; Tsujimoto, Y.; Fujimura, T.; Shinshi, H. Identification of Genes of the Plant-Specific Transcription-Factor Families Cooperatively Regulated by Ethylene and Jasmonate in Arabidopsis Thaliana. *J. Plant Res.* **2006**, *119*, 407–413. [CrossRef]
10. Yanagisawa, S.; Izui, K. Molecular Cloning of Two DNA-Binding Proteins of Maize That Are Structurally Different but Interact with the Same Sequence Motif. *J. Biol. Chem.* **1993**, *268*, 16028–16036. [CrossRef]
11. Cai, X.; Zhang, Y.; Zhang, C.; Zhang, T.; Hu, T.; Ye, J.; Zhang, J.; Wang, T.; Li, H.; Ye, Z. Genome-wide Analysis of Plant-specific Dof Transcription Factor Family in Tomato. *J. Integr. Plant Biol.* **2013**, *55*, 552–566. [CrossRef] [PubMed]
12. Zou, Z.; Zhu, J.; Zhang, X. Genome-Wide Identification and Characterization of the Dof Gene Family in Cassava (*Manihot esculenta*). *Gene* **2019**, *687*, 298–307. [CrossRef] [PubMed]
13. Chen, M.; Liu, X.; Huan, L.; Sun, M.; Liu, L.; Chen, X.; Gao, D.; Li, L. Genome-Wide Analysis of Dof Family Genes and Their Expression during Bud Dormancy in Peach (*Prunus Persica*). *Sci. Hortic.* **2017**, *214*, 18–26. [CrossRef]

14. Wu, Z.; Cheng, J.; Cui, J.; Xu, X.; Liang, G.; Luo, X.; Chen, X.; Tang, X.; Hu, K.; Qin, C. Genome-Wide Identification and Expression Profile of Dof Transcription Factor Gene Family in Pepper (*Capsicum Annuum* L.). *Front. Plant Sci.* **2016**, *7*, 574. [CrossRef] [PubMed]
15. Yanagisawa, S. Structure, Function, and Evolution of the Dof Transcription Factor Family. In *Plant Transcription Factors*; Elsevier: Amsterdam, The Netherlands, 2016; pp. 183–197.
16. Fang, Z.; Jiang, W.; He, Y.; Ma, D.; Liu, Y.; Wang, S.; Zhang, Y.; Yin, J. Genome-Wide Identification, Structure Characterization, and Expression Profiling of Dof Transcription Factor Gene Family in Wheat (*Triticum Aestivum* L.). *Agronomy* **2020**, *10*, 294. [CrossRef]
17. Yanagisawa, S. The Dof Family of Plant Transcription Factors. *Trends Plant Sci.* **2002**, *7*, 555–560. [CrossRef]
18. Lijavetzky, D.; Carbonero, P.; Vicente-Carbajosa, J. Genome-Wide Comparative Phylogenetic Analysis of the Rice and Arabidopsis Dof Gene Families. *BMC Evol. Biol.* **2003**, *3*, 17. [CrossRef]
19. Moreno-Risueno, M.Á.; Martínez, M.; Vicente-Carbajosa, J.; Carbonero, P. The Family of DOF Transcription Factors: From Green Unicellular Algae to Vascular Plants. *Mol. Genet. Genom.* **2007**, *277*, 379–390. [CrossRef]
20. Ruta, V.; Longo, C.; Lepri, A.; De Angelis, V.; Occhigrossi, S.; Costantino, P.; Vittorioso, P. The DOF Transcription Factors in Seed and Seedling Development. *Plants* **2020**, *9*, 218. [CrossRef]
21. Kang, W.-H.; Kim, S.; Lee, H.-A.; Choi, D.; Yeom, S.-I. Genome-Wide Analysis of Dof Transcription Factors Reveals Functional Characteristics during Development and Response to Biotic Stresses in Pepper. *Sci. Rep.* **2016**, *6*, 33332. [CrossRef]
22. Corrales, A.-R.; Carrillo, L.; Lasierra, P.; Nebauer, S.G.; Dominguez-Figueroa, J.; Renau-Morata, B.; Pollmann, S.; Granell, A.; Molina, R.-V.; Vicente-Carbajosa, J.; et al. Multifaceted Role of Cycling DOF Factor 3 (CDF3) in the Regulation of Flowering Time and Abiotic Stress Responses in Arabidopsis. *Plant Cell Environ.* **2017**, *40*, 748–764. [CrossRef] [PubMed]
23. Goralogia, G.S.; Liu, T.-K.; Zhao, L.; Panipinto, P.M.; Groover, E.D.; Bains, Y.S.; Imaizumi, T. CYCLING DOF FACTOR 1 Represses Transcription through the TOPLESS Co-repressor to Control Photoperiodic Flowering in Arabidopsis. *Plant J.* **2017**, *92*, 244–262. [CrossRef]
24. Xu, X.; Li, F.; Wang, Y.; Tang, S.; Dai, Q.; Zhu, S.; Liu, T. Identification of Dof Transcription Factors in Ramie (*Boehmeria Nivea* L. Gaud) and Their Expression in Response to Different Nitrogen Treatments. *3 Biotech* **2018**, *8*, 496. [CrossRef] [PubMed]
25. Cai, M.; Lin, J.; Li, Z.; Lin, Z.; Ma, Y.; Wang, Y.; Ming, R. Allele Specific Expression of Dof Genes Responding to Hormones and Abiotic Stresses in Sugarcane. *PLoS ONE* **2020**, *15*, e0227716. [CrossRef]
26. Liu, Y.; Liu, N.; Deng, X.; Liu, D.; Li, M.; Cui, D.; Hu, Y.; Yan, Y. Genome-Wide Analysis of Wheat DNA-Binding with One Finger (Dof) Transcription Factor Genes: Evolutionary Characteristics and Diverse Abiotic Stress Responses. *BMC Genom.* **2020**, *21*, 276. [CrossRef] [PubMed]
27. Zhang, Z.; Yuan, L.; Liu, X.; Chen, X.; Wang, X. Evolution Analysis of Dof Transcription Factor Family and Their Expression in Response to Multiple Abiotic Stresses in Malus Domestica. *Gene* **2018**, *639*, 137–148. [CrossRef]
28. Ramirez-Parra, E.; Perianez-Rodriguez, J.; Navarro-Neila, S.; Gude, I.; Moreno-Risueno, M.A.; del Pozo, J.C. The Transcription Factor OBP4 Controls Root Growth and Promotes Callus Formation. *New Phytol.* **2017**, *213*, 1787–1801. [CrossRef] [PubMed]
29. Qi, X.; Li, S.; Zhu, Y.; Zhao, Q.; Zhu, D.; Yu, J. ZmDof3, a Maize Endosperm-Specific Dof Protein Gene, Regulates Starch Accumulation and Aleurone Development in Maize Endosperm. *Plant Mol. Biol.* **2017**, *93*, 7–20. [CrossRef]
30. Wu, J.; Chen, L.; Chen, M.; Zhou, W.; Dong, Q.; Jiang, H.; Cheng, B. The DOF-Domain Transcription Factor ZmDOF36 Positively Regulates Starch Synthesis in Transgenic Maize. *Front. Plant Sci.* **2019**, *10*, 465. [CrossRef] [PubMed]
31. Fornara, F.; Panigrahi, K.C.; Gissot, L.; Sauerbrunn, N.; Rühl, M.; Jarillo, J.A.; Coupland, G. Arabidopsis DOF Transcription Factors Act Redundantly to Reduce CONSTANS Expression and Are Essential for a Photoperiodic Flowering Response. *Dev. Cell* **2009**, *17*, 75–86. [CrossRef]
32. Su, Y.; Liang, W.; Liu, Z.; Wang, Y.; Zhao, Y.; Ijaz, B.; Hua, J. Overexpression of GhDof1 Improved Salt and Cold Tolerance and Seed Oil Content in Gossypium Hirsutum. *J. Plant Physiol.* **2017**, *218*, 222–234. [CrossRef]
33. Zhou, Y.; Cheng, Y.; Wan, C.; Li, J.; Yang, Y.; Chen, J. Genome-Wide Characterization and Expression Analysis of the Dof Gene Family Related to Abiotic Stress in Watermelon. *PeerJ* **2020**, *8*, e8358. [CrossRef]
34. Qin, H.; Wang, J.; Chen, X.; Wang, F.; Peng, P.; Zhou, Y.; Miao, Y.; Zhang, Y.; Gao, Y.; Qi, Y. Rice Os DOF 15 Contributes to Ethylene-inhibited Primary Root Elongation under Salt Stress. *New Phytol.* **2019**, *223*, 798–813. [CrossRef] [PubMed]
35. Cai, X.; Zhang, C.; Shu, W.; Ye, Z.; Li, H.; Zhang, Y. The Transcription Factor SlDof22 Involved in Ascorbate Accumulation and Salinity Stress in Tomato. *Biochem. Biophys. Res. Commun.* **2016**, *474*, 736–741. [CrossRef]
36. Li, Y.; Zhu, F.-L.; Zheng, X.-W.; Hu, M.-L.; Dong, C.; Diao, Y.; Wang, Y.-W.; Xie, K.-Q.; Hu, Z.-L. Comparative Population Genomics Reveals Genetic Divergence and Selection in Lotus, Nelumbo Nucifera. *BMC Genom.* **2020**, *21*, 146. [CrossRef]
37. Shen-Miller, J. Sacred lotus, the long-living fruits of China Antique. *Seed Sci. Res.* **2002**, *12*, 131–143. [CrossRef]
38. Yang, M.; Zhu, L.; Pan, C.; Xu, L.; Liu, Y.; Ke, W.; Yang, P. Transcriptomic Analysis of the Regulation of Rhizome Formation in Temperate and Tropical Lotus (*Nelumbo nucifera*). *Sci. Rep.* **2015**, *5*, 13059. [CrossRef] [PubMed]
39. Sharma, B.R.; Gautam, L.N.S.; Adhikari, D.; Karki, R. A Comprehensive Review on Chemical Profiling of Nelumbo Nucifera: Potential for Drug Development. *Phytother. Res.* **2017**, *31*, 3–26. [CrossRef]
40. Sohn, D.-H.; Kim, Y.-C.; Oh, S.-H.; Park, E.-J.; Li, X.; Lee, B.-H. Hepatoprotective and Free Radical Scavenging Effects of Nelumbo Nucifera. *Phytomedicine* **2003**, *10*, 165–169. [CrossRef]
41. Liu, L.; White, M.J.; MacRae, T.H. Transcription Factors and Their Genes in Higher Plants: Functional Domains, Evolution and Regulation. *Eur. J. Biochem.* **1999**, *262*, 247–257. [CrossRef]

42. Zou, H.-F.; Zhang, Y.-Q.; Wei, W.; Chen, H.-W.; Song, Q.-X.; Liu, Y.-F.; Zhao, M.-Y.; Wang, F.; Zhang, B.-C.; Lin, Q. The Transcription Factor AtDOF4. 2 Regulates Shoot Branching and Seed Coat Formation in Arabidopsis. *Biochem. J.* **2013**, *449*, 373–388. [CrossRef]
43. Skirycz, A.; Jozefczuk, S.; Stobiecki, M.; Muth, D.; Zanor, M.I.; Witt, I.; Mueller-Roeber, B. Transcription Factor AtDOF4; 2 Affects Phenylpropanoid Metabolism in Arabidopsis Thaliana. *New Phytol.* **2007**, *175*, 425–438. [CrossRef]
44. Riechmann, J.L.; Heard, J.; Martin, G.; Reuber, L.; Jiang, C.-Z.; Keddie, J.; Adam, L.; Pineda, O.; Ratcliffe, O.J.; Samaha, R.R. Arabidopsis Transcription Factors: Genome-Wide Comparative Analysis among Eukaryotes. *Science* **2000**, *290*, 2105–2110. [CrossRef]
45. Qu, L.-J.; Zhu, Y.-X. Transcription Factor Families in Arabidopsis: Major Progress and Outstanding Issues for Future Research. *Curr. Opin. Plant Biol.* **2006**, *9*, 544–549. [CrossRef]
46. Lynch, M.; Force, A. The Probability of Duplicate Gene Preservation by Subfunctionalization. *Genetics* **2000**, *154*, 459–473. [CrossRef]
47. Jourda, C.; Cardi, C.; Mbéguié-A.-Mbéguié, D.; Bocs, S.; Garsmeur, O.; D'Hont, A.; Yahiaoui, N. Expansion of Banana (*Musa acuminata*) Gene Families Involved in Ethylene Biosynthesis and Signalling after Lineage-Specific Whole-Genome Duplications. *New Phytol.* **2014**, *202*, 986–1000. [CrossRef] [PubMed]
48. Koszul, R.; Fischer, G. A Prominent Role for Segmental Duplications in Modeling Eukaryotic Genomes. *Comptes Rendus Biol.* **2009**, *332*, 254–266. [CrossRef] [PubMed]
49. Qiao, X.; Yin, H.; Li, L.; Wang, R.; Wu, J.; Wu, J.; Zhang, S. Different Modes of Gene Duplication Show Divergent Evolutionary Patterns and Contribute Differently to the Expansion of Gene Families Involved in Important Fruit Traits in Pear (*Pyrus bretschneideri*). *Front. Plant Sci.* **2018**, *9*, 161. [CrossRef] [PubMed]
50. Guo, Y.; Qin, G.; Gu, H.; Qu, L.-J. Dof5.6/HCA2, a Dof Transcription Factor Gene, Regulates Interfascicular Cambium Formation and Vascular Tissue Development in Arabidopsis. *Plant Cell* **2009**, *21*, 3518–3534. [CrossRef]
51. Konishi, M.; Yanagisawa, S. Sequential Activation of Two Dof Transcription Factor Gene Promoters during Vascular Development in Arabidopsis Thaliana. *Plant Physiol. Biochem.* **2007**, *45*, 623–629. [CrossRef]
52. Rymen, B.; Kawamura, A.; Schäfer, S.; Breuer, C.; Iwase, A.; Shibata, M.; Ikeda, M.; Mitsuda, N.; Koncz, C.; Ohme-Takagi, M. ABA Suppresses Root Hair Growth via the OBP4 Transcriptional Regulator. *Plant Physiol.* **2017**, *173*, 1750–1762. [CrossRef] [PubMed]
53. Feng, B.-h.; Han, Y.-c.; Xiao, Y.-y.; Kuang, J.-f.; Fan, Z.-q.; Chen, J.-y.; Lu, W.-j. The banana fruit Dof transcription factor MaDof23 acts as a repressor and interacts with MaERF9 in regulating ripening-related genes. *J. Exp. Bot.* **2016**, *67*, 2263–2275. [CrossRef]
54. Lohani, N.; Babaei, S.; Singh, M.B.; Bhalla, P.L. Genome-Wide in Silico Identification and Comparative Analysis of Dof Gene Family in Brassica napus. *Plants* **2021**, *10*, 709. [CrossRef]
55. Jiao, Z.; Tang, D.; Fan, K.; Zhang, Q.; Liu, M.-Y.; Ruan, J. Genome-wide identification of the DNA-binding one zinc finger (Dof) transcription factor gene family and their potential functioning in nitrogen remobilization in tea plant (*Camellia sinensis* L.). *Sci. Hortic.* **2022**, *292*, 110615. [CrossRef]
56. Liu, J.-H.; Peng, T.; Dai, W. Critical Cis-Acting Elements and Interacting Transcription Factors: Key Players Associated with Abiotic Stress Responses in Plants. *Plant Mol. Biol. Report.* **2014**, *32*, 303–317. [CrossRef]
57. Hernandez-Garcia, C.M.; Finer, J.J. Identification and Validation of Promoters and Cis-Acting Regulatory Elements. *Plant Sci.* **2014**, 217–218. [CrossRef]
58. Cheng, L.; Li, S.; Hussain, J.; Xu, X.; Yin, J.; Zhang, Y.; Chen, X.; Li, L. Isolation and functional characterization of a salt responsive transcriptional factor, LrbZIP from lotus root (*Nelumbo nucifera* Gaertn). *Mol. Biol. Rep.* **2013**, *40*, 4033–4045. [CrossRef] [PubMed]
59. Liu, R.; Shi, H.; Wang, Y.; Chen, S.; Deng, J.; Liu, Y.; Li, S.; Chan, Z. Comparative physiological analysis of lotus (*Nelumbo nucifera*) cultivars in response to salt stress and cloning of NnCIPK genes. *Sci. Hortic.* **2014**, *173*, 29–36. [CrossRef]
60. Wu, P.; Zhang, Y.; Zhao, S.; Li, L. Comprehensive analysis of evolutionary characterization and expression for monosaccharide transporter family genes in Nelumbo nucifera. *Front. Ecol. Evol.* **2021**, *9*, 537398. [CrossRef]
61. Li, J.; Xiong, Y.; Li, Y.; Ye, S.; Yin, Q.; Gao, S.; Yang, D.; Yang, M.; Palva, E.T.; Deng, X. Comprehensive analysis and functional studies of WRKY transcription factors in Nelumbo nucifera. *Int. J. Mol. Sci.* **2019**, *20*, 5006. [CrossRef]
62. Chen, Y.; Cao, J. Comparative Analysis of Dof Transcription Factor Family in Maize. *Plant Mol. Biol. Report.* **2015**, *33*, 1245–1258. [CrossRef]
63. Li, H.; Yang, X.; Zhang, Y.; Gao, Z.; Liang, Y.; Chen, J.; Shi, T. Nelumbo Genome Database, an Integrative Resource for Gene Expression and Variants of Nelumbo Nucifera. *Sci. Data* **2021**, *8*, 38. [CrossRef] [PubMed]
64. Chen, C.; Chen, H.; Zhang, Y.; Thomas, H.R.; Frank, M.H.; He, Y.; Xia, R. TBtools: An integrative toolkit developed for interactive analyses of big biological data. *Mol. Plant* **2020**, *13*, 1194–1202. [CrossRef] [PubMed]

plants

MDPI

Article

Identification and Expression Analysis of NAC Gene Family in Weeping Trait of *Lagerstroemia indica*

Cuihua Gu [1,2,3,†], Linxue Shang [1,2,3,†], Guozhe Zhang [1,2,3,†], Qun Wang [1,2,3], Qingqing Ma [1,2,3], Sidan Hong [1,2,3], Yu Zhao [1,2,3,*] and Liyuan Yang [1,2,3,*]

1 College of Landscape and Architecture, Zhejiang Agriculture & Forestry University, Hangzhou 311300, China
2 Zhejiang Provincial Key Laboratory of Germplasm Innovation and Utilization for Garden Plants, Zhejiang Agriculture & Forestry University, Hangzhou 311300, China
3 Key Laboratory of National Forestry and Grassland Administration on Germplasm Innovation and Utilization for Southern Garden Plants, Zhejiang Agriculture & Forestry University, Hangzhou 311300, China
* Correspondence: zhaoyu@stu.zafu.edu.cn (Y.Z.); lyyang@zafu.edu.cn (L.Y.)
† These authors contributed equally to this work.

Abstract: *Lagerstroemia indica* is a widely used ornamental plant in summer gardens because of its desirable plant shape. The weeping traits of plants are related to secondary cell wall thickness and hormone signaling. NAC (NAM-ATAF1/2-CUC2), as one of the plant-specific transcription factors, is a switch for the secondary cell wall and also involved in leaf senescence, phytohormone signaling, and other growth processes. We identified a total of 21 *LiNAC* genes from the transcriptome data, which we divided into 14 subgroups and 2 groups. The physicochemical characteristics of amino acids, subcellular localization, transmembrane structure, GO and KEGG enrichment, and expression patterns were also examined. The qRT-PCR analysis showed that the expressions of *LiNAC8* and *LiNAC13* in upright *L. indica* 'Shaoguifei' and weeping *L. indica* 'Xiariwuniang' were significantly higher from the beginning to the end of growth stage (S1–S3), and the expressions of 'Shaoguifei' were always higher than those of 'Xiariwuniang'. However, *LiNAC2* showed a downward trend in S1–S3 and the relative expression level of 'Shaoguifei' was lower than that of 'Xiariwuniang'. It is hypothesized that these *LiNAC* genes may be involved in the regulation of weeping traits in *L. indica*. The results of this study provide a basis for analyzing the functions of *LiNAC* genes and help to explore the molecular regulatory mechanisms related to the weeping traits in *L. indica*.

Keywords: *Lagerstroemia indica*; NAC gene family; qRT-PCR analysis; weeping trait

Citation: Gu, C.; Shang, L.; Zhang, G.; Wang, Q.; Ma, Q.; Hong, S.; Zhao, Y.; Yang, L. Identification and Expression Analysis of NAC Gene Family in Weeping Trait of *Lagerstroemia indica*. *Plants* **2022**, *11*, 2168. https://doi.org/10.3390/plants11162168

Academic Editors: Aiping Song and Yu Chen

Received: 31 July 2022
Accepted: 15 August 2022
Published: 21 August 2022

1. Introduction

Weeping is a result of drooping branch development. The buds first develop upward, then as the tree grows, the apex of the branches shifts to growing downward. The reasons for the formation of the drooping branches are relatively convoluted, nevertheless. The development of drooping branches may be influenced by genetic make-up, hormones, and secondary growth, but it may also be intimately linked to certain external conditions, such as gravity, light, and supporting forces [1]. The weep gene mutation that led to the gravity anomaly in *Prunus persica* is what causes the weeping characteristic [2]. The weeping characteristics of *Salix babylonica* were brought on by a lack of mechanical support, brought on by an excessive trunk elongation [3]. For the weeping of *P. mume*, the secondary cell walls of branches in the pendant extension zone are thinner than those of upright branches [4]. To modulate the weeping phenotype, IAA and GA_3 biosynthetic-pathway-related genes were discovered in *P. mume* [5,6]. Recently, Zheng et al. found that genes related to cell division, cell development, and plant hormones played an important role in the tortuous-branch phenotype of *P. mume* [7]. Overexpression of a secondary-wall-associated cellulose synthase gene (*PtdCesA8*) was shown to result in a weeping phenotype in *Populus tremuloides* [8]. Ornamental plants often attract attention because of their specific traits. A certain plant architecture, known as weeping, has significant aesthetic appeal. Therefore, exploring the causes of plant drooping is potentially valuable for cultivating ornamental plants.

The ability to synthesize proteins via biological activities is controlled by transcription factors and some cis-acting elements, which also govern the spatiotemporal expression of downstream genes and eventually have an impact on the growth and development of organisms [9]. The genes NAM (no apical meristematic tissue) from *Petunia hybrida* [10], *ATAF1/2* (*Arabidopsis* transcription activation factor) found in *Arabidopsis thaliana* [11], and *CUC2* (cup2 shaped cotyledon) from *A. thaliana* are the sources of the name *NAC* [11]. At the N-terminal end of the protein, 150 amino acids make up its conserved structural domain, including five substructural components (A, B, C, D, and E) [12]. According to multiple studies, *NAC* has a wide spectrum of regulatory effects on plant growth and development [13–17]. Therefore, it is potentially possible to probe the weeping properties of plants from *NAC*.

Many studies show that the *NAC* gene family affects the weeping of plants by controlling the synthesis of secondary cell walls [15,18]. Inhibiting the expression of *SND1* and *NST1* in *Arabidopsis* at the same time can limit the synthetic gene expression in the three components of secondary cell walls (cellulose, xylem, and lignin), causing the flower stem cells to droop and fail to create secondary walls [15]. Studies on *SND4* and *SND5* revealed similar functions [19]. While the secondary cell wall of the stem of the overexpressed *PtrWND1B-s* plants greatly thickened and allowed them to grow upright, the secondary cell wall of the overexpressed *PtrWND1B-l* plants was unable to do so and displayed a phenotype similar to that of *PtrWND1B-RNAi* [18]. *NAC15* was significantly expressed in the poplar xylem and is most likely the primary factor for transgenic tobacco to develop tall stems [20]. Lignin is an important component in secondary cell walls and *EjNAC1* is associated with fruit lignification by activating genes involved in lignin biosynthesis [21]. The main regulatory mechanism in melatonin in enhancing flower stem strength was also to promote lignin accumulation and changed S/G lignin ratio [22]. Co-expression analysis of *Pm024213* showed that most of the related genes were involved in auxin and lignin biosynthesis [23].

Because of its lovely blossoms and exquisite branches, the crape myrtle (*Lagerstroemia indica* L.), a deciduous shrub and small tree, has considerable ornamental value [24]. *L. indica* branches were classified into flat, upright, and weeping branches by Tian [25]. The study of the molecular mechanisms of *L. indica's* weeping trait will benefit from an understanding of the *NAC* gene family function. However, nothing is known about the *NAC* gene studies in *L. indica*. In this study, the *NAC* genes in *L. indica* were discovered using a variety of bioinformatics techniques. Additionally, the physiological and biochemical properties of the LiNAC proteins were examined and a phylogenetic tree of the LiNAC proteins in *L. indica* and *A. thaliana* was created. To clarify whether *LiNAC* genes are involved in regulating the weeping trait of *L. indica*, we analyzed the relative expression levels of branches at different growth stages. Overall, the findings might support a theoretical hypothesis of *NAC* gene family in *L. indica*.

2. Results

2.1. Identification of NAC Genes in L. indica

Using AtNAC proteins as queries, 60 *LiNAC* genes were discovered by BLASTP searches in the *L. indica* transcriptome database and 43 putative *LiNAC* genes were discovered using an HMMER3 search in the *L. indica* protein database using the NAC-type NAM model (PF02365). Finally, using SMART and the NCBI Conserved Domain to remove duplicated genes and proteins without NAM-conserved domains, we were able to obtain 21 members of the *LiNAC* family. For further searching, all genes were given the designations *LiNAC1-LiNAC21* in the order of their assembled IDs (Table 1).

Table 1. *LiNAC* gene and protein characterization.

Transcriptome Gene ID (Gene Name)	Amino Acid Number/aa	Molecular Weight/Da	Isoelectric Point	Percentage of the Amino Acids with Highest Content%	Aliphatic Index	Instability Index	GRAVY	Subcellular Localization
TRINITY_DN1513_c0_g1_i2 (*LiNAC1*)	566	61,920.54	4.6	Ser(S)12.2%	65.81	46.22	−0.592	Nucleus
TRINITY_DN1513_c0_g1_i4 (*LiNAC2*)	433	48,711.53	4.49	Ser(S)10.2%	62.84	43.93	−0.758	Nucleus
TRINITY_DN1552_c0_g1_i1 (*LiNAC3*)	373	41,299	4.58	Ala(A)9.7%	69.87	36.19	−0.53	Nucleus
TRINITY_DN1552_c0_g2_i1 (*LiNAC4*)	406	44,609.42	4.42	Gly(G)9.9%	66.6	39.37	−0.525	Nucleus
TRINITY_DN17248_c0_g1_i1 (*LiNAC5*)	286	32,581.02	8.41	Leu(L)10.5%	73.01	48.22	−0.634	Nucleus
TRINITY_DN17248_c0_g1_i5 (*LiNAC6*)	288	32,682.93	8.67	Leu(L)10.4%	69.79	42.84	−0.619	Nucleus
TRINITY_DN1795_c0_g1_i3 (*LiNAC7*)	668	74,336.43	5.15	Ser(S)10.5%	68.64	51.19	−0.691	Nucleus
TRINITY_DN2758_c0_g1_i5 (*LiNAC8*)	414	47,164.68	5.39	Leu(L)8.7%	68.79	59.01	−0.731	Nucleus
TRINITY_DN419_c0_g1_i2 (*LiNAC9*)	266	30,251.89	9.36	Ser(S)10.9%	57.89	52.67	−0.85	Nucleus
TRINITY_DN419_c0_g1_i4 (*LiNAC10*)	264	29,841.31	8.9	Ser(S)12.1%	59.39	56.49	−0.795	Nucleus
TRINITY_DN419_c0_g1_i6 (*LiNAC11*)	270	30,776.37	8.89	Asn(N)& Ser (S) 9.3%	60.63	47.02	−0.905	Nucleus
TRINITY_DN419_c0_g1_i8 (*LiNAC12*)	270	30,781.24	8.89	Ser(S)10.0%	59.92	49.93	−0.942	Nucleus
TRINITY_DN4293_c0_g3_i1 (*LiNAC13*)	341	37,460.37	6.89	Ser(S)10.9%	68.94	49.69	−0.415	Nucleus
TRINITY_DN894_c0_g1_i10 (*LiNAC14*)	234	26,491.05	7.73	Leu(L)11.1%	75.47	29.93	−0.671	Nucleus
TRINITY_DN894_c0_g1_i11 (*LiNAC15*)	328	37,132.65	4.83	Asp(D)8.5%	68.69	38.22	−0.666	Nucleus
TRINITY_DN894_c0_g1_i22 (*LiNAC16*)	531	59,644.15	5.3	Ser(S)8.7%	71.62	48.59	−0.63	Nucleus
TRINITY_DN894_c0_g1_i31 (*LiNAC17*)	311	34,893.12	4.92	Pro(P)8.4%	67.46	47.53	−0.7	Nucleus
TRINITY_DN894_c0_g1_i4 (*LiNAC18*)	444	49,822.84	5.02	Pro(P)8.1%	69.19	47.22	−0.703	Nucleus
TRINITY_DN909_c0_g1_i1 (*LiNAC19*)	245	27,455.47	9.93	Arg(R)10.2%	71.18	62.67	−0.638	Nucleus
TRINITY_DN909_c0_g1_i3 (*LiNAC20*)	239	26,992.33	9.16	Ser(S)11.7%	63.18	58.10	−0.743	Nucleus
TRINITY_DN909_c0_g2_i1 (*LiNAC21*)	255	28,368.91	9.3	Ser(S)11.4%	59.65	55.76	−0.747	Nucleus

Proteins encoded by the 21 *LiNAC* genes contained 234 (LiNAC14) to 668 (LiNAC7) amino acids and their molecular weights (MWs) ranged from 26,491.05 (LiNAC14) to 74,336.43 (LiNAC7) Da, with an average molecular weight of 39,677 Da. Their predicted isoelectric point (pI) ranged from 4.42 (LiNAC4) to 9.93 (LiNAC19), with an average value of 6.90. Instability index calculation predicted that 17 (81%) of the NAC proteins were unstable. The LiNAC3, LiNAC4, LiNAC14, and LiNAC15 predicted proteins had instability indices less than 40 and were classified as stable proteins. Aliphatic amino acid indices showed that the thermal stability of the proteins ranged from 55.89 (LiNAC9) to 75.47 (LiNAC14). All LiNACs predicted protein grand average of hydropathicity (GRAVY) values as negative, demonstrating that they were predominantly hydrophilic. The predicted transmembrane structure results showed that LiNAC7 and LiNAC16 proteins had one transmembrane structure at the C-terminal of the 21 LiNAC proteins (Figure S1). Cell-PLoc subcellular localization predictions suggested that almost all LiNAC proteins were located in the Nucleus (Table 1).

2.2. Phylogenetic Analysis and Classification of LiNAC Proteins

We used the neighbor-joining (NJ) method to create a phylogenetic tree with 1000 bootstrap replications to shed light on the evolutionary relationships between proteins from *L. indica* and *A. thaliana*. Based on well-established *Arabidopsis* family classification and their homology with NAC proteins in *Arabidopsis*, 21 LiNAC proteins were classified into two groups, Group I and Group II, including 14 subgroups (Figure 1): OsNAC7, ANAC011, NAM, NAC1, ATAF, ANAC3/NAP, SENU5, ONAC022, TIP, NAC2, OsNAC8, ANAC001, ONAC003, and ANAC063. In our analysis, no LiNAC members from the subgroups ANAC011, NAM, ATAF, AtNAC3/NAP, OsNAC8, ONAC003, and ANAC063 were identified. Subgroup TIP had the most members (6) of *L. indica* and subgroups NAC1 and OsNAC7 had the fewest (1).

The LiNAC proteins clustered in the same subgroup, according to phylogenetic relationships, and may have related roles. For instance, NAC1 members are crucial for auxin signaling and the growth of lateral roots in plants [26]. The *NTL9* gene belonging to the TIP subgroup is heavily implicated in plant immune response [27]. The NAC family members in the ATAF, NAP, and AtNAC3 subgroups have a conserved role in stress response [28] and leaf senescence [17,29]. There are two distinct subgroups of NAC proteins that are involved in the formation of vascular vessels and the creation of secondary plant cell walls [30]. *SND2*, *SND3*, and other participants were placed into subgroup ONAC003, while *NST1*, *NST2*, *NST3/SND1*, and *VNDs* were grouped into subgroup OsNAC7.

2.3. Conserved Sequence Alignment and Motif Analysis

Multiple sequence alignments were created by Jalview for the 21 LiNAC proteins to investigate the presence and locations of conserved protein domains (Figure 2) and Weblogo displayed the conserved domains with five subdomains (A–E) (Figure 3). All LiNAC family members contained a NAM domain (Figure 4C), which contains a highly conserved A–E subdomain. To further analyze the structural diversity in LiNAC proteins, conserved motifs were searched using the MEME program. In total, 20 distinct motifs were identified (Figure 4B). Most LiNAC proteins contained motif 3 (representing subdomain A), motif 4 (representing subdomain B), motif 2 (representing subdomain C), motif 1 (representing subdomain D), and motif 5 (representing subdomain E) (Figure S2). Additionally, the majority of the members in the phylogenetic tree that were closely linked displayed comparable motifs in the same alignment and location (Figure 4A). Members in the TERN subgroup had nearly identical motifs. Interestingly, some motifs were identified in a certain subgroup. For example, compared with Group I, the number of motifs in Group II was small. Motif 14 was found only in the NAC1 subgroup and motif 16 in the ONAC022 subgroup. The results revealed that LiNAC proteins clustering in the same subgroup may have comparable biological roles and that specific motifs may be associated with specific functions of different subgroups.

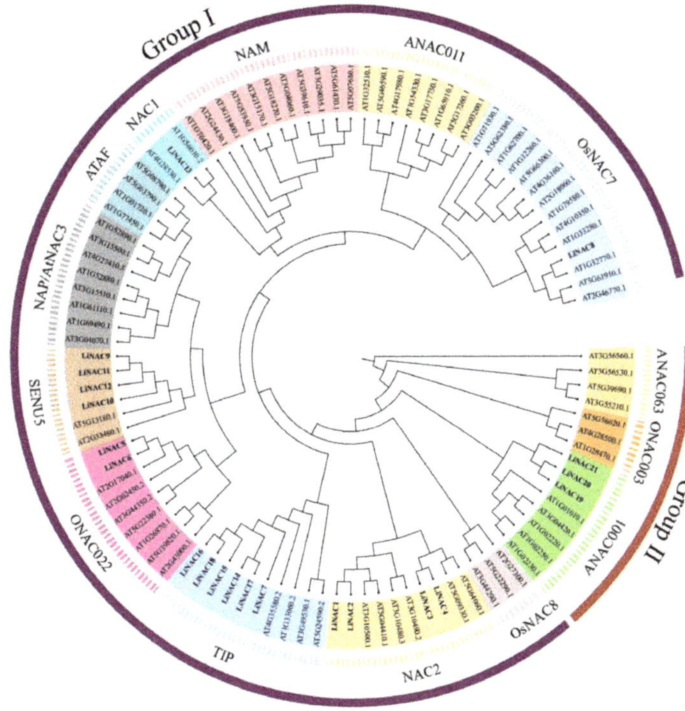

Figure 1. NAC phylogenetic tree of *L. indica* and *A. thaliana*. Each subgroup is distinguished by a different color.

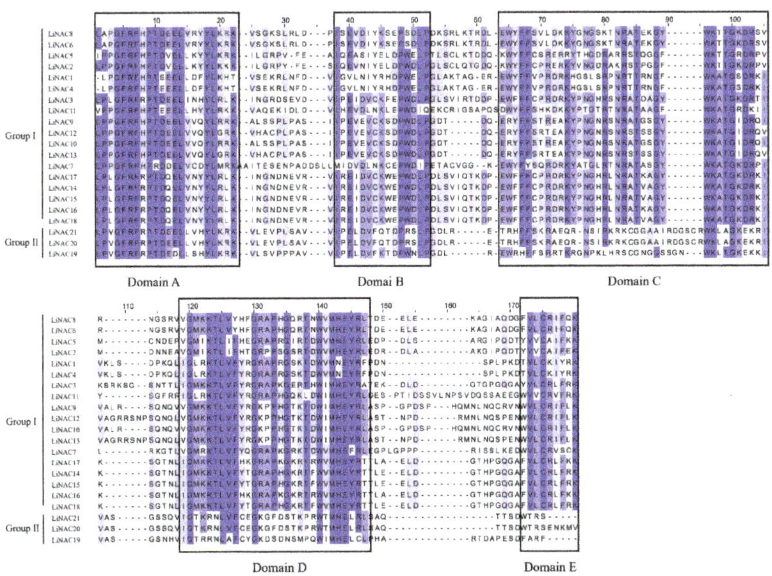

Figure 2. Amino acid sequence alignment of *L. indica*. The black box lines represent the five subdomains (**A**–**E**) of LiNAC.

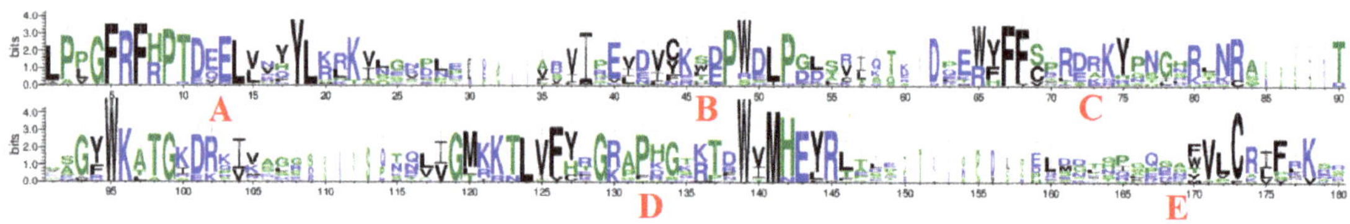

Figure 3. Conserved domain of LiNAC family by Jalview software and Weblogo. (A–E) represent five subdomains.

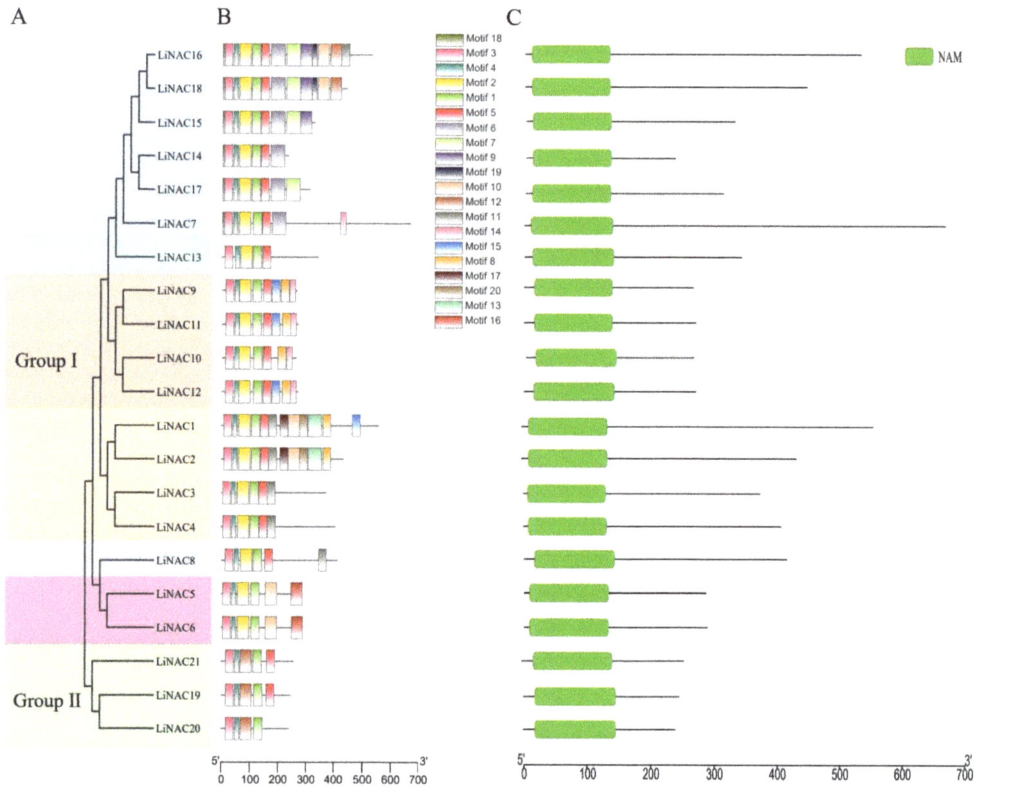

Figure 4. Evolutionary trees (**A**), conserved motifs (**B**), and conserved domain (**C**) of *LiNAC* genes.

2.4. Annotation and Enrichment Analysis of LiNAC Genes

GO functional annotation of 21 *LiNAC* genes revealed that 18 of them are involved in biological processes (BPs), cellular components (CCs), and molecular functions (MFs). We examined the GO enrichment data of 18 *LiNAC* genes to forecast their biological roles. The immunological effector mechanism has the highest degree of enrichment, as shown in Figure 5A, scoring 54.4, followed by the response of cells to cold stress, with an enrichment score of 51.72. Additionally, we found that many genes have roles in a variety of developmental processes (such as leaf development and shoot development).

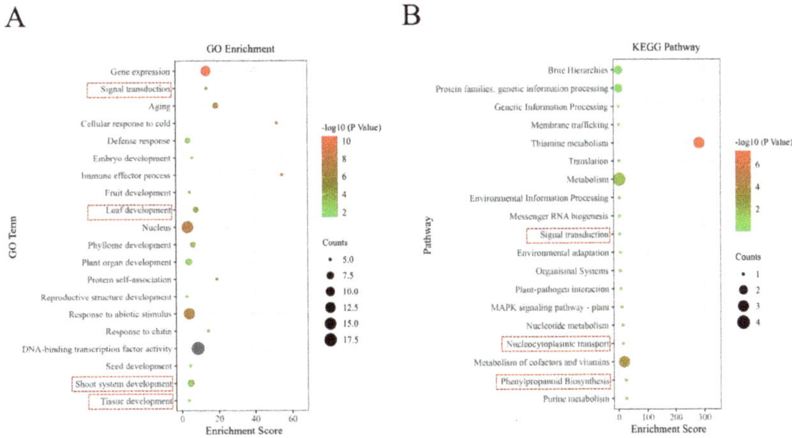

Figure 5. The GO terms (**A**) and KEGG pathways (**B**) enriched of *LiNAC* genes. The black circles indicate the number of target genes and different colors indicate the p−value.

As shown in Figure 5B, five *LiNAC* genes were enriched in the Nucleocytoplasmic Transport (ko03013), Phenylpropanoid Biosynthesis Pathway (ko00940), and Signal Pathway (ko04020), according to a KEGG pathway analysis of 21 *LiNAC* genes.

2.5. Determination and Analysis of Lignin Content

The lignin content in *L. indica* was measured by collecting the branches of 'Shaoguifei' and 'Xiariwuniang' at the early growth stage, the growth stage, and the end of the growth stage. The lignin content in the branches was analyzed separately for different stages and plants. Figure 6C shows that the lignin content in the branches of both 'Shaoguifei' and 'Xiariwuniang' increased gradually with the maturity of the growth period and the lignin content in 'Shaoguifei' was always higher than that in 'Xiariwuniang'. We found that the difference in lignin content between the two branches was not obvious at the early growth stage, but there was a significant difference between the lignin content in 'Shaoguifei' and 'Xiariwuniang' at the growth stage and the end of the growth stage.

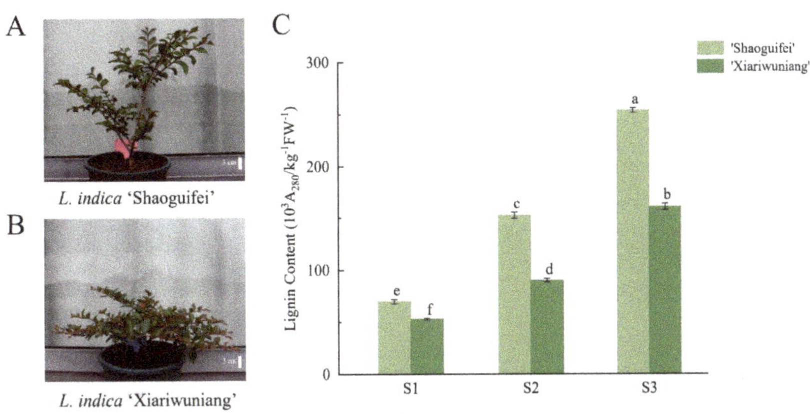

Figure 6. The trait of *L. indica* 'Shaoguifei' (**A**) and *L. indica* 'Xiariwuniang' (**B**) and lignin content of the two (**C**) at different growth stages. S1, S2, and S3 indicate the early growth stage, the growth stage, and the end of the growth stage. Different letters a–f indicate statistically significant differences in expression.

2.6. Expression Analysis of LiNAC Genes in Regulating the Weeping Trait

To identify which *LiNAC* genes were involved in regulating the weeping trait of *L. indica*, we used real-time PCR to analyze the expression patterns of *LiNAC* genes in branches of 'Shaoguifei' and 'Xiariwuniang' during the early growth stage, the growth stage, and the end of the growth stage. The expression pattern analysis is shown in Figure 7. The relative expression levels of *LiNAC3* and *LiNAC20* showed the same trend in the early growth stage (S1), growth stage (S2), and the end of the growth stage (S3) for 'Shaoguifei' and 'Xiariwuniang', which showed an increase first and then a decrease, and the expression level was higher in S2. The relative expression levels of five genes, *LiNAC7*, *LiNAC8*, *LiNAC14*, *LiNAC19*, and *LiNAC11*, in 'Shaoguifei' gradually increased from the S1 stage and reached the highest in S3. We found that, except for *LiNAC8* and *LiNAC11*, the relative expression of *LiNAC7*, *LiNAC14*, and *LiNAC19* showed a trend of increasing and then decreasing in 'Xiariwuniang', with the highest expression in the S2. Interestingly, the expression trend of *LiNAC8* in 'Xiariwuniang' is consistent with that in 'Shaoguifei', while the relative expression level of 'Shaoguifei' is slightly higher than that of 'Xiariwuniang' in three stages and the relative expression level is about twice that of 'Xiariwuniang'. We noticed that the relative expression of nine genes (*LiNAC1*, *LiNAC9*, *LiNAC10*, *LiNAC16*, *LiNAC4*, *LiNAC5*, *LiNAC9*, *LiNAC21*, and *LiNAC15*) decreased significantly from S1 to S2 in 'Shaoguifei' and increased gradually in S3. Except for four genes, *LiNAC1*, *LiNAC10*, *LiNAC16*, and *LiNAC15*, whose relative expression decreased from S1 to S3 in 'Xiariwuniang', all the other five genes showed an increasing and then decreasing trend and their relative expression peaked at S2. The relative expression levels of *LiNAC2*, *LiNAC6*, *LiNAC12*, *LiNAC18*, and *LiNAC13* in 'Shaoguifei' showed a trend of gradual decline and the lowest expression level in S3 (except *LiNAC13*), especially *LiNAC12*, in S3 period, the expression level of *LiNAC12* was 0.1-times that of 'Shaoguifei'. Interestingly, *LiNAC2*, *LiNAC18*, and *LiNAC13* also showed the same expression trend as that of 'Shaoguifei' in 'Xiariwuniang'. Notably, the relative expression of *LiNAC2* in 'Xiariwuniang' was always higher than that of 'Shaoguifei', especially in the S1–S2 stages. Further, the expression of *LiNAC13* was always higher than that of 'Xiariwuniang' in all stages; the relative expression was 9.7-times higher than that of 'Xiariwuniang' in S3.

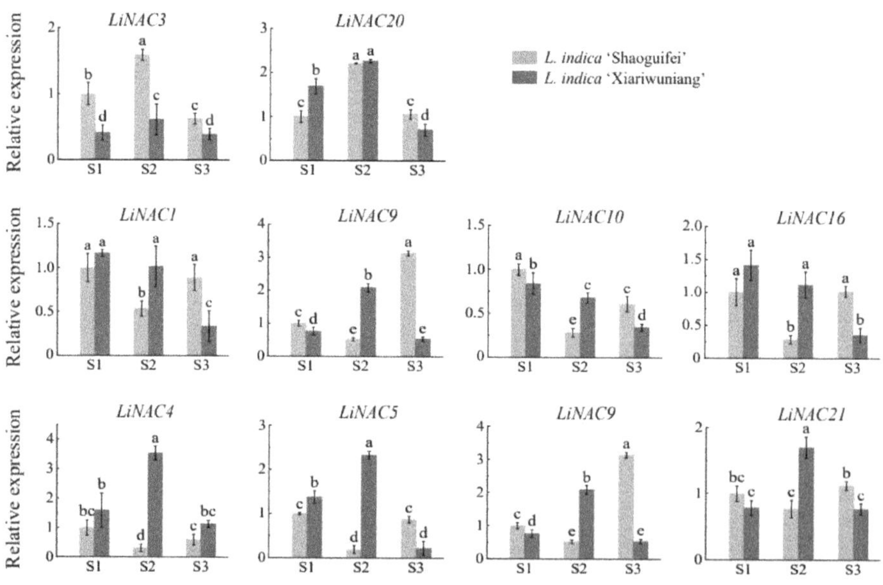

Figure 7. *Cont.*

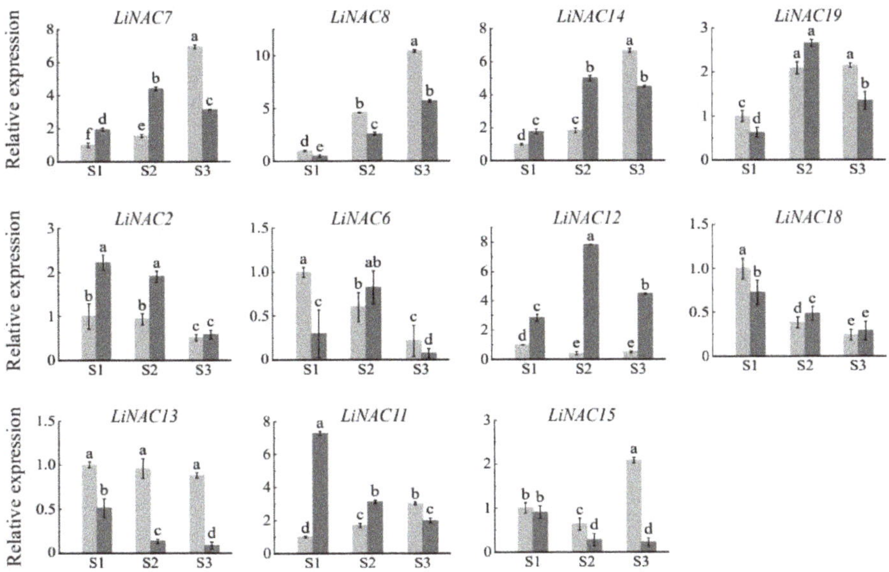

Figure 7. The qRT-PCR analysis of 21 *LiNAC* genes in Upright and Weeping *L. indica* at different growth stages. S1, S2, and S3 indicate the early growth stage, the growth stage, and the end of the growth stage; significant differences are identified by SPSS with Duncan's test ($p < 0.05$) and are represented by different letters above the error bars.

3. Discussion

3.1. The Characteristics of NAC Gene Family in L. indica

In this study, we identified 21 complete *LiNAC* genes with Open Reading Frames based on *L. indica* transcriptome data and analyzed their basic information. We predicted and analyzed 21 LiNAC proteins, such as isoelectric power, relative molecular weight, and transmembrane structure, LiNAC7, and LiNAC16 have a transmembrane structure at the C-terminal, and it is speculated that they are likely to be NAC-membrane-binding transcription factors. Almost all *LiNAC* genes are predicted to be localized in the nucleus (Table 1); *LiNACs* are probably mostly functional in the nucleus. The conserved structure and motif of *LiNACs* were also examined (Figure 4B) and it was shown that all of them possessed the normal NAM structure (Figure 4C). Nearly all *LiNAC* genes have motif 3, motif 4, and motif 1, although motif 2 is missing from Group II, which may be related to the loss of assembled data. The motifs found in the gene members of various evolutionary branches vary and motif 3, motif 4, motif 2, motif 1, and motif 5 may be the conserved motifs in the NAC family (Figure S2). Phylogenetic trees of *L. indica* and *Arabidopsis* divide into 2 groups with 14 subgroups (Figure 1) and *LiNAC* genes were not found in many subgroups (e.g., ANAC011, ATAF, and AtNAC3/NAP). Overexpression of *AtNAP* causes early senescence in *A. thaliana* and members of the ATAF subgroup have a significant role in leaf senescence and other characteristics of plants, indicating that these subgroups may be involved in the process of plant senescence. *LiNAC8* is closely connected to AT1G32770.1, AT3G61910.1, and AT2G46770, for instance, leading one to hypothesize about the function of the *LiNAC* gene based on genetic similarities. AT1G32770.1 (*SND1*) is a key gene that regulates the synthesis of cell secondary walls. In *Arabidopsis*, *SND1* is specifically expressed in xylem fibers, and functional silencing led to secondary wall thinning and the inability of pedicel fiber cells to form secondary walls and drooping. The other subgroups of the *LiNAC* gene may play different biological functions.

Further, 18 of the 21 *LiNAC* genes were shown to be engaged in the two biological processes of GO, leaf growth and branch creation, according to KEGG and GO enrichment analyses (Figure 5). While lignin, a phenylpropane-derived polymer, together with cellulose and hemicellulose, forms the cell wall of plant vascular tissue and provides mechanical support for plant upright growth, KEGG enrichment results revealed that *LiNAC* genes were involved in the phenylpropanoid biosynthesis pathway. As a result, it is hypothesized that the *LiNAC* genes may influence the production of secondary cell walls in plants to regulate the weeping trait.

3.2. *LiNACs Are Closely Associated with the Weeping Trait of* L. indica

It has been suggested that the weeping trait may be associated with abnormal GA signaling [31]. The drooping phenomenon also occurs when the gravity response changes [32]. Changes in the distribution of plant hormones often affect the formation of plant xylem and bast, which is another factor contributing to the phenomenon of weeping plants. Lignin is a phenylpropanoid-derived polymer that, together with cellulose and hemicellulose, forms the cell wall of plant vascular tissue and provides mechanical support for upright plant growth [33,34]. The NAC family of transcription factors is the master switch in the regulation of secondary cell wall thickening in plants and influences the plant phenotype by regulating the growth of secondary cell walls in plants. The cell wall thickness of upright branches in *P. mume* was significantly higher than that of weeping branches. In terms of overall expression, *PmWND1*(*PmNAC082*) [35] regulated downstream secondary-wall-synthesis-related genes, which were all significantly less expressed in the weeping branches than in the upright branches, indicating that *NAC* may play an important role in regulating the production of weeping traits in plants.

In this study, qRT-PCR analysis of upright *L. indica* 'Shaoguifei' and weeping *L. indica* 'Xiariwuniang' at three different growth stages showed that *LiNACs* played an important role in the weeping trait of *L. indica* and different *LiNACs* expression patterns were different during this process. The expression of *LiNAC13*, a member of the NAC1 subgroup, did not change significantly in the three growth stages of 'Shaoguifei', but decreased gradually in 'Xiariwuniang' and was much lower than that of 'Shaoguifei'. The expression of *LiNAC13* in 'Xiariwuniang' decreased gradually and was much lower than that of 'Shaoguifei'. NAC1 is normally involved in hormone signaling in plants to regulate plant growth and development [26] and it is speculated that *LiNAC13* may be involved in regulating weeping traits in plants by participating in the phytohormone signaling pathway, but this remains to be verified. The relative expression of *LiNAC2* in S1–S3 is always higher in 'Xiariwuniang' than in 'Shaoguifei', especially in the S1–S2 stages, and there is not much difference between them in the S3 stage. *LiNAC2* is closely related to AT3G105001/*NTL4*, so we hypothesized that *LiNAC2* began to negatively regulate the weeping trait of *L. indica* at S1. This still needs to be further verified by subsequent experiments. *LiNAC12* is closely related to AT5G13180.1 (*VNI2*), a VNI2 transcription factor that plays a molecular linkage role between plant response to environmental stress and regulation of leaf longevity [36]. In contrast, the expression of *LiNAC12* in 'Xiariwuniang' was always higher than that of 'Shaoguifei' and reached a peak during S2. It is speculated that 'Xiariwuniang' may have an advantage over 'Shaoguifei' in regulating plant growth and coping with environmental stress during development, which may also be a reason for its weeping branch formation. The relative expression of *LiNAC8* (which is closely related to AT1G32770.1/*SND1*) gradually increased in the three growth stages (S1–S3) of 'Shaoguifei' and 'Xiariwuniang' branches and the expression of 'Shaoguifei' was always higher than that of 'Xiariwuniang'. In our study, by measuring the lignin content in 'Shaoguifei' and 'Xiariwuniang' branches at different growth stages, we found that the lignin content in 'Shaoguifei' was always higher than that of 'Xiariwuniang'. Further, in the early stage of growth, there was not much difference in the content between the two, but with the development, especially at the end of the growth stage, the lignin content in 'Shaoguifei' was much higher than that of 'Xiariwuniang' (Figure 6). Inhibition of *SND1* expression in *Arabidopsis* leads to a dramatic decrease in the expression of all three secondary wall synthesis genes and pedicel fiber cells are unable to form secondary walls and droop grow [15]. In addition, dominant inhibition of *PtrWNDs* in poplar caused significant thinning of secondary walls of transgenic poplar stem fibers and failure of plants to grow upright [18]. Lignin is an important component in the secondary cell wall of plants and when the lignin content increases, the fruit of *Eriobotrya japonica* also becomes hard [21]. Therefore, we speculate that 'Shaoguifei' has more lignin content than 'Xiariwuniang' and the branches of 'Shaoguifei' provide more mechanical support, while 'Xiariwuniang' has less lignin content and, therefore, droops. *SND5* in *Arabidopsis* is closely connected to *SND2/3*, which is involved with the secondary wall and its direct homolog in poplar, according to Zhong et al. [19]. Similar findings were found in our study, *LiNAC8* showed a consistent trend with lignin content at S1–S3 in different *L. indica* species, which indicated that *LiNAC8* may be involved in affecting the synthesis of secondary walls and determining the weeping trait of *L. indica*. However, this conclusion has to be further investigated.

4. Materials and Methods

4.1. Identification and Sequence Analysis of NAC Genes from L. indica

As a reference sequence, the NAC protein sequence was acquired from the *Arabidopsis* TAIR database (https://www.arabidopsis.org/) (accessed on 17 March 2022) and members of the putative *LiNAC* genes were sought for using the *L. indica* transcriptome database (the raw sequence data were obtained from the website https://bigd.big.ac.cn/gsa) (accessed on 17 March 2022) and a BLASTP search. Then, we used the local HMMER3.0 software(Robert, D.F.; Ashburn, VA, USA) and the Hidden Markov (HMM) profile of the NAC protein (PF02365), which was downloaded from the Pfam database (http://pfam.xfam.org) (accessed on 17 March 2022). The potential gene members discovered using the two search techniques were pooled. The conserved NAM domain was discovered in all potential *LiNAC* genes using the Batch Web CD-Search Tool (https://www.ncbi.nlm.nih.gov/cdd) (accessed on 17 March 2022) and SMART (http://smart.embl.de/) (accessed on 17 March 2022), ensuring the correctness of the results.

4.2. Characterization of LiNAC Proteins

The website ExPasy (https://www.expasy.org/) (accessed on 18 March 2022) predicted and examined the physicochemical characteristics of all LiNAC potential proteins, including molecular weight and theoretical points. TMHMM-2.0, a website that can be accessed online (https://services.healthtech.dtu.dk/service.php?TMHMM-2.0) (accessed on 18 March 2022), examined the transmembrane structure of LiNAC protein. The online program MEME (https://meme-suite.org/meme/tools/meme) (accessed on 20 March 2022) was used to find conserved motifs. The following parameters were utilized: maximum number of motifs = 20; search model = zero or one occurrence per sequence; default values for the remainder. TBtools software v1.098661(Chen, C.J.; Guangzhou, China) was used to display the results [37]. Cell-PLoc2.0 was used to create predictions for subcellular localization (http://www.csbio.sjtu.edu.cn/bioinf/plant-multi/#) (accessed on 20 March 2022).

4.3. Sequence Alignment and Phylogenetic Tree Construction

Using MUSCLE, the sequences of the found 21 LiNAC proteins and 80 AtNAC protein sequences were aligned. In MEGA 6.0, the outcome was utilized to create a neighbor-joining (NJ) phylogenetic tree using 1000 bootstrap repetitions. P-distance and pairwise were the predetermined parameters. Phylogenetic tree were illustrations created with iTOL (https://itol.embl.de/) (accessed on 20 March 2022). Using the Jalview software 2.11.2.4 (Andrew, M.W.; Cambridge, MA, USA) (http://www.jalview.org/)(accessed on 20 March 2022), the amino acid sequences of conserved domains were compared and modified and conserved motif Logos were produced using the WebLogo program (http://weblogo.threeplusone.com) (accessed on 20 March 2022).

4.4. Annotation and Enrichment Analysis in GO and KEGG Databases

Online software EggNOG-Mapper (http://eggnog-mapper.embl.de/) (accessed on 20 March 2022) [38] and KEGG database (https://www.kegg.jp/) (accessed on 20 March 2022) were used to annotate the GO and KEGG functions of *LiNAC* gene. The results were collated using the eggNOG-mapper Helper function of TBtools v1.098661 and the text files for downstream analysis were output into GO Enrichment and KEGG Enrichment Analysis for enrichment analysis, respectively. Finally, the data were seen and examined using the online charting tool HIPLOT (https://hiplot.com.cn/) (accessed on 20 March 2022).

4.5. Determination of Lignin Content

The plant materials were collected from the greenhouse of Zhejiang Agriculture and Forestry University, which is situated at 30°15′2″ N/119°43′37″ E, by upright *L. indica* 'Shaoguifei' (Figure 6A) and weeping *L. indica* 'Xiariwuniang' (Figure 6B). They were relocated into a habitat with stable temperature and favorable growing circumstances. By keeping track of processes since the time of budding, the growth condition of branches was studied.

The whole branches of 'Shaoguifei' and 'Xiariwuniang' were weighed 0.5 g at the early growth stage, growth stage, and the end of the growth stage, respectively. Further, three samples of each were collected and ground in liquid nitrogen and the ground powder was packed into 5 mL of an eluent. We then, centrifuged at 12,000 rpm for 20 min after 30 min of shaking the sample at 28 °C, then discarded the supernatant. After adding 100% methanol, the mixture was shaken for 30 min before being centrifuged at 12,000 rpm for 20 min, with the supernatant being discarded. There were four iterations of this stage. After that, they were dried in an oven set to 80 °C for an overnight period.

Accurately weigh 10 mg (can be recorded repeatedly) of the powder (washed and dried) into a 10 mL tube (the total weight of the powder after drying should also be recorded): first add 1 mL of 2 M HCL, then add 0.1 mL of thioglycolic acid, mix it upside down and evenly, then place it in a boiling water bath and heat it for 8 h. Then it was cooled on ice and centrifuged at 12,800 rpm/4 °C for 20 min and the supernatant was discarded. The precipitate was washed twice with distilled water and the precipitate was dried overnight after centrifugation. Then the precipitate was resuspended in 2 mL of 1 M NaOH, mixed evenly, and slightly shaken at 28 °C to react for 18 h. The precipitate was centrifuged at 12,800 rpm for 20 min. Next, 0.5 mL of the supernatant was put into a new glass test tube, 100 μL of concentrated hydrochloric acid was added to each tube, and the solution was placed in a refrigerator at 4 °C for 4 h (this operation was to precipitate the thioglycolate-bound lignin). The solution was centrifuged at 12,800 rpm/4 °C for 20 min to precipitate 1 mL of 1 M NaOH.

After dilution, a UV spectrophotometer was used to determine the absorbance at A280 nm. Each sample underwent three biological replicates, with NaOH solution serving as the blank control. This lignin determination method is referenced by Xu et al. [21].

4.6. Plant Materials, RNA Extraction, and qRT-PCR Analysis

Take the entire branches of 'Xiariwuniang' and 'Shaoguifei' in the early growth stage, growth stage, and the end of the growth stage to examine the expression patterns of *LiNAC* genes. Three samples of each were collected and the sampling time was 10:00. All samples were immediately frozen in liquid nitrogen and stored at −80 °C until needed for RNA isolation.

Total RNA was extracted from plants according to the Instructions of FastPure® Plant Total RNA Isolation Kit (Vazyme, Nanjing, China). Reverse transcription reference HiScript® III All-in-one RT SuperMix perfect for qPCR (Vazyme, Nanjing, China) was used. Then, quantitative real-time PCR (qRT-PCR) analysis was performed on an ABI 7300 real-time PCR instrument (Applied Biosystems, Foster City, CA, USA). The reaction system is as follows: SYBR® Premix Ex TaqTM (TaKaRa, Dalian, China) 5 μL, cDNA 2 μL, forward and reverse primers 0.4 μL each, ddH_2O 2.2 μL. qRT-PCR reaction procedure is:

pre-denaturation at 95 °C for 30 s, denaturation at 95 °C for 5 s, denaturation at 60 °C for 30 s, 40 cycles. Three biological replicates were performed for each sample. The relative expression levels of *LiNACs* genes were analyzed by the $2^{-\Delta\Delta Ct}$ method [39] and the experimental data were analyzed by Excel 2010 and SPSS Statistics 20.0 software (IBM Corporation, Armonk, NY, USA). Finally, 21 pairs of gene-specific primers are shown in Table S1 and *LiEF-1α* [40] was used as the internal reference gene.

5. Conclusions

In this study, we identified a total of 21 *LiNAC* genes from the transcriptome data of *L. indica*. We analyzed the physicochemical properties of the 21 LiNAC proteins and all *LiNAC* genes were localized in the nucleus, and *LiNAC7* and *LiNAC16* also had a transmembrane structure at the C-terminus. Amino acid sequence alignment showed that almost all *LiNAC* genes contain a conserved NAM structure, consisting of five substructures, A, B, C, D, and E. The phylogenetic tree constructed with *A. thaliana* classified LiNAC proteins into 2 groups and 14 subgroups. GO and KEGG enrichment analysis also indicated that *LiNAC* genes are involved in plant growth development and metabolic pathways. The lignin content in the branches of upright *L. indica* 'Shaoguifei' and weeping *L. indica* 'Xiariwuniang' were also measured and it was found that the lignin content in the two branches differed at different growth and development stages. The qRT-PCR analysis showed that the expression of *LiNAC13* and *LiNAC8* in 'Shaoguifei' was always higher than that in 'Xiariwuniang' during S1–S3, while the relative expression of *LiNAC2* in 'Shaoguifei' was lower than that in 'Xiariwuniang' in S1–S3, which indicated that *LiNAC2*, *LiNAC13*, and *LiNAC8* might regulate the weeping traits of *L. indica* through their respective regulatory pathways. However, the specific molecular mechanism of regulation and the downstream genes that jointly regulate the weeping traits need to be further verified. In conclusion, the current study improved our understanding of the role of the *NAC* gene family in weeping traits in *L. indica*.

Supplementary Materials: The following supporting information can be downloaded at: https://www.mdpi.com/article/10.3390/plants11162168/s1, Figure S1: The transmembrane structure of *LiNAC7* and *LiNAC16*; Figure S2: Conserved motifs in five subdomains of *LiNACs*; Table S1: Specific primers for RT-PCR of *LiNAC* in this study.

Author Contributions: Conceptualization, L.Y., C.G. and L.S.; methodology, C.G.; software, L.S.; validation, Q.W., Y.Z. and G.Z.; investigation, C.G. and S.H.; data curation, L.S. and Q.M.; writing—original draft preparation, C.G. and L.S.; writing—review and editing, L.Y., C.G., L.S., Q.W. and G.Z.; funding acquisition, L.Y. and C.G. All authors have read and agreed to the published version of the manuscript.

Funding: This work was financially supported by the Research Development Fund of Zhejiang A&F University (2019FR028); Open Fund of Zhejiang Provincial Key Laboratory of Germplasm Innovation and Utilization for Garden Plants (2020E10013-K202104); Zhejiang Provincial Natural Science Foundation of China (No. LY21C160001); and Zhejiang Science and Technology Major Program on Agricultural New Variety Breeding (No. 2021C02071-4).

Institutional Review Board Statement: Not applicable.

Informed Consent Statement: Not applicable.

Data Availability Statement: All data in this study can be found in the manuscript or in the Supplementary Materials.

Conflicts of Interest: The authors declare no conflict of interest.

References

1. Hill, J.L.; Hollender, C.A. Branching out: New insights into the genetic regulation of shoot architecture in trees. *Curr. Opin. Plant Biol.* **2019**, *47*, 73–80. [CrossRef]
2. Hollender, C.A.; Pascal, T.; Tabb, A.; Hadiarto, T.; Srinivasan, C.; Wang, W.; Liu, Z.; Scorza, R.; Dardick, C. Loss of a highly conserved sterile alpha motif domain gene (*WEEP*) results in pendulous branch growth in peach trees. *Proc. Natl. Acad. Sci. USA* **2018**, *115*, e4690–e4699. [CrossRef]
3. Liu, J.; Zeng, Y.; Yan, P.; He, C.; Zhang, J. Transcriptional and hormonal regulation of weeping trait in *Salix matsudana*. *Genes* **2017**, *8*, 359. [CrossRef]
4. Hou, D.; Zhang, J.; Zheng, T.C.; Zhang, Q.X. A study about branch physical structure of upright/weeping Mei (*Prunus mume* Sieb. et Zucc). *Adv. Ornament Hortic. China* **2017**, *2017*, 552–559.
5. Mao, T.Y.; Zhu, H.H.; Liu, Y.; Bao, M.Z.; Zhang, J.W.; Fu, Q.; Xiong, C.F.; Zhang, J. Weeping candidate genes screened using comparative transcriptomic analysis of weeping and upright progeny in an F1 population of *Prunus mume*. *Physiol. Plant* **2020**, *170*, 318–334. [CrossRef]
6. Li, L.L.; Zhang, Y.C.; Zheng, T.C.; Zhuo, X.K.; Li, P.; Qiu, L.K.; Liu, W.C.; Wang, J.; Cheng, T.R.; Zhang, Q.X. Comparative gene expression analysis reveals that multiple mechanisms regulate the weeping trait in *Prunus mume*. *Sci. Rep.* **2021**, *11*, 2675. [CrossRef]
7. Zheng, T.C.; Li, P.; Zhuo, X.K.; Liu, W.C.; Qiu, L.K.; Li, L.L.; Yuan, C.Q.; Sun, L.D.; Zhang, Z.Y.; Wang, J.; et al. The chromosome-level genome provides insight into the molecular mechanism underlying the tortuous-branch phenotype of *Prunus mume*. *New Phytol.* **2022**, *235*, 141–156. [CrossRef]
8. Joshi, C.P.; Thammannagowda, S.; Fujino, T.; Gou, J.Q.; Avci, U.; Haigler, C.H.; McDonnell, L.M.; Mansfield, S.D.; Mengesha, B.; Carpita, N.C.; et al. Perturbation of wood cellulose synthesis causes pleiotropic effects in transgenic aspen. *Mol. Plant* **2011**, *4*, 331–345. [CrossRef]
9. Jensen, M.K.; Kjaersgaard, T.; Nielsen, M.M.; Galberg, P.; Petersen, K.; O'Shea, C.; Skriver, K. The *Arabidopsis thaliana* NAC transcription factor family: Structure-function relationships and determinants of *ANAC019* stress signaling. *Biochem. J.* **2010**, *426*, 183–196. [CrossRef]
10. Souer, E.; Van, H.A.; Kloos, D.; Mol, J.; Koes, R. The no apical meristem gene of *Petunia* is required for pattern formation in embryos and flowers and is expressed at meristem and primordia boundaries. *Cell* **1996**, *85*, 159–170. [CrossRef]
11. Aida, M.; Ishida, T.; Fukaki, H.; Fujisawa, H.; Tasaka, M. Genes involved in organ separation in *Arabidopsis*: An analysis of the cup-shaped cotyledon mutant. *Plant Cell* **1997**, *9*, 841–857. [CrossRef] [PubMed]
12. Ooka, H.; Satoh, K.; Doi, K.; Nagata, T.; Otomo, Y.; Murakami, K.; Matsubara, K.; Osato, N.; Kawai, J.; Carninci, P.; et al. Comprehensive analysis of NAC family genes in *Oryza sativa* and *Arabidopsis thaliana*. *DNA Res.* **2003**, *10*, 239–247. [CrossRef] [PubMed]
13. Melo, B.P.; Lourenço-Tessutti, I.T.; Fraga, O.T.; Pinheiro, L.B.; de Jesus Lins, C.B.; Morgante, C.V.; Engler, J.A.; Reis, P.A.B.; Grossi-de-Sá, M.F.; Fontes, E.P.B. Contrasting roles of *GmNAC065* and *GmNAC085* in natural senescence, plant development, multiple stresses and cell death responses. *Sci. Rep.* **2021**, *11*, 11178. [CrossRef]
14. Hu, P.; Zhang, K.M.; Yang, C.P. BpNAC012 positively regulates abiotic stress responses and secondary wall biosynthesis. *Plant Physiol.* **2019**, *179*, 700–717. [CrossRef] [PubMed]
15. Zhong, R.Q.; Richardson, E.A.; Ye, Z.H. Two NAC domain transcription factors, *SND1* and *NST1*, function redundantly in regulation of secondary wall synthesis in fibers of *Arabidopsis*. *Planta* **2007**, *225*, 1603–1611. [CrossRef] [PubMed]
16. Tusong, K.; Mamat, A.; Xu, J. Genome-wide investigation of NAC family genes potentially related to SCW biogenesis in korla pear fruit. *Plant Mol. Biol. Rep.* **2022**. [CrossRef]
17. Vargas-Hernández, B.Y.; Núñez-Muñoz, L.; Calderón-Pérez, B.; Xoconostle-Cázares, B.; Ruiz-Medrano, R. The NAC transcription factor *ANAC087* induces aerial rosette development and leaf senescence in *Arabidopsis*. *Front. Plant Sci.* **2022**, *13*, 818107. [CrossRef]
18. Zhao, Y.; Sun, J.Y.; Xu, P.; Zhang, R.; Li, L.G. Intron-mediated alternative splicing of wood-associated NAC transcription factor1B regulates cell wall thickening during fiber development in *Populus* species. *Plant Physiol.* **2014**, *164*, 765–776. [CrossRef]
19. Zhong, R.Q.; Lee, C.; Haghighat, M.; Ye, Z.H. Xylem vessel-specific *SND5* and its homologs regulate secondary wall biosynthesis through activating secondary wall NAC binding elements. *New Phytol.* **2021**, *231*, 1496–1509. [CrossRef]
20. Yao, W.; Zhang, D.W.; Zhou, B.; Wang, J.P.; Li, R.H.; Jiang, T.B. Over-expression of poplar *NAC15* gene enhances wood formation in transgenic tobacco. *BMC Plant Biol.* **2020**, *20*, 12. [CrossRef]
21. Xu, Q.; Wang, W.Q.; Zeng, J.K.; Zhang, J.; Grierson, D.; Li, X.; Yin, X.R.; Chen, K.S. A NAC transcription factor, *EjNAC1*, affects lignification of loquat fruit by regulating lignin. *Postharvest Biol. Tec.* **2015**, *102*, 25–31. [CrossRef]
22. Zhao, D.Q.; Luan, Y.T.; Shi, W.B.; Tang, Y.H.; Huang, X.Q.; Tao, J. Melatonin enhances stem strength by increasing lignin content and secondary cell wall thickness in herbaceous peony. *J. Exp. Bot.* **2022**. [CrossRef] [PubMed]
23. Zhuo, X.K.; Zheng, T.C.; Li, S.Z.; Zhang, Z.Y.; Zhang, M.; Zhang, Y.C.; Ahmad, S.; Sun, L.D.; Wang, J.; Cheng, T.R.; et al. Identification of the *PmWEEP* locus controlling weeping traits in *Prunus mume* through an integrated genome-wide association study and quantitative trait locus mapping. *Hort. Res.* **2021**, *8*, 131. [CrossRef] [PubMed]

24. Ju, Y.Q.; Feng, L.; Wu, J.Y.; Ye, Y.J.; Zheng, T.C.; Cai, M.; Cheng, T.R.; Wang, J.; Zhang, Q.X.; Pan, H.T. Transcriptome analysis of the genes regulating phytohormone and cellular patterning in *Lagerstroemia* plant architecture. *Sci. Rep.* **2018**, *8*, 15162. [CrossRef] [PubMed]
25. Zhang, Q.X.; Tian, M. Investigation on crape myrtle varieties in China. *Adv. Ornament Hortic. China* **2008**, *2008*, 654–658.
26. Zhu, Q.K.; Zou, J.X.; Zhu, M.L.; Liu, Z.B.; Feng, P.C.; Fan, G.T.; Wang, W.J.; Liao, H. In silico analysis on structure and DNA binding mode of *AtNAC1*, a NAC transcription factor from *Arabidopsis thaliana*. *J. Mol. Model.* **2014**, *20*, 2117. [CrossRef]
27. Guo, T.T.; Mao, X.G.; Zhang, H.; Zhang, Y.; Fu, M.D.; Sun, Z.F.; Kuai, P.; Lou, Y.G.; Fang, Y.D. Lamin-like proteins negatively regulate plant immunity through NAC with transmembrane motif1-like 9 and nonexpresser of pr genes1 in *Arabidopsis thaliana*. *Mol. Plant* **2017**, *10*, 1334–1348. [CrossRef]
28. Li, W.H.; Zeng, Y.L.; Yin, F.L.; Wei, R.; Mao, X.F. Genome-wide identification and comprehensive analysis of the NAC transcription factor family in sunflower during salt and drought stress. *Sci. Rep.* **2021**, *11*, 19865. [CrossRef]
29. Wang, X.; Basnayake, B.M.; Zhang, H.; Li, G.; Li, W.; Virk, N.; Mengiste, T.; Song, F. The Arabidopsis *ATAF1*, a NAC transcription factor, is a negative regulator of defense responses against necrotrophic fungal and bacterial pathogens. *Mol. Plant Microbe Interact.* **2009**, *22*, 1227–1238. [CrossRef]
30. Hu, R.B.; Qi, G.; Kong, Y.Z.; Kong, D.J.; Gao, Q.; Zhou, G.K. Comprehensive analysis of NAC domain transcription factor gene family in *Populus trichocarpa*. *BMC Plant Biol.* **2010**, *10*, 145. [CrossRef]
31. Fideghelli, C.; Sartori, A.; Grassi, F. Fruit tree size and architecture. *Acta. Hort.* **2003**, *622*, 279–293. [CrossRef]
32. Weise, E.S.; Kiss, Z.J. Gravitropism of inflorescence stems in starch-deficient mutants of *Arabidopsis*. *Int. J. Plant Sci.* **1999**, *160*, 521–527. [CrossRef] [PubMed]
33. Hoson, T.; Wakabayashi, K. Role of the plant cell wall in gravity resistance. *Phytochemistry* **2015**, *112*, 84–90. [CrossRef] [PubMed]
34. Zhao, Q. Lignification: Flexibility, Biosynthesis and Regulation. *Trends Plant Sci.* **2016**, *21*, 713–721. [CrossRef] [PubMed]
35. Li, X.; Li, P.; Zheng, T.C.; Zhang, Q.X. Genomic insights into the important ornamental and stress resistance traits of *Prunus mume*. *Sci. Hortic.* **2022**, *302*, 111179. [CrossRef]
36. Yang, S.D.; Seo, P.J.; Yoon, H.K.; Park, C.M. The *Arabidopsis* NAC transcription factor *VNI2* integrates abscisic acid signals into leaf senescence via the *COR/RD* genes. *Plant Cell* **2011**, *23*, 2155–2168. [CrossRef] [PubMed]
37. Chen, C.J.; Chen, H.; Zhang, Y.; Thomas, H.R.; Frank, M.H.; He, Y.H.; Xia, R. TBtools: An integrative toolkit developed for interactive analyses of big biological data. *Mol. Plant* **2020**, *13*, 1194–1202. [CrossRef] [PubMed]
38. Huerta-Cepas, J.; Szklarczyk, D.; Heller, D.; Hernández-Plaza, A.; Forslund, S.K.; Cook, H.; Mende, D.R.; Letunic, I.; Rattei, T.; Jensen, L.J.; et al. eggNOG 5.0: A hierarchical, functionally and phylogenetically annotated ortholog resource based on 5090 organisms and 2502 viruses. *Nucleic Acids Res.* **2018**, *47*, D309–D314. [CrossRef]
39. Livak, K.J.; Schmittgen, T.D. Analysis of relative gene expression data using real-time quantitative PCR and the $2^{-\Delta\Delta Ct}$ method. *Methods* **2001**, *25*, 402–408. [CrossRef]
40. Zheng, T.C.; Chen, Z.L.; Ju, Y.Q.; Zhang, H.; Cai, M.; Pan, H.T.; Zhang, Q.X. Reference gene selection for qRT-PCR analysis of flower development in *Lagerstroemia indica* and *L. speciosa*. *PLoS ONE* **2018**, *13*, e0195004. [CrossRef]

plants

MDPI

Article

Genome-Wide Identification and Expression Analysis of HSF Transcription Factors in Alfalfa (*Medicago sativa*) under Abiotic Stress

Jin Ma [1,2,3,†], Guozhe Zhang [1,2,3,†], Yacheng Ye [1,2,3,†], Linxue Shang [1,2,3], Sidan Hong [1,2,3], Qingqing Ma [1,2,3], Yu Zhao [1,2,3,*] and Cuihua Gu [1,2,3,*,†]

[1] College of Landscape and Architecture, Zhejiang Agriculture & Forestry University, Hangzhou 311300, China
[2] Zhejiang Provincial Key Laboratory of Germplasm Innovation and Utilization for Garden Plants, Zhejiang Agriculture & Forestry University, Hangzhou 311300, China
[3] Key Laboratory of National Forestry and Grassland Administration on Germplasm Innovation and Utilization for Southern Garden Plants, Zhejiang Agriculture & Forestry University, Hangzhou 311300, China
* Correspondence: zhaoyu@stu.zafu.edu.cn (Y.Z.); gucuihua@zafu.edu.cn (C.G.)
† These authors contributed equally to this work.

Abstract: Alfalfa (*Medicago sativa*) is one of the most important legume forage species in the world. It is often affected by several abiotic stressors that result in reduced yields and poor growth. Therefore, it is crucial to study the resistance of *M. sativa* to abiotic stresses. Heat shock transcription factors (HSF) are key players in a number of transcriptional regulatory pathways. These pathways play an essential role in controlling how plants react to different abiotic stressors. Studies on the HSF gene family have been reported in many species but have not yet undergone a thorough analysis in *M. sativa*. Therefore, in order to identify a more comprehensive set of *HSF* genes, from the genomic data, we identified 16 members of the *MsHSF* gene, which were unevenly distributed over six chromosomes. We also looked at their gene architectures and protein motifs, and phylogenetic analysis allowed us to divide them into 3 groups with a total of 15 subgroups. Along with these aspects, we then examined the physicochemical properties, subcellular localization, synteny analysis, GO annotation and enrichment, and protein interaction networks of amino acids. Finally, the analysis of 16 *MsHSF* genes' expression levels across all tissues and under four abiotic stresses using publicly available RNA-Seq data revealed that these genes had significant tissue-specific expression. Moreover, the expression of most *MsHSF* genes increased dramatically under abiotic stress, further validating the critical function played by the MsHSF gene family in abiotic stress. These results provided basic information about MsHSF gene family and laid a foundation for further study on the biological role of MsHSF gene in response to stress in *M. sativa*.

Keywords: *Medicago sativa*; HSF gene family; expression profile; abiotic stress

Citation: Ma, J.; Zhang, G.; Ye, Y.; Shang, L.; Hong, S.; Ma, Q.; Zhao, Y.; Gu, C. Genome-Wide Identification and Expression Analysis of HSF Transcription Factors in Alfalfa (*Medicago sativa*) under Abiotic Stress. *Plants* **2022**, *11*, 2763. https://doi.org/10.3390/plants11202763

Academic Editor: Małgorzata Nykiel

Received: 23 September 2022
Accepted: 17 October 2022
Published: 19 October 2022

1. Introduction

Normal plant growth is often affected by a variety of adverse environmental factors such as drought, high salinity, temperature extremes and other abiotic stresses [1]. Abiotic stress inhibits normal plant growth, development, and function by speeding up chlorophyll degradation, disrupting chloroplast membrane activities, and decreasing photosynthetic efficiency, as shown by a number of studies [2–4]. In response to external stimuli, plants' bodies produce signals that trigger the phosphorylation of downstream proteins, which in turn trigger a number of transcription factors [5]. Ultimately, plant-associated resistance genes and associated defense systems are induced, thus altering the ability of the plant to adapt to its environment [6–8].

Plants have developed multiple defense mechanisms and strategies to cope with adverse conditions and respond accordingly [9–11]. Under abiotic stress, induction of numerous proteins, including transcription factors (TF), can regulate the expression of specific functional genes and enhance plant resistance through signal transduction pathways [12]. Reactive oxygen species (ROS) scavenger enzymes and HSP are important functional proteins induced by HS, and their corresponding genes are targets of several HS-responsive TFs [13]. Previous studies have shown that HSFA6b is essential in *Arabidopsis thaliana* as a downstream regulator of the ABA-mediated heat stress response (HSR) [14,15].

HSF is a class of transcription factors that are widely present in eukaryotes [16]. Since the first *HSF* gene was isolated from *Solanum lycopersicum* in 1990, HSF has been reported in *A. thaliana*, *Oryza sativa*, *Glycine max* and other plants with the continuous improvement of genome sequencing technology [17–21]. The number of HSF gene family members varies widely among plants, with the highest number in *Triticum aestivum* containing 56 [22]. *G. max* [23], *Zea mays* [24], and *A. thaliana* [25] contained 52, 30, and 21, respectively. HSF plays a crucial role in the transmission and receipt of signals, the detection of heat shock components, the regulation of downstream genes, and the induction of plant stress responses [26]. The important role of HSF in plant responses to abiotic stresses can be well established.

Alfalfa (*Medicago sativa*) is one of the most economically valuable crops in the world and the most widely farmed legume fodder grass [27]. However, as it matures and flourishes, *M. sativa* is regularly damaged by a number of abiotic factors, such as salinity, cold and drought, which have a detrimental effect on *M. sativa* quality and output [28–31]. Therefore, it is particularly important to breed *M. sativa* germplasm resources with high resistance to stress. HSF is a class of transcription factors that are widely found in eukaryotes. They are crucial in the transmission and receipt of signals, the identification of heat shock components and the control of downstream genes, and the induction of plant responses to abiotic stresses [26,32–34]. Studies on the HSF gene family have been reported in many species but have not yet undergone a thorough analysis in *M. sativa*. Therefore, in this study, we identified the MsHSF family gene in *M. sativa* by integrating conserved motifs, gene structure, chromosome mapping, promoter cis-elements, and their phylogenetic relationships, and we analyzed the expression of HSF genes in *M. sativa* under four abiotic stresses. These results will deepen our current understanding of the evolutionary relationships and functional differentiation of *M. sativa HSF* genes and provide valuable information for further studies on the role of HSF genes in *M. sativa* for resistance under abiotic stresses.

2. Results

2.1. Identification of HSF Genes in M. sativa

In this study, we used the *A. thaliana* HSF protein (AtHSF) as a query to search the *M. sativa* genome database for 25 *MsHSF* candidate genes. We then used the HMMER3 search to search the *M. sativa* genome database using the HSF-type DBD model to retrieve 22 putative *MsHSF* candidate genes (PF00447). Finally, we used SMART and NCBI conserved structural domains to delete duplicate genes and proteins without DBD conserved structural domains to obtain 16 MsHSF family members. Based on these genes' chromosomal positions, they were given the new names *MsHSF01–MsHSF16*.

The predicted physicochemical properties of the amino acid sequences indicated that the 16 *HSF* genes encode proteins containing 211 (MsHSF06) to 543 (MsHSF09) amino acids with molecular weights (MWs) ranging from 24,541.86 (*MsHSF06*) to 60,982.41 Da (MsHSF06), with an average molecular weight of 42,922.38 Da. The predicted isoelectric points (pI) ranged from 4.72 (MsHSF05) to 8.16 (MsHSF06), with a mean value of 5.88. Instability index calculations predicted that 14 (87.5%) of the HSF proteins were unstable in vitro, and only MsHSF08 and MsHSF12 predicted proteins with an instability index of less than 40 were classified as stable proteins. In addition, the aliphatic amino acid index (A.I.) ranged from 62.79 (MsHSF08) to 83.61 (MsHSF12), indicating that their thermal stability was less different. The overall mean of the hydrophilic (GRAVY) scores of all HSF proteins was negative, indicating that they are all hydrophilic proteins. Finally, subcellular localization predictions showed that all MsHSF proteins were located in the nucleus (Table 1).

Table 1. Prediction of physicochemical properties of MsHSF.

Protein Name	Gene ID	Chromosome Location	Size (aa)	MW (kDa)	pI	Stability	A.I	GRAVY	Predicted Location
HSF01	MsG0180003410.01.T01	S1:61939996–61942587	496	55,750.22	4.91	U	69.91	−0.591	Nuclear
HSF02	MsG0180005479.01.T01	S1:92250440–92250760	297	32,972.9	6.26	U	71.28	−0.613	Nuclear
HSF03	MsG0380017150.01.T01	S3:92635955–92637670	372	42,094.06	8.16	U	62.82	−0.715	Nuclear
HSF04	MsG0380017281.01.T01	S3:94430904–94433609	381	42,838.88	4.93	U	78.5	−0.547	Nuclear
HSF05	MsG0480021648.01.T01	S4:61250000–61252171	480	53,424.6	4.85	U	69.62	−0.55	Nuclear
HSF06	MsG0480022400.01.T01	S4:72054039–72055708	211	24,541.86	6.02	U	69.72	−0.708	Nuclear
HSF07	MsG0480023951.01.T01	S4:91618543–91621035	401	45,748.67	4.99	U	80.7	−0.498	Nuclear
HSF08	MsG0580024826.01.T01	S5:10417724–10420833	287	32,168.91	6.9	S	62.79	−0.784	Nuclear
HSF09	MsG0580025495.01.T01	S5:19600912–19606188	543	60,982.41	5.1	U	71.47	−0.624	Nuclear
HSF10	MsG0680032444.01.T01	S6:41477418–41479676	295	33,653.64	5.66	U	68.14	−0.802	Nuclear
HSF11	MsG0680032449.01.T01	S6:41520725–41522690	319	36,439.77	5.27	U	70.34	−0.717	Nuclear
HSF12	MsG0680032451.01.T01	S6:41583472–41590098	527	59,504.68	6.96	S	83.61	−0.474	Nuclear
HSF13	MsG0680033612.01.T01	S6:69476487–69478031	359	39,836.88	5.54	U	76.3	−0.439	Nuclear
HSF14	MsG0680035659.01.T01	S6:110235176−110236554	351	40,233.48	5.58	U	63.85	−0.664	Nuclear
HSF15	MsG0780040480.01.T01	S7:78201225–78202956	226	26,119.55	7.56	U	71.55	−0.755	Nuclear
HSF16	MsG0780040676.01.T01	S7:80984782–80994274	527	60,447.57	5.43	U	73.76	−0.61	Nuclear

2.2. Phylogenetic Analysis of HSF Genes in M. sativa

We created a NJ phylogenetic tree using MEGA6.0 for the amino acid sequences of 16 *M. sativa* HSF, 21 *A. thaliana* HSF, and 25 *O. sativa* HSF in order to study the evolutionary relationships of *MsHSF* genes (Figure 1). Based on the well-established *A. thaliana* family classification, HSF are clearly classified into three groups, namely HSF A (green), HSF B (yellow) and HSF C (blue). Members of MsHSFs were identified in all three groups. The largest was MsHSF A, which made up 52.2% of the overall MsHSF, which was broken down into 9 subgroups (A1–A9). HSF A consists of nine proteins, namely MsHSF01, MsHSF04, MsHSF05, MsHSF07, MsHSF09, MsHSF10, MsHSF11, MsHSF12 and MsHSF16. MsHSF proteins were not aggregated into the three subgroups, A4, A7 and A9. MsHSF B was divided into 5 subgroups (B1–B5), accounting for 39.1%, and consisted of 6 members (MsHSF02, MsHSF03, MsHSF06, MsHSF08, MsHSF13, and MsHSF15). MsHSF C was the smallest group. Also, it contained only one MsHSF14. *M. sativa* HSF proteins clustered more closely with *A.thaliana* HSF proteins, according to the phylogenetic tree, which is interesting. This finding shows that the HSF proteins of dicotyledons and monocotyledons have distinct evolutionary differences.

2.3. Multiple Sequence Alignment and Protein Modeling Analysis of the HSF Gene in M. sativa

To check for the presence and position of conserved protein structural domains, we used Jalview to perform a multiple sequence alignment on 16 MsHSF proteins. We found that the DBD structural domain, which contains roughly 100 amino acids, is substantially conserved among all members of the MsHSF family (Figure 2). The DBD structural domain has three helix bundles (1–3) and four reverse parallel folds, as predicted from protein secondary structure (1–4). Additionally, we predicted the protein 3D structure of HSF of *M. sativa* and labeled the starting position of the 1 DBD structural domain in the figure (Figure 3).

2.4. Gene Structure and Conserved Motif Analysis of the HSF Gene in M. sativa

HSF proteins were split into three categories, HSF A, HSF B, and HSF C, based on the phylogenetic analyses mentioned above (Figure 4A). Our analysis of the MsHSF gene structure revealed that *HSF* genes from the same group typically have a similar number of introns in their structures. There is just one intron and two exons shared by all *HSFB* and *HSFC* genes in *M. sativa* (Figure 4B). Except for *MsHSF12*, which had eight introns, the *M. sativa HSF A* genes ranged in intron count from one to four. Three genes included two introns; three genes (*MsHSF01*, *MsHSF07*) contained four introns; two genes (*MsHSF10*, *MsHSF11*) contained one intron. *MsHSF09* and *MsHSF016* contain four introns.

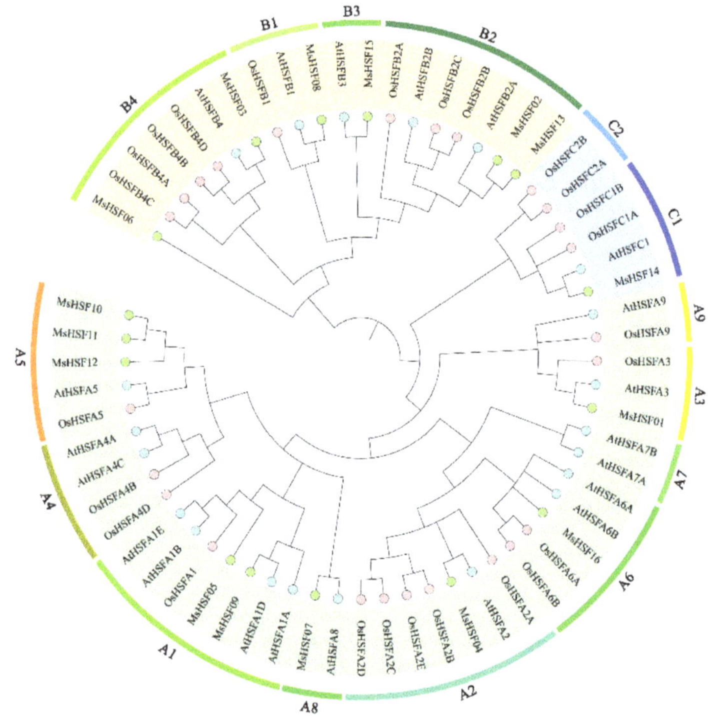

Figure 1. Model for phylogenetic analysis of HSF in *M. sativa*. Each subgroup is distinguished by a different color.

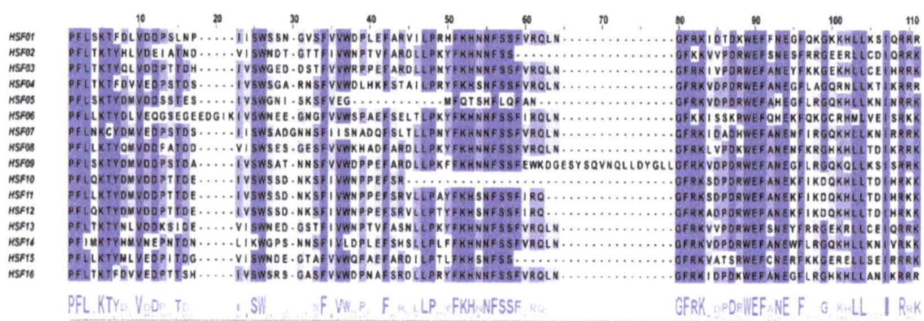

Figure 2. Conserved domain alignment of MsHSF members.

We also used MEME to identify up to 10 highly conserved motifs in each HSF protein (Figure 4C). The results show that the relative positions of most of the sequence motifs of the same group are similar. All MsHSF contain motifs 1 and 3, which constitute the most highly conserved part of the DBD, and the absence of motif 2 in some genes indicates that motif 2 is not necessary in the highly conserved part of the DBD. Motifs 7 and 9 were found only in HSF A, and motif 8 was present only in HSF B. These results suggest that the biological functions of the MsHSF proteins that are grouped together may be similar, and that different patterns may be related to different functions in separate subgroups.

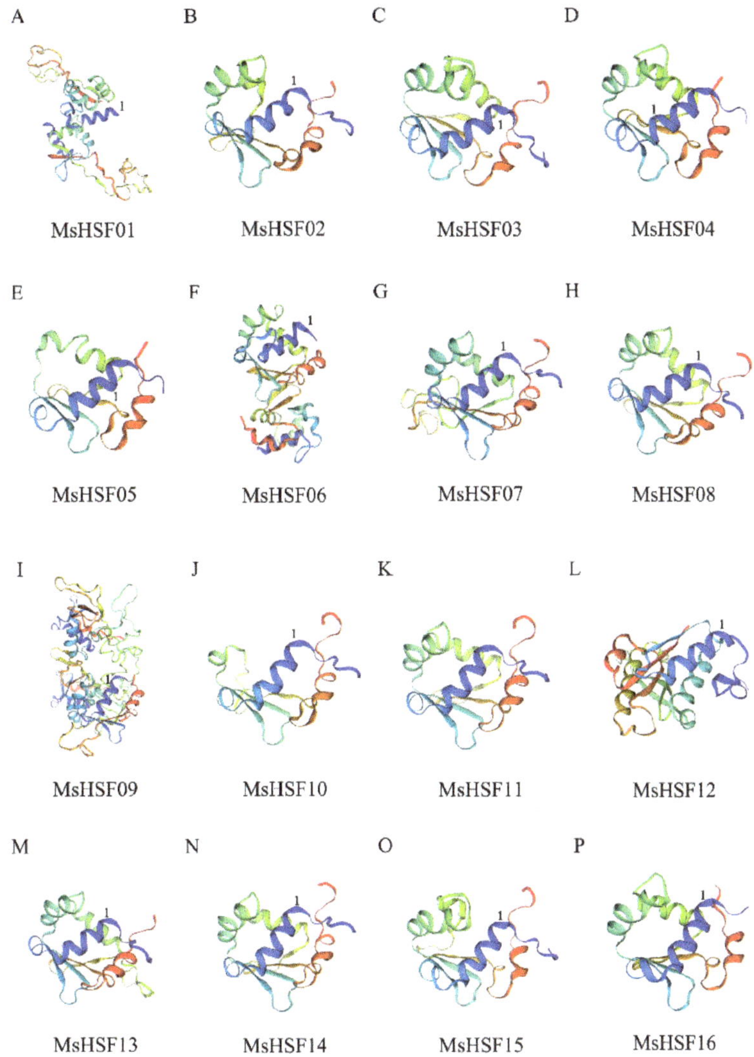

Figure 3. Three-dimensional model of the protein of MsHSF family members, with the starting position of the $\alpha1$ DBD structural domain indicated by 1. (**A–P**) in the figure represent the 16 proteins of the *M. sativa* HSF family, respectively.

2.5. Chromosome Distribution and Covariance Analysis of HSF Genes in M. sativa

M. sativa HSF genes were unevenly distributed on the six chromosomes (Figure 5). This included chromosomes 1, 3, 4, 5, 6, and 7, with the largest number of genes found on chromosome 6 with 5 genes, followed by 3 genes found on chromosome 4. All the remaining chromosomes contained only 2 *MsHSF* genes each.

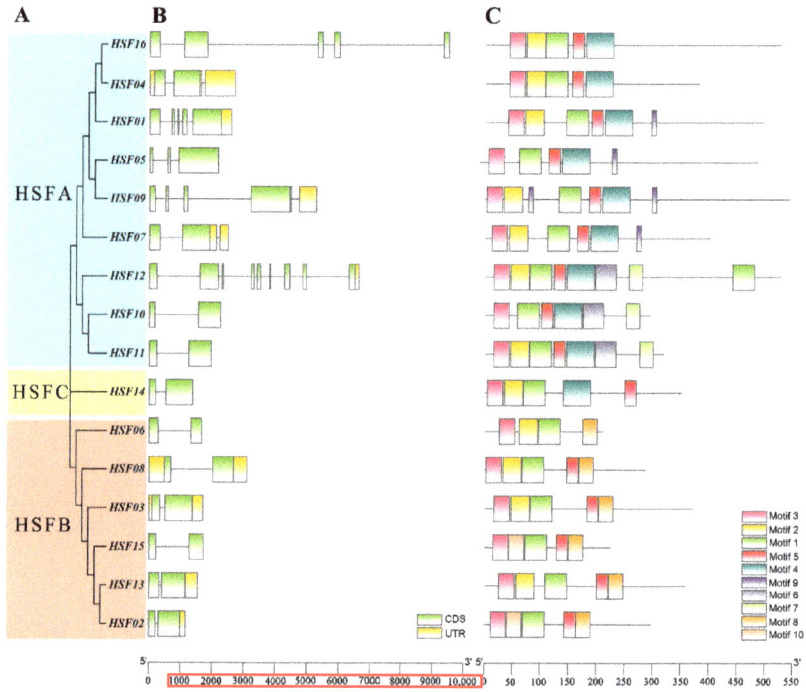

Figure 4. Phylogenetic relationship tree (**A**), gene structure (**B**) and conserved patterns (**C**) of HSF in *M. sativa*.

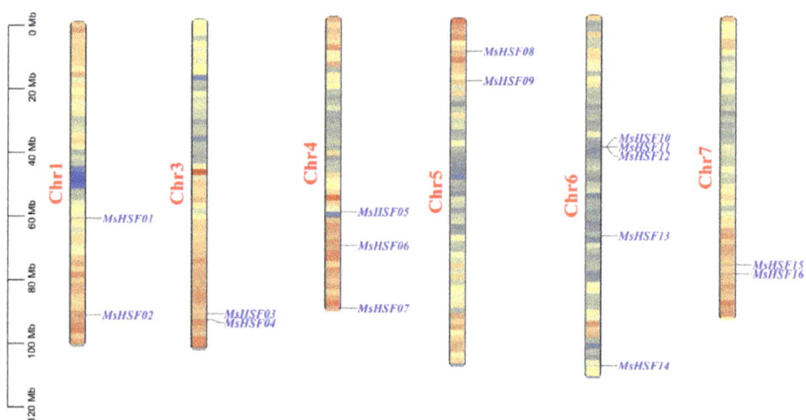

Figure 5. Distribution of *MsHSF* genes on the chromosomal scaffolds of *M. sativa*.

Gene duplication events are common in all species; they can generate new functional genes and drive species evolution. Therefore, we used MCScanX genomic homozygosity analysis to explore duplications in the *M. sativa* HSF gene family (Figure 6). Three (*MsHSF10*, *MsHSF11*, and *MsHSF12*) tandem duplicated genes and two pairs (*MsHSF01*, *MsHSF02*, *MsHSF15*, and *MsHSF16*) of codominant genes were detected in the MsHSF family.

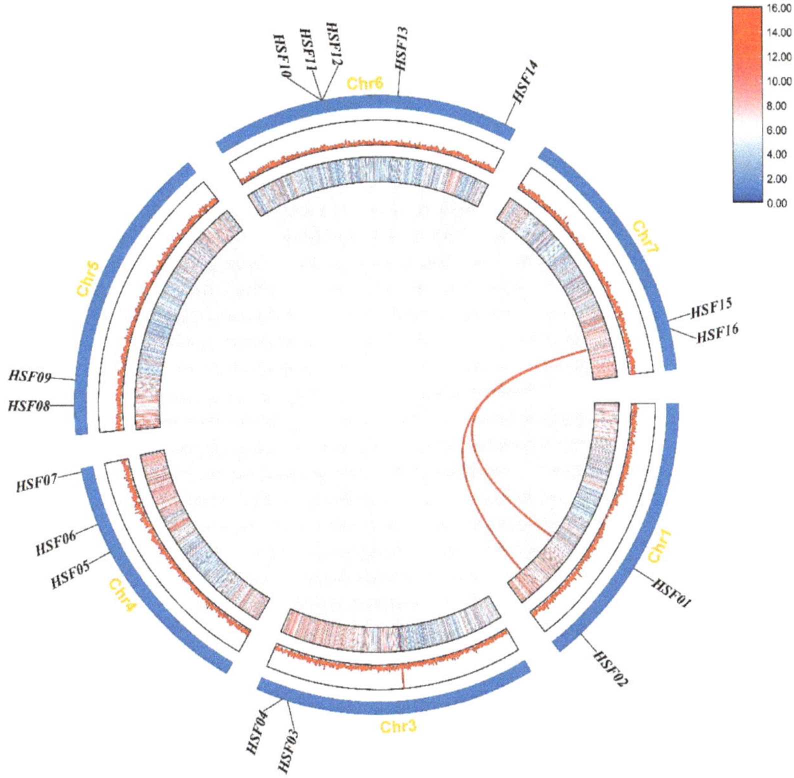

Figure 6. Chromosome distribution and interchromosomal relationships of *MsHSF* genes. Gray lines indicate synthetic blocks within the *M. sativa* genome, and red lines indicate duplicated *MsHSF* gene pairs.

Additionally, the proportion of *HSF* genes shared by *M. sativa* and other species may reflect the evolutionary relationships of the HSF gene family among *M. sativa*, dicotyledons, and monocotyledons. Therefore, in this study, we constructed a comparative homozygous map of three *HSF* genes from *M. sativa* and three dicots (*A. thaliana*, *G. max*, and *Medicago truncatula*) and one monocot (*Z. mays*) (Figure 7). A total of 44 direct homologous pairs (including one *M. sativa* gene corresponding to more than one *G. max* gene) were studied between *M. sativa* and *G. max HSF* genes. 20, 11, and 7 pairs of direct homologous genes were presented in *M. truncatula*, *A. thaliana*, and *Z. mays*, respectively. During evolution, most of the *HSF* genes of *M. sativa* have more than two direct homologs in *A. thaliana*, which further suggests that *M. sativa* has experienced more whole gene duplication events.

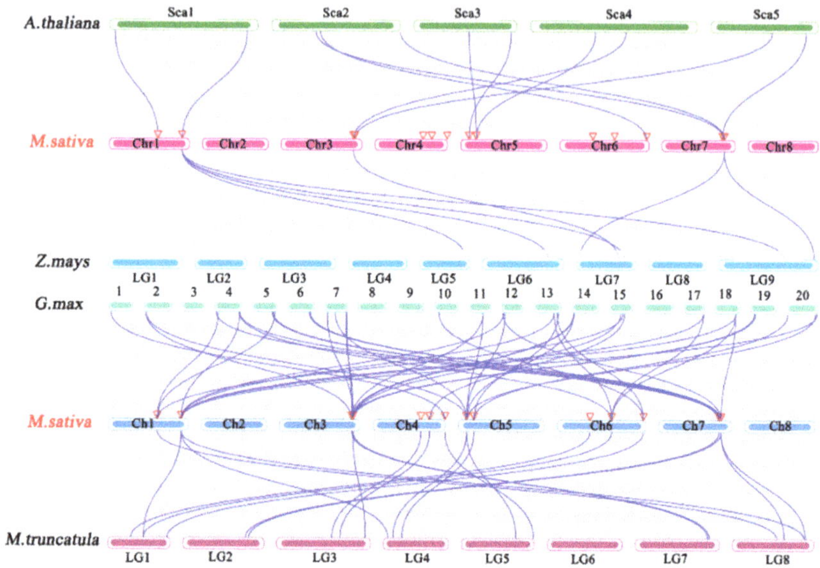

Figure 7. Synthetic analysis of the *M. sativa* genome with the genomes of one monocotyledon and three dicotyledon plants. Gray lines represent alignment blocks between paired genomes, and blue lines indicate synthetic *HSF* gene pairs.

2.6. Analysis of the Promoter Cis-Element of the HSF Gene in M. sativa

In order to learn more about how the *MsHSF* gene is regulated, this study examined the cis-acting components of the promoter region. We searched the PlantCARE database for potential cis-acting elements in the 2000 bp upstream sequence of the *MsHSF* translation start codon (Figure 8). As a result, most are stress response elements, and all promoters contain at least one TATA-box, CAAT-box, and short_function element. The next most common element is the ABRE element (contained in all promoters except *MsHSF11*), which is associated with the abscisic acid response. Other elements present in MsHSF promoters include the MBS element (involved in drought, high salt, and low temperature responses), TC-rich repeat sequences (involved in defense and stress responses), AE-box, and TGA erythromycin response elements. These findings suggest that *MsHSF* genes may be involved in multiple transcriptional regulatory mechanisms of plant growth and stress responses.

2.7. Expression Profiling of the M. sativa HSF Gene in Different Tissues

To determine the expression patterns of individual *MsHSF* genes in various tissues, a hierarchical clustering heat map was also constructed from public RNA-seq data obtained from NCBI for exploring the transcriptional patterns of *MsHSF* genes in this study (Figure 9). The blue color in the graph indicates low transcript abundance and the red color indicates high transcript abundance. The analysis showed that there were different *HSF* genes showing a trend of high expression in six tissues of *M. sativa*. *MsHSF06* and *MsHSF15* were highly expressed in the roots. The expression of each gene was not significant in the root nodules. In pre-elongating stems, *MsHSF05*, *MsHSF09* and *MsHSF14* showed higher expression. However, the gene with higher expression in elongating stems was *MsHSF03*. In *M. sativa* leaves, *MsHSF07*, *MsHSF08* and *MsHSF16* showed higher expression. Interestingly, *MsHSF10* was not expressed in any other tissue parts of alfalfa but was expressed in the flowers, a phenomenon that suggests that different *MsHSF* genes have more obvious specificity in different tissues.

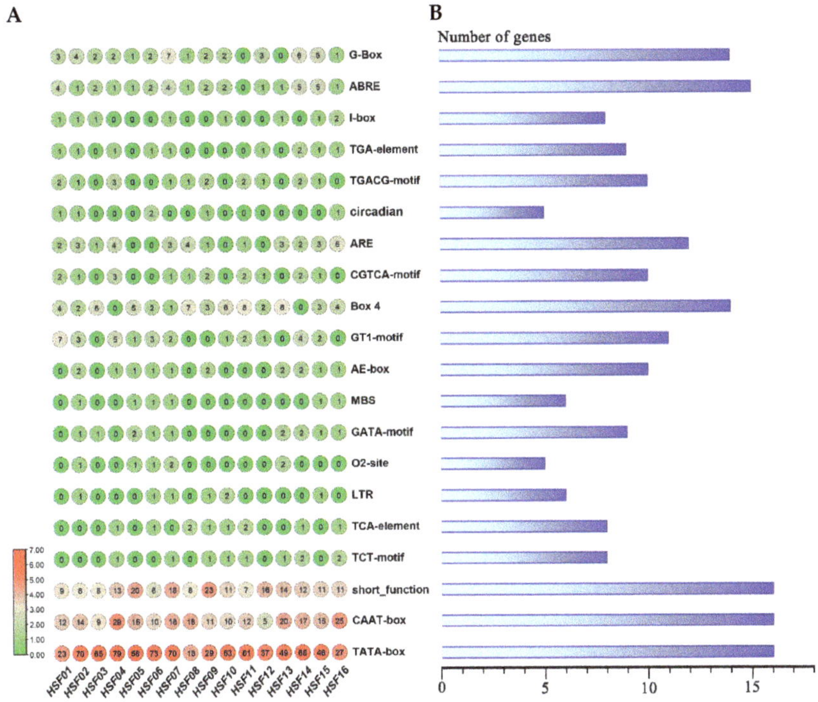

Figure 8. A schematic representation of the cis-acting elements identified in the 2000 bp promoter region upstream of the *MsHSF* gene. The different colors represent the number of cis-acting elements contained (**A**). The number of *MsHSF* genes corresponding to the cis-acting element (**B**).

2.8. Expression Analysis of HSF Genes in M. sativa in Response to Abiotic Stresses

To investigate the expression levels of *MsHSF* genes under abiotic stress, we analyzed the expression patterns of *MsHSF* genes under treatment with cold stress, abscisic acid (ABA), drought, and salt using published transcriptome data (Figure 10). The results of the analysis showed that the genes that functioned in *M. sativa* under different stress treatments were slightly different compared to the control. However, the genes that appeared to be the first to function and to be highly expressed in the immediate period of abiotic stress were *MsHSF09* and *MsHSF15*, while *MsHSF07* and *MsHSF08* gradually became more highly expressed as the time of stress increased to counteract the external stimuli. By the later stages of stress, it is *MsHSF04*, *MsHSF05* and *MsHSF13* that play a role. Interestingly, most of the *MsHSF* genes were induced to increase in response to abiotic stress, except for *MsHSF02* and *MsHSF16*, which were suppressed in response to cold stress, which may be related to cellular trauma in *M. sativa* during cold stress.

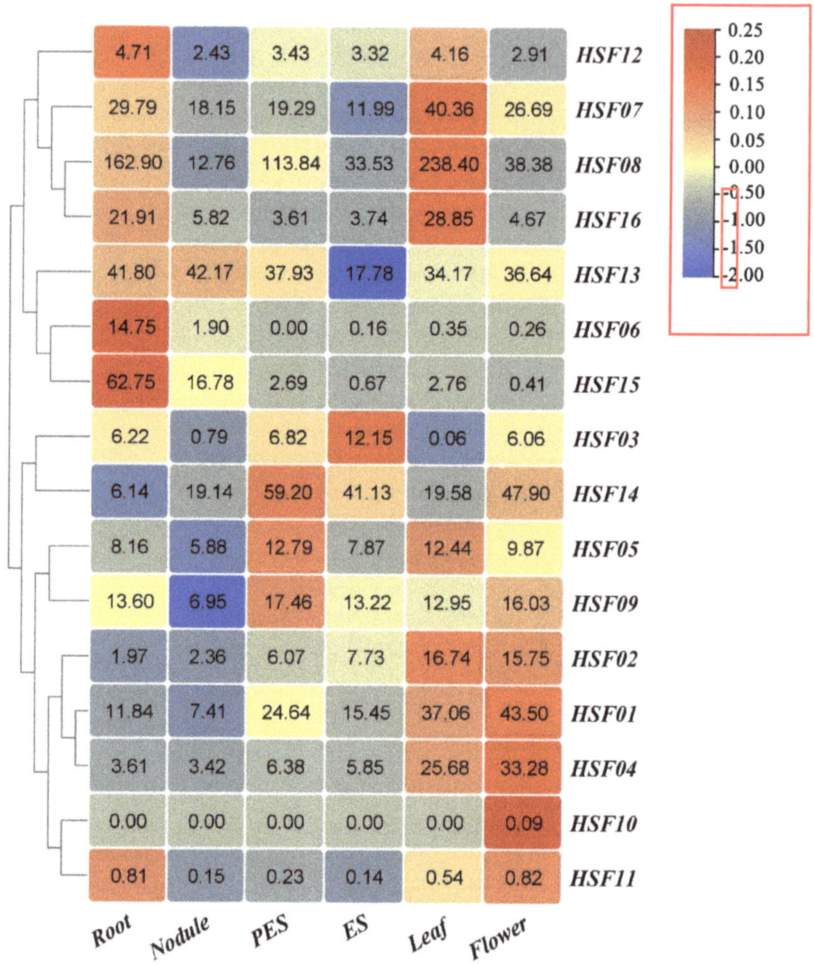

Figure 9. A heat map representation of *MsHSF* expression between different tissues. The values in the rectangle represent the magnitude of the gene expression.

2.9. GO Annotation and Enrichment Analysis of *M. sativa* HSF Protein

Plants have evolved complex mechanisms to sense and respond to biotic and abiotic stresses, and HSF is an important component of these defense systems. We carried out GO annotation and enrichment analysis on 16 MsHSF proteins in order to learn more about the biological functions of this protein (Figure 11). MsHSF was enriched for 55 biological processes, 1 cellular component, and 2 molecular functions in comparison to the entire GO database. According to the GO enrichment data, MsHSF transcription factors are primarily involved in biological processes including responding to abiotic stimuli, responding to temperature stimuli, responding to heat, and responding to xenobiotic stimuli. The findings again suggest that HSF genes play an extremely important role in resisting abiotic stresses.

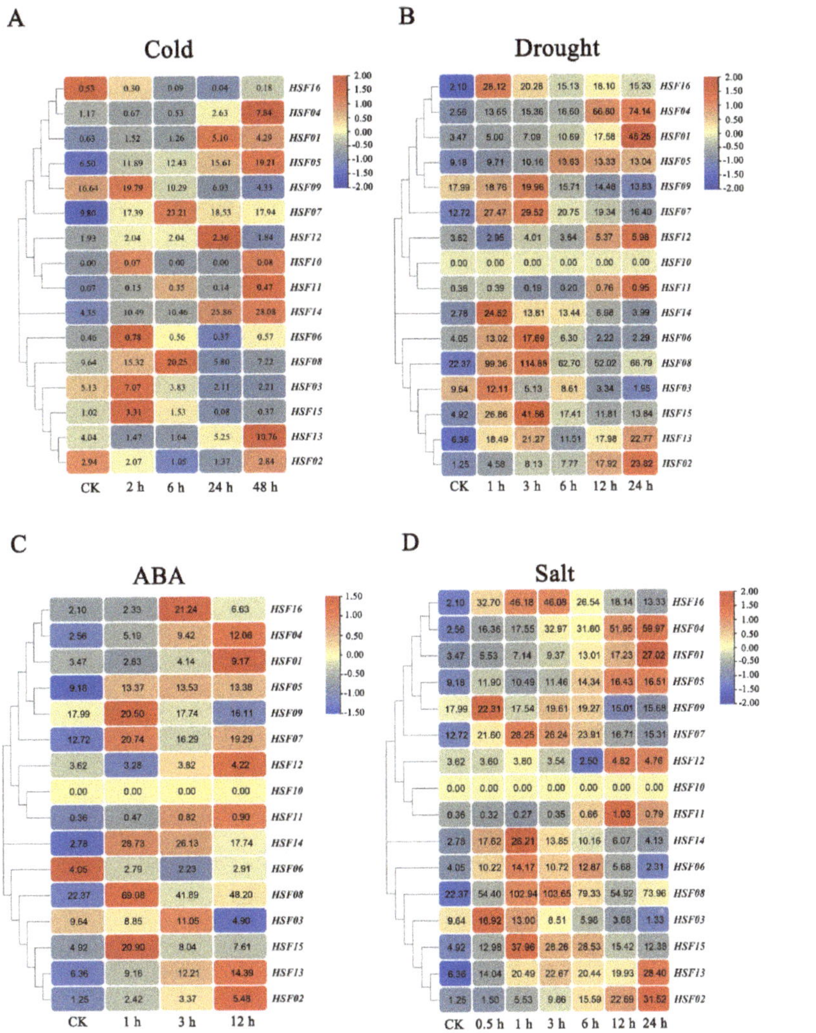

Figure 10. Expression of 16 *MsHSF* genes in cold (**A**), drought (**B**), ABA (**C**), and salt (**D**) treatments. The values in the rectangle represent the magnitude of the gene expression.

2.10. Interaction Network Analysis of HSF Proteins in M. sativa

We predicted probable interactions between MsHSF proteins using the STRING database in order to better comprehend MsHSF protein interactions (Figure 12). The findings demonstrate that while certain proteins, like MsHSF10 and MsHSF16, exhibit direct connections, others, including MsHSF10, MsHSF16, and MsHSF08, exhibit more complex multigene interactions. where it is projected that the major nodes MsHSF01, MsHSF08, and MsHSF13 radiate a significant number of connections to additional nodes.

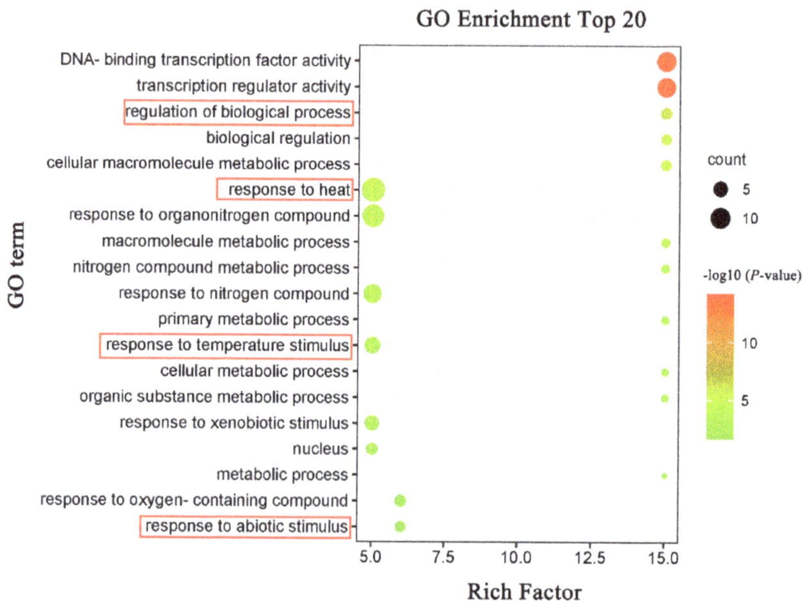

Figure 11. GO enrichment analysis of the MsHSF proteins relative to the GO database. The horizontal axis indicates the enrichment factor, and the size of the circle indicates the number of genes annotated with a given GO term.

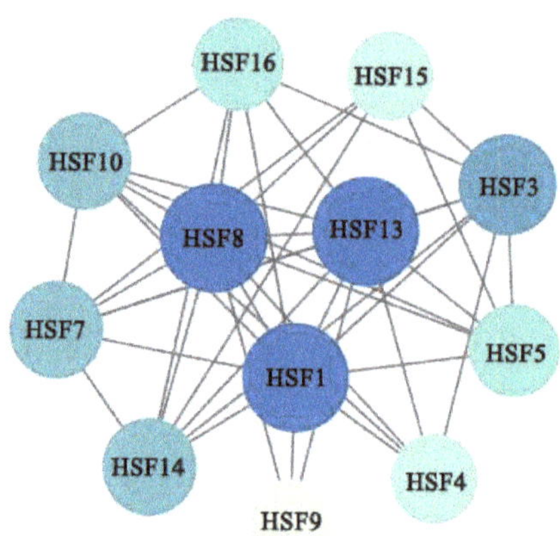

Figure 12. Interaction network of HSF proteins in *M. sativa*. Nodes represent proteins; central nodes are indicated in blue, and black lines indicate interactions between nodes.

3. Discussion

3.1. The Characteristics of HSF Gene Family in M. sativa

HSF is a particular sort of transcription factor that is crucial for plants' ability to resist diverse stressors [35]. The highly conserved plant HSF DBD is located at the N-terminal end where it is able to precisely locate and recognize the heat stress element (HSE) in the promoter of the target gene [36–38]. All MsHSF proteins comprised DBD with three helices and four folds, according to multiple sequence alignment and secondary structure prediction (Figure 2). It's interesting to note that some MsHSF proteins also have additional conserved structural domains; further experimental confirmation is required to determine whether this is a sign of the gene family's functional diversification. Tertiary structural analysis showed that the portion of transcription factors interacting with nucleic acids is conserved in the subfamily. HSF B and HSF C members may not have transcriptional activation because they lack AHA motifs. which is consistent with the results of previous studies [39–42].

Similar gene architectures and conserved protein motifs among members of the same phylogenetic group typically indicate a tight phylogenetic relationship [43]. Short sequences involved in significant biological processes are typically referred to as motif [44]. The presence of Motifs 1–2 in every MsHSF raises the possibility that they may have significant biological roles, however this has not yet been established. Because motif 9 is specific to the HSF A subgroup, it is possible that the *HSF* genes in this subgroup perform a particular role. Additionally, we discovered that the majority of MsHSF share exon-intron architectures and motif distributions within the same evolutionary tree grouping, indicating that genes within a subfamily frequently have comparable biological activities.

Only 16 HSF genes have been found in *M. sativa*, which is less than other plant species and may reflect the lack of expansion of the MsHSF family. MsHSF genes might be further grouped into three categories using homology matching and multispecies matching: A, B, and C. They are clustered in the same way as the members of the *A. thaliana* HSF gene family, with group A having the most genes and group C having the fewest [26]. Three subgroups, A4, A7, and A9, are missing from *M. sativa*, demonstrating that despite HSF family proteins sharing a common ancestor, they have evolved separately in different species. The majority of the alfalfa HSF proteins grouped with the *A. thaliana* HSF proteins but not with the *O. sativa* HSF proteins, suggesting that MsHSF and AtHSF have a tight evolutionary relationship. The HSF proteins of dicotyledons and monocotyledons have evolved in quite different ways.

Gene duplication events have a big impact on how gene families are formed. By supplying the necessary building blocks for the creation of new genes, gene duplication aids in the development of new, functional genes [45]. The majority of gene duplication occurs as tandem and fragmental duplication [46]. The 16 *MsHSF* genes in the *M. sativa* genome contained four homologous gene pairs, all of which underwent WGD (whole genome duplication) or fragmental duplication events and intense purifying selection pressure. These findings imply that WGD or fragmental replication is essential for *MsHSF* gene amplification.

Cis-acting elements are nucleotide sequences found upstream or downstream of genes that regulate their transcriptional levels [47]. When plants react to numerous developmental processes and stressors, they work by binding to certain transcription factors [48]. According to studies, cis-acting elements are present in plant-inducible promoters in response to adverse stress. There are numerous hormone-responsive core promoter elements and binding sites spread across the 16 *M. sativa* HSF promoter regions. According to this, MsHSF might be involved in the communication between several hormone signaling pathways. Aside from the heat stress element, the majority of MsHSF also contained the drought response element MBS, the anaerobic induction response element ARE, and the low temperature response element LTR. This shows that this gene family may control the effects of a variety of abiotic stimuli. These findings imply that MsHSF may interact with hormone signaling pathways that control growth and development as well as stress responses in *M. sativa*.

According to GO enrichment analysis, 15 of the 16 *MsHSF* genes are involved in two biological processes of GO resistance to abiotic stress and synthesis of abiotic stress factors (Figure 11). Several studies have shown that HSF regulates the expression of stress-related proteins, such as heat shock protein (HSP), which plays an important role in the plant stress response, especially heat stress [49–51]. Therefore, we speculate that the *MsHSF* gene may play a key role in plant resistance to abiotic stresses.

3.2. The Potential Roles of Differentially Expressed MsHSF Genes

In order to protect plants from heat stress, HSPs can raise the denaturation temperature of their proteins. They can also fix damaged proteins, enabling plants to withstand high temperatures [52]. HSFs play a major role in transcriptionally controlling the expression of HSPs. Additionally, the role of the HSFs signaling pathway encompasses many stresses, including cold, osmosis, drought, and salt, in addition to the response to heat stress [53]. It's interesting to note that osmotic pressures such as drought, salinity disruption, and other stresses result in the buildup of ROS, ABA, and H_2O_2 as well as alterations to cell walls. Ca^{2+} and ROS are the key factors causing abiotic stress response processes. Therefore, we examined the *M. sativa* transcriptome under ABA, salt, drought, and low temperature stressors. It was discovered that the expression of a considerable number of HSF genes was elevated under these stressful circumstances. This suggests that these *HSF* genes may be involved in some processes in the response of plants to external stresses. In conclusion, *MsHSF* genes are an important class of regulatory genes that control the plant's growth, development, and response to stress.

HSF plays an important role in the plant's response to abiotic stresses because it can achieve resistance to abiotic stresses by regulating the expression of different genes [54,55]. Among the *HSF* genes in plants, HSFA is a major transcriptional activator because it is essential to awaken HSR [56–58]. Although there were some subtle differences in the responses of *MsHSF* genes to different stresses in this study, the first to play a role in resistance was the *MsHSF09* gene in HSFA, which reinforces the important role of HSFA in resistance to abiotic stresses. Unlike HSFA, a considerable number of HSFB and HSFC have not been reported as transcriptional activators, but interestingly, in *M. sativa*, many genes in HSFB, such as *MsHSF07*, *MsHSF08*, and *MsHSF15*, also play important roles in resistance to abiotic stresses, and based on previous studies and analyses, it is clear that genes in HSFB in alfalfa function as transcriptional co-activators of HSFA. Taken together, several MsHSF genes are differentially expressed under abiotic stresses (including heat, salt, or ABA stress), and these results suggest that they may be involved in plant responses to abiotic stresses.

4. Materials and Methods

4.1. Identification and Sequence Analysis of HSF Genes in M. sativa

HSF protein sequences from the *A. thaliana* TAIR database (https://www.arabidopsis.org/) (accessed on 7 June 2022) were used as a reference sequence, and members of the putative *MsHSF* gene were sought using *M. sativa* genomic data from the National Genomics Data Center (https://ngdc.cncb.ac.cn/) (accessed on 7 June 2022) and a BLASTP search. We then used the native HMMER 3.0 software (Robert, D.F.; Ashburn, VA, USA) [59], using Hidden Markov (HMM) mapping of the HSF protein (PF0047), downloaded from the pfam database (http://pfam.xfam.org) (accessed on 9 July 2022). The potential gene members of *M. sativa* HSF identified were pooled using these two search techniques. WebCD-search (https://www.ncbi.nlm.nih.gov/cdd) and SMART (http://smart.embl.de/) were used to identify conserved HSF structural domains in all potential *MsHSF* genes (accessed on 9 July 2022). We finally identified 16 *MsHSF* genes and renamed them according to their position on the *M. sativa* chromosome.

We predicted and examined the physicochemical characteristics of all MsHSF potential proteins, including amino acid numbers, molecular weights, and theoretical sites, through the website ExPasy (https://www.expasy.org/) (accessed on 11 August 2022). Cell-PLoc2.0 was used to create subcellular localizations (http://www.csbio.sjtu.edu.cn/bioinf/plant-multi/#) (accessed on 11 August 2022).

4.2. Construction of Phylogenetic Tree and Sequence Comparison

Whole genome information for *A.thaliana* and *O.sativa* was downloaded from the NCBI database (https://www.ncbi.nlm.nih.gov/) (accessed 2 August 2022). Using MUSCLE technology, the 16 MsHSF protein sequences, 21 AtHSF proteins, and 25 OsHSF sequences found were compared to multiple sequences. The comparison parameters were in multiple comparison mode (other parameters were in default mode), and the obtained comparison results were used to construct a neighbor-joining (NJ) phylogenetic tree that we created in MEGA 7.0 using 1000 bootstrap replications, and the phylogenetic tree was created as an illustration using iTOL (https://itol.embl.de/) (accessed on 2 August 2022).

Intraspecific classification of *M. sativa* HSF sequences was performed based on interspecific phylogenetic trees, and amino acid sequences of conserved structural domains were compared and modified using Jalview software 2.11.2.4 (Andrew, M.W.; Cambridge, MA, USA) (http://www.jalview.org/) (accessed on 2 August 2022), and conserved thematic WebLogo were produced.

We also submitted the Jalview output to SOPMA (https://npsa-prabi.ibcp.fr/cgi-bin/npsa_automat.pl?page=npsa_sopma.html) (accessed on 3 August 2022) for protein secondary structure prediction by using default parameters. For tertiary structure prediction, we completed the analysis via the online website SWISS-MODEL (https://swissmodel.expasy.org/interactive) (accessed on 3 August 2022).

4.3. Gene Structure and Motif Identification

The conserved amino acid sequences of HSF proteins were analyzed using the online MEME tool (https://meme-suite.org/meme/) (accessed on 5 August 2022) with the parameters of minimum width of $\geq$ for 6, maximum width of 50, and number of parentheses of 10; all other parameters were set to default values. Then we obtained the intron-exon distribution of the *MsHSF* gene from the *M. sativa* genome using the GFF annotation method, and finally the results were displayed using TBtools software v1.098661 (Chen, C.J.; Guangzhou, China) [60].

4.4. Chromosome Location and Covariance Analysis

Chromosome lengths and gene positions were obtained from *M. sativa* genome annotation files, and MG2C v.2 (http://mg2c.iask.in/mg2c_v2.0/) (accessed on 8 August 2022) was used to visualize gene positions on chromosomes. *M. sativa* protein sequences were aligned to each other or to those of *A. thaliana*, *G. max*, *Z. mays*, or *M. truncatula* using TBtools software. MCScanX was used along with default parameters to identify homozygous relationships between gene replication events and HSF proteins, and results were visualized using Circos and Dual Synteny Plot in TBtools.

4.5. Identification of Cis-Acting Elements

A 2000 bp promoter region upstream of the MsHSF transcriptional start site was extracted from the *M. sativa* genome (http://www.genoscope.cns.fr/brassicanapus/) (accessed on 25 August 2022) and submitted to PlantCARE (http://bioinformatics.psb.ugent) (accessed on 25 August 2022) to identify the three types of regulatory cis-acting elements and finally visualize the results using TBtools.

4.6. Analysis of Tissue-Specific Expression and Abiotic Stress Transcriptome Data

Relevant transcriptome data were downloaded from the public NCBI database to investigate the expression patterns of *MsHSF* genes in different tissues and under different biotic stresses. *M. sativa* RNA-Seq data for different tissues (pre-elongated stems, elongated stems, flowers, leaves, roots, and rhizomes) can be downloaded from accession number SRP055547. The NCBI Short Read Long Archive database contains abiotic stress-related transcriptome data (cold, at SRR7091780-SRR7091794; drought, salt, and ABA, at SRR7160313-SRR7160357). The raw data was filtered and the SRA files were converted to FASTQ files by the SRA to Fastq program of TBtools. Gene expression levels were calculated by fragment number per kilobase per million mapped reads (FPKM values). Finally, the Hetmap program of TBtools was imported to generate the associated heat map.

4.7. Protein Interaction Network Prediction and GO Enrichment Analysis

HSF protein sequences were uploaded to the STRING database (https://string-db.org/) (accessed on 19 August 2022) for node comparison, and relationships between important proteins were predicted based on *A. thaliana* protein interactions. Finally, Cytoscape (Shannon, P.; California, USA) [61] was used to visualize the generated networks.

The online software EggNOG-Mapper (http://eggnog-mapper.embl.de/) (accessed on 18 August 2022) was used to annotate the GO function of the *MsHSF* gene. The results were collated using the eggNOG-mapper Helper function of TBtools and the text files used for downstream analysis were exported separately to GO Enrichment for enrichment analysis. Finally, use the online charting tool HIPLOT (https://hiplot.com.cn/) (accessed on 19 August 2022) to view and examine the data.

5. Conclusions

From the genomic information of *M. sativa*, we discovered a total of 16 *MsHSF* genes in this study. We analyzed the physicochemical properties of the 16 MsHSF proteins and found that all the *MsHSF* genes were localized in the nucleus. Amino acid sequence comparison showed that all *MsHSF* genes contain conserved DBD structures, after which we classified MsHSF proteins into 2 groups and 15 subgroups based on evolutionary relationships and showed that they are mostly similar within the same group but differ significantly between subgroups. Cis-acting element and GO enrichment analyses suggest that *MsHSF* genes may be involved in multiple transcriptional regulatory mechanisms for plant growth and stress response. In addition, expression profiling indicated that *MsHSF* genes could show significant specificity during tissue development and that *MsHSF* genes play a very important role in response to abiotic stresses. Overall, bioinformatic analysis and expression profiling studies of HSF can help to understand the important role of HSF in abiotic stress responses in alfalfa and provide a basis for exploring ways to understand and regulate these stress responses.

Author Contributions: Conceptualization, C.G., J.M., G.Z. and Y.Z.; methodology, J.M.; software, G.Z.; validation, Y.Y. and Y.Z.; investigation, J.M. and Y.Y.; data curation, G.Z. and Q.M.; writing original draft preparation, J.M. and G.Z.; writing—review and editing, C.G., J.M., G.Z., S.H. and L.S.; funding acquisition, C.G. All authors have read and agreed to the published version of the manuscript.

Funding: This work was financially supported by Zhejiang Provincial Natural Science Foundation of China (No. LY21C160001); National Natural Science Foundation of China (31272494); and Zhejiang Provincial Natural Science Foundation of China (No. LY16C170003).

Institutional Review Board Statement: Not applicable.

Informed Consent Statement: Not applicable.

Data Availability Statement: All data in this study can be found in the manuscript.

Conflicts of Interest: The authors declare no conflict of interest.

References

1. Pu, J.; Li, M.; Mao, P.; Zhou, Q.; Liu, W.; Liu, Z. Genome-Wide Identification of the Q-type C2H2 Transcription Factor Family in Alfalfa (*Medicago sativa*) and Expression Analysis under Different Abiotic Stresses. *Genes* **2021**, *12*, 1906. [CrossRef] [PubMed]
2. Song, Y.; Lv, J.; Ma, Z.; Dong, W. The mechanism of alfalfa (*Medicago sativa* L.) response to abiotic stress. *Plant Growth Regul.* **2019**, *89*, 239–249. [CrossRef]
3. Mao, P.; Jin, X.; Bao, Q.; Mei, C.; Zhou, Q.; Min, X.; Liu, Z. WRKY Transcription Factors in *Medicago sativa* L.: Genome-Wide Identification and Expression Analysis Under Abiotic Stress. *DNA Cell Biol.* **2020**, *39*, 2212–2225. [CrossRef] [PubMed]
4. Yuan, Y.; Yu, J.; Kong, L.; Zhang, W.; Hou, X.; Cui, G. Genome-wide investigation of the PLD gene family in alfalfa (*Medicago sativa* L.): Identification, analysis and expression. *BMC Genom.* **2022**, *23*, 243. [CrossRef] [PubMed]
5. Yang, Y.; Qi, L.; Nian, L.; Zhu, X.; Yi, X.; Jiyu, Z.; Qiu, J. Genome-Wide Identification and Expression Analysis of the SRS Gene Family in *Medicago sativa*. *DNA Cell Biol.* **2021**, *40*, 1539–1553. [CrossRef]
6. Dong, X.; Deng, H.; Ma, W.; Zhou, Q.; Liu, Z. Genome-wide identification of the MADS-box transcription factor family in autotetraploid cultivated alfalfa and expression analysis under abiotic stress. *BMC Genom.* **2021**, *22*, 603. [CrossRef]

7. Sheng, S.; Guo, X.; Wu, C.; Xiang, Y.; Duan, S.; Yang, W.; Li, W.; Cao, F.; Liu, L. Genome-wide identification and expression analysis of DREB genes in alfalfa (*Medicago sativa*) in response to cold stress. *Plant Signal. Behav.* **2022**, *17*, 2081420. [CrossRef]
8. He, F.; Wei, C.; Zhang, Y.; Long, R.; Li, M.; Wang, Z.; Yang, Q.; Kang, J.; Chen, L. Genome-Wide Association Analysis Coupled With Transcriptome Analysis Reveals Candidate Genes Related to Salt Stress in Alfalfa (*Medicago sativa* L.). *Front. Plant Sci.* **2022**, *12*, 826584. [CrossRef]
9. Westerheide, S.D.; Raynes, R.; Powell, C.; Xue, B.; Uversky, V.N. HSF Transcription Factor Family, Heat Shock Response, and Protein Intrinsic Disorder. *Curr. Protein Pept. Sci.* **2012**, *13*, 86–103. [CrossRef]
10. Nian, L.; Zhang, X.; Yi, X.; Liu, X.; Ain, N.U.; Yang, Y.; Li, X.; Haider, F.U.; Zhu, X. Genome-wide identification of ABA receptor PYL/RCAR gene family and their response to cold stress in *Medicago sativa* L. *Physiol. Mol. Biol. Plants* **2021**, *27*, 1979–1995. [CrossRef]
11. Nian, L.; Liu, X.; Yang, Y.; Zhu, X.; Yi, X.; Haider, F.U. Genome-wide identification, phylogenetic, and expression analysis under abiotic stress conditions of LIM gene family in *Medicago sativa* L. *PLoS ONE* **2021**, *16*, e0252213. [CrossRef] [PubMed]
12. Panzade, K.P.; Kale, S.S.; Kapale, V.; Chavan, N.R. Genome-Wide Analysis of Heat Shock Transcription Factors in *Ziziphus jujuba* Identifies Potential Candidates for Crop Improvement Under Abiotic Stress. *Appl. Biochem. Biotechnol.* **2021**, *193*, 1023–1041. [CrossRef] [PubMed]
13. Zhang, L.; Chen, W.; Shi, B. Genome-wide analysis and expression profiling of the heat shock transcription factor gene family in Physic Nut (*Jatropha curcas* L.). *PeerJ* **2020**, *8*, e8467. [CrossRef] [PubMed]
14. Wang, P.; Song, H.; Li, C.; Li, P.; Li, A.; Guan, H.; Hou, L.; Wang, X. Genome-wide dissection of the heat shock transcription factor family genes in Arachis. *Front. Plant Sci.* **2017**, *8*, 106. [CrossRef]
15. Wan, X.; Yang, J.; Guo, C.; Bao, M.; Zhang, J. Genome-wide identification and classification of the Hsf and sHsp gene families in *Prunus mume*, and transcriptional analysis under heat stress. *PeerJ* **2019**, *7*, e7312. [CrossRef]
16. Li, M.; Xie, F.; Li, Y.; Gong, L.; Luo, Y.; Zhang, Y.; Chen, Q.; Wang, Y.; Lin, Y.; Zhang, Y.; et al. Genome-Wide Analysis of the Heat Shock Transcription Factor Gene Family in *Brassica juncea*: Structure, Evolution, and Expression Profiles. *DNA Cell Biol.* **2020**, *39*, 1990–2004. [CrossRef] [PubMed]
17. Gong, C.; Pang, Q.; Li, Z.; Li, Z.; Chen, R.; Sun, G.; Sun, B. Genome-Wide Identification and Characterization of Hsf and Hsp Gene Families and Gene Expression Analysis under Heat Stress in Eggplant (*Solanum melongema* L.). *Horticulturae* **2021**, *7*, 149. [CrossRef]
18. Rehman, A.; Atif, R.M.; Azhar, M.T.; Peng, Z.; Li, H.; Qin, G.; Jia, Y.; Pan, Z.; He, S.; Qayyum, A.; et al. Genome wide identification, classification and functional characterization of heat shock transcription factors in cultivated and ancestral cottons (*Gossypium* spp.). *Int. J. Biol. Macromol.* **2021**, *182*, 1507–1527. [CrossRef]
19. Huang, B.; Huang, Z.; Ma, R.; Chen, J.; Zhang, Z.; Yrjälä, K. Genome-wide identification and analysis of the heat shock transcription factor family in moso bamboo (*Phyllostachys edulis*). *Sci. Rep.* **2021**, *11*, 16492. [CrossRef]
20. Song, X.; Liu, G.; Duan, W.; Liu, T.; Huang, Z.; Ren, J.; Li, Y.; Hou, X. Genome-wide identification, classification and expression analysis of the heat shock transcription factor family in Chinese cabbage. *Mol. Genet. Genom.* **2014**, *289*, 541–551. [CrossRef]
21. Zhang, J.; Liu, B.; Li, J.; Zhang, L.; Wang, Y.; Zheng, H.; Lu, M.; Chen, J. Hsf and Hsp gene families in Populus: Genome-wide identification, organization and correlated expression during development and in stress responses. *BMC Genom.* **2015**, *16*, 181. [CrossRef] [PubMed]
22. Xue, G.-P.; Sadat, S.; Drenth, J.; McIntyre, C.L. The heat shock factor family from *Triticum aestivum* in response to heat and other major abiotic stresses and their role in regulation of heat shock protein genes. *J. Exp. Bot.* **2014**, *65*, 539–557. [CrossRef] [PubMed]
23. Lin, Y.; Cheng, Y.; Jin, J.; Jin, X.; Jiang, H.; Yan, H.; Cheng, B. Genome Duplication and Gene Loss Affect the Evolution of Heat Shock Transcription Factor Genes in Legumes. *PLoS ONE* **2014**, *9*, e102825. [CrossRef] [PubMed]
24. Li, H.-C.; Li, G.-L.; Liu, Z.-H.; Zhang, H.-M.; Zhang, Y.-M.; Guo, X.-L. Cloning, Localization and Expression Analysis of ZmHsf-like Gene in Zea mays. *J. Integr. Agric.* **2014**, *13*, 1230–1238. [CrossRef]
25. Zhao, L.; Pan, H.; Sun, M.; Zhang, Q. Molecular cloning of *Arabidopsis thaliana* HSFA2 gene and agrobacterium-mediated genetic transformation of chrysanthemum morifolium ramat. *Adv. Intell. Soft Comput.* **2010**, *134*, 827–834. [CrossRef]
26. Guo, M.; Liu, J.-H.; Ma, X.; Luo, D.-X.; Gong, Z.-H.; Lu, M.-H. The Plant Heat Stress Transcription Factors (HSFs): Structure, Regulation, and Function in Response to Abiotic Stresses. *Front. Plant Sci.* **2016**, *7*, 114. [CrossRef]
27. Wang, T.; Ren, L.; Li, C.; Zhang, D.; Zhang, X.; Zhou, G.; Gao, D.; Chen, R.; Chen, Y.; Wang, Z.; et al. The genome of a wild Medicago species provides insights into the tolerant mechanisms of legume forage to environmental stress. *BMC Biol.* **2021**, *19*, 96. [CrossRef]
28. Zhang, L.; Jia, X.; Zhao, J.; Hasi, A.; Niu, Y. Molecular characterisation and expression analysis of NAC transcription factor genes in wild *Medicago falcata* under abiotic stresses. *Funct. Plant Biol.* **2020**, *47*, 327–341. [CrossRef]
29. Xu, C.; He, C.G.; Wang, Y.J.; Bi, Y.F.; Jiang, H. Effect of drought and heat stresses on photosynthesis, pigments, and xanthophyll cycle in alfalfa (*Medicago sativa* L.). *Photosynthetica* **2020**, *58*, 1226–1236. [CrossRef]
30. Guiza, M.; Benabdelrahim, M.A.; Brini, F.; Haddad, M.; Saibi, W. Assessment of Alfalfa (*Medicago sativa* L.) Cultivars for Salt Tolerance Based on Yield, Growth, Physiological, and Biochemical Traits. *J. Plant Growth Regul.* **2021**, *10*, 1435-8107. [CrossRef]
31. He, K.; Li, C.; Zhang, Z.; Zhan, L.; Cong, C.; Zhang, D.; Cai, H. Genome-wide investigation of the ZF-HD gene family in two varieties of alfalfa (*Medicago sativa* L.) and its expression pattern under alkaline stress. *BMC Genom.* **2022**, *23*, 150. [CrossRef] [PubMed]

32. Tang, M.; Xu, L.; Wang, Y.; Cheng, W.; Luo, X.; Xie, Y.; Fan, L.; Liu, L. Genome-wide characterization and evolutionary analysis of heat shock transcription factors (HSFs) to reveal their potential role under abiotic stresses in radish (*Raphanus sativus* L.). *BMC Genom.* **2019**, *20*, 772. [CrossRef] [PubMed]
33. Guo, M.; Lu, J.-P.; Zhai, Y.-F.; Chai, W.-G.; Gong, Z.-H.; Lu, M.-H. Genome-wide analysis, expression profile of heat shock factor gene family (CaHsfs) and characterisation of CaHsfA2 in pepper (*Capsicum annuum* L.). *BMC Plant Biol.* **2015**, *15*, 151. [CrossRef] [PubMed]
34. Ma, G.; Shen, J.; Yu, H.; Huang, X.; Deng, X.; Hu, Z.; Amee, M.; Chen, L.; Cao, L. Genome-wide identification and functional analyses of heat shock transcription factors involved in heat and drought stresses in ryegrass. *Environ. Exp. Bot.* **2022**, *201*, 104968. [CrossRef]
35. Liu, X.; Meng, P.; Yang, G.; Zhang, M.; Peng, S.; Zhai, M.Z. Genome-wide identification and transcript profiles of walnut heat stress transcription factor involved in abiotic stress. *BMC Genom.* **2020**, *21*, 474. [CrossRef]
36. Zhang, H.Z.; Yang, J.L.; Chen, Y.L.; Mao, X.L.; Wang, Z.C.; Li, C.H. Identification and expression analysis of the heat shock transcription factor (HSF) gene family in Populus trichocarpa. *PLANT OMICS* **2013**, *6*, 415–424. [CrossRef]
37. Fan, K.; Mao, Z.; Ye, F.; Pan, X.; Li, Z.; Lin, W.; Zhang, Y.; Huang, J.; Lin, W. Genome-wide identification and molecular evolution analysis of the heat shock transcription factor (HSF) gene family in four diploid and two allopolyploid Gossypium species. *Genomics* **2021**, *113*, 3112–3127. [CrossRef]
38. Li, W.; Wan, X.-L.; Yu, J.-Y.; Wang, K.-L.; Zhang, J. Genome-wide Identification, classification, and expression analysis of the HSF gene family in carnation (*Dianthus caryophyllus*). *Int. J. Mol. Sci.* **2019**, *20*, 5233. [CrossRef]
39. Xue, H.; Slavov, D.; Wischmeyer, P. Glutamine activates heat shock transcription factor-1 (HSF1) gene transcription. *FASEB J.* **2012**, *26*, 646.27. [CrossRef]
40. Duan, S.; Liu, B.; Zhang, Y.; Li, G.; Guo, X. Genome-wide identification and abiotic stress-responsive pattern of heat shock transcription factor family in *Triticum aestivum* L. *BMC Genom.* **2019**, *20*, 257. [CrossRef]
41. Li, P.-S.; Yu, T.-F.; He, G.-H.; Chen, M.; Zhou, Y.-B.; Chai, S.-C.; Xu, Z.-S.; Ma, Y.-Z. Genome-wide analysis of the Hsf family in soybean and functional identification of GmHsf-34 involvement in drought and heat stresses. *BMC Genom.* **2014**, *15*, 1009. [CrossRef] [PubMed]
42. Zhang, Q.; Geng, J.; Du, Y.; Zhao, Q.; Zhang, W.; Fang, Q.; Yin, Z.; Li, J.; Yuan, X.; Fan, Y.; et al. Heat shock transcription factor (Hsf) gene family in common bean (*Phaseolus vulgaris*): Genome-wide identification, phylogeny, evolutionary expansion and expression analyses at the sprout stage under abiotic stress. *BMC Plant Biol.* **2022**, *22*, 33. [CrossRef] [PubMed]
43. Yu, X.-Y.; Yao, Y.; Hong, Y.-H.; Hou, P.-Y.; Li, C.-X.; Xia, Z.-Q.; Geng, M.-T.; Chen, Y.-H. Differential expression of the Hsf family in cassava under biotic and abiotic stresses. *Genome* **2019**, *62*, 563–569. [CrossRef] [PubMed]
44. Shyamli, P.S.; Pradhan, S.; Panda, M.; Parida, A. De novo Whole-Genome Assembly of Moringa oleifera Helps Identify Genes Regulating Drought Stress Tolerance. *Front. Plant Sci.* **2021**, *12*, 766999. [CrossRef]
45. Shen, C.; Yuan, J. Genome-wide characterization and expression analysis of the heat shock transcription factor family in pumpkin (*Cucurbita moschata*). *BMC Plant Biol.* **2020**, *20*, 471. [CrossRef]
46. Agarwal, P.; Khurana, P. Functional characterization of HSFs from wheat in response to heat and other abiotic stress conditions. *Funct. Integr. Genom.* **2019**, *19*, 497–513. [CrossRef]
47. Ducy, P.; Karsenty, G. Two distinct osteoblast-specific cis-acting elements control expression of a mouse osteocalcin gene. *Mol. Cell. Biol.* **1995**, *15*, 1858–1869. [CrossRef]
48. Wang, R.; Zhong, Y.; Liu, X.; Zhao, C.; Zhao, J.; Li, M.; Hassan, M.U.; Yang, B.; Li, D.; Liu, R.; et al. Cis- regulation of the amino acid transporter genes *ZmAAP2* and *ZmLHT1* by ZmPHR1 transcription factors in maize ear under phosphate limitation. *J. Exp. Bot.* **2021**, *72*, 3846–3863. [CrossRef]
49. Binder, R.J.; Blachere, N.E.; Srivastava, P.K. Heat shock protein-chaperoned peptides but not free peptides introduced into the cytosol are presented efficiently by major histocompatibility complex I molecules. *J. Biol. Chem.* **2001**, *276*, 17163–17171. [CrossRef]
50. Yang, D.; Tu, Y.; Wang, X.; Cao, C.; Hu, Y.; Shao, J.; Weng, L.; Mou, X.; Dong, X. A photo-triggered antifungal nanoplatform with efflux pump and heat shock protein reversal activity for enhanced chemo-photothermal synergistic therapy. *Biomater. Sci.* **2021**, *9*, 3293–3299. [CrossRef]
51. Altaff, K.; Radha, V. Influence of heat shock protein (hsp-70) enhancing compound from red alga (Porphyridium purpureum) for augmenting egg production in copepod culture–A new in silico report. *Mar. Sci. Technol. Bull.* **2021**, *10*, 186–192. [CrossRef]
52. Simon, S.; Aissat, A.; Degrugillier, F.; Simonneau, B.; Fanen, P.; Arrigo, A.P. Small Hsps as Therapeutic Targets of Cystic Fibrosis Transmembrane Conductance Regulator Protein. *Int. J. Mol. Sci.* **2021**, *22*, 4252. [CrossRef] [PubMed]
53. Tian, F.; Hu, X.-L.; Yao, T.; Yang, X.; Chen, J.-G.; Lu, M.-Z.; Zhang, J. Recent advances in the roles of HSFs and HSPs in heat stress response in woody plants. *Front. Plant Sci.* **2021**, *12*, 704905. [CrossRef] [PubMed]
54. Zhou, L.; Yu, X.; Wang, D.; Li, L.; Zhou, W.; Zhang, Q.; Wang, X.; Ye, S.; Wang, Z. Genome-wide identification, classification and expression profile analysis of the HSF gene family in *Hypericum perforatum*. *PeerJ* **2021**, *9*, e11345. [CrossRef] [PubMed]
55. Tang, R.; Zhu, W.; Song, X.; Lin, X.; Cai, J.; Wang, M.; Yang, Q. Genome-Wide Identification and Function Analyses of Heat Shock Transcription Factors in Potato. *Front. Plant Sci.* **2016**, *7*, 490. [CrossRef]
56. Samtani, H.; Sharma, A.; Khurana, J.P.; Khurana, P. The heat stress transcription factor family in *Aegilops tauschii*: Genome-wide identification and expression analysis under various abiotic stresses and light conditions. *Mol. Genet. Genom.* **2022**, *2*, 11–15. [CrossRef]

57. Saha, D.; Mukherjee, P.; Dutta, S.; Meena, K.; Sarkar, S.K.; Mandal, A.B.; Dasgupta, T.; Mitra, J. Genomic insights into HSFs as candidate genes for high-temperature stress adaptation and gene editing with minimal off-target effects in flax. *Sci. Rep.* **2019**, *9*, 5581. [CrossRef]
58. Tan, B.; Yan, L.; Li, H.; Lian, X.; Cheng, J.; Wang, W.; Zheng, X.; Wang, X.; Li, J.; Ye, X.; et al. Genome-wide identification of HSF family in peach and functional analysis of PpHSF5 involvement in root and aerial organ development. *PeerJ* **2021**, *9*, e10961. [CrossRef]
59. Eddy, S.R. What is a hidden Markov model? *Nat. Biotechnol.* **2004**, *22*, 1315–1316. [CrossRef]
60. Chen, C.; Chen, H.; Zhang, Y.; Thomas, H.R.; Frank, M.H.; He, Y.; Xia, R. TBtools: An Integrative Toolkit Developed for Interactive Analyses of Big Biological Data. *Mol. Plant* **2020**, *13*, 1194–1202. [CrossRef]
61. Shannon, P.; Markiel, A.; Ozier, O.; Baliga, N.S.; Wang, J.T.; Ramage, D.; Amin, N.; Schwikowski, B.; Ideker, T. Cytoscape: A Software Environment for Integrated Models of Biomolecular Interaction Networks. *Genome Res.* **2003**, *13*, 2498–2504. [CrossRef] [PubMed]

plants

MDPI

Article

Genome-Wide Analysis of the *Rhododendron* AP2/ERF Gene Family: Identification and Expression Profiles in Response to Cold, Salt and Drought Stress

Zhenhao Guo [1,2], Lisi He [1,2], Xiaobo Sun [1,2], Chang Li [1,2], Jiale Su [1,2], Huimin Zhou [1,2] and Xiaoqing Liu [1,2,*]

1 Institute of Leisure Agriculture, Jiangsu Academy of Agricultural Sciences, Nanjing 210014, China
2 Jiangsu Key Laboratory for Horticultural Crop Genetic Improvement, Nanjing 210014, China
* Correspondence: lxqjaas@aliyun.com

Abstract: The AP2/ERF gene family is one of the most conserved and important transcription factor families mainly occurring in plants with various functions in regulating plant biological and physiological processes. However, little comprehensive research has been conducted on the AP2/ERF gene family in *Rhododendron* (specifically, *Rhododendron simsii*), an important ornamental plant. The existing whole-genome sequence of *Rhododendron* provided data to investigate the AP2/ERF genes in *Rhododendron* on a genome-wide scale. A total of 120 *Rhododendron* AP2/ERF genes were identified. The phylogenetic analysis showed that RsAP2 genes were classified into five main subfamilies, AP2, ERF, DREB, RAV and soloist. Cis-acting elements involving plant growth regulators, response to abiotic stress and MYB binding sites were detected in the upstream sequences of RsAP2 genes. A heatmap of RsAP2 gene expression levels showed that these genes had different expression patterns in the five developmental stages of *Rhododendron* flowers. Twenty RsAP2 genes were selected for quantitative RT-PCR experiments to clarify the expression level changes under cold, salt and drought stress treatments, and the results showed that most of the RsAP2 genes responded to these abiotic stresses. This study generated comprehensive information on the RsAP2 gene family and provides a theoretical basis for future genetic improvement.

Keywords: *Rhododendron*; transcription factor; genome-wide; abiotic stress; gene duplication; gene expression

Citation: Guo, Z.; He, L.; Sun, X.; Li, C.; Su, J.; Zhou, H.; Liu, X. Genome-Wide Analysis of the *Rhododendron* AP2/ERF Gene Family: Identification and Expression Profiles in Response to Cold, Salt and Drought Stress. *Plants* **2023**, *12*, 994. https://doi.org/10.3390/plants12050994

Academic Editors: Aiping Song and Yu Chen

Received: 2 February 2023
Revised: 20 February 2023
Accepted: 20 February 2023
Published: 22 February 2023

1. Introduction

As a widely distributed ornamental genus around the world, it is difficult to deny that *Rhododendron* has more dimensions of ornamental diversity than other popular woody horticultural plants. In China, *Rhododendron simsii* is the most widely cultivated *Rhododendron* species in landscaping, and numerous cultivars have been obtained through hybridization [1]. *Rhododendron* flower buds generally undergo a period of dormancy between flower bud differentiation and anthesis [2] to cope with the environmental conditions of autumn and winter. Previous studies have shown that *Rhododendron* flower buds are more sensitive to cold than are leaf buds, leaves and stems, and that the cold-tolerance gap between flower tissues and vegetative tissues is more obvious in plants with strong cold tolerance [3]. The critical low temperature that flower buds can tolerate before damage occurs is regarded as an important trait indicator [4]. Therefore, cold tolerance is one of the most important prerequisites for maintaining ornamental characteristics. In addition to temperature, various abiotic stress factors such as drought and high soil salt concentrations, not only affect the overall health of plants but also cause damage to flower buds [5,6]. Gardeners formerly used cross-breeding to obtain varieties with high abiotic stress tolerance. However, this method was less efficient in terms of time and the resulting varieties were often difficult to combine ornamental traits together with abiotic stresses resistance [4]. It is necessary to understand the molecular mechanism related to flower stress resistance to meet the demand. Recent research on the rose APERALA2/Ethylene Responsive Factor (AP2/ERF) gene family showed that the isolated member gene *RcDREB2B* is repressed under drought stress, while the overexpression of *RcDREB2B* was found to promote sensitivity to salt, Abscisic Acid (ABA) and poly (ethylene glycol) [7]. Enhancing abiotic stress tolerance through molecular breeding can enable the barriers of traditional breeding methods to be overcome and can influence the regional distribution of cultivation [4].

The AP2/ERF superfamily is one of the most conserved and important Transcription Factor (TF) families and is mainly found plants [8]. This TF family has various functions for regulating plant biological and physiological processes, including responses to abiotic stress and contributions to developmental processes and plant morphogenesis [9]. AP2/ERF proteins are recognized with AP2 binding domains, and the APE/ERF family can be divided into five subfamilies: AP2, Dehydration-Responsive Element-Binding (DREB) protein, ERF, ABI3/VP1 (RAV) and soloist, according to the features of the domains [8]. Different structural characteristics determine the differences in function among these subfamilies, while the AP2 subfamily members contribute to the regulation of anthesis and development of flower organs [10,11]. The ERF and DREB subfamily members mainly regulate the response toward abiotic stresses such as cold and salt [12,13], and the RAV subfamily members participate in the regulation of plant hormones such as ethylene and brassinolide [14]. The soloist subfamily function requires more research to clarify.

It has been reported in *Arabidopsis thaliana* that exogenously applied abscisic acid, cold, dehydration and high salinity will induce the expression of genes with cis-acting elements sharing a conserved 'A/GCCGAC' core sequence [15–18]. One of the cis-elements, TACCGACAT, was named the Dehydration-Responsive Element (DRE) and can bind to the TFs named DRE-Binding Protein 1/C-repeat Binding Factor (DREB1/CBF) and DREB2 [19,20]. Overexpression of DREB1/CBF ultimately improves the tolerance to cold, drought and high salinity in *Arabidopsis*, while DREB2 can help counter the stress caused by drought, high salinity and heat shock through a series of expression changes in downstream genes [21–23]. AP2/ERF transcription factors have been widely detected in plants, and their functions in flower physiology and development have been studied, including in *Arabidopsis* [24], tomato [25] and maize [26], while their functions in abiotic stress response have also been studied, mainly in model and crop plants such as rice [27] and wheat [28]. However, in ornamental plants, where research on flower development is more important and prominent, there is no comprehensive understanding of the AP2/ERF transcription factor family as there is in food-crop plants. In flowers, an important reproductive organ, there is still a lack of research on the AP2/ERF gene family and the mechanism toward abiotic stress resistance, and the need for obtaining more bioinformatics data. The achievement of the *Rhododendron* whole-genome sequence provides opportunities to comprehensively study the *Rhododendron* AP2/ERF gene family.

In this study, we identified AP2/ERF genes from the *Rhododendron* genome using bioinformatics methods and obtained complete information such as sequence features, classification, and chromosomal and promoter locations. We generated more understanding of the RsAP2 genes' function through determining the expression patterns of the AP2/ERF genes in different *Rhododendron* flower developmental stages and the expression levels in closed buds under different abiotic stresses, including cold, salt and drought. This study provides a theoretical basis for the further validation of RsAP2 gene function and protein interaction.

2. Results

2.1. Identification and Analysis of RsAP2 Genes

A total of 120 AP2 genes were obtained through Hidden Markov Model (HMM) searches with the *Rhododendron*-specific AP2 model, and their AP2 domains were confirmed by the Simole Modular Architecture Research Tool (SMART), the National Center for Biotechnology Information Conserved Domains Database (NCBI CDD) and Pfam. The sequences were named according to their chromosomal location. Gene characteristics, including gene names, IDs, subfamilies, length of proteins, molecular weight of proteins (MW), isoelectric points (pI), chromosomal locations and main subcellular localizations are displayed in Table S1. Among all of the RsAP2 proteins, RsAP2-59 was identified as the largest protein with 906 aa and 100.7 kDa, while RsAP2-16 was the smallest, with 129 aa and 13.6 kDa in protein length and MW, respectively. The results of subcellular localization prediction showed that RsAP2 proteins were mainly located in the nucleus, followed by the chloroplast.

2.2. Phylogenetic Analysis and Classification of RsAP2 Genes

A phylogenetic tree of AP2 genes in *Rhododendron* and *Arabidopsis* was constructed to analyze the evolutionary relationship. A total of 195 AP2 genes were classified into 5 subfamilies, and the DREB and ERF subfamilies were further divided into 11 subgroups, as shown in Figure 1. Seventeen RsAP2 genes were classified into the AP2 subfamily with two tandem-duplicated AP2 domains. The DREB and ERF subfamilies both have only one AP2 domain. Forty-four RsAP2 genes with the 14th and 19th amino acids of the domain being valine and glutamic acid, respectively, were confirmed to be DREB subfamily members. Fifty-three RsAP2 genes with the 14th and 19th amino acids of the domain being alanine and aspartic acid, respectively, were confirmed to be ERF subfamily members. Four RsAP2 genes with an AP2 domain at the C-terminus and a B3 domain at the N-terminus were recognized as RAV subfamily members. The remaining two RsAP2 genes with an AP2 domain but with large differences in gene structure from other subfamily members were classified into the soloist subfamily. As shown in the figure, AP2 genes with the same structural features were accurately clustered into a certain subfamily, which indicated that the AP2 gene family has high homology across different species.

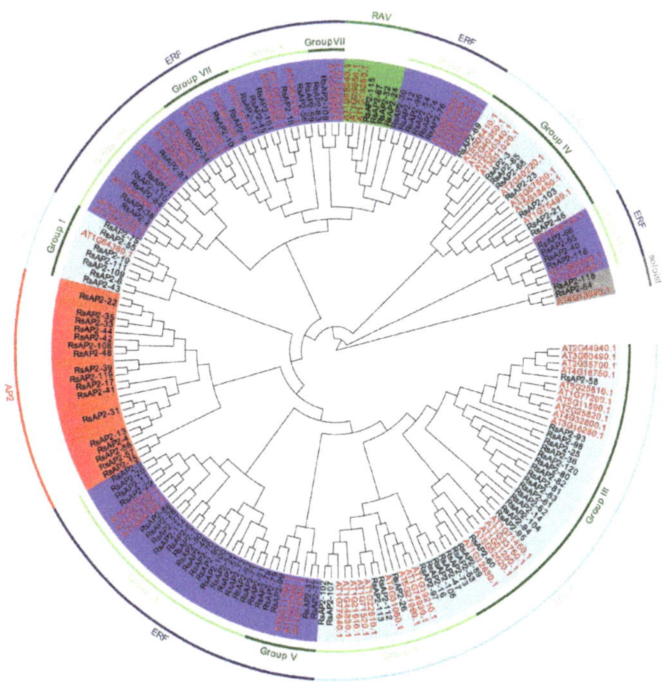

Figure 1. Phylogenetic tree of AP2 genes from *Rhododendron* and *Arabidopsis*. The identifiers colored in black are RsAP2 genes and the AP2 genes from *Arabidopsis* are colored in red. Subfamilies of AP2, DREB, ERF, RAV and soloist are distinguished by backgrounds shown in red, light blue, dark blue, green and gray, respectively.

2.3. *RsAP2 Protein Motifs and Gene Structure Analysis*

Fifteen conserved motifs of RsAP2 proteins were identified with Multiple EM for Motif Elicitation (MEME) software. These motifs were composed of 8–60 amino acids as shown in Figure 2. The types and distribution of motifs were different in RsAP2 genes, but several combined patterns of motifs, such as motif 11-1-8 and motif 3-2, appear rather frequently. Gene structures of introns and exons were acquired from alignment of genome sequence data. Among the RsAP2 genes, 79 genes had no introns, 17 genes had 1 intron and 24 genes had 2 or more introns. To intuitively compare the features of the RsAP2 genes, Figure 3 was constructed by using a phylogenetic tree, the distribution of motifs and the gene structure of RsAP2 genes.

Figure 2. Fifteen main conserved motifs of RsAP2 genes.

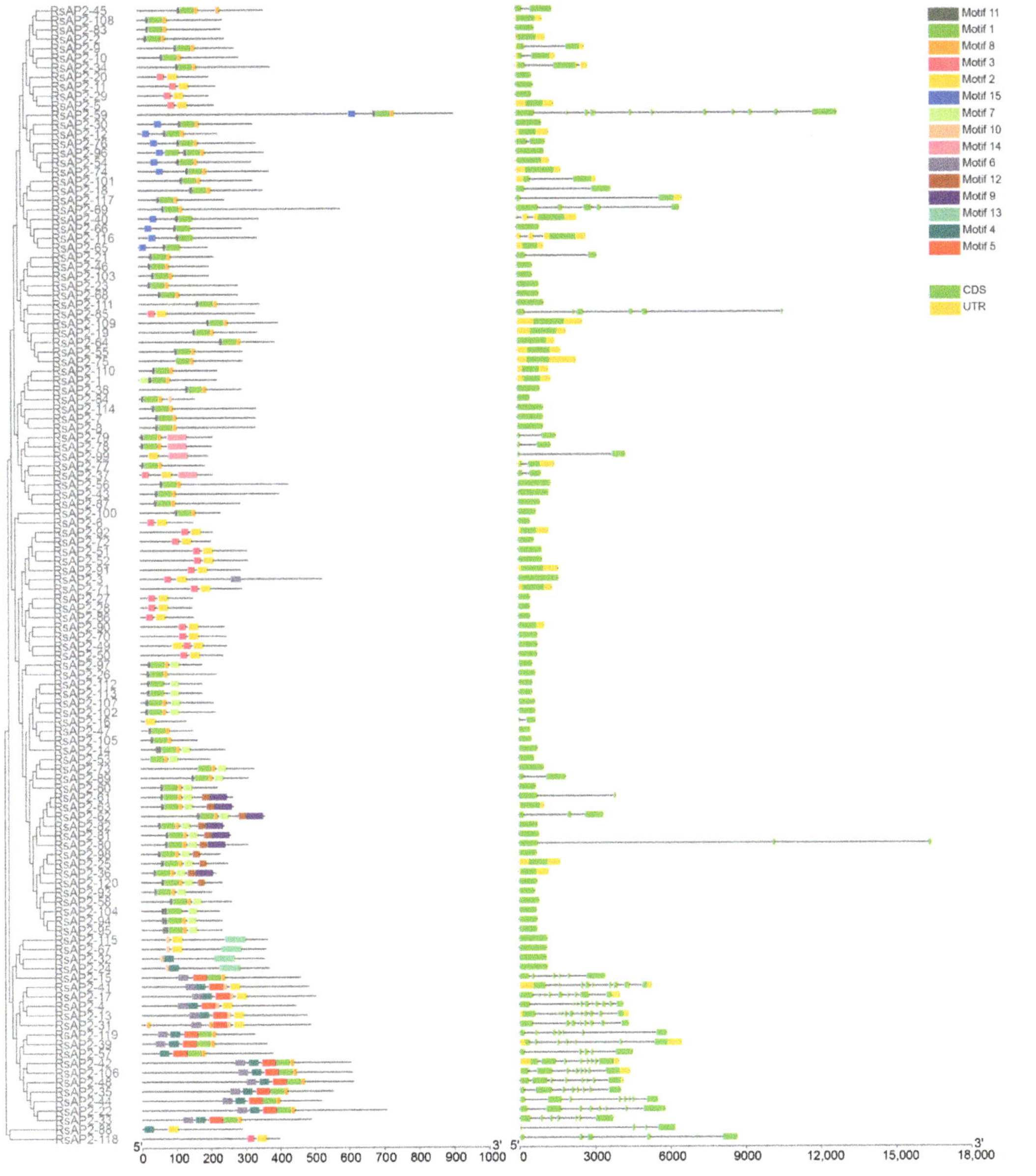

Figure 3. Chromosomal distribution and gene duplication analysis of RsAP2 genes.

2.4. Chromosome Location and Gene Duplication of RsAP2 Genes

The 120 RsAP2 genes were distributed on the thirteen chromosomes unevenly, with the chromosome 7 containing fifteen of the RsAP2 genes and the chromosome 11 contained only three RsAP2 genes, which were the largest and smallest numbers, respectively (Figure 4).

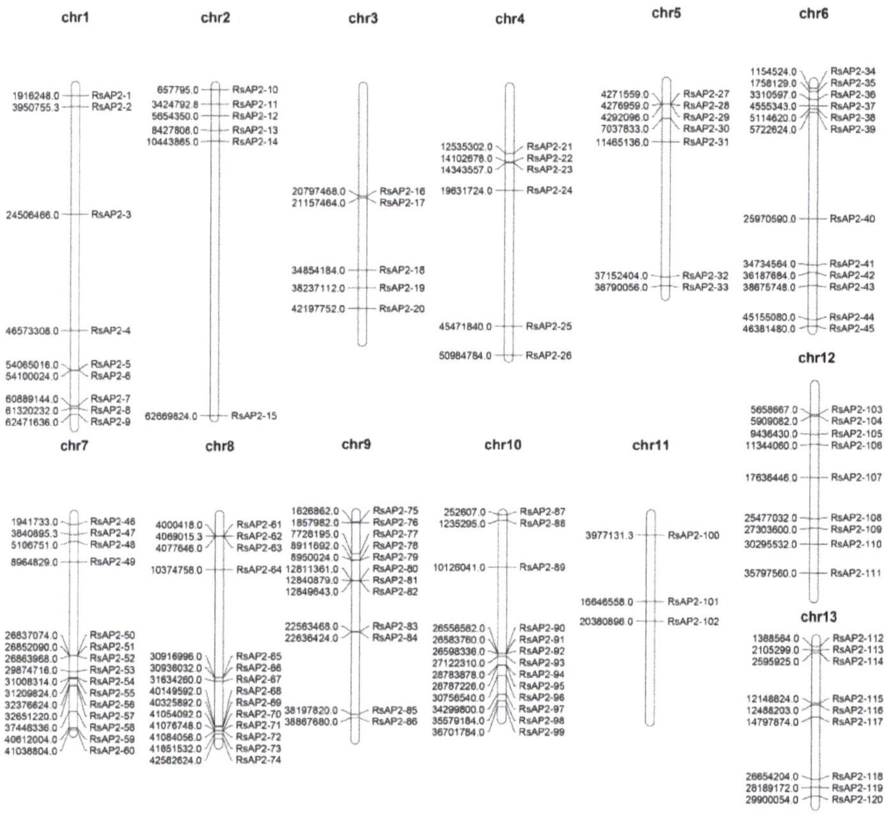

Figure 4. Chromosomal distribution and location of RsAP2.

The evolution of genomes and genetic systems is considered to be driven by gene duplications and could be distinguished as segmental duplication and tandem duplication [29]. A tandem duplication event is described as a chromosomal region within 200 kb containing two or more genes, according to previous research [30]. As shown in Figure 5, thirty RsAP2 genes (RsAP2-5/6, RsAP2-27/28/29, RsAP2-50/51/52, RsAP2-61/62/63, RsAP2-65/66, RsAP2-68/69, RsAP2-70/71/72, RsAP2-78/79, RsAP2-80/81/82, RsAP2-83/84, RsAP2-90/91/92 and RsAP2-94/95) were clustered into twelve tandem duplication event regions on *Rhododendron* chromosomes 1, 5, 7, 8, 9 and 10. The most frequent AP2/ERF gene tandem duplication event occurred on chromosome 8 with four clusters. Fifty segmental duplication events involving seventy-one RsAP2 genes were identified with Multiple Collinearity Scan toolkit X version software (MCScanX). The tandem and segmental duplication contributed greatly to the expansion and evolution process of RsAP2, especially for the latter pattern.

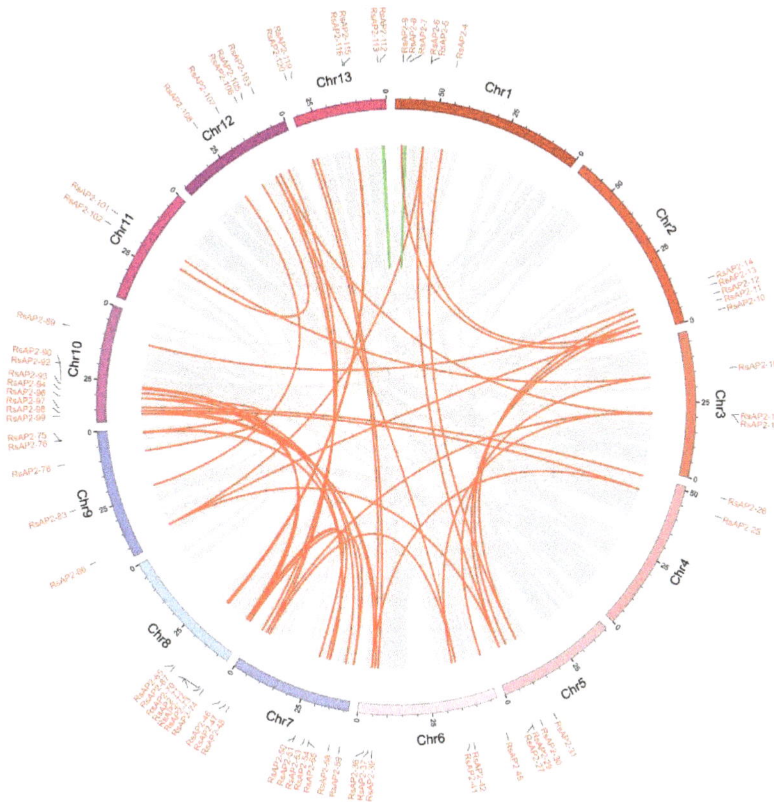

Figure 5. Synteny analysis and chromosomal relationships of *Rhododendron* RsAP2 genes. Red lines indicate duplicated RsAP2 gene pairs between different chromosomals, green lines indicate duplicated RsAP2 gene pairs in the same chromosomal and gray lines indicate all synteny blocks in the *Rhododendron* genome.

2.5. Important Cis-Acting Elements in RsAP2 Promoters

Upstream sequences (2 kb) were extracted according to the chromosomal locations of RsAP2 genes. Cis-acting elements were detected by using the Plant Cis-Acting Regulatory Element (PlantCARE) and the results showed elements with several putative functions. These elements were divided into three main categories according to their putative functions, including plant growth regulation, responses to abiotic stress and MYB binding sites (Figure 6). The cis-acting elements ABRE and LTR, which respond to abscisic acid and are involved in low-temperature responsiveness, respectively, were widely found as promoters of the 106 and 68 RsAP2 genes, respectively. These findings indicate the vital function of RsAP2 in counteracting environmental stress, especially cold stress.

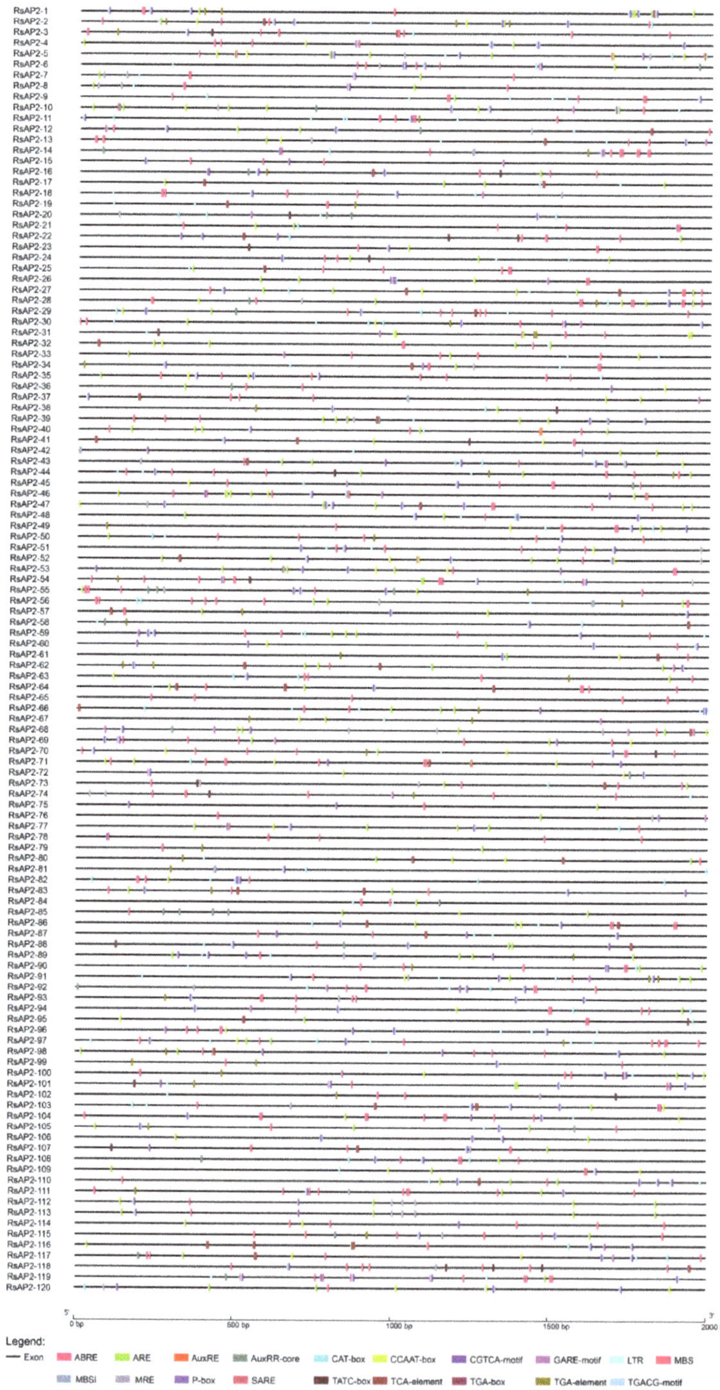

Figure 6. Cis-acting elements in *RsAP2* promoters.

2.6. Expression Patterns of RsAP2 Genes in Different Stages of the Flowering Process

The heatmap of 115 RsAP2 genes (with 5 RsAP2 genes eliminated through the results of local Basic Local Alignment Search Tool, BLAST) was generated by the Fragments per Kilobase of Transcript per Million Mapped reads (FPKM) values, transformed by log2 (Figure 7). The RsAP2 genes had different expression patterns across the five development stages of *Rhododendron* flowers. Most of the RsAP2 genes had an especially high expression stage that was distinct from the others. Almost all of the RsAP2 gene clusters with tandem duplication events, such as RsAP2-5/6 and RsAP2-50/51/52, showed similar expression patterns, and they all had higher expression levels in the last flowering stage. Additionally, the cluster of RsAP2-80/81/82 had higher expression levels in the S1 and S2 stages but lower expression levels in other stages.

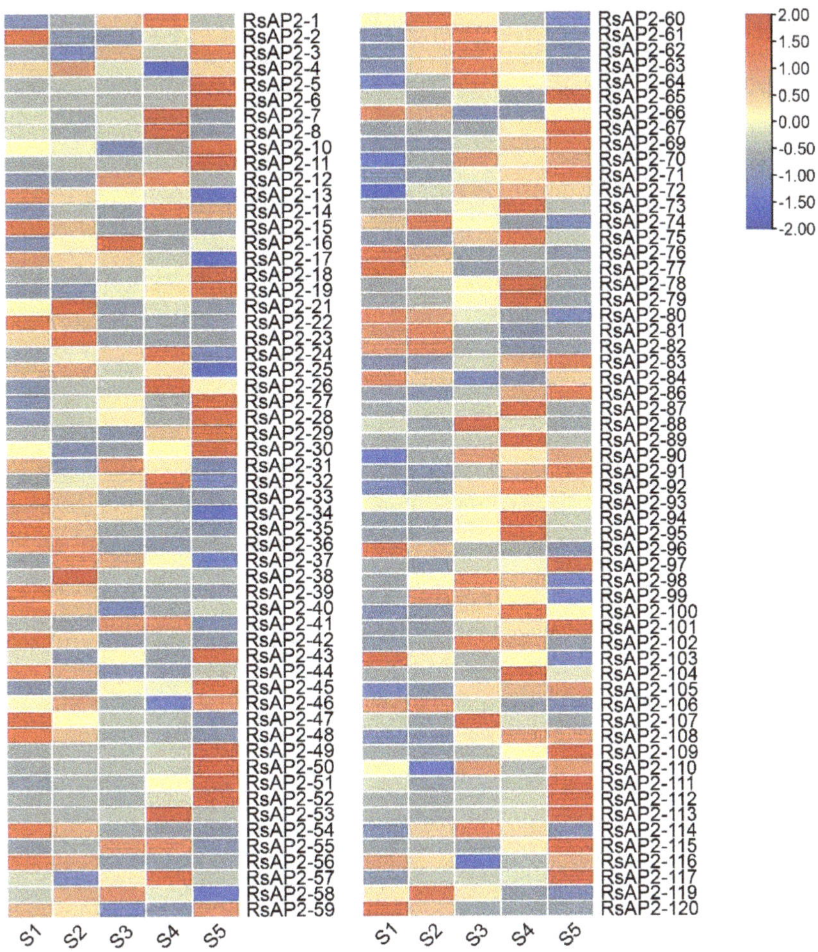

Figure 7. Heatmap of expression levels of RsAP2 genes in different development stages of *Rhododendron* flowers. The FPKM levels are transformed by log2 and labels of S1, S2, S3, S4 and S5 represent closed buds, buds showing color at the top but with the scales still present, initial flowering stage, full flowering stage and last flowering stage, respectively.

2.7. *Interaction Network Analysis of RsAP2 Proteins*

The protein interaction network referring to *Arabidopsis* orthologous proteins was constructed to speculate the relationship and synergy of RsAP2 proteins (Figures 8 and S1). According to the result, several RsAP2 genes represented to be core regulators to interact with other proteins. RsAP2-4 had high homology with AP2 protein in *Arabidopsis* and may be upstream from the regulatory network to affect various proteins. The Cytokinin Response Factor 2 (CRF2), CRF3 and CRF4 were determined to have pairwise interaction through the yeast two-hybrid system [31], and the corresponding RsAP2-54, RsAP2-76 and RsAP2-74 may have similar function in response to abiotic stress.

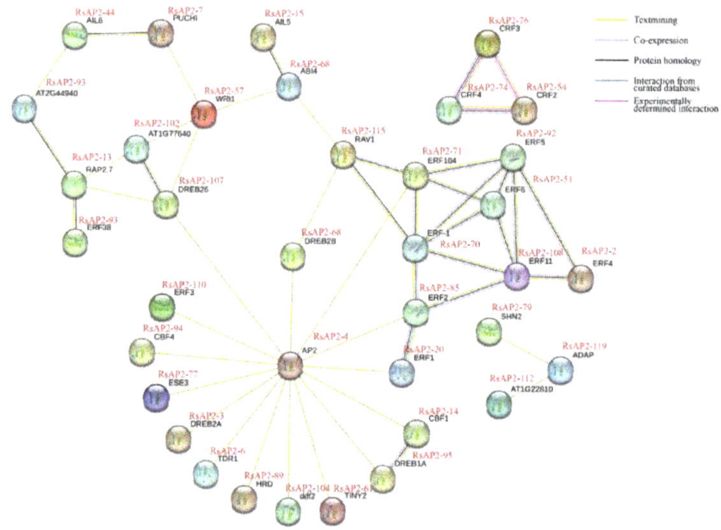

Figure 8. Partial interaction network of RsAP2 protein referring to orthologs in *Arabidopsis*.

2.8. *Expression Profiles of RsAP2 Genes under Abiotic Stress*

Quantitative Real-Time Fluorescent Polymerase Chain Reaction (qRT-PCR) was conducted to clarify the expression level change in RsAP2 genes with cold, salt and drought stress treatments in closed buds with three biological replicates and two technical replicates. Twenty RsAP2 genes were selected as examples according to the conditions of being confirmed as AP2-like ethylene-responsive transcription factors with annotations of the Pfam, SwissProt and NCBI Non-Redundant (NR) protein databases and had relatively higher FPKM values at the closed bud stage, while their promoters contained cis-acting elements response to abiotic stress and abscisic acid. Some RsAP2 genes showed significant changes in expression levels in response to abiotic stress treatments.

Overall, most of the RsAP2 genes were sensitive to cold stress, as 16 RsAP2 genes had significantly upregulated relative expression with cold treatment for 4 and 24 h compared to the control treated at room temperature and shielded from light (Figure 9). The peak RsAP2-42, 66, 102, 106 and 119 expression levels occurred at 4 h of cold stress, which may indicate a fast response situation.

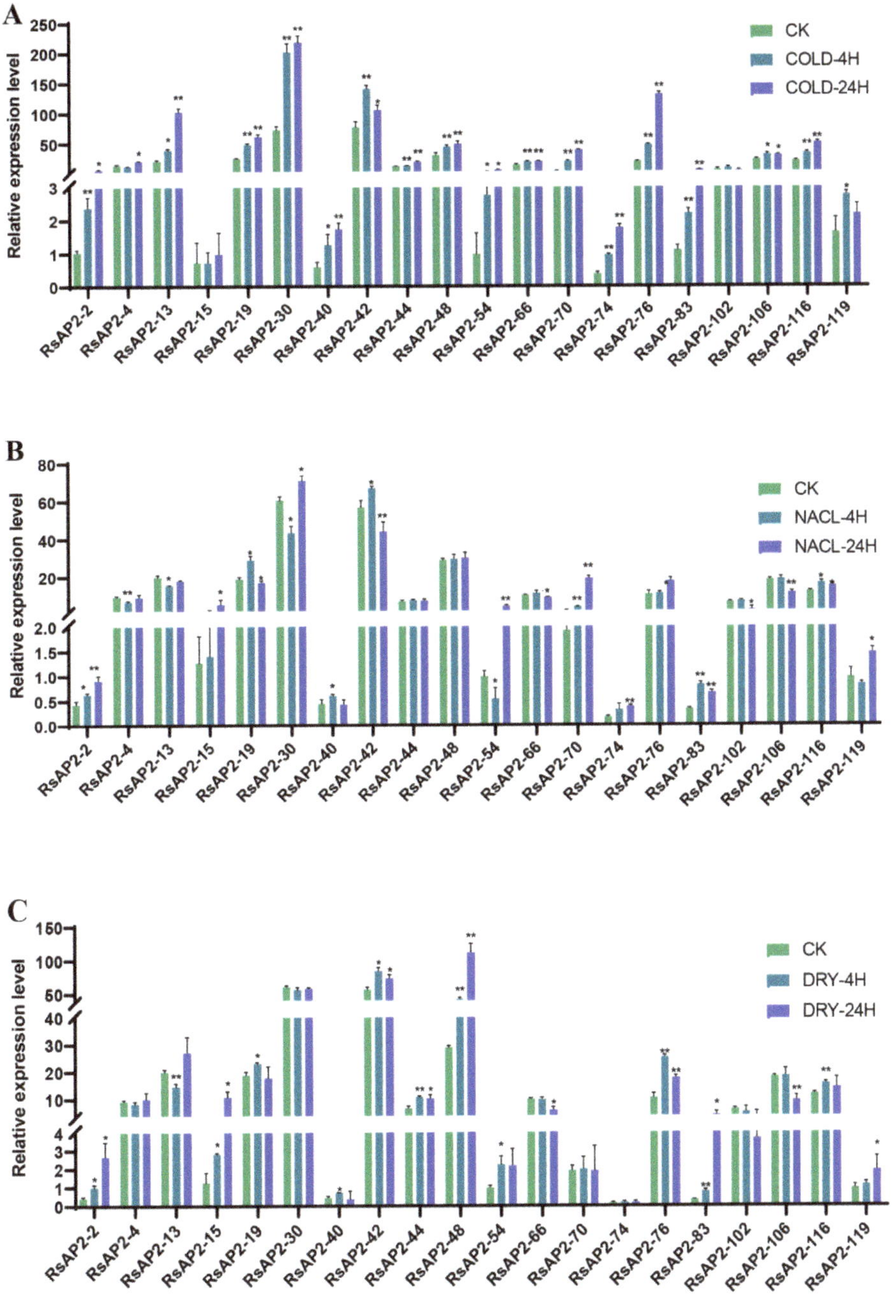

Figure 9. Relative expression level of 20 selected RsAP2 genes in response to cold (**A**), salt (**B**) and drought (**C**) treatments. Actin gene was taken as the internal reference gene. The relative expression levels are presented with mean and standard deviation of qRT-PCR. The significance test is presented with asterisks: * $p < 0.05$, ** $p < 0.01$, Student's t-test.

Although salt treatment may also cause drought, several RsAP2 genes, such as RsAP2-30 and RsAP2-48, had obvious differences in expression patterns. RsAP2-30 expression had barely any volatility with the drought treatment, but the salt treatment resulted in this gene being significantly downregulated at 4 h but significantly upregulated at 24 h. As for RsAP2-30, the expression was mostly stable with the salt treatment, but the drought treatment resulted in extreme upregulation of the expression at 4 and 24 h. However, the expression patterns were rather similar in RsAP2-2, 15, 19, 40, 66, 106 and 116 with both the salt and drought treatments.

3. Discussion

With a large number of family members and diversified functions including abiotic stress resistance, hormone signal transduction and plant metabolite regulation, the AP2/ERF transcription family is important in many plant species. Many species specific AP2/ERF transcripts have been identified with the study of whole-genome sequencing data of many plants such as *Arabidopsis* [32], rice [33] and poplar [34]. Most studies have concentrated on model and crop plants, while only a few studies have detected transcription factor families in ornamental plants. Therefore, we performed a comprehensive and overall study of the AP2/ERF transcript family in *Rhododendron*, which has not been done before.

We identified 120 RsAP2 genes and carried out comprehensive studies on their phylogeny, sequence structure, chromosomal location, collinearity and expression patterns in flower development stages and abiotic stresses. This information can provide a theoretical basis and data support for further research on the functional pathways in stress resistance and flowering physiology of *Rhododendron* and other ornamental plants.

Previous research has reported that the number of AP2/ERF genes was not directly related to genome size, as the numbers of AP2/ERF genes in *Arabidopsis*, rice, poplar and rose were 147, 164, 200 and 135, respectively, while their genome sizes were 125, 430, 485 and 600 Mb, respectively. With 120 AP2/ERF genes identified and a 609 Mb genome size, the findings for *Rhododendron* could support this conclusion.

Phylogenetic trees can be used to study visually the relationships of sequences. In our research, the sequences were not clustered by species firstly but, rather, according to the classification of structure. The AP2 subfamily genes were grouped into an independent cluster, while the DREB and ERF subfamilies had a staggered distribution and formed different subgroups. This may be due to the tandem double AP2 domains, a relatively obvious specific structure.

The conserved motifs are conserved amino acid sequences with various biological functions involved in biological processes such as protein cointeraction, transcriptional activation and nuclear localization [32]. The tandem motif 11-1-8 and motif 3-2 occurred in all of the ERF and DREB subfamily members, while in AP2 subfamily members the tandem motif 6-4-5 was a common structure, suggesting that sequence structures have a close relationship with certain functions.

Some characteristics and commonalities can be found through the analysis of gene structures. There were 79 RsAP2 genes with no introns, which accounted for 65.83% of all the family members, similar to the situation in rose [7]. Interestingly, the RAV subfamily members all had no introns, while the AP2 subfamily members contained more than four introns without exception.

According to many studies, the evolution of genomes and genetic systems cannot occur without gene duplications. Gene families could therefore be the product of segmental duplication and tandem duplication [35]; in these ways, plants can achieve rapid adaption to environmental change [36]. In the present study, the duplication event of RsAP2 genes was analyzed to complete the understanding of the expansion pattern. The distribution of tandem duplications was uneven, with 30 RsAP2 genes clustered into 12 tandem duplication event regions on *Rhododendron* chromosomes 1, 5, 7, 8, 9 and 10, while 50 segmental duplication events involving 71 RsAP2 genes were identified. This indicated that tandem duplication and segmental duplication both contributed greatly to the evolution of *Rhodo-*

dendron AP2/ERF genes. Many of the RsAP2 genes associated with tandem duplication events tended to have similar expression patterns in different developing stages of *Rhododendron* flowers, as shown in Figure 7, which is evidence of the close relationship between structure and function.

Cis-acting elements not only determine the tissue-specific expression of genes, but also have close relationship with stress resistance [37]. The cis-acting elements were mainly distributed in the 2 kb upstream sequence and could be analyzed to clarify the mechanisms and potential functions of RsAP2 transcriptional regulation. Among the elements, ABREs had the largest variety of sequence types according to data for four species (Table S2). The ABREs in *Rhododendron* were widely distributed upstream of 108 RsAP2 genes, and 78 of them appeared more than once. The high frequency and wide range of ABRE elements suggest that the function involved in abscisic acid responsiveness to resist abiotic stresses including cold, drought and salt [38], played an important role and was guaranteed in quantity. Another important cis-acting element is mainly involved in low temperature responsiveness [39], which occurred in the promoter of 68 RsAP2 genes. Three kinds of elements with MYB binding sites for MYB TFs, MBS, MBSI and MRE, were detected for their functions of drought-inducibility, flavonoid biosynthetic genes regulation and light responsiveness, respectively. This may imply the relation and interaction of AP2/ERF and MYB transcription factors and the complex regulatory network.

The expression patterns of genes were analyzed to achieve deeper understanding of the AP2/ERF genes in terms of overall trends and individual differences. Transcriptome data were the main data source to reveal the expression level of genes in different states of plant tissue. There is a period of dormancy with special biological significance for the *Rhododendron* flower bud before anthesis. According to the RNA-seq data for *R. simsii*, the RsAP2 genes presented different expression patterns in the five developmental stages of flowers. At the closed buds stage (S1), several RsAP2 genes had peak expression levels, while others had the lowest expression levels. Additionally, combined with the gene duplication events mentioned above, the clustered tandem duplicated genes exhibited high similarities in expression patterns among all five stages. This phenomenon may indicate the important roles of duplicated genes in developmental physiology and stress response because their function may be enhanced through an increase in quantity.

Quantitative RT-PCR can intuitively illustrate the RsAP2 genes in *Rhododendron* flower buds under different abiotic stress treatments. Twenty RsAP2 genes were selected as examples according to conditions, including phylogenetic analysis, annotations, FPKM values, cis-acting elements and protein interaction networks. With cold hardiness being an important breeding goal for researchers around the world, we focused on the gene response to cold stress, which significantly affected most of the selected RsAP2 genes. The accumulation of expression levels was positively correlated with the increase in time treated with cold stress for the overall trend. The functions of RsAP2-30/54/74/76 were supposed to be similar to those of the AT4G27950 gene, with a close phylogenetic relationship. AT4G27950 was regarded as CRF4 and showed transcriptional induction after exposure to cold [40]. Several special cases, such as RsAP2-42/106/119, had their expression peak with 4 h of cold treatment, which may indicate the characteristics of a rapid response. Although the application of higher salt concentrations may cause cellular osmotic water loss resulting in a drought-like effect, some AP2/ERF genes show specific patterns in response to salt treatment compared with mannitol-induced drought treatment, such as RsAP2-70 and RsAP2-48. RsAP2-70 had a close structural relationship with AT1G06160, which was named AtERF15. AtERF15 was reported to be the binding factor of Coupling Element 1 and was a positive regulator of the ABA response [41]. It can be assumed from this that RsAP2-70 was induced with salt stress and regulated by ABA as a downstream response factor. Meanwhile, RsAP2-48 exhibited no significant change in expression level after salt treatment. The function of RsAP2-48 may be similar to that of AT3G20840, the PLETHORA 1 gene in *Arabidopsis*, induced by auxin to control the alteration of stomatal apertures under drought stress [42]. Abiotic stresses, plant growth regulators and TFs can form regulatory networks, and the AP2 gene family has extensive involvement in the process of environmental adaptation. Some of the RsAP2 genes appear to be redundant under abiotic stress and show similar expression patterns; they can be multiple types of insurance in terms of mechanisms or can be functionally cumulative in terms of quantity.

The above analysis was performed to systematically mine the genome-wide AP2 gene family of *Rhododendron*. Precise studies of regulatory systems and molecular mechanisms can provide a theoretical basis for future genetic improvement.

4. Materials and Methods

4.1. Identification of AP2 Family Members in the Rhododendron Genome

The whole *Rhododendron* protein sequence was generated from the *Rhododendron* genome sequencing data. (The data have been submitted to NCBI genebank under the BioProject number of PRJNA902133). HMM was used as the main method to identify *Rhododendron* AP2 candidates. We downloaded the model file (PF00847) from the Protein Family (Pfam) database (http://pfam.xfam.org/, accessed on 23 June 2022). The first domain search was conducted using the model file with the standard of E-value < 1×10^{-10} to achieve a species-specific model in the *Rhododendron* genome protein sequence. Putative protein sequences were achieved with the second search using the species-specific model with the standard of E-value < 1×10^{-10}. The putative sequences were submitted to the SMART (http://smart.embl-heidelberg.de/, accessed on 28 June 2022), NCBI CDD (https://www.ncbi.nlm.nih.gov/Structure/bwrpsb/bwrpsb.cgi, accessed on 28 June 2022) and Pfam databases to verify the AP2 domain. The resulting sequences with AP2 domains confirmed by all three databases were clarified by removing redundant sequences, and the other sequences were categorized as RsAP2 genes. The RsAP2 genes were sorted according to their chromosome number and their location on the chromosome, and they were named in this order at the same time. The subcellular localization of RsAP2 proteins was predicted with WoLF PSORT (https://wolfpsort.hgc.jp/, accessed on 12 October 2022). The RsAP2 sequences were submitted to the Expert Protein Analysis System (ExPASy, http://web.expasy.org/protparam/, accessed on 28 June 2022) to analyze their molecular weights and theoretical isoelectric points (pI).

4.2. Phylogenetic and Classification Analysis

The phylogenetic tree of AP2 genes from *Rhododendron* and *Arabidopsis* was conducted using neighbor-joining method with 1000 bootstrap replications to show their evolutionary relationship with MEGA software (version 5.1) [43]. The phylogenetic tree was further modified with Evolview (https://www.evolgenius.info/evolview/#/treeview, accessed on 29 June 2022) [44]. The *Rhododendron* AP2 genes were classified into five subfamilies, AP2, ERF, DREB, RAV and soloist, depending on the number of AP2 domains and other features in sequence according to previous studies [45]. The ERF and DREB subfamilies were further classified into eleven subgroups according to the topology of phylogenetic tree and classification of the *Arabidopsis* AP2 gene family.

4.3. Sequence Analysis

The conserved motifs of RsAP2 genes were identified with MEME (http://meme-suite.org/, accessed on 29 June 2022) [46]. The information of gene structures describing the exon and intron was extract from the gff file of genome sequence data. Visualization of the RsAP2 gene structures was conducted by Gene Structure Display Server (GSDS, http://gsds.gao-lab.org/, accessed on 29 June 2022) and Toolbox for Biologists (TBtools, version 1.098774) [47,48].

4.4. Chromosomal Localization and Colinearity Analysis

We used MapChart (version 2.32) to present the localization information for RsAP2 genes on chromosomes [49]. The duplication and expansion of the RsAP2 gene family was analyzed with MCScanX (http://chibba.pgml.uga.edu/mcscan2/MCScanX.zip, accessed on 30 June 2022), and we used CIRCOS (https://circos.ca/, accessed on 5 July 2022) to complete the figure of RsAP2 gene family collinearity [50].

4.5. Main Cis-Acting Elements Analysis

We extracted 2 kb upstream sequence of the RsAP2 gene sequences from the genome sequencing data and submitted it to PlantCARE (http://bioinformatics.psb.ugent.be/webtools/plantcare/html/, accessed on 30 June 2022) [51]. Cis-acting elements involved in different kinds of plant growth regulators, response to abiotic stresses, anaerobic induction, meristem expression and flavonoid biosynthetic were identified.

4.6. Plant Materials and Abiotic Stress Treatments

The *Rhododendron* cultivar 'Yanzhi Mi' grown in the greenhouse at the Jiangsu Academy of Agricultural Sciences (Nanjing, China, 118.881 E, 32.039 N) was used in this study. In November 2022, four-year-old plants with closed buds that were grown in separate pots were grouped according to the treatments to be subjected to different abiotic stresses. For cold stress, the plants were placed in a paper box with a black plastic light-shielding bag. Two groups were placed in a 4 °C artificial cold storage room, while the control group was placed at room temperature. For salt and drought stress, the plants were placed in a greenhouse and treated with 200 mM NaCl or 260 mM mannitol, respectively, while the control group was treated with pure water. Closed buds were collected from every group at 4 and 24 h after treatment, frozen in liquid nitrogen immediately after sampling and then stored in a freezer at −80 °C.

4.7. RNA-Seq Data Analysis of RsAP2 Genes

RNA-seq data were obtained from a previous study using *Rhododendron* cultivar 'Yanzhi Mi' flowers at 5 different developmental stages, including closed buds (S1), buds showing color at the top but with scales still present (S2), initial flowering stage (S3), full flowering stage (S4) and last flowering stage (S5), to analyze the expression patterns of RsAP2 genes during the flower development process (PRJNA700000) [52]. We created a local BLAST database with the Unigenes in the RNA-seq data and processed the tblastn search program with the RsAP2 protein sequence using BioEdit software (version 7.2.5). The corresponding relations were confirmed for the match of query–subject pairs with the lowest E-value, less than 1×10^{-50}. We used TBtools (version 1.098774) to construct the heatmap with the FPKM values transformed by log2. The RsAP2 protein interaction network was constructed by STRING (https://cn.string-db.org/, accessed on 4 January 2023) with the reference organism of *A. thaliana*.

4.8. Total RNA Extractions and Expression Analysis of RsAP2 Genes

Total RNA extraction was conducted with a FastPure Universal Plant Total RNA Isolation Kit (Vazyme, Nanjing, China). The cDNA reverse-transcription process was conducted with HiScript III RT SuperMix for qPCR (Vazyme, Nanjing, China). The primers were designed with Primer Premier software (version 5.0) and synthesized by General Biol Company (Anhui, China). The quantitative real-time PCR process was conducted with ChamQ Universal SYBR qPCR Master Mix (Vazyme, Nanjing, China) on the platform of ROCHE (LightCycler® 480 II, Swiss). The RT-qPCR reaction was conducted in a 96-well plate containing 10.0 μL SYBR Mix, 0.4 μL 10 μM solution of primer, respectively, 0.18 μL undiluted cDNA and 9.02 μL double distilled H2O. The qPCR program was conducted in 3 steps: pre-denaturation stage had 1 cycle of 95 °C for 30 s; recirculation reaction stage had 40 cycles of 95 °C for 10 s and 60 °C for 30 s; melting curve had 1 cycle of 95 °C for 15 s, 60 °C for 60 s and 95 °C for 15 s, in order. Three repeats were performed for each sample. Housekeeping gene *Actin* was taken as the internal reference gene [53]. Relative expression levels of the genes were calculated using the $2^{-\Delta\Delta CT}$ formula [54].

5. Conclusions

This research focused on the analysis of *Rhododendron* AP2/ERF gene family. A total of 120 AP2/ERF genes were identified and their structural features described. The genes were classified into five subfamilies according to their domain and sequence structural characteristics. The phylogenetic comparison of AP2/ERF genes from *Arabidopsis* provided a valuable reference for the evolutionary characteristics of *Rhododendron* AP2/ERF genes. Many of the RsAP2 genes may participate in the development of *Rhododendron* and could be response factors to abiotic stress, such as cold, salt and drought. The expression patterns in different developmental stages and under different abiotic stress treatments indicated the different functions of RsAP2 members. This research generated relatively comprehensive information on the RsAP2 gene family and provides a theoretical basis for future further validation of RsAP2 gene function and protein interaction.

Supplementary Materials: The following supporting information can be downloaded at: https://www.mdpi.com/article/10.3390/plants12050994/s1, Figure S1: Interaction network of RsAP2 protein referring to orthologs in *Arabidopsis*. Table S1: Features of RsAP2. Table S2: Cis-acting elements of RsAP2 promoters.

Author Contributions: Conceptualization, L.H. and Z.G.; methodology, X.S.; software, Z.G.; validation, C.L., X.L. and J.S.; formal analysis, Z.G.; resources, H.Z.; data curation, Z.G. and X.S.; writing—original draft preparation, Z.G.; writing—review and editing, L.H. and C.L.; visualization, Z.G.; supervision, H.Z.; project administration, X.L.; funding acquisition, J.S. and L.H. All authors have read and agreed to the published version of the manuscript.

Funding: This research was funded by the 'JBGS' Project of Seed Industry Revitalization in Jiangsu Province (JBGS(2021)095); Special Fund for Jiangsu Provincial Forestry Innovation and Promotion (LYKJ(2021)06); Special Fund for Forestry Development of Jiangsu Province (sczh (2022) No. 18); Jiangsu Agricultural industrial Technology system construction project (JATS(2022)101); Jiangsu Provincial Agricultural Science and Technology Innovation Project (CX (22)3061); the Central Financial Services Demonstration Funds for Forestry Science and Technology Promotion Project (Su (2021) TG06).

Institutional Review Board Statement: Not applicable.

Informed Consent Statement: Not applicable.

Data Availability Statement: Data are contained within the article and Supplementary Materials.

Acknowledgments: The authors gratefully acknowledge the financial support provided by the Natural Science Foundation of Jiangsu Province. We are thankful for the support provided by Jiangsu Academy of Agricultural Sciences.

Conflicts of Interest: The authors declare no conflict of interest.

References

1. Yang, F.S.; Nie, S.; Liu, H.; Shi, T.L.; Tian, X.C.; Zhou, S.S.; Bao, Y.; Jia, K.; Guo, J.; Zhao, W.; et al. Chromosome-level genome assembly of a parent species of widely cultivated azaleas. *Nat. Comm.* **2020**, *11*, 5269. [CrossRef]
2. Christiaens, A.; Pauwels, E.; Gobin, B.; Van Labeke, M. Flower differentiation of azalea depends on genotype and not on the use of plant growth regulators. *Plant Growth Regul.* **2015**, *75*, 245–252. [CrossRef]
3. Sakai, A.; Fuchigami, L.; Weiser, C.J. Cold hardiness in the genus Rhododendron. *J. Am. Soc. Hortic. Sci.* **1986**, *111*, 273–280. [CrossRef]
4. Krebs, S. Rhododendron. In *Ornamental Crops*, 2nd ed.; Van Huylenbroeck, J., Ed.; Springer International Publishing: Cham, Switzerland, 2018; Volume 11, pp. 673–718.
5. Mizoi, J.; Shinozaki, K.; Yamaguchi-Shinozaki, K. AP2/ERF family transcription factors in plant abiotic stress responses. *Biochim. Biophys. Acta* **2012**, *1819*, 86–96. [CrossRef]
6. Agarwal, P.; Agarwal, P.; Joshi, A.; Sopory, S.; Reddy, M. Overexpression of PgDREB2A transcription factor enhances abiotic stress tolerance and activates downstream stress-responsive genes. *Mol. Biol. Rep.* **2010**, *37*, 1125–1135. [CrossRef]
7. Li, W.; Geng, Z.; Zhang, C.; Wang, K.; Jiang, X. Whole-genome characterization of Rosa chinensis AP2/ERF transcription factors and analysis of negative regulator RcDREB2B in Arabidopsis. *BMC Genom.* **2021**, *22*, 90. [CrossRef]

8. Licausi, F.; Ohme-Takagi, M.; Perata, P. APETALA 2/Ethylene Responsive Factor (AP2/ERF) transcription factors: Mediators of stress responses and developmental programs. *New Phytol.* **2013**, *199*, 639–649. [CrossRef]
9. Feng, K.; Hou, X.; Xing, G.; Liu, J.; Duan, A.; Xu, Z.; Li, M.; Zhuang, J.; Xiong, A. Advances in AP2/ERF super-family transcription factors in plant. *Crit. Rev. Bio.* **2020**, *40*, 750–776. [CrossRef]
10. Aukerman, M.J. Regulation of flowering time and floral organ identity by a MicroRNA and its APETALA2-like target genes. *Plant Cell* **2003**, *15*, 2730–2741. [CrossRef]
11. Klucher, K.M.; Chow, H.; Reiser, L.; Fischer, R.L. The AINTEGUMENTA gene of Arabidopsis required for ovule and female gametophyte development is related to the floral homeotic gene APETALA2. *Plant Cell* **1996**, *8*, 137–153. [CrossRef]
12. Ito, Y.; Katsura, K.; Maruyama, K.; Taji, T.; Kobayashi, M.; Seki, M.; Yamaguchi-Shinozaki, K. Functional analysis of rice DREB1/CBF-type transcription factors involved in cold-responsive gene expression in transgenic rice. *Plant Cell Physiol.* **2006**, *47*, 141–153. [CrossRef]
13. Wu, L.; Zhang, Z.; Zhang, H. Transcriptional modulation of ethylene response factor protein JERF3 in the oxidative stress response enhances tolerance of tobacco seedlings to salt, drought and freezing. *Plant Physiol.* **2008**, *148*, 1953–1963. [CrossRef]
14. Hu, Y.X.; Wang, Y.X.; Liu, X.F. Arabidopsis RAV1 is downregulated by brassino steroid and may act as a negative regulator during plant development. *Cell Res.* **2004**, *14*, 8–15. [CrossRef]
15. Yamaguchi-Shinozaki, K.; Shinozaki, K. Characterization of the expression of a desiccation-responsive rd29 gene of Arabidopsis thaliana and analysis of its promoter in transgenic plants. *Mol. Gen. Genet.* **1993**, *236*, 331–340. [CrossRef]
16. Yamaguchi-Shinozaki, K.; Shinozaki, K. A novel Arabidopsis DNA binding protein contains the conserved motif of HMG-box proteins. *Nucleic Acids Res.* **1992**, *20*, 6737. [CrossRef]
17. Yamaguchi-Shinozaki, K.; Shinozaki, K. A novel cis-acting element in an Arabidopsis gene is involved in responsiveness to drought, low-temperature, or high-salt stress. *Plant Cell* **1994**, *6*, 251–264. [CrossRef]
18. Baker, S.S.; Wilhelm, K.S.; Thomashow, M.F. The 5′-region of Arabidopsis thaliana cor15a has cis-acting elements that confer cold-, drought- and ABA-regulated gene expression. *Plant Mol. Biol.* **1994**, *24*, 701–713. [CrossRef]
19. Stockinger, E.J.; Gilmour, S.J.; Thomashow, M.F. Arabidopsis thaliana CBF1 encodes an AP2 domain-containing transcriptional activator that binds to the C-repeat/DRE, a cis-acting DNA regulatory element that stimulates transcription in response to low temperature and water deficit. *Proc. Natl. Acad. Sci. USA* **1997**, *94*, 1035–1040. [CrossRef]
20. Liu, Q.; Kasuga, M.; Sakuma, Y.; Abe, H.; Miura, S.; Yamaguchi-Shinozaki, K.; Shinozaki, K. Two transcription factors, DREB1 and DREB2, with an EREBP/AP2 DNA binding domain separate two cellular signal transduction pathways in drought and low-temperature-responsive gene expression, respectively, in Arabidopsis. *Plant Cell* **1998**, *10*, 1391–1406. [CrossRef]
21. Jaglo-Ottosen, K.R.; Gilmour, S.J.; Zarka, D.G.; Schabenberger, O.; Thomashow, M.F. Arabidopsis CBF1 overexpression induces COR genes and enhances freezing tolerance. *Science* **1998**, *280*, 104–106. [CrossRef]
22. Nakashima, K.; Shinwari, Z.K.; Sakuma, Y.; Seki, M.; Miura, S.; Shinozaki, K.; Yamaguchi-Shinozaki, K. Organization and expression of two Arabidopsis DREB2 genes encoding DRE-binding proteins involved in dehydration- and highsalinity- responsive gene expression. *Plant Mol. Biol.* **2000**, *42*, 657–665. [CrossRef]
23. Sakuma, Y.; Maruyama, K.; Qin, F.; Osakabe, Y.; Shinozaki, K.; Yamaguchi-Shinozaki, K. Dual function of an Arabidopsis transcription factor DREB2A in water-stressresponsive and heat-stress-responsive gene expression. *Proc. Natl. Acad. Sci. USA* **2006**, *103*, 18822–18827. [CrossRef] [PubMed]
24. Jofuku, K.D.; Den Boerb, G.; Van Montagu, M. Control of Arabidopsis flower and seed development by the homeotic gene APETALA2. *Plant Cell* **1994**, *6*, 1211. [CrossRef] [PubMed]
25. Nakano, T.; Fujisawa, M.; Shima, Y. The AP2/ERF transcription factor SlERF52 functions in flower pedicel abscission in tomato. *J. Exp. Bot.* **2014**, *65*, 3111–3119. [CrossRef] [PubMed]
26. Chuck, G.; Meeley, R.; Hake, S. Floral meristem initiation and meristem cell fate are regulated by the maize AP2 genes ids1 and sid1. *Development* **2008**, *135*, 3013–3019. [CrossRef] [PubMed]
27. Jung, S.E.; Bang, S.W.; Kim, S.H. Overexpression of OsERF83, a vascular tissue-specific transcription factor gene, confers drought tolerance in rice. *Int. J. Mol. Sci.* **2021**, *22*, 7656. [CrossRef] [PubMed]
28. Xu, Z.S.; Xia, L.Q.; Chen, M. Isolation and molecular characterization of the Triticum aestivum L. ethylene-responsive factor 1(TaERF1) that increases multiple stress tolerance. *Plant Mol. Biol.* **2007**, *65*, 719–732. [CrossRef] [PubMed]
29. Zhu, Y.; Wu, N.; Song, W.; Yin, G.; Qin, Y.; Yan, Y.; Hu, Y. Soybean (Glycine max) expansin gene superfamily origins: Segmental and tandem duplication events followed by divergent selection among subfamilies. *BMC Plant Biol.* **2014**, *14*, 93. [CrossRef]
30. Holub, E.B. The arms race is ancient history in Arabidopsis, the wildflower. *Nat. Rev. Genet.* **2001**, *2*, 516. [CrossRef]
31. James, W.C.; Eva, H.; Alexander, H.; Aaron, M.R. CRFs form protein–protein interactions with each other and with members of the cytokinin signalling pathway in Arabidopsis via the CRF domain. *J. Exp. Bot.* **2011**, *62*, 4995–5002. [CrossRef]
32. Nakano, T.; Suzuki, K.; Fujimura, T. Genome-wide analysis of the ERF gene family in Arabidopsis and rice. *Plant Physiol.* **2006**, *140*, 411–432. [CrossRef]
33. Sharoni, A.M.; Nuruzzaman, M.; Satoh, K. Gene structures, classification and expression models of the AP2/EREBP transcription factor family in rice. *Plant Cell Physiol.* **2011**, *52*, 344–360. [CrossRef] [PubMed]
34. Zhuang, J.; Cai, B.; Peng, R.H. Genome-wide analysis of the AP2/ERF gene family in Populus trichocarpa. *Biochem. Biophys. Res. Commun.* **2008**, *371*, 468–474. [CrossRef] [PubMed]

35. Flagel, L.E.; Wendel, J.F. Gene duplication and evolutionary novelty in plants. *New Phytol.* **2009**, *183*, 557–564. [CrossRef] [PubMed]
36. Fraser, J.A.; Huang, J.C.; Pukkila-Worley, R.; Alspaugh, J.A.; Mitchell, T.G.; Heitman, J. Chromosomal translocation and segmental duplication in Cryptococcus neoformans. *Eukaryot Cell* **2013**, *4*, 401–406. [CrossRef] [PubMed]
37. Le, D.T.; Nishiyama, R.; Watanabe, Y. Identification and expression analysis of cytokinin metabolic genes in soybean under normal and drought conditions in relation to cytokinin levels. *PLoS ONE* **2012**, *7*, e42411. [CrossRef] [PubMed]
38. Narusaka, Y.; Nakashima, K.; Shinwari, Z.K.; Sakuma, Y.; Furihata, T.; Abe, H.; Narusaka, M.; Shinozaki, K.; Yamaguchi-Shinozaki, K. Interaction between two cis-acting elements, ABRE and DRE, in ABA-dependent expression of Arabidopsis rd29A gene in response to dehydration and high-salinity stresses. *Plant J.* **2010**, *34*, 137–148. [CrossRef]
39. He, C.; Yu, Z.; Teixeira, J.A.; Zhang, J.; Liu, X.; Wang, X.; Zhang, X.; Zeng, S.; Wu, K.; Tan, J.; et al. DoGMP1 from Dendrobium officinale contributes to mannose content of water-soluble polysaccharides and plays a role in salt stress response. *Sci. Rep.* **2017**, *7*, 41010. [CrossRef]
40. Zwack, P.J.; Compton, M.A.; Adams, C.I. Cytokinin response factor 4 (CRF4) is induced by cold and involved in freezing tolerance. *Plant Cell Rep.* **2016**, *35*, 573–584. [CrossRef]
41. Lee, S.; Lee, S.; Kim, S.Y. AtERF15 is a positive regulator of ABA response. *Plant Cell Rep.* **2015**, *34*, 71–81. [CrossRef]
42. Lin, Q.; Wang, S.; Dao, Y.; Wang, J.; Wang, K. Arabidopsis thaliana trehalose-6-phosphate phosphatase gene TPPI enhances drought tolerance by regulating stomatal apertures. *J. Exp. Bot.* **2020**, *71*, 4285–4297. [CrossRef]
43. Tamura, K.; Stecher, G.; Peterson, D.; Filipski, A.; Kumar, S. MEGA6: Molecular evolutionary genetics analysis version 6.0. *Mol. Biol. Evol.* **2013**, *30*, 2725–2729. [CrossRef] [PubMed]
44. Subramanian, B.; Gao, S.; Lercher, M.; Hu, S.; Chen, W. Evolview v3: A webserver for visualization, annotation, and management of phylogenetic trees. *Nucleic Acids Res.* **2019**, *47*, W270–W275. [CrossRef] [PubMed]
45. Gou, Y.; Zhang, L.; Guo, H.; Ma, H.; Bao, A. Research progress on the AP2/ERF transcription factor in plants. *Pratacult. Sci.* **2020**, *37*, 1150–1159. [CrossRef]
46. Bailey, T.L.; Elkan, C. Fitting a mixture model by expectation maximization to discover motifs in biopolymers. *Proc. Int. Conf. Intell. Syst. Mol. Biol.* **1994**, *2*, 28–36. [PubMed]
47. Hu, B.; Jin, J.; Guo, A.; Zhang, H.; Luo, J.; Gao, G. GSDS 2.0: An upgraded gene feature visualization server. *Bioinformatics* **2015**, *31*, 1296–1297. [CrossRef]
48. Chen, C.J.; Chen, H.; Zhang, Y.; Thomas, H.R.; Frank, M.H.; He, Y.H.; Xia, R. TBtools: An integrative toolkit developed for interactive analyses of big biological data. *Mol. Plant* **2020**, *13*, 1194–1202. [CrossRef]
49. Voorrips, R.E. MapChart: Software for the graphical presentation of linkage maps and QTLs. *J. Hered.* **2002**, *93*, 77–78. [CrossRef]
50. Krzywinski, M.; Schein, J.; Birol, I.; Connors, J.; Gascoyne, R.; Horsman, D.; Jones, S.; Marra, M. Circos: An information aesthetic for comparative genomics. *Genome Res.* **2009**, *19*, 1639–1645. [CrossRef]
51. Lescot, M.; Déhais, P.; Thijs, G.; Marchal, K.; Moreau, Y.; Peer, Y.; Rouzé, P.; Rombauts, S. PlantCARE, a database of plant cis-acting regulatory elements and a portal to tools for in silico analysis of promoter sequences. *Nucleic Acids Res.* **2002**, *30*, 325–327. [CrossRef]
52. Sun, X.; He, L.; Guo, Z.; Xiao, Z.; Su, J.; Liu, X.; Zhou, H.; Li, C.; Gao, H. Comparative transcriptome analyses reveal genes related to pigmentation in the petals of a flower color variation cultivar of *Rhododendron obtusum*. *Mol. Biol. Rep.* **2022**, *49*, 2641–2653. [CrossRef]
53. Xiao, Z.; Sun, X.; Liu, X.; Li, C.; He, L.; Chen, S.; Su, J. Selection of reliable reference genes for gene expression studies on Rhododendron molle G. Don. *Front. Plant Sci.* **2016**, *7*, 1547. [CrossRef] [PubMed]
54. Livak, K.J.; Schmittgen, T.D. Analysis of relative gene expression data using real-time quantitative PCR and the 2-ΔΔCT method. *Methods* **2001**, *25*, 402–408. [CrossRef]

plants

Article

Genome-Wide Identification and Analysis of *MYB* Transcription Factor Family in *Hibiscus hamabo*

Dina Liu [1,2,†], Chunsun Gu [1,2,3,†], Zekai Fu [1,2] and Zhiquan Wang [1,3,*]

1 Institute of Botany, Jiangsu Province and Chinese Academy of Sciences, Nanjing 210014, China
2 College of Forest Sciences, Nanjing Forestry University, Nanjing 210037, China
3 Jiangsu Key Laboratory for the Research and Utilization of Plant Resources, Nanjing 210014, China
* Correspondence: wangzhiquan@cnbg.net
† These authors contributed equally to this work.

Abstract: MYB transcription factors constitute one of the largest gene families in plants and play essential roles in the regulation of plant growth, responses to stress, and a wide variety of physiological and biochemical processes. In this study, 204 MYB proteins (HhMYBs) were identified in the *Hibiscus hamabo* Sieb. et Zucc (*H. hamabo*) genome and systematically analyzed based on their genomic sequence and transcriptomic data. The candidate HhMYB proteins and MYBs of *Arabidopsis thaliana* were divided into 28 subfamilies based on the analysis of their phylogenetic relationships and their motif patterns. Expression analysis using RNA-seq and quantitative real-time PCR (qRT-PCR) indicated that most *HhMYB*s are differentially regulated under drought and salt stresses. qRT-PCR analysis of seven selected *HhMYB*s suggested that the HhMYB family may have regulatory roles in the responses to stress and hormones. This study provides a framework for a more comprehensive analysis of the role of MYBs in the response to abiotic stress in *H. hamabo*.

Keywords: MYB family; transcription factors; abiotic stress; semi-mangrove plant; *Hibiscus hamabo*

Citation: Liu, D.; Gu, C.; Fu, Z.; Wang, Z. Genome-Wide Identification and Analysis of *MYB* Transcription Factor Family in *Hibiscus hamabo*. *Plants* **2023**, *12*, 1429. https://doi.org/10.3390/plants12071429

Academic Editor: Rebecca Grumet

Received: 29 January 2023
Revised: 18 March 2023
Accepted: 20 March 2023
Published: 23 March 2023

1. Introduction

Hibiscus hamabo Sieb.et Zucc. (*H. hamabo*) is a deciduous shrub native to the coastlines of Korea, Japan, and China that is currently considered an endangered plant in the Zhejiang Province of China [1–3]. As a semi-mangrove species, it is adaptable to pronounced changes in ecological characteristics in land–sea transitions [2,4], with a strong adaptability to saline alkali soil [5]. Therefore, *H. hamabo* plays an important role in the improvement of saline–alkali lands and has been widely applied in the afforestation of coastal beaches in environmental governance efforts in recent years [5]. *H. hamabo* is an excellent material for the study of plant salt tolerance. At present, studies on salt tolerance in *H. hamabo* have mainly been based on morphological observations: investigations of the physiological and molecular mechanisms involved [6–9]. Previously, we presented the genome sequencing of *H. hamabo* using PacBio, Illumina, and Hi-C sequencing technology and analyzed its salt tolerance mechanism based on genome and transcriptome data [2].

The MYB transcription factor family is one of the largest in plants and plays key roles in the regulation of plant growth and stress [10,11]. MYB family members have highly conserved domains forming three α-helices in three-dimensional space [10]. The second and third helices form a "helix-turn-helix" structure with a hydrophobic core containing three tryptophan residues, which is the DNA binding site [10]. According to the number of adjacent imperfect repeats, MYB genes can be divided into four gene subfamilies, namely 1R-MYB, 2R-MYB, 3R-MYB, and 4R-MYB [12].

MYB transcription factors have been reported to alleviate or eliminate plant stress-related damage through their signal transduction network, which regulates targeted genes to initiate physiological and biochemical responses [13,14]. Abscisic acid (ABA) and salt can induce transcription levels of *AmMYB1*, and the overexpression of *AmMYB1* in tobacco improves the tolerance to salt stress [15]. Overexpression of the wheat R2R3-MYB gene members *TaMYB32*, *TaMYB33*, *TaMYB56-B*, and *TaMYB73* in *Arabidopsis thaliana* (*A. thaliana*) significantly improved salt tolerance [11]. All the genome data of *A. thaliana* provided the possibility for the description and classification of the MYB gene family in plants [16]. Since then, scientists have identified the MYB family from different species and clarified their biological function and that of related genes by molecular biological and other means [12,17,18]. In this study, all MYB genes detected in the genome of *H. hamabo* and their chromosomal distribution, phylogenetic relationships, gene structural organization, and protein motif content were analyzed. Furthermore, their expression in response to stress stimuli was analyzed using previously published transcriptomic data of *H. hamabo*, which was validated for selected MYB members by comparative RT-qPCR. This work provides a preliminary framework for the identification of candidate MYB genes mediating stress responses, especially to salt stress in *H. hamabo*. The candidate genes identified could be useful both in future studies of the underlying mechanisms of salt tolerance in *H. hamabo* and as potential markers in subsequent *H. hamabo* breeding.

2. Materials and Methods

2.1. Identification of MYB Gene Members in the H. hamabo Genome

HMMER software (v. 3.1) was used with the hidden Markov model (HMM) of the MYB domain (PF00249), obtained from the Pfam database (http://pfam.xfam.org/, accessed on 18 March 2021), to identify MYB genes in the *H. hamabo* genome (Bioproject accession number: PRJNA759075) with E-values < 0.001 [2,4]. ClustalX software (v2.0) was used for alignment of protein sequences and eliminating redundant sequences. The Pfam website (http://pfam.xfam.org/search#tabview=tab1, accessed on 18 March 2021) was used to check the MYB domain structure, and sequences with a typical conserved domain were identified as *H. hamabo* MYB members (*HhMYB*), which were used in further analyses.

2.2. Phylogenetic, Intron-Exon Structure, and Motif Compositional Analyses of HhMYBs

The phylogenetic tree of MYB proteins of *H. hamabo* and *A. thaliana* was constructed using MEGA 7.0 [19] with the nearest neighbor (NJ) method and 1000 bootstraps. The gene structure was visualized using Gene Structure Display Server (GSDS) [20]. MEME (http://meme.nbcr.net/meme, accessed on 18 March 2021) was used for the analysis of conserved motifs with the maximum number of motifs set at 10.

2.3. Chromosome Mapping of HhMYBs

The mapping of *HhMYB*s to chromosomes was based on their annotation information from the H. hamabo genome and was visualized with CIRCOS [21].

2.4. Expression Pattern Analysis of HhMYBs Obtained from Transcriptome Sequencing

The expression profile data for expression analysis of *H. hamabo* in different periods of salt and drought treatment were obtained from RNA-Seq data [22]. Fragments per kb per million reads (FPKM) were used as the standard for expression pattern, and the omicshare tool was used to generate the associated heatmap figures (https://www.omicshare.com/tools/Home/Soft/heatmap, accessed on 18 March 2021).

2.5. Plant Materials and Stress Treatment

The plant materials used in this study were collected from Institute of Botany, Jiangsu Province, and the Chinese Academy of Sciences (Nanjing Botanical Garden Mem. Sun Yat-Sen). Healthy and well-developed *H. hamabo* seeds were placed in a refrigerator at 4 °C for vernalization for 20 days. The seeds were then treated with concentrated sulfuric acid for 15 min and then rinsed thoroughly with running water before planting in plastic pots with a substrate composed of peat and vermiculite (1:1) and placing in a phytotron with a photoperiod of 16/8 h and day/night temperatures of 24/20 °C. Seedlings with 8–10 true leaves were transplanted into 1/2 MS solution for 7 days, followed with abiotic stress and hormone treatments.

High and low temperature stress treatments consisted of exposing the seedlings to 42 °C or 4 °C, respectively. A certain number of seedlings were exposed in 400 mM NaCl or 500 mM mannitol for high salt or drought treatment, respectively. Other seedlings were sprayed with 200 μM salicylic acid (SA), 1 mM methyl jasmonate (MeJA), or 50 μM abscisic acid (ABA) until the leaves were completely moist. The youngest fully expanded leaf from each plant was collected at 1, 2, 6, 12, and 24 h after treatment and then frozen in liquid nitrogen before storage at −80 °C. The youngest fully expanded leaf from each plant of untreated seedlings was similarly collected as control material. Three biological replicates with nine different plants, three plants in a group, were produced for each treatment.

2.6. Expression Analysis of the HhMYB Genes by qRT-PCR

Total RNA was extracted with a total RNA extraction kit (Shanghai PuDi Biotech Co., Ltd., Shanghai, China), and cDNA was synthesized with HiScript® III 1st Strand cDNA Synthesis kit (Vazyme, Nanjing, China), both procedures following their respective manufacturer recommendations. The Genscript tool (https://www.genscript.com/tools/pcr-primers-designer, accessed on 20 March 2021) was used for designing primers for qRT-PCR (Table S1). qRT-PCR was performed using a StepOnePlus TM Real-Time PCR system (Applied Biosystems, Foster City, CA, USA). The reaction mixture contained 2 μL diluted cDNA, 10 μL 2 × SYBR Green Master Mix (Bimake, Houston, TX, USA), 0.4 μL ROX (Dye I), 1 μL each of forward and reverse primer, and ddH_2O to a total volume of 20 μL. The PCR procedure utilized an initial denaturation step set as 95 °C for 10 min, followed by 40 cycles of 95 °C for 15 s, 60 °C for 30 s, and 72 °C for 30 s. Melting curve analyses of the amplified products were conducted at 60–95 °C. The *Actin* was selected as a reference gene: forward primer: 5′-GGCACCTCTCAACCCCAAGG-3′, reverse primer: 3′-GAGAGAACGGCCTGGATGGC-5′ [5]. Three technical replicates were prepared for each extract, and the quantitative results were analyzed using the $2^{-\Delta\Delta CT}$ method.

3. Results

3.1. Identification of MYB Gene Family Members in H. hamabo

In order to identify MYB family genes in the genome of *H. hamabo*, we used the hidden Markov model profile (HMM) of the MYB domain (PF00249) to conduct an HMM-search based against the predicted gene coding sequences. A total of 204 genes were obtained with E-values < 0.001 and labeled as *HhMYB1-HhMYB204* (Table S2). The length range of the HhMYB sequences was from 300 aa to 1731 aa, of which HhMYB186 had the longest amino acid sequence. In addition, the molecular weights of HhMYBs ranged from 32.81 kDa (HhMYB4) to 196.98 kDa (HhMYB186), and the range of theoretical isoelectric points was 4.51 (HhMYB195) to 10.01 (HhMYB30) (Table S2).

3.2. Phylogenetic Analysis of HhMYBs

In order to analyze the phylogenetic relationships of HhMYBs, a phylogenetic tree was constructed from their predicted coding sequences. MYBs of *A. thaliana* (*AtMYBs*) were included for comparison (Figure 1). The candidate HhMYBs and AtMYBs were clustered within 28 subfamilies but showed differing clustering patterns among the subfamilies, suggesting the divergence of the MYB family during the evolution of these species. The XX subfamily had the most HhMYBs (22), whereas the XIV and XXVII subfamilies each contained only one member, and VII, VIII, X, XXII, and XXVI subfamilies contained no HhMYBs. It is worth noting that several MYB subfamilies were found to contain a greater number of HhMYBs relative to AtMYBs. For example, the III subfamily contains 20 HhMYBs but only 5 AtMYBs. Conversely, some subfamilies consisted exclusively of AtMYBs, indicating that these genes may be lost during the evolution of *H. hamabo*.

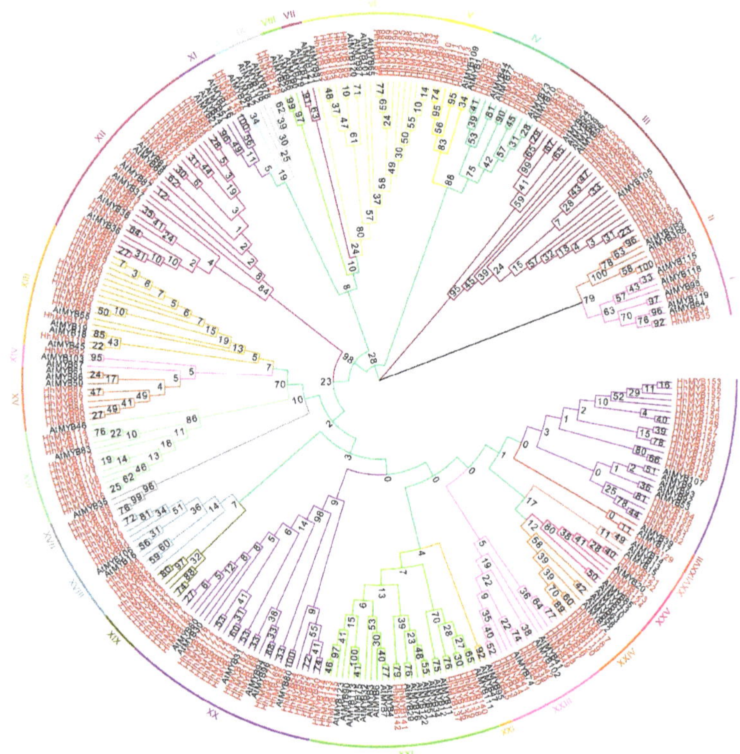

Figure 1. Phylogenetic tree of MYB family members of *H. hamabo* and *A. thaliana* using the NJ method. The prefix At was used before the names of the *Arabidopsis* MYBs. The subfamilies were represented by different colors.

3.3. Conserved Motif and Gene Structure Analysis

In order to determine if the hierarchical clustering in the phylogenetic tree reflected differences in conserved motifs and/or gene structure, the candidate HhMYBs and AtMYBs were analyzed for motif content and their genes for intron/exon organization. The results (Figure S1) showed that motif 1 was highly conserved and present in most MYB members in *H. hamabo* and *Arabidopsis*, while other motifs showed a larger variability. Notably, the profiles of the HhMYB motif contents differed between the branches of the evolutionary tree and were relatively consistent within each branch. Only a few HhMYB proteins located in the same branch contained different motif profiles. Structural analysis revealed that the majority of HhMYB members (147) had a typical structure containing two introns (Figure S1). However, HhMYBs 1, 6–8, and 10–13 had no introns, whereas other members contained 1 to 15 introns, and HhMYBs with the same number of introns were grouped into a similar subclade.

3.4. Chromosome Mapping Analysis of MYBs in H. hamabo

The 204 *HhMYB* genes were mapped onto the 46 chromosomes using their annotated information provided from the earlier genomic analysis of *H. hamabo*. This showed that *HhMYB*s were unevenly distributed across the 46 chromosomes in a manner independent of chromosomal length (Figure S2). The largest number of *HhMYB*s were located on chromosome 3, with a total of 11 *HhMYB* genes. Some chromosomes contained only one *HhMYB*, such as chromosomes 27, 33, 38, 43, and 45.

3.5. Analysis of HhMYB Gene Expression Patterns under Salt and Drought Stress

In order to gain insight into the potential function of *HhMYBs* in response to saline and drought stress, we analyzed the expression patterns of *HhMYBs* using transcriptomic data obtained from *H. hamabo* subjected to these conditions (Figure 2) [22]. *HhMYB* genes in subfamilies XIX, XVIII, XVII, and I were not detected in any sample, whereas other subfamilies showed significant differences in expression patterns in response to the stress treatments. Several *HhMYBs* showed similar response trends to the two stresses. For example, the expression levels of *HhMYB46* of XXIII and *HhMYB117* of XXIV were up-regulated under both stresses, while the expression levels of *HhMYB140* of the XXI subgroup was down-regulated under both stresses, suggesting that these genes may play roles in both drought and salt stress. However, some genes, such as *HhMYB96* of the subgroup XIII, displayed different expression patterns under the two stresses. The expression level of *HhMYB96* increased under drought stress but decreased under salt stress, suggesting the differential deployment of this gene in response to temperature and drought stresses in *H. hamabo*. In addition, *HhMYB* genes in the same evolutionary branch showed different expression characteristics, indicating their functional differentiation.

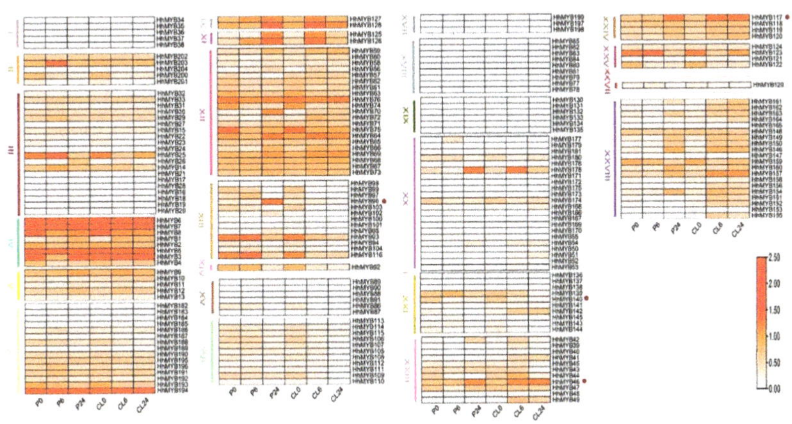

Figure 2. Expression pattern of *MYB* genes in *H. hamabo* under salt and drought stress as determined from RNA-Seq data. The expression level was represented by color: red, higher expression; white, lower expression. HhMYBs with a dot were specifically mentioned in the manuscript. P0, P6, P24, CL0, CL6, and CL24 represent treatments for 0 h, 6 h, and 24 h under PEG treatment and 0 h, 6 h, and 24 h under NaCl treatment, respectively.

3.6. qRT-PCR Analysis of Expression Patterns of HhMYBs under Temperature and Hormonal Treatments

Seven selected *HhMYB* genes with relatively large fold changes under the PEG and NaCl treatments were tested for their response to salt and drought stress; high and cold temperatures; and SA, ABA, and MeJA treatments by qRT-PCR analysis. It can be seen from Figure 3 that the expression of these genes underwent significant changes under the different stress treatments. The expression trends of HhMYB75 were essentially opposite under low and high temperature conditions. All the selected *HhMYBs* were responsive to the various hormone treatments but showed differing expression trends, indicating that *HhMYB* genes are differentially deployed in response hormones.

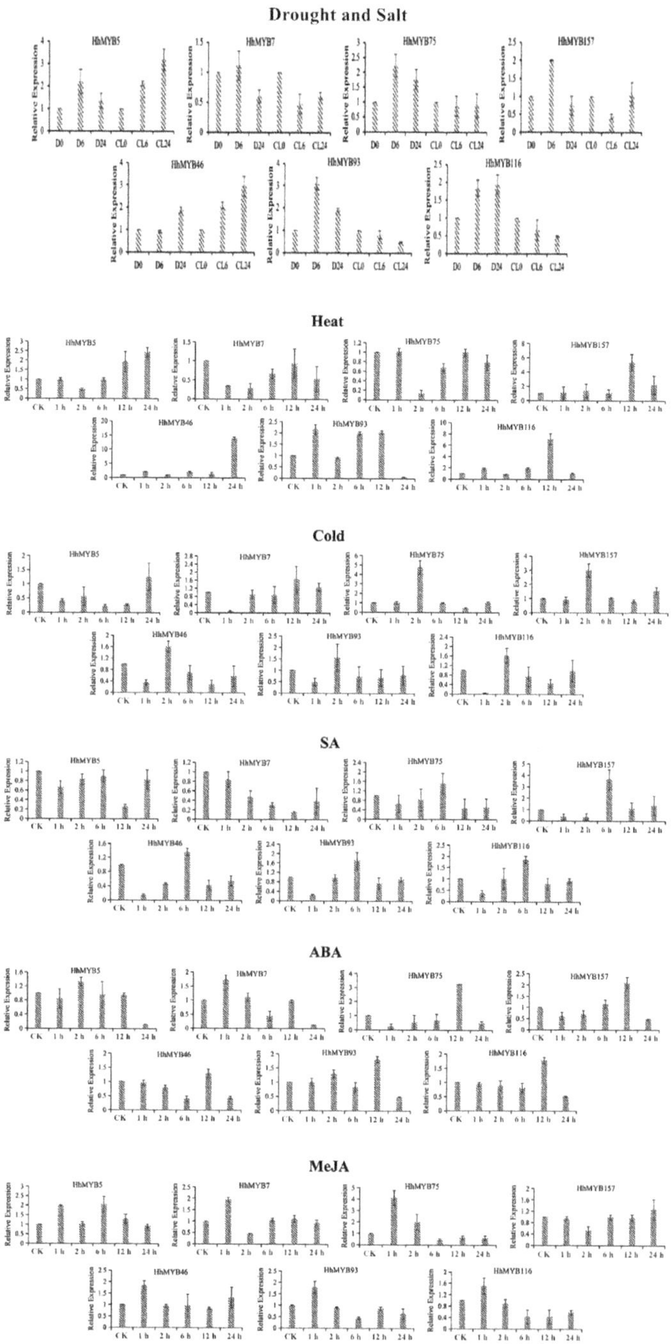

Figure 3. Expression pattern of *HhMYB* genes in *H. hamabo* under 500 mM mannitol (Drought), 400 mM NaCl (Salt), 42 °C (Heat), 4 °C (Cold), 200 μM salicylic acid (SA), 50 μM abscisic acid (ABA), or 1 mM methyl jasmonate (MeJA) treatments. D0, D6, D24, CL0, CL6, and CL24 represent treatments for 0 h, 6 h, and 24 h under drought stress and 0 h, 6 h, and 24 h under salt stress, respectively.

4. Discussion

MYBs constitute one of the largest transcription factor families in plants, with a conserved MYB–DNA binding domain composed of about 52 amino acids at the N-terminus [23]. MYB transcription factors play a key role in plant development, secondary metabolism, hormone signal transduction, disease resistance, and abiotic stress tolerance [11,24,25]. With the increasing number of sequenced plant genomes, members of the MYB family have been systematically identified in various plant species [17,26]. For example, 197 candidate MYBs were identified in *Arabidopsis*, and a different number of candidate MYBs has also been identified in other species, such as 155 in rice and 245 in *Helianthus annuus* L. [27,28]. However, the MYB gene family in *H. hamabo* has not been reported, and its functions remain unclear. *H. hamabo*, with strong adaptability to saline alkali soil and seawater immersion, can adapt to the land–sea transition zone. Because *H. hamabo* is a good material for studying plant resistance, MYB is an important plant regulator. Genome-wide identification and analysis of MYB transcription factor family in *H. hamabo* were performed for the first time in this study. The results of this study showed that *H. hamabo* has a family of 204 *MYB*s, which may be due to large-scale fragment replication in the genome of *H. hamabo* [2].

Based on a phylogenetic analysis, the 204 candidate HhMYB proteins and AtMYBs were divided into 28 subfamilies. Almost all of these subfamilies contained both HhMYBs and AtMYBs, but in differing proportions. HhMYBs clustered in the same subfamily, in many cases, shared similar and highly conserved profiles of MYB motifs, suggesting that they may be functionally related. Only a few HhMYB proteins located in the same branch contained different motif profiles, indicating that these HhMYB members may have arisen as the result of functional differentiation during *H. hamabo* evolution. Additionally, HhMYB members have introns varying from 0 to 15, which are similar to those found in studies on, for example, *Hedychium coronarium* [29]. The results again indicated complex differentiation during evolution.

As a semi-mangrove plant, *H. hamabo* has a high resistance to saline and alkaline soils and drought and barren environments [8]. Some *Arabidopsis* MYB transcription factors have been reported to be involved in plant resistance to salt and drought stresses [30,31]. *AtMYB60* can influence the drought response by regulating the ABA signal transduction pathway [32]. The wheat *MYB33* was reported to enhance the salt and drought tolerance of transgenic *Arabidopsis* through its restoration of osmotic balance and increase in ROS-scavenging capabilities [33]. Here, the analysis of transcriptomic data indicated that several *HhMYB*s were involved in the plant responses to both saline and drought stresses. The expression patterns of the genes were different under drought or saline stress, suggesting that these *HhMYB*s may have regulatory functions in the tolerance to saline and drought stress and deserve further investigation.

Other biotic stresses, such as high and low temperatures, also seriously affect the growth of *H. hamabo*. MYBs have been reported to regulate the response and resistance to temperature stress. Transgenic rice overexpressing *OsMYB3R-2* exhibited an enhanced cold tolerance as well as an increased cell mitotic index [34]. Therefore, the expression levels of *HhMYB*s under temperature stress were also analyzed in this study. The expression trends of *HhMYB75* were essentially opposite under low and high temperature conditions. Similarly, AtMYB104 has been reported to be down-regulated by cold stress but up-regulated under high temperature, while the expression trend of *AtMYB81* displayed the opposite [12].

The involvement of MYBs in plant responses to environmental factors may be mediated by hormones [35]. *AtMYB*s *2*, *13*, *15*, and *101* have been shown to respond to ABA, and the ABA regulation of *AtMYB2* can influence the response to saline stress [36]. Many potato *StMYB* genes were induced under ABA, GA, and IAA treatments [36]. The results in this study showed that a high proportion of *HhMYB*s are responsive to ABA, SA, or MeJa, but with differing levels of intensity and at different time points after hormonal stimulation.

This study provides a reference for future research into the role of MYB transcription factors in the response of *H. hamabo* to abiotic stresses.

5. Conclusions

Based on the importance of the MYB family in plants and the strong adaptability of *H. hamabo* to stressed environments, genome-wide identification and analysis of the MYB transcription factor family in H. hamabo were performed for the first time in this study. The search for MYB family members in the *H. hamabo* genome led to the identification of 204 putative HhMYB family members. The HhMYB family was clustered into 28 subfamilies with AtMYBs according to phylogenetics and their motif patterns. In addition, we found that the 204 *HhMYB* genes were unevenly distributed across the 46 chromosomes of *H. hamabo*. Expression analysis with transcriptomic data and qRT-PCR showed that the candidate HhMYB family displays a wide range of sensitivities to abiotic stress and hormones. Several *HhMYBs* were involved in the plant responses to both saline and drought stresses to varying degrees, indicating that these *HhMYBs* may have regulatory functions in tolerance to saline and drought stress and deserve further investigation. Our results provide a foundation for further researching MYB gene functions in *H. hamabo*.

Supplementary Materials: The following are available online at https://www.mdpi.com/article/10.3390/plants12071429/s1. Figure S1: Phylogeny (a), conserved motif (b), and gene structure (c) analyses of HhMYBs; Figure S2: Chromosome distribution of *HhMYBs*; Table S1: Primer sequences for qRT-PCR; Table S2: List of the MYB family genes in *H. hamabo*.

Author Contributions: Conceptualization, C.G. and Z.W.; investigation, D.L. and Z.F.; writing—original draft preparation, D.L. and Z.W.; writing—review and editing, Z.W. and C.G. All authors have read and agreed to the published version of the manuscript.

Funding: This research was funded by Six Talent Peaks Project of Jiangsu Province [NY-042].

Institutional Review Board Statement: Not applicable.

Informed Consent Statement: Not applicable.

Data Availability Statement: No new data were created or analyzed in this study. Data sharing is not applicable to this article.

Conflicts of Interest: The authors declare no conflict of interest.

References

1. Sakhanokho, H.F.; Islam-Faridi, N.; Babiker, E.M.; Nelson, C.D.; Stringer, S.J.; Adamczyk, J.J., Jr. Determination of nuclear DNA content, ploidy, and fish location of ribosomal DNA in hibiscus hamabo. *Sci. Hortic.* **2020**, *264*, 109167. [CrossRef]
2. Wang, Z.; Xue, J.-Y.; Hu, S.-Y.; Zhang, F.; Yu, R.; Chen, D.; Van de Peer, Y.; Jiang, J.; Song, A.; Ni, L. The genome of *Hibiscus hamabo* reveals its adaptation to saline and waterlogged habitat. *Hortic. Res.* **2022**, *9*, uhac067. [CrossRef]
3. Yan, A.L.; Qi, Y.T.; Li, D.W. In Current status of Hangjiahu plain wetlands resources and proposals for protection and management. *Adv. Mater. Res.* **2014**, *955*, 3683–3686.
4. Ni, L.; Wang, Z.; Liu, X.; Wu, S.; Hua, J.; Liu, L.; Yin, Y.; Li, H.; Gu, C. Genome-wide study of the GRAS gene family in *Hibiscus hamabo* sieb. Et zucc and analysis of HhGRAS14-induced drought and salt stress tolerance in *Arabidopsis*. *Plant Sci.* **2022**, *319*, 111260. [CrossRef] [PubMed]
5. Ni, L.; Wang, Z.; Liu, L.; Guo, J.; Li, H.; Gu, C. Selection and verification of candidate reference genes for gene expression by quantitative RT-PCR in *Hibiscus hamabo* sieb. Et zucc. *Trees* **2019**, *33*, 1591–1601. [CrossRef]
6. Wang, Z.; Ni, L.; Guo, J.; Liu, L.; Li, H.; Yin, Y.; Gu, C. Phylogenetic and transcription analysis of *Hibiscus hamabo* sieb. Et zucc. WRKY transcription factors. *DNA Cell Biol.* **2020**, *39*, 1141–1154. [CrossRef]
7. Ni, L.; Wang, Z.; Fu, Z.; Liu, D.; Yin, Y.; Li, H.; Gu, C. Genome-wide analysis of basic helix-loop-helix family genes and expression analysis in response to drought and salt stresses in *Hibiscus hamabo* sieb. Et zucc. *Int. J. Mol. Sci.* **2021**, *22*, 8748. [CrossRef]
8. Ni, L.; Wang, Z.; Liu, X.; Wu, S.; Hua, J.; Yin, Y.; Li, H.; Gu, C. Transcriptome analysis of salt stress in *Hibiscus hamabo* sieb. Et zucc based on pacbio full-length transcriptome sequencing. *Int. J. Mol. Sci.* **2021**, *23*, 138. [CrossRef] [PubMed]
9. Shi, Q.; Bao, X.W.; Hua, J.F.; Yu, C.G.; Yin, Y.L.; Lu, Z.G. Effects of drought stress and recovery on photosynthesis and physiological characteristics of *Hibiscus hamabo*. *J. Appl. Ecol.* **2019**, *30*, 2600–2606.
10. Du, H.; Zhang, L.; Liu, L.; Tang, X.-F.; Yang, W.-J.; Wu, Y.-M.; Huang, Y.-B.; Tang, Y.-X. Biochemical and molecular characterization of plant MYB transcription factor family. *Biochemistry* **2009**, *74*, 1–11. [CrossRef]

11. Li, J.; Han, G.; Sun, C.; Sui, N. Research advances of MYB transcription factors in plant stress resistance and breeding. *Plant Signal. Behav.* **2019**, *14*, 1613131. [CrossRef] [PubMed]
12. Wang, Y.; Zhang, Y.; Fan, C.; Wei, Y.; Meng, J.; Li, Z.; Zhong, C. Genome-wide analysis of MYB transcription factors and their responses to salt stress in *Casuarina equisetifolia*. *BMC Plant Biol.* **2021**, *21*, 328. [CrossRef]
13. Baldoni, E.; Genga, A.; Cominelli, E. Plant MYB transcription factors: Their role in drought response mechanisms. *Int. J. Mol. Sci.* **2015**, *16*, 15811–15851. [CrossRef] [PubMed]
14. Smita, S.; Katiyar, A.; Chinnusamy, V.; Pandey, D.M.; Bansal, K.C. Transcriptional regulatory network analysis of MYB transcription factor family genes in rice. *Front. Plant Sci.* **2015**, *6*, 1157. [CrossRef] [PubMed]
15. Ganesan, G.; Sankararamasubramanian, H.; Harikrishnan, M.; Ashwin, G.; Parida, A. A MYB transcription factor from the grey mangrove is induced by stress and confers NaCl tolerance in tobacco. *J. Exp. Bot.* **2012**, *63*, 4549–4561. [CrossRef] [PubMed]
16. Dubos, C.; Stracke, R.; Grotewold, E.; Weisshaar, B.; Martin, C.; Lepiniec, L. MYB transcription factors in *Arabidopsis*. *Trends Plant Sci.* **2010**, *15*, 573–581. [CrossRef]
17. Arce-Rodríguez, M.L.; Martínez, O.; Ochoa-Alejo, N. Genome-wide identification and analysis of the MYB transcription factor gene family in chili pepper (*Capsicum* spp.). *Int. J. Mol. Sci.* **2021**, *22*, 2229. [CrossRef]
18. Wang, J.; Liu, Y.; Tang, B.; Dai, X.; Xie, L.; Liu, F.; Zou, X. Genome-wide identification and capsaicinoid biosynthesis-related expression analysis of the R2R3-MYB gene family in *Capsicum annuum* L. *Front. Genet.* **2020**, *11*, 598183. [CrossRef]
19. Kumar, S.; Stecher, G.; Tamura, K. MEGA7: Molecular evolutionary genetics analysis version 7.0 for bigger datasets. *Mol. Biol. Evol.* **2016**, *33*, 1870–1874. [CrossRef]
20. Hu, B.; Jin, J.; Guo, A.-Y.; Zhang, H.; Luo, J.; Gao, G. GSDS 2.0: An upgraded gene feature visualization server. *Bioinformatics* **2015**, *31*, 1296–1297. [CrossRef]
21. Krzywinski, M.; Schein, J.; Birol, I.; Connors, J.; Gascoyne, R.; Horsman, D.; Jones, S.J.; Marra, M.A. Circos: An information aesthetic for comparative genomics. *Genome Res.* **2009**, *19*, 1639–1645. [CrossRef] [PubMed]
22. Wang, Z.; Ni, L.; Hua, J.; Liu, L.; Yin, Y.; Li, H.; Gu, C. Transcriptome analysis reveals regulatory framework for salt and drought tolerance in *Hibiscus hamabo* siebold & zuccarini. *Forests* **2021**, *12*, 454.
23. Wang, X.; Niu, Y.; Zheng, Y. Multiple functions of MYB transcription factors in abiotic stress responses. *Int. J. Mol. Sci.* **2021**, *22*, 6125. [CrossRef] [PubMed]
24. Ambawat, S.; Sharma, P.; Yadav, N.R.; Yadav, R.C. MYB transcription factor genes as regulators for plant responses: An overview. *Physiol. Mol. Biol. Plants* **2013**, *19*, 307–321. [CrossRef] [PubMed]
25. Cao, Y.; Li, K.; Li, Y.; Zhao, X.; Wang, L. MYB transcription factors as regulators of secondary metabolism in plants. *Biology* **2020**, *9*, 61. [CrossRef]
26. Salih, H.; Gong, W.; He, S.; Sun, G.; Sun, J.; Du, X. Genome-wide characterization and expression analysis of MYB transcription factors in *Gossypium hirsutum*. *BMC Genet.* **2016**, *17*, 129. [CrossRef]
27. Li, J.; Liu, H.; Yang, C.; Wang, J.; Yan, G.; Si, P.; Bai, Q.; Lu, Z.; Zhou, W.; Xu, L. Genome-wide identification of MYB genes and expression analysis under different biotic and abiotic stresses in *Helianthus annuus* L. *Ind. Crops Prod.* **2020**, *143*, 111924. [CrossRef]
28. Katiyar, A.; Smita, S.; Lenka, S.K.; Rajwanshi, R.; Chinnusamy, V.; Bansal, K.C. Genome-wide classification and expression analysis of MYB transcription factor families in rice and *Arabidopsis*. *BMC Genom.* **2012**, *13*, 544. [CrossRef]
29. Abbas, F.; Ke, Y.; Zhou, Y.; Yu, Y.; Waseem, M.; Ashraf, U.; Wang, C.; Wang, X.; Li, X.; Yue, Y.; et al. Genome-wide analysis reveals the potential role of MYB transcription factors in floral scent formation in *Hedychium coronarium*. *Front. Plant Sci.* **2021**, *12*, 623742. [CrossRef]
30. Zhang, P.; Wang, R.; Yang, X.; Ju, Q.; Li, W.; Lü, S.; Tran, L.S.P.; Xu, J. The R2R3-MYB transcription factor *AtMYB49* modulates salt tolerance in *Arabidopsis* by modulating the cuticle formation and antioxidant defence. *Plant Cell Environ.* **2020**, *43*, 1925–1943. [CrossRef]
31. Park, M.Y.; Kang, J.-y.; Kim, S.Y. Overexpression of *AtMYB52* confers ABA hypersensitivity and drought tolerance. *Mol. Cells* **2011**, *31*, 447–454. [CrossRef]
32. Hussain, S.S.; Kayani, M.A.; Amjad, M. Transcription factors as tools to engineer enhanced drought stress tolerance in plants. *Biotechnol. Prog.* **2011**, *27*, 297–306. [CrossRef] [PubMed]
33. Qin, Y.; Wang, M.; Tian, Y.; He, W.; Han, L.; Xia, G. Over-expression of *TaMYB33* encoding a novel wheat MYB transcription factor increases salt and drought tolerance in *Arabidopsis*. *Mol. Biol. Rep.* **2012**, *39*, 7183–7192. [CrossRef] [PubMed]
34. Ma, Q.; Dai, X.; Xu, Y.; Guo, J.; Liu, Y.; Chen, N.; Xiao, J.; Zhang, D.; Xu, Z.; Zhang, X. Enhanced tolerance to chilling stress in *OsMYB3R-2* transgenic rice is mediated by alteration in cell cycle and ectopic expression of stress genes. *Plant Physiol.* **2009**, *150*, 244–256. [CrossRef]
35. Abe, H.; Urao, T.; Ito, T.; Seki, M.; Shinozaki, K.; Yamaguchi-Shinozaki, K. Arabidopsis *AtMYC2* (*bHLH*) and *AtTMYB2* (*MYB*) function as transcriptional activators in abscisic acid signaling. *Plant Cell* **2003**, *15*, 63–78. [CrossRef] [PubMed]
36. Sun, W.; Ma, Z.; Chen, H.; Liu, M. MYB gene family in potato (*Solanum tuberosum* L.): Genome-wide identification of hormone-responsive reveals their potential functions in growth and development. *Int. J. Mol. Sci.* **2019**, *20*, 4847. [CrossRef]

plants

MDPI

Article

Genome-Wide Identification of the Rose *SWEET* Gene Family and Their Different Expression Profiles in Cold Response between Two Rose Species

Xiangshang Song, Yaping Kou, Mingao Duan, Bo Feng, Xiaoyun Yu, Ruidong Jia, Xin Zhao, Hong Ge * and Shuhua Yang *

State Key Laboratory of Vegetable Biobreeding, Key Laboratory of Biology and Genetic Improvement of Flower Crops (North China), Ministry of Agriculture and Rural Affairs, Institute of Vegetables and Flowers, Chinese Academy of Agricultural Sciences, Beijing 100081, China; fengb1107@163.com (B.F.)

* Correspondence: gehong@caas.cn (H.G.); yangshuhua@caas.cn (S.Y.); Tel.: +86-10-8210-9542 (S.Y.)

Abstract: Sugars Will Eventually be Exported Transporter (SWEET) gene family plays indispensable roles in plant physiological activities, development processes, and responses to biotic and abiotic stresses, but no information is known for roses. In this study, a total of 25 *RcSWEET* genes were identified in *Rosa chinensis* 'Old Blush' by genome-wide analysis and clustered into four subgroups based on their phylogenetic relationships. The genomic features, including gene structures, conserved motifs, and gene duplication among the chromosomes of *RcSWEET* genes, were characterized. Seventeen types of *cis*-acting elements among the *RcSWEET* genes were predicted to exhibit their potential regulatory roles during biotic and abiotic stress and hormone responses. Tissue-specific and cold-response expression profiles based on transcriptome data showed that *SWEETs* play widely varying roles in development and stress tolerance in two rose species. Moreover, the different expression patterns of cold-response *SWEET* genes were verified by qRT-PCR between the moderately cold-resistant species *R. chinensis* 'Old Blush' and the extremely cold-resistant species *R. beggeriana*. Especially, *SWEET2a* and *SWEET10c* exhibited species differences after cold treatment and were sharply upregulated in the leaves of *R. beggeriana* but not *R. chinensis* 'Old Blush', indicating that these two genes may be the crucial candidates that participate in cold tolerance in *R. beggeriana*. Our results provide the foundation for function analysis of the *SWEET* gene family in roses, and will contribute to the breeding of cold-tolerant varieties of roses.

Keywords: rose; *SWEET* gene family; sugar transport; gene expression; cold response

Citation: Song, X.; Kou, Y.; Duan, M.; Feng, B.; Yu, X.; Jia, R.; Zhao, X.; Ge, H.; Yang, S. Genome-Wide Identification of the Rose *SWEET* Gene Family and Their Different Expression Profiles in Cold Response between Two Rose Species. *Plants* **2023**, *12*, 1474. https://doi.org/10.3390/plants12071474

Academic Editors: Aiping Song and Yu Chen

Received: 18 December 2022
Revised: 2 March 2023
Accepted: 3 March 2023
Published: 28 March 2023

1. Introduction

Sugars Will Eventually be Exported Transporter (SWEET) is a new type of sugar transporter that was first identified in *Arabidopsis thaliana* in 2010 [1]. Comparing with monosaccharide transporters (MST) and sucrose transporters (SUT), SWEET proteins are capable of transporting sugar bidirectionally without energy dependence, and the types of substrates for sugar transport are more extensive [2]. In plants, SWEET proteins usually contain seven transmembrane helices (TMHs); two MtN3/saliva domains with three TMHs are connected by the fourth TMH to form a "3-1-3" structure [3]. The SWEET family is divided into four Clades. Different clades have different preferences for monosaccharides or disaccharides, in which Clades I and II prefer transporting hexose and Clade III prefer transporting sucrose, while SWEET proteins of Clade IV tend to transport fructose on the tonoplast [4–7].

Due to the advances in whole-genome sequencing in plants, the genome-wide identification of *SWEET* genes has been reported in many crops, vegetables, and fruits, such as rice (*Oryza sativa*) [8], soybean (*Glycine max*) [9], sorghum (*Sorghum bicolor*) [10], tomato (*Solanum lycopersicum*) [11], potato (*Solanum tuberosum*) [12], cabbage (*Brassica oleracea*) [13], tea plant (*Camellia sinensis*) [14], apple (*Malus domestica*) [15], grape (*Vitis vinifera*) [16], and banana (*Musa acuminate*) [17]. Previous studies show that *SWEET* genes in plants are involved in many biological processes, including the regulation of pollen development, nectar secretion, seed development, phloem loading, and leaf senescence. In Arabidopsis, loss of function of *AtSWEET8/RPG1* led to male sterility, indicating a key role in the maintenance of pollen viability [1,18]. *AtSWEET13* is also demonstrated to have the function for pollen development in Arabidopsis [19]. *AtSWEET9/Nec1* and its homologous genes in tobacco (*Nicotiana attenuata*) and rape (*Brassica rapa*) are proven to be essential for nectar secretion [20]. AtSWEET10 is reported to be the downstream of FLOWERING LOCUS T (FT) during floral transition in Arabidopsis [21]. AtSWEET11, 12, and 15 are beneficial to seed filling by mediating sucrose transfer from the seed coat to the endosperm [22]. AtSWEET11 and 12 are also responsible for phloem loading for long-distance transport of sucrose [7,23]. OsSWEET4, 11, and 15 in rice (*Oryza sativa*) and ZmSWEET4 in maize (*Zea mays*) are involved in the transport of hexose or sucrose in the endosperm to promote seed filling [24,25]. AtSWEET13 and 14 may be involved in modulating the GA response in Arabidopsis [26]. The NAC transcription factor ORE1 could bind to the *AtSWEET15* promoter to positively regulate leaf senescence [27]. CmSWEET17 could be involved in the process of sucrose-induced axillary bud outgrowth in chrysanthemum (*Chrysanthemum moriflorum*), possibly via the auxin transport pathway [28].

SWEET genes also play very important roles in response to biotic and abiotic stresses. Overexpression of *AtSWEET16* and its homologous genes in tea and apples can enhance the cold resistance of the transgenic calli or plants [14,29,30]. Moreover, overexpression of *CsSWEET1a* and *CsSWEET17* from tea also improves the cold tolerance of the transgenic Arabidopsis plants [31]. Loss of function in *AtSWEET17* may affect lateral root development and lead to impaired drought resistance [32]. *AtSWEET2* inhibits *Pythium* infection by reducing the availability of sugars in the rhizosphere of Arabidopsis seedlings [33,34]. In grape, overexpression of the *VvSWEET4* increases hexose content in hair roots and enhances resistance to *Pythium* [35]. In rice, *OsSWEET11, 13,* and *14* are involved in resistance to bacterial blight disease by regulation of the upstream transcription factors [36–38]. However, no information about the *SWEET* gene family is known for the genus *Rosa* L.

Rose is one of the most popular flowers in the world and is widely used in cut flowers, potted flowers, garden cultivation, and essential oil production [39]. Cold stress affects the growth and distribution of plants, resulting in freezing injuries and even the death of plants [40–42]. The rose industry suffers great commercial losses every year due to cold stress [40–42]. *Rosa beggeriana* is a wild species that originated in cold and arid Central Asia, including northwest China. It has extreme cold resistance, which can be inherited through interspecific hybridization with modern rose cultivars [43]. Our previous study found that the significantly higher freezing tolerance in the shoot of *R. beggeriana* than that of *R. fortuneana*, which originated in southeastern China, might be due to the stronger ability of the spatial transfer of soluble sugars from leaves to shoots and consequently the storage of soluble sugars as starch in the shoots during overwintering [44]. The transcriptome analyses in *R. multiflora* and *R. xanthina* further indicated that the pathways of starch and sucrose metabolism were activated during cold stress [42,45]. However, it is not clear whether and which members of the SWEET gene family may participate in the cold response of a rose. Furthermore, it is not sure if there are species differences in these cold-response SWEET genes among the cold-sensitive and cold-tolerant species in *Rosa* L.

In this study, the *SWEET* family genes in rose were identified, and their gene structures, motif compositions, phylogenetic relationships, chromosome locations, gene duplications, and *cis*-acting elements were analyzed on the basis of the genome sequence of *R. chinensis* 'Old blush'. The comparison of expression patterns of the *SWEET* family genes was further investigated between *R. chinensis* 'Old blush' and *R. beggeriana* during cold treatment. Finally, the analysis of subcellular localization was conducted on the screened cold-response *RbSWEET* genes in *R. beggeriana*. Our results will contribute to a better understanding of the *SWEET* gene family in rose and provide valuable information for further functional analysis of *RbSWEET* genes in response to cold tolerance.

2. Results

2.1. Genome-Wide Identification of the SWEET Family Genes in R. chinensis

A total of 29 SWEET proteins were identified on the basis of HMM profiles of the SWEET domain (PF03083) search results. Meanwhile, 28 SWEET homologous proteins were obtained by using 17 protein sequences of AtSWEETs to BLASTP in the *R. chinensis* proteome with an e-value threshold of 1×10^{-5}. After the intersection of these two results, 25 *SWEET* genes were finally identified in *R. chinensis* according to the conserved domains in the relevant database (Table 1). These genes were named from *RcSWEET1* to *RcSWEET17b* according to their homology with *A. thaliana*. Protein characterization analysis showed that the ranges of genomic DNA (gDNA) and coding sequences (CDS) are 717~3289 bp and 708~933 bp, respectively. The amino acid length of RcSWEET family members ranged from 235 (RcSWEET2a, RcSWEET2b, and RcSWEET5c) to 310 (RcSWEET12), and the average molecular weight (MW) ranged from 26,284.25 (RcSWEET2b) to 34,702.73 (RcSWEET12). The range of the theoretical isoelectric point (pI) is between 5.93 (RcSWEET15b) and 10.11 (RcSWEET10c). The TMHMM predicted results showed that all 25 RcSWEET proteins contained 7 transmembrane domains, indicating that the transmembrane structure of RcSWEET proteins was relatively conservative (Figure S1).

Table 1. Characteristics of the *RcSWEET* gene family members in *R. chinensis* 'Old Blush'.

Gene Name	Accession No.	Clade	Chr. No.	Location	gDNA (bp)	CDS (bp)	Protein (aa)	MW (Da)	PI
RcSWEET1	LOC112172425	I	6	NC_037093.1 (39918414 ... 39920220)	1807	744	247	27,128.17	9.58
RcSWEET2a	LOC112201532	I	5	NC_037092.1 (8689160...8691537)	2378	708	235	26,347.62	9.57
RcSWEET2b	LOC112166177	I	5	NC_037092.1 (22992540 ... 22995828)	3289	708	235	26,284.25	8.95
RcSWEET3	LOC112181844	I	1	NC_037088.1 (65904502 ... 65906291)	1790	756	251	28,320.45	8.90
RcSWEET4a	LOC112185269	II	2	NC_037089.1 (84804695 ... 84806493)	1799	732	243	26,958.32	9.71
RcSWEET4b	LOC112175327	II	7	NC_037094.1 (6173469 ... 6176058)	2590	762	253	28,128.47	9.58
RcSWEET5a	LOC112177410	II	7	NC_037094.1 (12047796 ... 12049179)	1384	750	249	28,162.82	8.92
RcSWEET5b	LOC112164442	II	5	NC_037092.1 (89007600 ... 89008331)	732	732	243	27,298.40	9.12
RcSWEET5c	LOC112184760	II	2	NC_037089.1 (64595639 ... 64598570)	2932	708	235	26,374.52	9.41
RcSWEET5d	LOC112177411	II	7	NC_037094.1 (12068260 ... 12070141)	1882	714	237	26,541.70	9.62
RcSWEET5e	LOC112196855	II	4	NC_037091.1 (37884667 ... 37885398)	732	732	243	27,172.30	8.91
RcSWEET5f	LOC112167162	II	1	NC_037088.1 (42566318 ... 42567034)	717	717	238	26,881.01	8.96
RcSWEET5g	LOC112186521	II	1	NC_037088.1 (51365641 ... 51366372)	732	732	243	27,261.37	8.96
RcSWEET5h	LOC112190416	I	1	NC_037088.1 (51946076 ... 51946820)	745	732	243	27,273.32	8.96
RcSWEET9	LOC112171608	III	6	NC_037093.1 (55529800 ... 55531112)	1313	801	266	29,697.38	9.30
RcSWEET10a	LOC112176691	III	7	NC_037094.1 (8669485 ... 8671441)	1957	867	288	32,706.42	8.18
RcSWEET10b	LOC112174578	III	6	NC_037093.1 (55634260 ... 55636316)	2057	882	293	32,633.41	6.51
RcSWEET10c	LOC112173222	III	6	NC_037093.1 (55593310 ... 55595334)	2025	915	304	33,868.54	10.11
RcSWEET11a	LOC112173492	III	6	NC_037093.1 (55588084 ... 55590577)	2494	882	293	32,435.99	9.67

Table 1. *Cont.*

Gene Name	Accession No.	Clade	Chr. No.	Location	gDNA (bp)	CDS (bp)	Protein (aa)	MW (Da)	PI
RcSWEET11b	LOC112173793	III	6	NC_037093.1 (55578837 ... 55580213)	1377	861	181	31,799.93	9.23
RcSWEET12	LOC112180477	III	7	NC_037094.1 (8655605 ... 8657888)	2284	933	310	34,702.73	7.50
RcSWEET15a	LOC112196569	III	4	NC_037091.1 (54285212 ... 54287511)	2300	918	305	34,033.00	6.51
RcSWEET15b	LOC112198953	III	4	NC_037091.1 (20120784 ... 20122566)	1783	909	302	33,623.13	5.93
RcSWEET17a	LOC112176490	IV	1	NC_037088.1 (36716888 ... 36718685)	1798	732	243	26,871.80	9.19
RcSWEET17b	LOC112200151	IV	4	NC_037091.1 (48596784 ... 48598877)	2094	726	241	27,071.79	7.24

2.2. Gene Structure, Motif, and Phylogenetic Analysis of the RcSWEET Genes

The gene structures, including untranslated regions (UTRs), exons, and introns, of each *RcSWEET* gene were predicted based on gDNA and CDS sequences (Table S2). The majority of the *RcSWEET* genes (*RcSWEET1, RcSWEET2a, RcSWEET2b, RcSWEET3, RcSWEET4a, RcSWEET5a, RcSWEET5c, RcSWEET5d, RcSWEET9, RcSWEET10a, RcSWEET10b, RcSWEET10c, RcSWEET11a, RcSWEET12, RcSWEET15a, RcSWEET15b, RcSWEET17a,* and *RcSWEET17b*) contained six exons. Two *RcSWEET* genes (*RcSWEET4b* and *RcSWEET11b*) had five exons, whereas five *RcSWEET* genes (*RcSWEET5b, RcSWEET5e, RcSWEET5f, RcSWEET5g,* and *RcSWEET5h*) had only one exon and no intron (Figure 1B). The conserved motifs were investigated based on the protein sequences of the *RcSWEET* genes. Ten consensus motifs were detected in the *RcSWEET* genes (Figure 1C). Each of the 25 members contained two MtN3_slv domains that were composed of three transmembrane helical domains. To investigate the phylogenetic relationship of the *SWEET* genes in *R. chinensis*, the identified *SWEET* family members of rose were constructed alone and combined with the *SWEET* family members of rice and Arabidopsis. The results showed that 25 *RcSWEETs* were divided into four clades, including 4 members in Clade I, 10 members in Clade II, 9 members in Clade III, and 2 members in Clade IV (Figures 1A and 2).

2.3. Chromosomal Location and Gene Duplication Analysis of the RcSWEET Genes

The analysis of chromosomal location showed that 25 *RcSWEET* genes were distributed on six chromosomes of *R. chinensis* 'Old Blush'. Chr6 contained the greatest number of six *RcSWEET* genes, including *RcSWEET1, RcSWEET9, RcSWEET10b, RcSWEET10c, RcSWEET11b,* and *RcSWEET11b*. Both Chr1 and Chr7 contained five genes, which were *RcSWEET3, RcSWEET5f, RcSWEET5g, RcSWEET5h,* and *RcSWEET17a* on Chr1 and *RcSWEET4b, RcSWEET5a, RcSWEET5d, RcSWEET10a,* and *RcSWEET12* on Chr7. There were four genes, including *RcSWEET5e, RcSWEET15a, RcSWEET15b,* and *RcSWEET17b*, on Chr4. Three genes, including *RcSWEET2a, RcSWEET2b,* and *RcSWEET5b*, are located on Chr5. Only two genes, *RcSWEET4a* and *RcSWEET5c*, are distributed on Chr2. However, there was no *RcSWEET* gene on Chr3.

Gene duplication events usually play an indispensable role in gene expansion and the evolution of gene families. A pair of tandemly duplicated genes (*RcSWEET10a* and *RcSWEET12*) located on Chr7, and another cluster of five tandemly duplicated genes (*RcSWEET9, RcSWEET10b, RcSWEET10c, RcSWEET11a,* and *RcSWEET11b*) located on Chr6 (Figure 3A). In addition, two segmental duplication events, including *RcSWEET4a*/*RcSWEET5a* and *RcSWEET12*/*RcSWEET15a*, were detected on three different chromosomes (Figure 3B). The selection pressure was further estimated by the Ka (nonsynonymous) and Ks (synonymous) substitution ratios [46]. Only one pair of segmentally duplicated *RcSWEETs*, namely *RcSWEET4a* and *RcSWEET5a*, had a Ka/Ks ratio of 0.15, which was significant, and indicated a synonymous change that has been selected during plant genome evolution (Table S4).

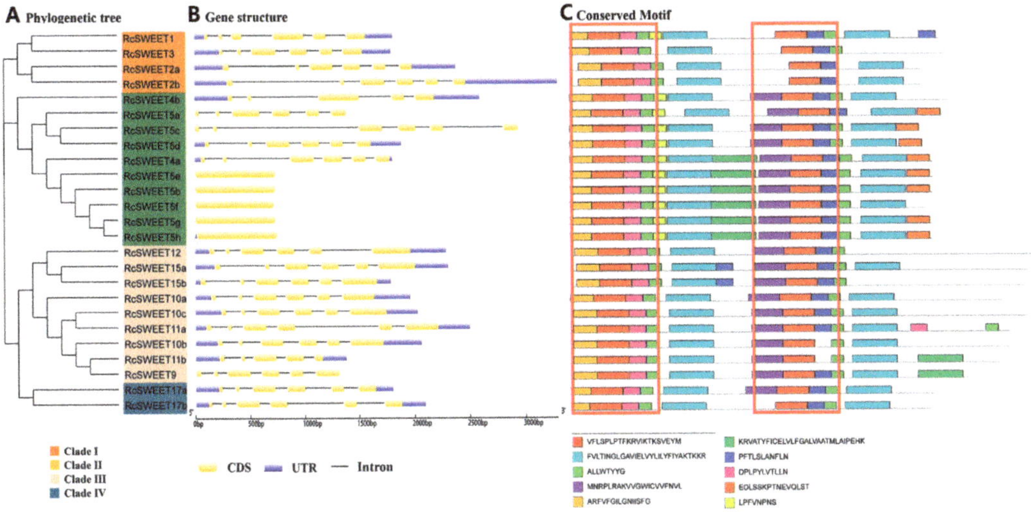

Figure 1. Phylogenetic relationship, gene structure, and conserved motif analysis of the *RcSWEET* genes. (**A**) Phylogenetic tree of 25 RcSWEET proteins. The unrooted neighbor-joining (NJ) phylogenetic tree was constructed with MEGA7 using amino acid sequences of 25 RcSWEET proteins, and the bootstrap test replicate was set at 1000 times. (**B**) Gene structure of *RcSWEET* genes. Yellow boxes represent CDS, and black lines represent introns. The upstream/downstream regions of the *RcSWEET* genes are indicated in blue boxes. (**C**) Distributions of conserved motifs in *RcSWEET* genes. Ten putative motifs are indicated in different colored boxes. The MtN3_slv domains were marked in two red frames.

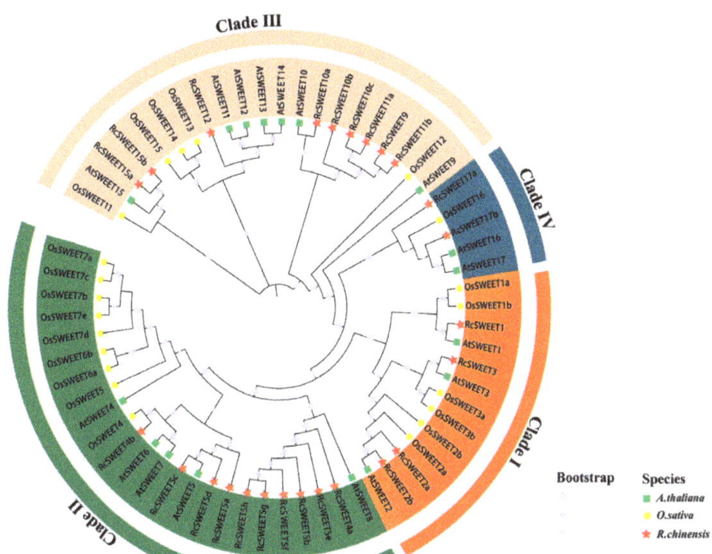

Figure 2. Phylogenetic tree of SWEET proteins of Arabidopsis, rice, and rose. The phylogenetic tree was constructed using the NJ (neighbor joining) method with 1000 bootstrap replications. The 4 subfamilies were distinguished by different colors.

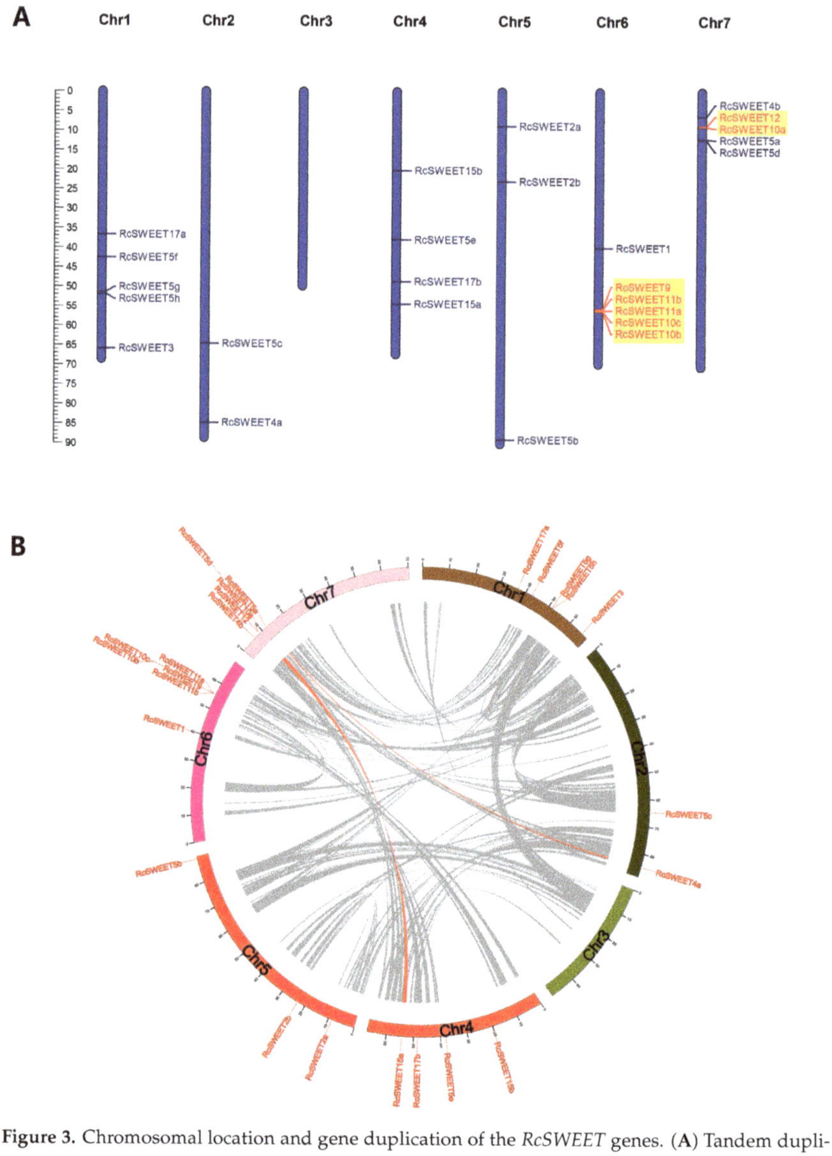

Figure 3. Chromosomal location and gene duplication of the *RcSWEET* genes. (**A**) Tandem duplication of *RcSWEET* genes. The tandemly duplicated genes are marked by laurel-green rectangles. (**B**) Segmental duplication of the *RcSWEET* genes. The segmentally duplicated genes of the *R. chinensis* genome are connected by gray lines, and the segmentally duplicated genes of the *RcSWEET* family are highlighted.

2.4. Cis-Elements Analysis in the Promoters of the RcSWEET Genes

To further study the potential regulatory mechanisms of the *RcSWEET* genes in response to abiotic stress, the 2000 bp upstream sequences from the translation start sites of the *RcSWEET* genes were submitted into PlantCARE to detect the cis-elements. Seventeen cis-acting elements associated with stress and hormones were identified, including ABRE, ARE, AuxRR core, Box 4, CGTCA-motif, circadian, ERE, GARE-motif, LTR, MBS, P-box, TATC-box, TCA-element, TC-rich, TGACG-motif, TGA-element, and WUN-motif (Figure 4).

The details of the cis-acting elements contained in each *RcSWEET* gene have been listed in Table S5. All the *RcSWEET* genes possessed at least one stress-relevant cis-element; ABRE, ARE, and Box 4 were the most abundant genes in the *SWEET* family, indicating that abscisic acid, light, and anaerobic conditions had a great influence on the expression of *RcSWEET* genes, and these cis-acting elements were also related to stress response. In particular, 13 elements (52%) contained one or more LTR, suggesting a potential cold-stress response under low temperature conditions.

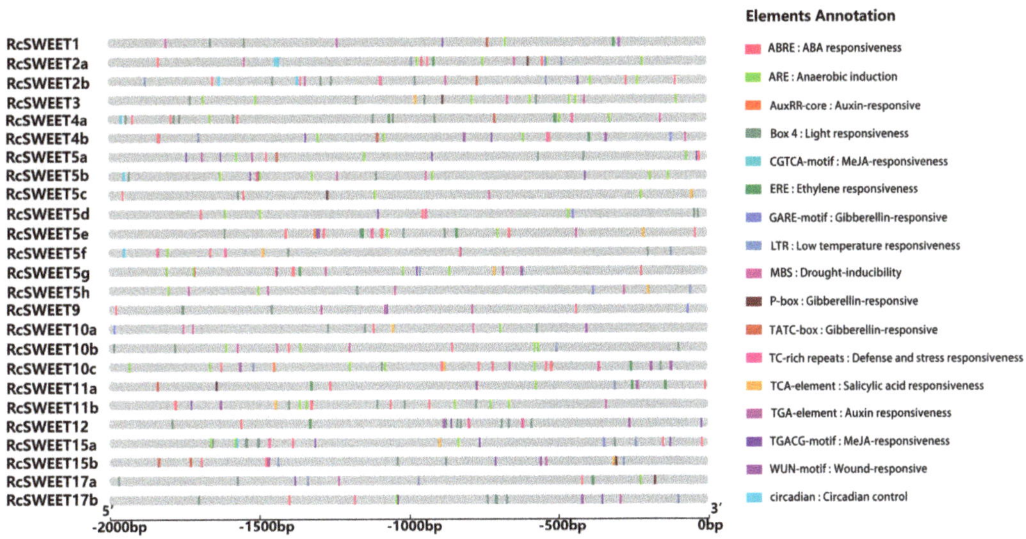

Figure 4. Predicted cis-elements in the promoters of the *RcSWEET* genes. Promoter sequences (−2000 bp) of 25 *RcSWEET* genes were analyzed by PlantCARE. Seventeen stress-relevant cis-acting elements are indicated in different colored boxes.

2.5. Expression Patterns of RcSWEET Genes in Different Tissues

The spatial expression patterns of *RcSWEET* genes in different tissues have been analyzed according to the published transcriptome sequencing (RNA-seq) data in *R. chinensis* 'Old Blush' under the growth condition of a 12 h (25 °C) day/12 h (18 °C) night cycle [47]. There were 13 *RcSWEET* genes (52%) expressed in roots, 10 genes (40%) in stems, 9 genes (36%) in prickles and leaves, 18 genes (72%) in stamens, and 15 genes (60%) in pistils and ovaries (Figure 5). Six *RcSWEET* genes, including *RcSWEET1*, *RcSWEET2a*, *RcSWEET2b*, *RcSWEET4b*, *RcSWEET5h*, and *RcSWEET17a*, were expressed in all tissues. Especially, *RcSWEET1* was highly expressed (FPKM > 20) in prickles, leaves, stamens, pistils, and ovaries. *RcSWEET4b* was highly expressed (FPKM > 20) in roots, stems, prickles, stamens, pistils, and ovaries. Some of the *RcSWEET* genes exhibited similar expression patterns in the same tissues. For example, *RcSWEET5d* and *RcSWEET10a* showed extremely high expression levels (FPKM > 95) in stamen. Moreover, the highly expressed levels (FPKM > 20) were presented in *RcSWEET2a* and *RcSWEET3* in leaves, as well as *RcSWEET12* and *RcSWEET15a* in roots.

2.6. Expression Patterns of Cold-Response SWEET Genes in Two Rosa Species

Identically as in *R. chinensis*, 25 *RbSWEET* genes have been identified in *R. beggeriana* based on the transcriptome analysis of leaves and shoots in response to cold stress (Figure 6). At least 12 genes were statistically detected under the growth temperature of 23/18 °C, including *RbSWEET1*, *RbSWEET2a*, *RbSWEET2b*, *RbSWEET3*, *RbSWEET4b*, *RbSWEET10c*, *RbSWEET11a*, *RbSWEET15a*, *RbSWEET17a*, and *RbSWEET17b* in both tissues, as well as *RbSWEET5f* and *RbSWEET5h* in shoots. The above results presented very similar expression patterns in the leaves and shoots of *R. chinensis* 'Old Blush'. After cold stress at 4 °C, the differentially expressed genes (DEGs) were screened out of the leaves and shoots of *R. beggeriana*. In leaves, *RbSWEET2a* and *RbSWEET10c* were upregulated, while *RbSWEET4b* was downregulated during cold stress. *RbSWEET1* and *RbSWEET15a* were downregulated at 1 h and then upregulated until 24 h of cold stress. On the contrary, *RbSWEET3* and *RbSWEET2b* presented the upregulated expression level at 1 h and then the downregulated expression level at 6 h and 24 h during cold stress. In shoots, *RbSWEET10c*, *RbSWEET4b*, *RbSWEET1*, and *RbSWEET2b* exhibited very similar expression trends as those genes in leaves. In contrast to leaves, there were much higher expression levels of *RbSWEET10c* and *RbSWEET4b* in shoots. *RbSWEET2a* and *RbSWEET15a* exhibited the downregulated expression levels, but *RbSWEET3* fluctuated in shoots during cold stress. In addition, *RbSWEET17a* did not show a different effect on growth temperature but maintained the relatively higher expression levels in leaves and shoots under both growth temperatures.

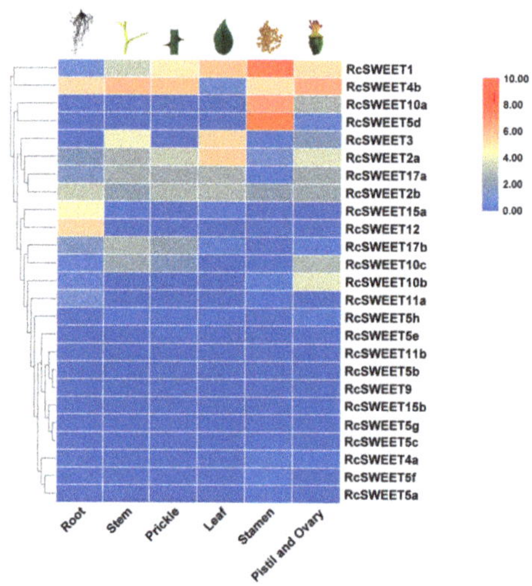

Figure 5. Expression patterns of *RcSWEET* genes in different tissues. Expression levels are shown as the $\log_2$ FPKM values obtained from the RNA-Seq data.

The expression patterns of DEGs or highly expressed *SWEET* genes in leaves and shoots of *R. chinensis* 'Old Blush' and *R. beggeriana* during cold treatment are shown in Figure 7. In the leaves of *R. chinensis* 'Old Blush', significantly increased relative expression levels were found in *RcSWEET1* at 6 h and *RcSWEET15a* at 24 h during cold stress. *RcSWEET2a* and *RcSWEET10c* showed a significant decline at 1 h but recovered at 6 h and 24 h. *RcSWEET2b* did not significantly change during cold stress. The relative expression levels of *RcSWEET3* and *RcSWEET17a* fluctuated during cold stress, significantly decreasing at 1 h, increasing at 6 h, and declining at 24 h. *RcSWEET4b* exhibited the exact opposite trend as *RcSWEET3*, but maintained significantly higher relative expression levels during cold treatment than those at 0 h. In shoots of *R. chinensis* 'Old Blush', *RcSWEET1* was significantly upregulated at 1 h and 24 h, but *RcSWEET2a* was unchanged during cold treatment. The increased expression levels were observed in *RcSWEET2b* at 6 h and in *RcSWEET10c* at 1 h. *RcSWEET3* was dramatically upregulated at 1 h, then significantly downregulated at 6 h, but recovered at 24 h. The relative expression levels of *RcSWEET4b*, *RcSWEET15a*, and *RcSWEET17a* significantly increased at 1 h and 6 h but sharply declined at 24 h.

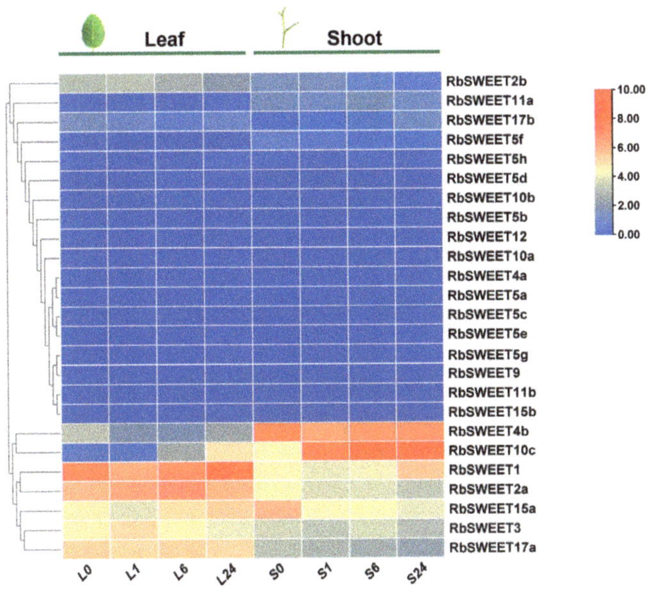

Figure 6. Transcriptional levels of *RbSWEET* genes after cold stress in *R. beggeriana*. Heatmap showing expression patterns of *RbSWEET* genes in the leaves and shoots treated at 4 °C for 0 h, 1 h, 6 h, and 24 h, represented by L0, L1, L6, L24, and S0, S1, S6, S24, respectively. Expression levels are shown as the $\log_2$ FPKM values obtained from the RNA-Seq data.

As for the leaves and shoots of *R. beggeriana*, the expression patterns of most *RbSWEETs* determined using qRT-PCR detection were similar to the transcriptome results. However, the qRT-PCR analysis showed the expression levels of *RbSWEET4b* increased at 1 h and 24 h in leaves and enhanced at 1 h and 6 h in shoots, instead of the decreased results in the RNA-seq data. In addition, *RbSWEET17a* did not alter in the transcriptome data but significantly declined in leaves by qRT-PCR analysis during cold stress.

The further comparison analysis disclosed that *SWEET* genes showed various expression patterns in two species during cold treatment (Figure 7). For instance, *SWEET1* and *SWEET4b* exhibited upregulated trends in leaves and shoots of both species. The expression patterns of *SWEET2a* and *SWEET10c* were quite similar in the shoots of both species. In leaves, *RcSWEET2a* and *RcSWEET10c* sharply declined at 1 h, then recovered to the same expression levels as those genes at 0 h, while *RbSWEET2a* and *RbSWEET10c* stably enhanced and reached the significant differences after 6 h and 24 h of cold treatment, respectively. Moreover, *SWEET2b*, *SWEET3*, and *SWEET15a* exhibited upregulated trends in the shoots of *R. chinensis* 'Old Blush' at the beginning of cold stress, but downregulated in the shoots of *R. beggeriana* during the whole cold treatment.

2.7. Subcellular Localization of Three RbSWEET Genes

Based on the expression patterns of the *SWEET* genes in both species, three *RbSWEET* genes were selected for subcellular localization. *RbSWEET1*, *RbSWEET2a*, and *RbSWEET10c* were transiently overexpressed in tobacco leaf epidermal cells. As shown in Figure 8, *RbSWEET1* and *RbSWEET10c* were found to be localized to the plasma membrane, and *RbSWEET2a* was found to be localized to the tonoplast, whereas the control (GFP alone) was observed clearly in the plasma membrane, cytoplasm, and nucleus.

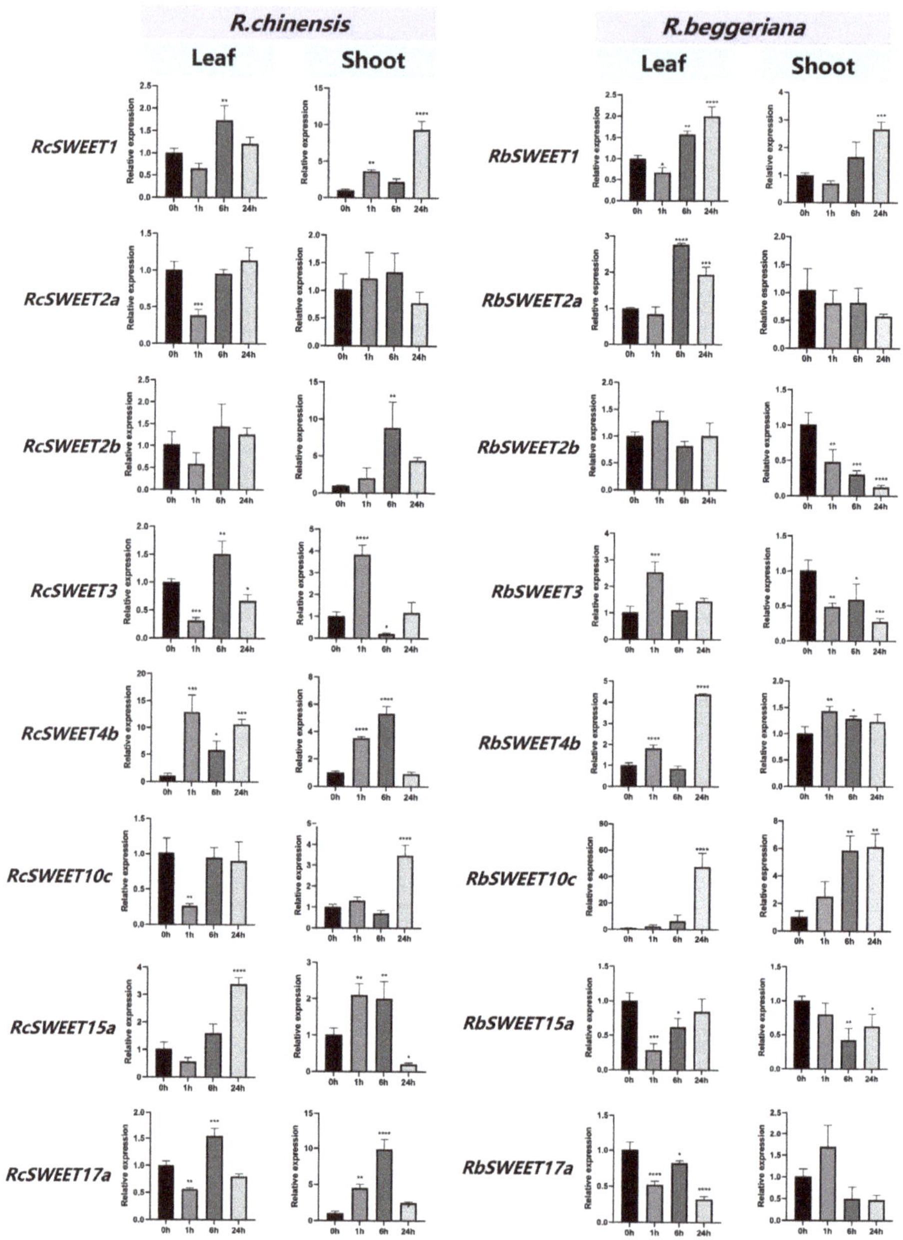

Figure 7. Expression levels of DEGs or highly expressed *SWEET* genes in *R. chinensis* 'Old Blush' and *R. beggeriana* via qRT-PCR. Error bars indicate standard deviation, and asterisks indicate significant differences between the control (0 h) and 4 °C treatment for 1 h, 6 h, 24 h, * $p < 0.05$, ** $p < 0.01$, *** $p < 0.001$, and **** $p < 0.0001$.

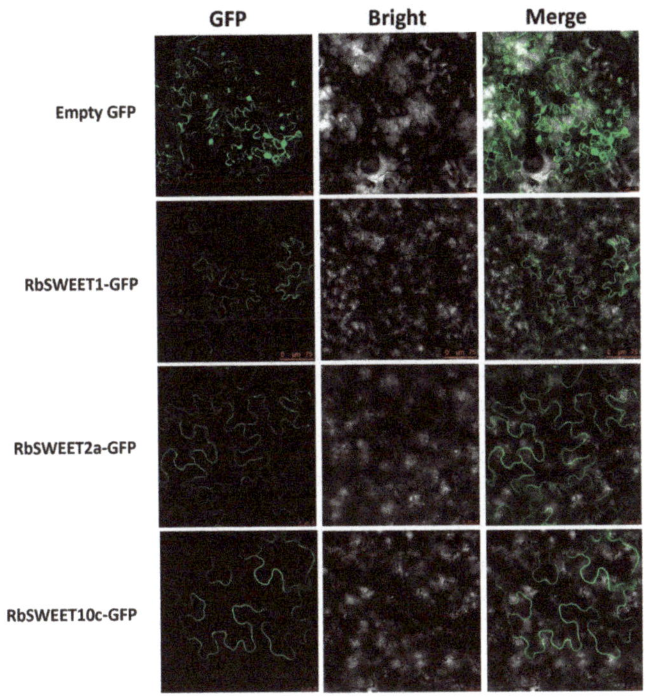

Figure 8. Subcellular localization of RbSWEET-GFP fusion proteins in tobacco leaves.

3. Discussion

The SWEET gene family has been identified in many major crops such as rice [8], sorghum [10], and soybean [9], as well as important horticultural plants such as cabbage [13], potato [12], tomato [11], and apple [15]. However, little information has been reported on roses and other ornamental plants, especially their molecular regulatory mechanisms in cold response. The genome sequencing of *R. chinensis* 'Old Blush' laid the foundation for the whole genome identification and analysis of the rose gene family [48]. In this study, 25 *RcSWEET* genes were identified and classified into four subfamilies (Clades I–IV) according to their phylogenetic evolutionary relationship, including four members in Clade I, ten members in Clade II, nine members in Clade III, and two members in Clade IV (Figure 1A). The phylogenetic tree constructed with the SWEET family proteins of Arabidopsis, rice, and rose showed that the SWEET proteins of rose were more closely related to Arabidopsis, which also belongs to dicotyledons (Figure 2). Analysis of the *RcSWEET* gene structure indicated that most genes contained five or six exons, which was consistent with the analysis results in other species such as cabbage [13] and banana [17]. Moreover, similar intron and extron arrangements were found in the same subfamily members (Figure 1B), which implied a different function in each clade. In plants, SWEET proteins usually consist of seven transmembrane helices (TMHs), and two MtN3/saliva domains with three TMHs connected by the fourth transmembrane helix to form a "3-1-3" structure [3]. Compared with plants and other eukaryotes, SWEET protein in prokaryotes usually has only three TMHs, which is named SemiSWEET protein, suggesting that eukaryotic SWEET protein is likely to evolve due to gene replication or fusion [6]. Ten consensus motifs were detected in the RcSWEET proteins, and all the RcSWEET proteins contained two MtN3_slv domains. This result suggested that the structure of RcSWEET proteins was highly conserved (Figure 1C).

Gene duplication events, including whole-genome duplication (WGD), tandem duplication, and segmental duplication events, usually result in functional segregation during gene expansion and evolution [46]. A total of 11 *RcSWEET* genes were identified as being tandemly or segmentally duplicated, which included a pair of tandemly duplicated genes (*RcSWEET10a* and *RcSWEET12*) located on Chr7, another cluster of five tandemly duplicated genes (*RcSWEET9*, *RcSWEET10b*, *RcSWEET10c*, *RcSWEET11a*, and *RcSWEET11b*) located on Chr6, and two pairs of genes (*RcSWEET4a-RcSWEET5a* and *RcSWEET12-RcSWEET15a*) were segmentally duplicated (Figure 3). These results indicated that tandem duplication and segmental duplication synergistically contribute to the expansion of the *RcSWEET* gene family, but the former is more influential than the latter. Although the duplicated *RcSWEET* genes can be derived from a common ancestor, their expression patterns and functions may have diverged. For example, the expression levels of the tandemly duplicated gene pair of *RcSWEET10a-RcSWEET12* identified in this work were very different in the various tissues of *R. chinensis* 'Old Blush'. In Arabidopsis, the unique *AtSWEET10* functions for early flowering [21], but there are three *RcSWEET10a/b/c* genes in rose. Moreover, *RcSWEET10a* exhibited higher expression levels in flowers, pistils, and ovaries, suggesting that *RcSWEET10a* may play an important role during the reproductive development. *AtSWEET12* was reported to be responsible for phloem loading and seed filling in Arabidopsis [11,25,28]. However, *RcSWEET12* was highly expressed in roots, suggesting that it might be involved in root development in roses.

The regulation of gene expression in higher plants usually occurs at the transcriptional level, which is coordinated by various *cis*-acting elements and trans-acting factors [49]. However, there are few studies on *SWEET* transcriptional regulation in plants. For example, the NAC transcription factor ORE1 and three abscisic acid (ABA)-responsive element (ABRE)-binding transcription factors, ABF2 (AREB1), ABF3, and ABF4 (AREB2), were found to be involved in directly regulating senescence-associated genes *AtSWEET15* by binding to their promoters in Arabidopsis [27,50]. A waterlogging-responsive ERF (MaRAP2-4) from mint (*Mentha spp.*) was reported to specifically target the DRE or GCC box in the *AtSWEEET10* promoter to regulate soluble sugar availability and waterlogging tolerance [51]. The rice transcription factor OsDOF1 is involved in the regulation of pathogen interactions by targeting the promoters of *SWEET11/14*. In this study, a wide variety of *cis*-acting elements have been identified in the promoters of *RcSWEET* genes, including *cis*-acting elements associated with stress and hormone response, suggesting that family members are involved in complex signaling pathway regulation (Figure 4). The number of ABRE elements was the largest in the whole family. Previous studies showed that the DREB transcription factor could bind to ABRE in the gene promoter region to respond to cold stress, suggesting that *RcSWEETs* containing these *cis* elements might be involved in the regulation of the cold response in roses [52]. Moreover, 13 *RcSWEETs* contained LTR elements, which might directly participate in the regulation of low temperature response.

Low temperature is one of the typical abiotic stress factors for plants, which has an important effect on the growth and development of plants and their geographical distribution [40]. Cold acclimation at non-freezing temperatures can enhance cold resistance and induce many physiological and biochemical changes in plants, such as the accumulation of osmotic regulatory substances, the removal of reactive oxygen species, and the expression of cold responsive genes (CORs) [53–56]. It is true that the cold-induced increase of the soluble sugar content should be directly associated with the redistribution and balance of sugar in plants through the regulation of sugar transporters [57]. Previous studies showed that *SWEET* genes could be induced by low temperatures and participate in cold stress responses in many plants. For example, the overexpression of *DlSWEET1* in the tropic fruit plant longan (*Dimocarpus longan*) improved cold tolerance in transgenic *Arabidopsis* plants [58]. *BoSWEET2b*, *BoSWEET4a*, and *BoSWEET15b* were significantly upregulated in the cabbage leaves after cold stress [13]. A hexose transporter, CsSWEET2, from cucumber (*Cucumis sativus*), can improve cold tolerance in *Arabidopsis* [59]. The double mutant of *sweet11 sweet12* exhibited significantly increased freezing tolerance in the leaves of Arabidopsis, indicating that AtSWEET11 and AtSWEET12 function negatively under cold stress [60]. The Arabidopsis transgenic plants with the overexpressed *AtSWEET16* gene exhibited higher cold resistance than wild type [29]. Moreover, overexpression of *MdSWEET16* in apple 'Orin' calli was able to increase their cold tolerance compared with *MdSWEET16* RNA interference calli [30]. Enhancing the expression of CsSWEET16/1a/17 in tea plants could improve cold tolerance in the transgenic Arabidopsis plants [14,31]. In addition, more than 90% of *MaSWEETs* in banana leaves were induced in response to cold stress in two different varieties [17].

In the present study, all eight candidate genes statistically presented significant differences in the leaves or shoots of two rose species during the different hours of cold treatment. Many of the SWEET genes that have been proven to be activated under cold stress in other species were also found in the two rose species studied. For instance, *SWEET1* has been recently reported to play an important role in cold resistance in longan [57], which is also enhanced in two rose species after cold stress. There were increased expression levels of the homologous genes *CsSWEET2* in cucumber [59], *BoSWEET2b* in cabbage [13], and *RbSWEET2a* in rose after cold treatment. Moreover, *BoSWEET4a* and *BoSWEET15b* in cabbage [13], as well as *RcSWEET4b* and *RcSWEET15a* in rose, could be activated under the cold growth temperature. We also found that the expression level of *SWEET10c* was significantly enhanced in the tissues of rose species. As far as we know, this is the first report that the *SWEET10* gene could be induced under low temperature in plants. In Arabidopsis, *AtSWEET10* acted downstream of FT to promote plant flowering [21]. *PbSWEET10* might also contribute to pollen development in the Chinese white pear [61]. Moreover, *StSWEET10b* was downregulated in the leaves of potatoes after drought treatment [62]. Recently, *GmSWEET10a/b* were reported to participate in the domestication of seed development in soybean [63].

Analysis using qRT-PCR further demonstrated that the expression levels of *SWEET* genes varied between the two species during cold treatment. The similar upregulated expression patterns of *SWEET1* and *SWEET4b* were observed in leaves and shoots of both species, implying that these two genes should be relatively conserved for the cold response in the genus *Rosa* L. Since *RbSWEET1* was localized in the plasma membrane, it may function on the cellular level for the accumulation of soluble sugars after cold stress. *RbSWEET2a* and *RbSWEET10c* were significantly upregulated in the leaves of *R. beggeriana*, while *RcSWEET2a* and *RcSWEET10c* sharply declined at 1 h and then recovered in the leaves of *R. chinensis* 'Old Blush' after cold treatment. The results suggested that *RcSWEET2a* and *RcSWEET10c* could be the crucial candidate genes for the cold-induced transportation of soluble sugars in the leaves of the deciduous and extremely cold-tolerant species *R. beggeriana*. As for *R. chinensis* 'Old Blush', *RcSWEET2a* and *RcSWEET10c* might be more sensitive to cold stress than their homologues in *R. beggeriana*, where low temperatures could initially impair their activities but then recover after the cold adaptation in this evergreen and moderately cold-tolerant species. Moreover, RcSWEET2a was localized in the tonoplast membrane, but RcSWEET10c was localized in the plasma membrane, implying that *RbSWEET2a* and *RbSWEET10c* may have different functions for the acquisition of cold tolerance in *R. beggeriana*.

4. Materials and Methods

4.1. Identification of the SWEET Family Genes in R. chinensis

The genome sequence and annotation of *R. chinensis* 'Old Blush' were downloaded from the NCBI genome database (https://www.ncbi.nlm.nih.gov/genome/?term=rosa+chinensis, accessed on 18 December 2022) [48]. The Hidden Markov Model (HMM) profiles of the SWEET domain (PF03083) were downloaded from the Pfam database (http://pfam.xfam.org/, accessed on 18 December 2022) and used to search the SWEET proteins of *R. chinensis* with HMMER (3.3.2) software (e-value < 0.01) [64]. In addition, the protein sequences of 17 AtSWEETs [1] were downloaded from TAIR (http://www.arabidopsis.org/, accessed on 18 December 2022) and used as queries to search the *R. chinensis* proteome by using BLASTP with an e-value threshold of 1×10^{-5}. Then the results of the two methods are intersected to obtain the candidate sequences, whose conserved domains were further identified by the Pfam database (https://pfam.xfam.org/search#tabview=tab1, accessed on 18 December 2022) and the Conserved Domain Database (http://www.ncbi.nlm.nih.gov/Structure/cdd/wrpsb.cgi/, accessed on 18 December 2022). Finally, the ProtParam tool (http://web.expasy.org/prot-param, accessed on 18 December 2022) was used to analyze the sequence length, molecular weight, and theoretical isoelectric point values of each RcSWEET protein. The transmembrane helix of RcSWEET proteins was predicted by the TMHMM Server version 2.0 (http://www.cbs.dtu.dk/services/TMHMM/, accessed on 18 December 2022) [65].

4.2. Gene Structure, Motif, and Phylogenetic Analysis

Based on the genomic sequence and coding sequence, the gene structure of *RcSWEET* genes was predicted by the Gene Structure Display Server (GSDS; http://gsds.gao-lab.org/, accessed on 18 December 2022) [66]. The conserved motifs in full-length SWEET proteins were identified by MEME (https://meme-suite.org/meme/tools/meme, accessed on 18 December 2022) [67]. Multiple sequence alignments were carried out considering the full-length SWEET protein sequences from *A. thaliana, O. sativa*, and *R. chinensis* (Table S3). Subsequently, an unrooted neighbor-joining (NJ) tree was constructed by MEGA 7.0 with 1000 bootstrap replications [68].

4.3. Chromosomal Location and Tandem Duplication Analysis

The information on location for each *RcSWEET* gene was obtained from the GFF genome annotation of *R. chinensis* 'Old Blush'. The chromosomal localization of the *RcSWEET* genes was mapped using MapChart (2.3.2) software [69]. The MCscanX software was used to search for tandem and duplicated genes. The Ka and Ks values for duplicated gene pairs (Table S4) were calculated based on the coding sequence alignments using the Ka/Ks calculator in TBtools [46].

4.4. Cis-Acting Element Analysis

The upstream sequences (2000 bp) of the 25 *RcSWEET* coding sequences were retrieved and then submitted to PlantCARE (http://bioinformatics.psb.ugent.be/webtools/plantcare/html/, accessed on 18 December 2022) to identify cis-acting elements [70]. The prediction results of cis-acting elements were illustrated by GSDS.

4.5. RNA-Seq Data Analysis

To investigate the expression patterns of *RcSWEET* genes in different tissues of *R. chinensis* 'Old Blush', organ specificity transcriptome data (PRJNA546486) of the rose, including roots, stems, leaves, stamens, prickles, pistils, and ovaries, was extracted from the NCBI SRA database (https://www.ncbi.nlm.nih.gov/sra/, accessed on 18 December 2022) [47]. The raw data have been listed in Table S6. FPKM (Fragments Per Kilobase of transcript per Million mapped reads) was used to examine the expression levels, and the heatmap of *RcSWEET* genes in different tissues was generated using TBtools based on the $\log_2$FPKM values.

The grafted seedlings of *R. beggeriana* 'Old Blush' with the rootstock of *R. multiflora* were placed in a growth chamber with an air temperature of 23 °C day/18 °C night and a photoperiod of 10 h light/14 h dark. After a week of acclimatization, the grafted seedlings were maintained at 4 °C for 0, 1, 6, and 24 h. Current-year shoots of similar thickness and intact leaves were selected as sampling materials, and all the samples were collected in three biological replicates (the data has not been published yet). The Illumina HiSeq2500 high-throughput sequencing platform was used to sequence the cDNA library and obtain the raw data. All high-quality reads of each sample that passed quality control were mapped to the *R. chinensis* 'Old Blush' RchiOBHm-V2 genome. Gene expression levels were calculated as FPKM. The *RbSWEET* genes have been identified, and the transcriptome data are listed in Table S7. The heatmap of *RbSWEET* gene expression profiles was generated using

TBtools based on the log_2FPKM values. DEGs were screened according to certain criteria (fold change ≥ 2 and *p*-value ≤ 0.05).

4.6. Quantitative Real-Time PCR (qRT-PCR)

The cuttage seedlings of *R. chinensis* 'Old Blush' and the grafted seedlings of *R. beggeriana* were used for qRT-PCR, and the cold treatment of the two species is the same as in Section 2.6. Total RNA was extracted using the Quick RNA Isolation Kit (Huayueyang Biotech, Beijing, China). The first-strand cDNA was synthesized using the PrimeScript™ RT reagent Kit with gDNA Eraser (Takara Bio Inc., Dalian, Liaoning Province, China). The qRT-PCR reactions were performed using TB Green™ Premix Ex Taq™ II (TaKaRa, Japan) and carried out on a CFX96 Real-Time PCR System (Bio-Rad Laboratories, Hercules, CA, USA). Conditions for the reaction were as follows: 95 °C for 30 s, followed by 40 cycles of 95 °C for 5 s and 60 °C for 30 s. The specific primers for *SWEET* genes were designed using the Real-time PCR (TaqMan) Primer and Probes Design Tool (https://www.genscript.com/tools/real-time-pcr-taqman-primer-design-tooland, accessed on 18 December 2022) and have been listed in Table S1. *Rba-Tubulin* was used as the reference gene. All reactions were performed in three biological replicates, and the $2^{-\Delta\Delta Ct}$ method was applied to calculate the relative expression. GraphPad Prism 8.4.0 (https://www.graphpad.com, accessed on 18 December 2022) software was used to generate graphs and to perform statistical analyses. A *p*-value of <0.05 was considered to determine the significance level of the data.

4.7. Subcellular Localization in Tobacco

The pCAMBIA1300-c-GFP vector was used for the subcellular localization of the *RbSWEET* genes in *Nicotiana benthamiana*. The CDS sequences of three *RbSWEET* genes were amplified using a forward primer containing a HindIII restriction site and a reverse primer containing a KpnI restriction site. The amplification products were recovered using the FastPure Gel DNA Extraction Mini Kit (Vazyme Bio Co., Nanjing, Jiangsu Province, China) and then recombined with the digested plasmid by using the ClonExpress® Ultra One Step Cloning Kit (Vazyme). The recombined plasmids were later transformed into *Agrobacterium tumefaciens* strain GV3101. The infection solution was prepared with sterile water containing 10 mM $MgCl_2$, 10 mM 2-Morpholinoethanesulfonic Acid (MES), and 0.1 M Acetosyringone (AS), and then the *A. tumefaciens* was resuspended to an optical density (OD600) of 1.0. After standing in the dark for 3 h, the infection solution was injected into the leaf of *N. benthamiana* with a syringe. Two days after infiltration, the transient expression position of the *RbSWEET* proteins was observed by a Leica TCS SP8 fluorescence microscope (Leica Microsystems, Inc., Buffalo Grove, IL, USA) using filter blocks to select for spectral emission at 488 nm, and the empty vector was used as a control.

5. Conclusions

In this study, 25 *SWEET* genes were firstly identified and analyzed in *R. chinensis* 'Old Blush'. Gene structure and motif analysis indicated that members of the SWEET family in the rose are highly conserved among the species. The prediction of *Cis*-acting elements suggested that some family members may be involved in complex signaling pathway regulation. The transcriptome data showed that there were at least eight DEGs of *SWEETs* in the leaves and shoots of *R. chinensis* 'Old Blush' and *R. beggeriana* after cold treatment. The qRT-PCR results further disclosed that there were various expression patterns of *SWEET* genes in two species. *SWEET1* and *SWEET4b* were relatively conserved in *Rosa* L., with similar increased trends both in tissues and species. However, *SWEET2a* and *SWEET10c* exhibited the species difference after cold treatment that sharply upregulated in the leaves of the deciduous and extremely cold-tolerant species *R. beggeriana* but not the evergreen and moderately cold-tolerant species *R. chinensis* 'Old Blush'. The results indicated that *SWEET2a* and *SWEET10c* might be the crucial candidates to participate in the acquisition of cold tolerance in rose. Our findings provide important information for further research on the function analysis of *SWEET* genes in cold response and their application in the breeding of rose varieties with strong cold tolerance.

Supplementary Materials: The following supporting information can be downloaded at: https://www.mdpi.com/article/10.3390/plants12071474/s1, Figure S1: The transmembrane structure of 25 RcSWEET proteins; Table S1: Primers used in this study; Table S2: Nucleotide sequences of 25 *RcSWEET* genes; Table S3: Amino acid sequences of 63 SWEET proteins from three species; Table S4: Estimated Ka/Ks ratios of the duplicated RcSWEET genes; Table S5: The numbers of predicted cis-elements in the promoters of 25 *RcSWEET* genes; Table S6: Transcriptome sequencing data of *RcSWEET* genes in different tissues (PRJNA546486); Table S7: Transcriptome sequencing data of *RbSWEET* genes in leaves and shoots during cold treatment.

Author Contributions: Conceptualization, H.G. and S.Y.; methodology, X.S., Y.K., M.D. and X.Y.; software and validation, X.S., B.F. and R.J.; investigation, S.Y., X.S., M.D. and X.Z.; formal analysis, X.S., Y.K. and S.Y.; resources, S.Y.; data curation, X.S. and S.Y.; writing—original draft preparation, X.S.; writing—review and editing, Y.K. and S.Y.; visualization, X.S., R.J. and X.Z.; supervision, H.G. and S.Y. All authors have read and agreed to the published version of the manuscript.

Funding: This study was supported by the National Key R&D Program of China (2018YFD1000401; 2019YFD1000401) and the Beijing Innovation Consortium of Agriculture Research System, China (BAIC09-2023).

Data Availability Statement: All data are included in the main text.

Conflicts of Interest: The authors declare no conflict of interest.

References

1. Chen, L.-Q.; Hou, B.-H.; Lalonde, S.; Takanaga, H.; Hartung, M.L.; Qu, X.-Q.; Guo, W.-J.; Kim, J.-G.; Underwood, W.; Chaudhuri, B.; et al. Sugar transporters for intercellular exchange and nutrition of pathogens. *Nature* **2010**, *468*, 527–532. [CrossRef] [PubMed]
2. Chen, L.Q.; Cheung, L.S.; Feng, L.; Tanner, W.; Frommer, W.B. Transport of sugars. *Annu. Rev. Biochem.* **2015**, *84*, 865–894. [CrossRef] [PubMed]
3. Xuan, Y.H.; Hu, Y.B.; Chen, L.Q.; Sosso, D.; Ducat, D.C.; Hou, B.H.; Frommer, W.B. Functional role of oligomerization for bacterial and plant SWEET sugar transporter family. *Proc. Natl. Acad. Sci. USA* **2013**, *110*, E3685–E3694. [CrossRef] [PubMed]
4. Zhou, Y.; Liu, L.; Huang, W.; Yuan, M.; Zhou, F.; Li, X.; Lin, Y. Overexpression of OsSWEET5 in rice causes growth retardation and precocious senescence. *PLoS ONE* **2014**, *9*, e94210. [CrossRef] [PubMed]
5. Chardon, F.; Bedu, M.; Calenge, F.; Klemens, P.A.W.; Spinner, L.; Clement, G.; Chietera, G.; Léran, S.; Ferrand, M.; Lacombe, B.; et al. Leaf fructose content is controlled by the vacuolar transporter SWEET17 in Arabidopsis. *Curr. Biol.* **2013**, *23*, 697–702. [CrossRef]
6. Guo, W.J.; Nagy, R.; Chen, H.Y.; Pfrunder, S.; Yu, Y.C.; Santelia, D.; Frommer, W.B.; Martinoia, E. SWEET17, a facilitative transporter, mediates fructose transport across the tonoplast of Arabidopsis roots and leaves. *Plant Physiol.* **2014**, *164*, 777–789. [CrossRef]
7. Chen, L.Q.; Qu, X.Q.; Hou, B.H.; Sosso, D.; Osorio, S.; Fernie, A.R.; Frommer, W.B. Sucrose efflux mediated by SWEET Proteins as a key step for phloem transport. *Science* **2012**, *335*, 207–211. [CrossRef]
8. Yuan, M.; Wang, S. Rice MtN3/saliva/SWEET family genes and their homologs in cellular organisms. *Mol. Plant* **2013**, *6*, 665–674. [CrossRef]
9. Patil, G.; Valliyodan, B.; Deshmukh, R.; Prince, S.; Nicander, B.; Zhao, M.; Sonah, H.; Song, L.; Lin, L.; Chaudhary, J.; et al. Soybean (Glycine max) SWEET gene family: Insights through comparative genomics, transcriptome profiling and whole genome re-sequence analysis. *BMC Genom.* **2015**, *16*, 520. [CrossRef]
10. Mizuno, H.; Kasuga, S.; Kawahigashi, H. The sorghum *SWEET* gene family: Stem sucrose accumulation as revealed through transcriptome profiling. *Biotechnol. Biofuels* **2016**, *9*, 127. [CrossRef]
11. Feng, C.Y.; Han, J.X.; Han, X.X.; Jiang, J. Genome-wide identification, phylogeny, and expression analysis of the *SWEET* gene family in tomato. *Gene* **2015**, *573*, 261–272. [CrossRef]
12. Manck-Gotzenberger, J.; Requena, N. Arbuscular mycorrhiza symbiosis induces a major transcriptional reprogramming of the potato SWEET sugar transporter family. *Front Plant Sci.* **2016**, *7*, 487. [CrossRef]
13. Zhang, W.; Wang, S.; Yu, F.; Tang, J.; Shan, X.; Bao, K.; Yu, L.; Wang, H.; Fei, Z.; Li, J. Genome-wide characterization and expression profiling of SWEET genes in cabbage (Brassica oleracea var. capitata L.) reveal their roles in chilling and clubroot disease responses. *BMC Genom.* **2019**, *20*, 93. [CrossRef]
14. Wang, L.; Yao, L.; Hao, X.; Li, N.; Qian, W.; Yue, C.; Ding, C.; Zeng, J.; Yang, Y.; Wang, X. Tea plant SWEET transporters: Expression profiling, sugar transport, and the involvement of CsSWEET16 in modifying cold tolerance in Arabidopsis. *Plant Mol. Biol.* **2018**, *96*, 577–592. [CrossRef]

15. Wei, X.; Liu, F.; Chen, C.; Ma, F.; Li, M. The Malus domestica sugar transporter gene family: Identifications based on genome and expression profiling related to the accumulation of fruit sugars. *Front Plant Sci.* **2014**, *5*, 569. [CrossRef] [PubMed]
16. Chong, J.; Piron, M.C.; Meyer, S.; Merdinoglu, D.; Bertsch, C.; Mestre, P. The SWEET family of sugar transporters in grapevine: VvSWEET4 is involved in the interaction with Botrytis cinerea. *J. Exp. Bot.* **2014**, *65*, 6589–6601. [CrossRef] [PubMed]
17. Miao, H.; Sun, P.; Liu, Q.; Miao, Y.; Liu, J.; Zhang, K.; Hu, W.; Zhang, J.; Wang, J.; Wang, Z.; et al. Genome-wide analyses of SWEET family proteins reveal involvement in fruit development and abiotic/biotic stress responses in banana. *Sci. Rep.* **2017**, *7*, 3536. [CrossRef] [PubMed]
18. Guan, Y.-F.; Huang, X.-Y.; Zhu, J.; Gao, J.-F.; Zhang, H.-X.; Yang, Z.-N. *RUPTURED POLLEN GRAIN1*, a member of the MtN3/saliva gene family, is crucial for exine pattern formation and cell integrity of microspores in Arabidopsis. *Plant Physiol.* **2008**, *147*, 852–863. [CrossRef]
19. Han, L.; Zhu, Y.P.; Liu, M.; Zhou, Y.; Lu, G.Y.; Lan, L.; Wang, X.P.; Zhao, Y.F.; Zhang, X.J.C. Molecular mechanism of substrate recognition and transport by the AtSWEET13 sugar transporter. *Proc. Natl. Acad. Sci. USA* **2017**, *114*, 10089–10094. [CrossRef]
20. Lin, I.W.; Sosso, D.; Chen, L.Q.; Gase, K.; Kim, S.G.; Kessler, D.; Klinkenberg, P.M.; Gorder, M.K.; Hou, B.H.; Qu, X.Q.; et al. Nectar secretion requires sucrose phosphate synthases and the sugar transporter SWEET9. *Nature* **2014**, *508*, 546–549. [CrossRef] [PubMed]
21. Andres, F.; Kinoshita, A.; Kalluri, N.; Fernandez, V.; Falavigna, V.S.; Cruz, T.M.D.; Jang, S.; Chiba, Y.; Seo, M.; Mettler-Altmann, T.; et al. The sugar transporter SWEET10 acts downstream of FLOWERING LOCUS T during floral transition of *Arabidopsis thaliana*. *BMC Plant Biol.* **2020**, *20*, 53. [CrossRef]
22. Chen, L.Q.; Lin, I.W.N.; Qu, X.Q.; Sosso, D.; McFarlane, H.E.; Londono, A.; Samuels, A.L.; Frommer, W.B. A cascade of sequentially expressed sucrose transporters in the seed coat and endosperm provides nutrition for the Arabidopsis embryo. *Plant Cell* **2015**, *27*, 607–619. [CrossRef]
23. Dinant, S.; Wolff, N.; De Marco, F.; Vilaine, F.; Gissot, L.; Aubry, E.; Sandt, C.; Bellini, C.; Le Hir, R. Synchrotron FTIR and Raman spectroscopy provide unique spectral fingerprints for Arabidopsis floral stem vascular tissues. *J. Exp. Bot.* **2019**, *70*, 871–883. [CrossRef] [PubMed]
24. Yang, J.L.; Luo, D.P.; Yang, B.; Frommer, W.B.; Eom, J.S. SWEET11 and 15 as key players in seed filling in rice. *New Phytol.* **2018**, *218*, 604–615. [CrossRef]
25. Sosso, D.; Luo, D.; Li, Q.B.; Sasse, J.; Yang, J.; Gendrot, G.; Suzuki, M.; Koch, K.E.; McCarty, D.R.; Chourey, P.S.; et al. Seed filling in domesticated maize and rice depends on SWEET-mediated hexose transport. *Nat. Genet.* **2015**, *47*, 1489–1493. [CrossRef] [PubMed]
26. Kanno, Y.; Oikawa, T.; Chiba, Y.; Ishimaru, Y.; Shimizu, T.; Sano, N.; Koshiba, T.; Kamiya, Y.; Ueda, M.; Seo, M. AtSWEET13 and AtSWEET14 regulate gibberellin-mediated physiological processes. *Nat. Commun.* **2016**, *7*, 13245. [CrossRef] [PubMed]
27. Matallana-Ramirez, L.P.; Rauf, M.; Farage-Barhom, S.; Dortay, H.; Xue, G.P.; Droge-Laser, W.; Lers, A.; Balazadeh, S.; Mueller-Roeber, B. NAC transcription factor ORE1 and senescence-induced BIFUNCTIONAL NUCLEASE1 (BFN1) constitute a regulatory cascade in Arabidopsis. *Mol. Plant* **2013**, *6*, 1438–1452. [CrossRef] [PubMed]
28. Liu, W.X.; Peng, B.; Song, A.P.; Jiang, J.F.; Chen, F.D. Sugar transporter, CmSWEET17, promotes bud outgrowth in *Chrysanthemum morifolium*. *Genes* **2020**, *11*, 26. [CrossRef]
29. Klemens, P.A.W.; Patzke, K.; Deitmer, J.; Spinner, L.; Le Hir, R.; Bellini, C.; Bedu, M.; Chardon, F.; Krapp, A.; Neuhaus, H.E. Overexpression of the vacuolar sugar carrier *AtSWEET16* modifies germination, growth, and stress tolerance in Arabidopsis. *Plant Physiol.* **2013**, *163*, 1338–1352. [CrossRef]
30. Yang, G.X.; Xu, H.F.; Zou, Q.; Zhang, J.; Jiang, S.H.; Fang, H.C.; Wang, Y.C.; Su, M.Y.; Wang, N.; Chen, X.S. The vacuolar membrane sucrose transporter MdSWEET16 plays essential roles in the cold tolerance of apple. *Plant Cell Tiss. Org.* **2020**, *140*, 129–142. [CrossRef]
31. Yao, L.N.; Ding, C.Q.; Hao, X.Y.; Zeng, J.M.; Yang, Y.J.; Wang, X.C.; Wang, L. CsSWEET1a and CsSWEET17 mediate growth and freezing tolerance by promoting sugar transport across the plasma membrane. *Plant Cell Physiol.* **2020**, *61*, 1669–1682. [CrossRef] [PubMed]
32. Valifard, M.; Le Hir, R.; Muller, J.; Scheuring, D.; Neuhaus, H.E.; Pommerrenig, B. Vacuolar fructose transporter SWEET17 is critical for root development and drought tolerance. *Plant Physiol.* **2021**, *187*, 2716–2730. [CrossRef] [PubMed]
33. Chen, H.Y.; Huh, J.H.; Yu, Y.C.; Ho, L.H.; Chen, L.Q.; Tholl, D.; Frommer, W.B.; Guo, W.J. The Arabidopsis vacuolar sugar transporter SWEET2 limits carbon sequestration from roots and restricts *Pythium* infection. *Plant J.* **2015**, *83*, 1046–1058. [CrossRef]
34. Rouina, H.; Tseng, Y.H.; Nataraja, K.N.; Shaanker, R.U.; Oelmueller, R. Arabidopsis restricts sugar loss to a colonizing trichoderma harzianum strain by downregulating SWEET11 and -12 and upregulation of SUC1 and SWEET2 in the roots. *Microorganisms.* **2021**, *9*, 1246. [CrossRef] [PubMed]
35. Meteier, E.; La Camera, S.; Goddard, M.L.; Laloue, H.; Mestre, P.; Chong, J. Overexpression of the VvSWEET4 transporter in grapevine hairy roots increases sugar transport and contents and enhances resistance to *Pythium irregulare*, a soilborne pathogen. *Front. Plant Sci.* **2019**, *10*, 884. [CrossRef] [PubMed]
36. Wang, X.; Zhang, L.Y.; Ji, H.T.; Mo, X.Y.; Li, P.; Wang, J.Z.; Dong, H.S. Hpa1 is a type III translocator in Xanthomonas oryzae pv. oryzae. *BMC Microbiol.* **2018**, *18*, 105. [CrossRef]
37. Wu, Y.F.; Lee, S.K.; Yoo, Y.; Wei, J.; Kwon, S.Y.; Lee, S.W.; Jeon, J.S.; An, G. Rice transcription factor OsDOF11 modulates sugar transport by promoting expression of sucrose transporter and *SWEET* genes. *Mol. Plant* **2018**, *11*, 833–845. [CrossRef]

38. Cheng, Q.; Mao, W.H.; Xie, W.Y.; Liu, Q.S.; Cao, J.B.; Yuan, M.; Zhang, Q.L.; Li, X.H.; Wang, S.P. Characterization of a disease susceptibility locus for exploring an efficient way to improve rice resistance against bacterial blight. *Sci. China Life Sci.* **2017**, *60*, 298–306. [CrossRef]
39. Saint-Oyant, L.H.; Ruttink, T.; Hamama, L.; Kirov, I.; Lakhwani, D.; Zhou, N.N.; Bourke, P.M.; Daccord, N.; Leus, L.; Schulz, D.; et al. A high-quality genome sequence of *Rosa chinensis* to elucidate ornamental traits. *Nat. Plants* **2018**, *4*, 473–484. [CrossRef]
40. Thomashow, M.F. Plant cold acclimation: Freezing tolerance genes and regulatory mechanisms. *Annu. Rev. Plant Phys.* **1999**, *50*, 571–599. [CrossRef]
41. Fürtauer, L.; Weiszmann, J.; Weckwerth, W.; Nägele, T. Dynamics of plant metabolism during cold acclimation. *Int. J. Mol. Sci.* **2019**, *20*, 5411. [CrossRef] [PubMed]
42. Zhang, X.; Zhang, J.; Zhang, W.; Yang, T.; Xiong, Y.; Che, D. Transcriptome sequencing and de novo analysis of *Rosa multiflora* under cold stress. *Acta Physiol. Plant.* **2016**, *38*, 164. [CrossRef]
43. Huang, S.W.; Hong, G. Study on the use of *Rosa beggeriana* Schrenk as a source of cold resistance in rose breeding. *Acta Hortic. Sinica.* **1989**, *16*, 237–240.
44. Liu, Q.; Yang, S.H.; Jia, R.D.; Zhao, X.; Hong, G. Analysis of freezing tolerances and its physiological differences of two *Rosa* species during the overwintering. *Acta Hortic. Sinica.* **2017**, *44*, 1344–1354. [CrossRef]
45. Zhuang, D.; Ma, C.; Xue, L.; Li, Z.; Wang, C.; Lei, J.; Yuan, X. Transcriptome and de novo analysis of Rosa xanthina f. spontanea in response to cold stress. *BMC Plant Biol.* **2021**, *21*, 472. [CrossRef] [PubMed]
46. Sharma, H.; Sharma, A.; Rajput, R.; Sidhu, S.; Dhillon, H.; Verma, P.C.; Pandey, A.; Upadhyay, S.K. Molecular characterization, evolutionary analysis, and expression profiling of BOR genes in important cereals. *Plants* **2022**, *11*, 911. [CrossRef]
47. Han, Y.; Yu, J.; Zhao, T.; Cheng, T.; Wang, J.; Yang, W.; Pan, H.; Zhang, Q. Dissecting the genome-wide evolution and function of R2R3-MYB transcription factor family in *Rosa chinensis*. *Genes* **2019**, *10*, 823. [CrossRef]
48. Raymond, O.; Gouzy, J.; Just, J.; Badouin, H.; Verdenaud, M.; Lemainque, A.; Vergne, P.; Moja, S.; Choisne, N.; Pont, C.; et al. The *Rosa* genome provides new insights into the domestication of modern roses. *Nat. Genet.* **2018**, *50*, 772–777. [CrossRef]
49. de Bruijn, F.J.; Felix, G.; Grunenberg, B.; Hoffmann, H.J.; Metz, B.; Ratet, P.; Simons-Schreier, A.; Szabados, L.; Welters, P.; Schell, J. Regulation of plant genes specifically induced in nitrogen-fixing nodules: Role of cis-acting elements and trans-acting factors in leghemoglobin gene expression. *Plant Mol. Biol.* **1989**, *13*, 319–325. [CrossRef]
50. Gao, S.; Gao, J.; Zhu, X.; Song, Y.; Li, Z.; Ren, G.; Zhou, X.; Kuai, B. ABF2, ABF3, and ABF4 promote ABA-mediated chlorophyll degradation and leaf senescence by transcriptional activation of chlorophyll catabolic genes and senescence-associated genes in Arabidopsis. *Mol. Plant* **2016**, *9*, 1272–1285. [CrossRef]
51. Phukan, U.J.; Jeena, G.S.; Tripathi, V.; Shukla, R.K. MaRAP2-4, a waterlogging-responsive ERF from *Mentha*, regulates bidirectional sugar transporter AtSWEET10 to modulate stress response in Arabidopsis. *Plant Biotechnol. J.* **2018**, *16*, 221–233. [CrossRef] [PubMed]
52. Sazegari, S.; Niazi, A.; Ahmadi, F.S. A study on the regulatory network with promoter analysis for Arabidopsis DREB-genes. *Bioinformation* **2015**, *11*, 101–106. [CrossRef] [PubMed]
53. Nishida, I.; Murata, N. Chilling sensitivity in plants and cyanobacteria: The crucial contribution of membrane lipids. *Annu. Rev. Plant Physiol.* **1996**, *47*, 541–568. [CrossRef]
54. Kaplan, F.; Kopka, J.; Haskell, D.W.; Zhao, W.; Schiller, K.C.; Gatzke, N.; Sung, D.Y.; Guy, C.L. Exploring the temperature-stress metabolome of Arabidopsis. *Plant Physiol.* **2004**, *136*, 4159–4168. [CrossRef] [PubMed]
55. Chinnusamy, V.; Zhu, J.; Zhu, J.K. Cold stress regulation of gene expression in plants. *Trends Plant Sci.* **2007**, *12*, 444–451. [CrossRef]
56. Chang, C.Y.; Brautigam, K.; Huner, N.P.A.; Ensminger, I. Champions of winter survival: Cold acclimation and molecular regulation of cold hardiness in evergreen conifers. *New Phytol.* **2021**, *229*, 675–691. [CrossRef]
57. Yamada, K.; Osakabe, Y.; Mizoi, J.; Nakashima, K.; Fujita, Y.; Shinozaki, K.; Yamaguchi-Shinozaki, K. Functional analysis of an Arabidopsis thaliana abiotic stress-inducible facilitated diffusion transporter for monosaccharides. *J. Biol. Chem.* **2010**, *285*, 1138–1146. [CrossRef]
58. Fang, T.; Rao, Y.; Wang, M.; Li, Y.; Liu, Y.; Xiong, P.; Zeng, L. Characterization of the SWEET gene family in longan (Dimocarpus longan) and the role of DlSWEET1 in cold tolerance. *Int. J. Mol. Sci.* **2022**, *23*, 8914. [CrossRef]
59. Hu, L.; Zhang, F.; Song, S.; Yu, X.; Ren, Y.; Zhao, X.; Liu, H.; Liu, G.; Wang, Y.; He, H. CsSWEET2, a hexose transporter from cucumber (*Cucumis sativus* L.), affects sugar metabolism and improves cold tolerance in Arabidopsis. *Int. J. Mol. Sci.* **2022**, *23*, 3886. [CrossRef]
60. Le Hir, R.; Spinner, L.; Klemens, P.A.; Chakraborti, D.; de Marco, F.; Vilaine, F.; Wolff, N.; Lemoine, R.; Porcheron, B.; Gery, C.; et al. Disruption of the sugar transporters AtSWEET11 and AtSWEET12 affects vascular development and freezing tolerance in Arabidopsis. *Mol. Plant* **2015**, *8*, 1687–1690. [CrossRef]
61. Li, J.; Qin, M.; Qiao, X.; Cheng, Y.; Li, X.; Zhang, H.; Wu, J. A new insight into the evolution and functional divergence of SWEET transporters in Chinese white pear (*Pyrus bretschneideri*). *Plant Cell Physiol.* **2017**, *58*, 839–850. [CrossRef] [PubMed]
62. Aliche, E.B.; Theeuwen, T.; Oortwijn, M.; Visser, R.G.F.; van der Linden, C.G. Carbon partitioning mechanisms in potato under drought stress. *Plant Physiol. Biochem.* **2020**, *146*, 211–219. [CrossRef]

63. Wang, S.D.; Liu, S.L.; Wang, J.; Yokosho, K.; Zhou, B.; Yu, Y.C.; Liu, Z.; Frommer, W.B.; Ma, J.F.; Chen, L.Q.; et al. Simultaneous changes in seed size, oil content and protein content driven by selection of SWEET homologues during soybean domestication. *Natl. Sci. Rev.* **2020**, *7*, 1776–1786. [CrossRef] [PubMed]
64. Finn, R.D.; Clements, J.; Eddy, S.R. HMMER web server: Interactive sequence similarity searching. *Nucleic Acids Res.* **2011**, *39*, W29–W37. [CrossRef] [PubMed]
65. Chen, Y.J.; Yu, P.; Luo, J.C.; Jiang, Y. Secreted protein prediction system combining CJ-SPHMM, TMHMM, and PSORT. *Mamm. Genome.* **2003**, *14*, 859–865. [CrossRef]
66. Hu, B.; Jin, J.; Guo, A.Y.; Zhang, H.; Luo, J.; Gao, G. GSDS 2.0: An upgraded gene feature visualization server. *Bioinformatics* **2015**, *31*, 1296–1297. [CrossRef]
67. Bailey, T.L.; Boden, M.; Buske, F.A.; Frith, M.; Grant, C.E.; Clementi, L.; Ren, J.Y.; Li, W.W.; Noble, W.S. MEME SUITE: Tools for motif discovery and searching. *Nucleic Acids Res.* **2009**, *37*, W202–W208. [CrossRef] [PubMed]
68. Kaur, A.; Sharma, A.; Madhu; Dixit, S.; Singh, K.; Upadhyay, S.K. OSCA Genes in Bread Wheat: Molecular characterization, expression profiling, and interaction analyses indicated their diverse roles during development and stress response. *Int. J. Mol. Sci.* **2022**, *23*, 14867. [CrossRef]
69. Voorrips, R.E. MapChart: Software for the graphical presentation of linkage maps and QTLs. *J. Hered.* **2002**, *93*, 77–78. [CrossRef]
70. Huanxin, Z.; Ning, C.; Chunjuan, D.; Qingmao, S. Genome-wide identification and expression of ARF gene family during adventitious root development in hot pepper (*Capsicum annuum*). *Hortic. Plant J.* **2017**, *3*, 151–164. [CrossRef]

plants

Article

Selection and Validation of Optimal RT-qPCR Reference Genes for the Normalization of Gene Expression under Different Experimental Conditions in *Lindera megaphylla*

Hongli Liu [1,†], Jing Liu [1,†], Peng Chen [1], Xin Zhang [1], Ke Wang [2], Jiuxing Lu [1,*] and Yonghua Li [1,*]

[1] International Union Laboratory of Landscape Architecture of Henan, College of Landscape Architecture and Arts, Henan Agricultural University, Zhengzhou 450003, China
[2] Zhengzhou Botanical Garden, Zhengzhou 450042, China
* Correspondence: lujiuxing@hotmail.com (J.L.); liyhhany@henau.edu.cn (Y.L.)
† These authors contributed equally to this work.

Abstract: *Lindera megaphylla*, a broad-leaved evergreen that is used as a landscape ornamental plant and medicinal plant, is an ecologically important and dominant tree species. However, little is known about the molecular mechanisms of its growth, development, and metabolism. The selection of suitable reference genes is critical for molecular biological analyses. To date, no research on reference genes as a foundation for gene expression analysis has been undertaken in *L. megaphylla*. In this study, 14 candidate genes were selected from the transcriptome database of *L. megaphylla* for RT-qPCR assay under different conditions. Results showed that *helicase-15* and *UBC28* were most stable in different tissues of seedlings and adult trees. For different leaf developmental stages, the best combination of reference genes was *ACT7* and *UBC36*. *UBC36* and *TCTP* were the best under cold treatment, while *PAB2* and *CYP20-2* were the best under heat treatment. Finally, a RT-qPCR assay of *LmNAC83* and *LmERF60* genes were used to further verify the reliability of selected reference genes above. This work is the first to select and evaluate the stability of reference genes for the normalization of gene expression analysis in *L. megaphylla* and will provide an important foundation for future genetic studies of this species.

Keywords: *Lindera megaphylla*; normalization genes; RT-qPCR; tissue specificity; temperature; transcript analysis

Citation: Liu, H.; Liu, J.; Chen, P.; Zhang, X.; Wang, K.; Lu, J.; Li, Y. Selection and Validation of Optimal RT-qPCR Reference Genes for the Normalization of Gene Expression under Different Experimental Conditions in *Lindera megaphylla*. *Plants* **2023**, *12*, 2185. https://doi.org/10.3390/plants12112185

Academic Editors: Aiping Song and Yu Chen

Received: 12 April 2023
Revised: 18 May 2023
Accepted: 29 May 2023
Published: 31 May 2023

1. Introduction

Lindera megaphylla is a predominant, broad-leaved, and aromatic evergreen tree species belonging to the Lauraceae family and is widely distributed in the subtropical and warm-temperate zones of China. *L. megaphylla* has not only ecological and ornamental value, but also medicinal and therapeutic value as a source of an essential oils, spices, and drugs [1–3]. For example, d-dicentrine, an aporphine alkaloid, is isolated from the root of *L. megaphylla* and has potential antitumor activity [4,5]. These trees are rich in terpenoids, alkaloids, and flavonoids, many of which could be used to make pesticides or industrial feedstocks, while its volatile compounds, mainly terpenoids, have strong bactericidal ability. These trees can also help to improve air quality [6], a property that could be improved through molecular breeding. To synthesize antibacterial compounds, it is necessary to first explore the related regulatory genes and to analyze their functions. To date, studies of *L. megaphylla* have largely focused on the cultivation of seedlings, various kinds of biotic and abiotic stress responses and the analysis of volatile substances and their potential applications [7–12], while few studies have focused on the molecular biology of *L. megaphylla* due to a lack of genomic information. Gene expression analysis circumvents that lack of a sequenced genome to explore the molecular mechanisms underlying transcriptional regulation of phenotype. Transcriptome datasets derived from different tissues and differently aged leaves of *L. megaphylla* have been obtained (unpublished data), which will greatly promote functional genetic studies in this species.

Real-time reverse transcriptase quantitative polymerase chain reaction (RT-qPCR) is an important tool for analyzing gene expression because of its high throughput, sensitivity and precision [13,14]. However, RT-qPCR data are affected by many factors, such as extraction protocols, the purity and integrity of the extracted RNA, the efficiency of reverse transcription and PCR amplification, and primer specificity [15,16]. Therefore, a stably expressed reference gene is essential to avoid unnecessary errors generated through confounding factors and to increase the accuracy of the RT-qPCR data analysis. In the pre-genomic era, traditional reference genes were chosen based on their known or suspected housekeeping roles in basic cellular processes, cell structure, or primary metabolism, including genes encoding actin (*ACT*), 18S rRNA (*18S*) and glyceraldehyde-3 phosphate dehydrogenase (*GAPDH*) [17–20]. Unfortunately, these housekeeping genes are not stably expressed in all tissues, under different experimental conditions, or between different species [21,22]. In addition, a growing number of studies have revealed that some novel genes can also be used as internal references and are better than some traditional housekeeping genes [23–26]. Thus, it is necessary to systematically select the most appropriate reference genes to ensure the accuracy of RT-qPCR analysis for specific conditions and in specific materials. For this reason, optimal reference genes for transcript normalization must be determined through statistical algorithms such as delta cycle threshold (ΔCt) [27], geNorm [28], NormFinder [29], BestKeeper [30], and RefFinder [31] for each type of sample. These algorithms are widely used to assess the transcript stability of candidate reference genes in various species [32–35]. The use of reference genes in expression analysis has greatly facilitated our understanding of the important information related to gene functions and complex biological processes in plants, such as the signaling and metabolic pathways that underlie developmental and cellular processes [36–39].

Gene expression databases of model plant species such as *Arabidopsis thaliana* and tomato (*Lycopersicon esculentum*) are important resources for identifying and searching for genes of interest and their expression patterns (http://www.ebi.ac.uk/arrayexpress/; accessed on 5 June 2022; http://www.ncbi.nlm.nih.gov/geo/, accessed on 5 June 2022) [17]. Czechowski et al. selected, verified, and recommended 18 new reference genes that were superior to traditional reference genes in terms of expression stability across an extensive sample series or under a range of environmental conditions through the publicly available AtGenExpress database (http://web.uni-frankfurt.de/fb15/botanik/mcb/AFGN/atgenex.htm, accessed on 5 June 2022) and the author's own ATH1 database [17]. Orthologues of known genes in Arabidopsis can serve the same purposes as in other species. In addition, Lin et al. validated suitable reference genes for reliable normalization of data from *Litsea cubeba* [40]. Moreover, both *L. megaphylla* and *L. cubeba* belong to the Lauraceae family, and their evolutionary relationship is relatively close.

Based on these previous results, we first selected 40 genes that have been used as internal reference from published model plants such as Arabidopsis thaliana and other species in the Lauraceae family as candidate reference genes for *L. megaphylla*. Second, we screened homologous sequences in the transcriptome database of *L. megaphylla* and obtained 20 candidate genes based on expression multiples less than 1.5 and FPKM value > 50 in different tissues.

In this study, 14 candidate genes were identified with E-values between 91.035% and 107.169% and R^2 values from 0.991 to 0.999, indicating that the primer pairs may be more accurate for standardized evaluation by RT-qPCR. The candidate genes included translationally controlled tumor protein (*TCTP*), *ACT7*, *GAPDH*, ubiquitin-conjugating enzyme E2 36/7 (*UB C36*, *UBC7*), elongation factor 2-like (*EF2*), peptidyl-prolyl cis-trans isomerase CYP20-2, chloroplastic (*CYP20-2*), polyubiquitin (*UBQ*), alpha-tubulin (*TUA*), ubiquitin-conjugating enzyme E2 28-like (*UBC28*), NADH dehydrogenase (*ubiquinone*), pentatricopeptide repeat-containing protein (PPR), eukaryotic initiation factor 4A-3-like (*EIF4A-3*), DEAD-box ATP-dependent RNA helicase 15 (*helicase-15*), and polyadenylate-binding protein 2-like (*PAB2*). These 14 candidate genes were then assessed for stability of expression under specific conditions, including different tissues of one-year-old seedlings (roots, stems, and leaves), tissues of 10-year-old trees (leaf buds, young stems, young seeds, young leaves, and mature leaves), 16 different leaf developmental stages and under different temperature stresses (cold and heat). Finally, the *NAC* and *ERF* genes, which are from an important family of transcription factors in plants, were used to verify the reliability of the selected reference genes in different samples. Our research identified the best reference genes for RT-qPCR analysis of *L. megaphylla* tissues under different conditions, laying a basis for further studies of the molecular mechanisms regulating gene expression in this important tree species.

2. Materials and Methods

2.1. Plant Materials and Treatments

L. megaphylla trees were grown in a field at the Zhengzhou Botanical Garden located in Zhengzhou city, Henan Province, China. Young seedlings were cultivated in plant growth chambers with LED lighting for temperature treatments. In total, 6 experimental sets were cultivated for RNA extraction. The first experimental set consisted of roots, stems, and leaves from 1-year-old seedlings that displayed robust and consistent growth. The second experimental set consisted of leaf buds, young stems, young seeds, young leaves, and mature leaves from adult trees of *L. megaphylla* that had been growing in the natural environment for approximately 10 years. The third experimental set included leaves at different developmental stages. Growing leaves were collected roughly every 3–7 or 3–15 days from the beginning of leaf bud growth in late March until the leaves were fully mature by the end of July. The specific sampling dates were 24 March, 29 March, 1 April, 4 April, 7 April, 15 April, 22 April, 29 April, 14 May, 23 May, 4 June, 16 June, 1 July, 15 July, and 31 July. The fourth set was exposed to cold stress [41]. Furthermore, 1-year-old seedlings were grown at 65% relative humidity and under 16 h/8 h light/dark conditions in an LED plant growth incubator (Shengyuan Instrument Co., Ltd., Zhengzhou, China). The seedlings were first treated at 25 °C for 7 days as control, and leaves were collected. Then, the seedlings were transferred to 4 °C for 7 days for long-term chilling acclimation (CA). Next, the seedlings were shifted to 0 °C for an additional 7 days for long-term freezing acclimation (FA). Then, the seedlings were again moved to control conditions (25 °C) for 7 days for long-term de-acclimation (DA). The fifth and sixth experimental groups consisted of leaves collected from 1-year-old seedlings that were treated with cold and heat, respectively. For cold treatments, the seedlings were cultivated in an LED plant growth incubator at 25 °C, 4 °C, 0 °C, −4 °C, or −6 °C for 24 h. For heat treatments, the seedlings were cultivated in an incubator at 25 °C, 30 °C, 35 °C, 40 °C, or 45 °C for 24 h. Data regarding all six sample sets described above are summarized in Table 1. All collected samples were immediately frozen in liquid nitrogen and stored at −80 °C. Three independent biological replicates were collected for each sample.

Table 1. Six experimental sets of *L. megaphylla*.

Experimental Sample Sets	Tissue Type	Biological Replicates	Sampling Dates	Total Number of Samples
Different tissues of oneyear-old seedlings	Roots, stems and leaves	3	1	3
Different tissues of adult trees	Leaf buds, young stems, young seeds, young leaves and mature leaves	3	1	15
Developmental stages	Leaves	3	15	45
Cold stress for 7 days	Leaves	3	4	12
Cold stress for 24 h	Leaves	3	5	15
Heat stress for 24 h	Leaves	3	5	15

2.2. Total RNA Extraction and cDNA Synthesis

Total RNA was extracted from 0.05 g samples using a Quick RNA isolation kit (HUAYUEYANG Biotechnology, Beijing, China) according to the manufacturer's protocol [42]. The RNA integrity, purity and concentration were assessed using 2% (*w*/*v*) agarose gel electrophoresis and a NanoDrop ND-1000 spectrophotometer (Thermo Scientific, Waltham, MA, USA). Total RNA (1 μg) with A_{260}/A_{280} and A_{260}/A_{230} ratios greater than 1.8 was used for first-strand cDNA synthesis using the Evo M-MLV RT Kit with gDNA Clean for qPCR according to the manufacturer's instructions (Accurate Biotechnology, Changsha, China). Specifically, a 10 μL reaction system (1 μg total RNA, 2 μL 5 × gDNA Clean Reaction Mix, to a total volume of 10 μL with RNase-free water) was subjected to 42 °C for 2 min. Then, a 20 μL reaction system (10 μL of first reaction solution, 4 μL 5 × *Evo M-MLV* RT Reaction Mix and 6 μL RNase-free water) was subjected to 37 °C for 15 min and 85 °C for 5 s. Five-fold diluted cDNA was used for subsequent RT-qPCR experiments. All cDNA samples were stored at −20 °C until use.

2.3. Selection of Candidate Reference Genes and Design of RT-qPCR Primers

In total, 20 candidate genes (*TCTP, ACT7, GAPDH, UBC36, UBC7, EF2, CYP20-2, UBQ, TUA, UBC28, ICln, ubiquinone, PPR, SDE2, EIF4A-3, helicase-15, PAB2, CYP9, RHA2A,* and *EF1α*) were selected from the transcriptome database of *L. megaphylla* based on reference genes reported in the literature. All primers were designed using the qPCR primer quest tool (https://sg.idtdna.com/pages/tools/primerquest?returnurl=%2Fprimerquest%2FHome%2FIndex, accessed on 9 June 2022) based on the coding sequences (CDS) in the transcriptome database of *L. megaphylla* (Supplementary Table S1). Details of these candidate reference genes and primers are shown in Table 2.

Table 2. Candidate reference genes and designed primers for RT-qPCR normalization in *L. megaphylla*.

Gene Symbol	Gene Name	Primer Sequence (5′ → 3′)	Product Length (bp)	Standard Curve	E (%)	R^2
TCTP	translationally controlled tumor protein	F:GTTTCTCACCCTCCAACTTAGG R:CATTTCGCCTCCAGGAACA	102	y = −2.4502x + 29.195	95.070	0.9992
ACT7	actin-related protein 7	F:AAGCCAACAGGGAGAAGATG R:CACCCGAGTCCAGAACAATAC	132	y = −2.3523x + 28.25	103.003	0.9971
GAPDH	Glyceraldehyde 3-phosphate dehydrogenase	F:CGGAGGATGATGTGGTTTCTAC R:GCGACAAGCTTGACAAAGTG	106	y = −2.3607x + 27.623	98.050	0.9994
UBC36	ubiquitin-conjugating enzyme E2 36	F:CCCGAAGGTTCGATTTCTCA R:TGAAGAGCAGGACTCCATTTATC	102	y = −2.3618x + 29.443	101.509	0.9978
UBC7	ubiquitin-conjugating enzyme E2 7	F:TCATGAGCTTCCCAGCAAATTA R:CGTCCGTCGGGATAAACATTAG	91	y = −2.442x + 29.477	96.932	0.9974
EF2	elongation factor 2-like	F:GCGGATAAGGGTAGGTTCTTT R:TTCTGGCCAGGAACATAGTTAG	104	y = −1.9725x + 27.627	96.045	0.9919
CYP20-2	peptidyl-prolyl cis-trans isomerase CYP20-2, chloroplastic	F:AACACCAACGGTAGCCAAT R:TCCAGAACCTGCCCAAATAC	86	y = −2.4129x + 27.719	101.307	0.9963
UBQ	polyubiquitin	F:CCTCGCCGACTACAATATTCA R:CACCTCCAGAGTAATCGTCTTC	115	y = −2.2086x + 23.796	85.947	0.9989
TUA	Alpha-tubulin	F:GCCTTACAACAGTGTGCTTTC R:ATCTAGAGATCGACGGCAGATA	106	y = −2.3673x + 27.37	101.414	0.9972
UBC28	ubiquitin-conjugating enzyme E2 28-like	F:ACAATTATGGGACCAGCAGATAG R:GGGTGGCTTGAATGGGTAAT	90	y = −2.3925x + 28.93	101.491	0.9972
ICln	chloride conductance regulatory protein ICln	F:TGAGCGACACCGATAGAGAA R:TAAATGCAAGGAGAGGCGTAAG	103	y = −2.6401x + 31.012	64.420	0.9953

Table 2. *Cont.*

Gene Symbol	Gene Name	Primer Sequence (5′ → 3′)	Product Length (bp)	Standard Curve	E (%)	R^2
ubiquinone	NADH dehydrogenase	F:ATCCGACGGGCGATTAAAG R:TCTAGCCTCTTCTTCCAGATACT	123	y = −2.1552x + 28.534	107.169	0.9975
PPR	pentatricopeptide repeatcontaining protein	F:CTTTAAGCCAGACCAGCAAATG R:TCCTCTTTCAGCCATCTTTCC	106	y = −2.3288x + 29.75	102.616	0.9976
SDE2	replication stress response regulator SDE2-like	F:TAGACGGGCGGACCAGAT R:GAGGAGGACGGTGCAGGAG	197	y = −2.7496x + 30.276	86.753	0.9912
EIF4A-3	eukaryotic initiation factor 4A-3like	F:TCTTTGTTGCGGTTGAGCG R:ACCAATCCACCTTTCTTTTCG	117	y = −2.8406x + 28.345	95.752	0.9918
helicase-15	DEAD-box ATP-dependent RNA helicase 15	F:CCTGGGAGAATACTGGCACTG R:GGCCTCGTCGTCCACATAA	249	y = −3.1364x + 30.269	92.361	0.9992
PAB2	polyadenylate-binding protein 2like	F:CCCAAGCTGTTGAGGATCTTA R:CCTTTCAGCTCCATCTCTCTTT	100	y = −2.4748x + 31.269	91.035	0.9919
CYP95	peptidyl-prolyl cis-trans isomerase CYP95 like	F:GGGTTCAGTCATCGTTACTCTT R:GCGTTCACTTCTTCCTCCATA	99	y = −2.4245x + 29.937	103.599	0.9898
RHA2A	E3 ubiquitin-protein ligase RHA2A	F:CTTTAGCGGGAGCGATGT R:CAAGCACTCTCTGTGGAAGA	112	y = −2.3815x + 31.29	93.904	0.9870
EF1α	Translation elongation factor EF1A	F:AAATGAGGAGGAGCGTGTAAAG R:CGCTGATCATGTTAGGGACATAG	128	y = −2.6481x + 30.807	83.491	0.8779

2.4. RT-PCR and RT-qPCR Data Analysis

To verify the accuracy of the designed primers, each pair was used for RT-PCR amplification. Each 20 µL reaction system contained 2 µL of 10 µM forward and reverse primers, 1 µL cDNA, 10 µL 2 × Rapid Taq Master Mix (Vazyme, Nanjing, China) and 7 µL ddH_2O. The reaction was carried out at 95 °C for 3 min, followed by 35 cycles of 95 °C for 30 s, 60 °C for 30 s and 72 °C for 15 s, with a final extension at 72 °C for 2 min. The PCR products were visualized by 1% (*w*/*v*) agarose gel electrophoresis.

To monitor the E-value, the cDNA templates from all samples were serially diluted five-fold (cDNA:water, v:v). RT-qPCR was performed for each pair of primers to obtain Ct values and to establish a standard curve; the R^2, slope and E-values were calculated with Microsoft Office Excel 2019 using the following formula: $E = (5^{-1/slope} - 1) \times 100\%$ [43].

All RT-qPCRs were performed with an Applied Biosystems™ (ABI) QuantStudio™ 5 real-time PCR system (ABI, Los Angeles, CA, USA) using the following amplification procedure: 95 °C for 30 s, followed by 40 cycles of 95 °C for 5 s and 60 °C for 30 s. A melting curve was generated at 60–95 °C. Each 20 µL RT-qPCR reaction (10 µL 2 × SYBR® Green *Pro Taq* HS Premix, 2 µL 5-fold diluted cDNA, 0.4 µL 10 µM forward primer, 0.4 µL 10 µM reverse primer, 0.4 µL 4 µM ROX Reference Dye, and 6.8 µL RNase-free water) was prepared according to the instructions for SYBR® Green Premix Pro Taq HS qPCR Kit (Rox Plus) (Accurate Biology, Shanghai, China). A negative control without the addition of cDNA was used to test for background amplification. Three technical replicates were performed for each sample, and the mean was used for RT-qPCR analysis.

2.5. Candidate Reference Gene Expression Stability Analysis

CT values were used to assess the expression levels of candidate reference genes in all samples by RT-qPCR. Four common algorithms, namely, delta Ct (ΔCt), geNorm (version 3.5), NormFinder (version 0.953) and BestKeeper (version 1.0), were used to evaluate the stability of the expression of the candidate reference genes in the different experimental groups. In geNorm and NormFinder, the M value reflects the stability of each candidate reference gene [28,29], with a smaller M value indicating higher stability. The geNorm package also determines the number of optimal reference genes based on the ratio $V_{n/n+1}$ by calculating pairwise variations in the normalized factor after introducing a new internal reference gene [28]. A cut-off value of 0.15 was used for pairwise variation. If the value of $V_{n/n+1}$ was less than 0.15, n was selected for the number of optimal internal reference genes; if the value of $V_{n/n+1}$ was greater than 0.15, n + 1 was selected. For BestKeeper, the values of CV and SD were used to evaluate the relative expression stability of each candidate gene [29]. The smaller the CV and SD values are, the more stable the gene. Finally, the RefFinder program (http://blooge.cn/RefFinder/, accessed on 25 September, 2022) was used to comprehensively rank the candidate reference genes by ΔCt, geNorm, NormFinder, and BestKeeper as previously described [31].

2.6. Validation of Candidate Reference Genes by RT-qPCR

The NAC and ERF transcription factors function as central switches of growth, development and various abiotic/biotic stress responses in plants [44–48]. Hence, *LmNAC83* and *LmERF60* genes were selected as targets to determine the reliability of the most stable and unstable reference genes. The primers for these two genes are shown in Supplementary Table S2. The expression levels of these genes were calculated using the $2^{-\Delta\Delta Ct}$ method [35]. The RT-qPCR analysis was carried out using RNA from three biological replicates.

3. Results

3.1. Verification of Amplicon Size, Primers Specificity and PCR Amplification Efficiency

A total of 20 candidate genes from *L. megaphylla* were selected as potential reference genes for the normalization of target gene transcript levels using RT-qPCR. Standard PCR amplification with primers (Table 2, Supplementary Table S1) targeting the candidate genes was performed with reverse-transcribed cDNA from each sample as templates. Agarose gel electrophoresis indicated that all PCR products were single bands of the expected sizes, indicating that the primers were specific (Supplementary Figure S1). The melting curves, obtained after 40 cycles of amplification by RT-qPCR, showed single peaks, which also verified that the primers for the 20 candidate reference genes had strong specificity (Supplementary Figure S2). Standard curves, generated using a five-fold serial dilution for each candidate gene, had linear correlation coefficients (R^2) greater than 0.99 for each specific primer pair, and the amplification efficiencies of the RT-qPCR reactions were 64.42–107.17% (Table 2). Because the amplification efficiencies of *ICln* (64.42%), *UBQ* (85.95%), *SDE2* (86.75%), and *EF1α* (83.49%) were less than 90%, and the R^2 values of *CYP95* and *RHA2A* were <0.99, these six candidate genes, based on the designed primers, were not appropriate for validating expression. Therefore, the remaining 14 candidate genes were used for subsequent experiments.

3.2. Transcript Abundance of Candidate Reference Genes

The transcript abundance of 14 candidate reference genes was estimated using the average cycle threshold (Ct) values for RNA extracted from different tissues or plants grown under different experimental conditions. All candidate reference genes were expressed at a wide range of transcript levels under the different experimental conditions. The gene expression level is negatively correlated with the Ct value, which means that a gene with a higher transcript level has a smaller Ct value. As shown in Figure 1, the minimum Ct value was 17.94, and the maximum was 27.51. Among the genes, *UBC36* exhibited the highest transcript abundance, with the minimum, median and maximum Ct values of 20.91, 22.90, and 27.51, respectively. *PAB2* showed the lowest expression abundance, with minimum, median and maximum Ct values being 17.94, 20.20, and 22.65, respectively. Compared with other genes, *GAPDH, UBC28,* and *TCTP* had obviously narrow range of Ct values (19.14–22.69, 20.74–23.95, and 21.10–24.98, respectively), indicating that they have a relatively stable expression level. While all the candidate reference genes exhibited significant expression abundance, none of the candidate genes were expressed stably in all samples. Thus, it is necessary to screen the appropriate internal reference genes for *L. megaphylla* under different experimental conditions or tissue types.

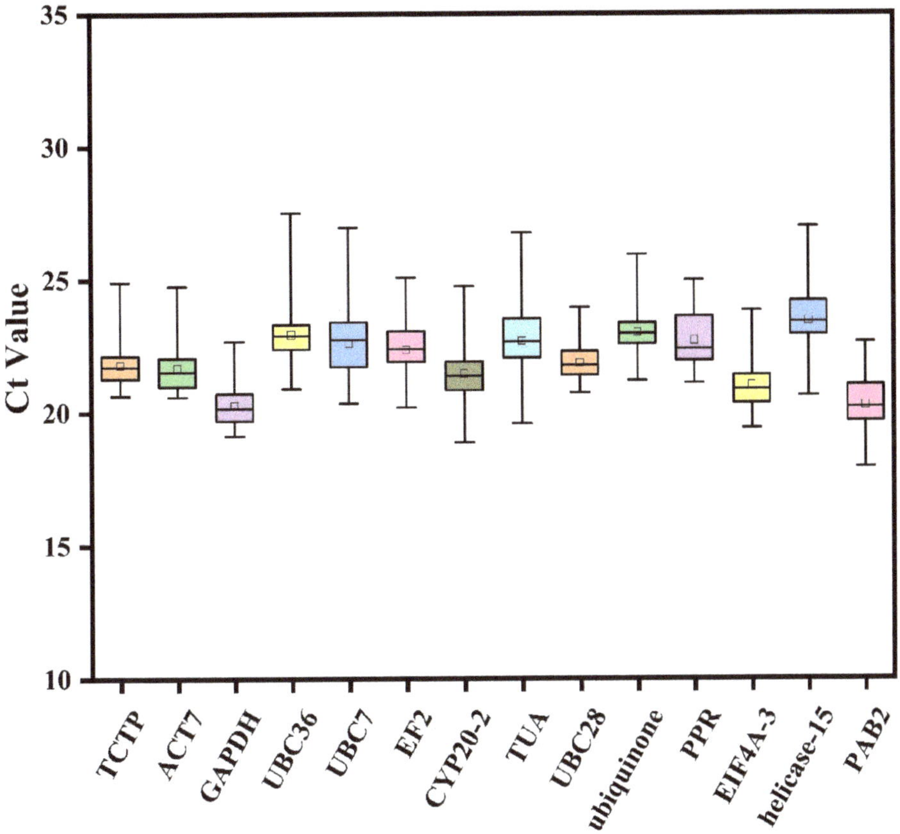

Figure 1. Distribution of average cycle threshold (Ct) values of the 14 candidate reference genes across all 38 samples from *L. megaphylla*. Boxes represent the interquartile range, with the middle solid lines indicating the median Ct values. The upper and lower boundaries of each box represent the 75th and 25th percentiles, respectively. The upper and lower bars represent the maximum and minimum Ct values, respectively, and the small squares represent the average values.

3.3. *Estimation of the Stability of the Reference Genes under Different Experimental Conditions*

To identify the optimal reference genes for the normalization of gene expression analysis in *L. megaphylla*, the stability of the 14 candidate genes was assessed by four different algorithms. The RefFinder software was used for overall ranking.

3.4. *Delta Ct Method Analysis*

The delta Ct (ΔCt) method ranks the stability of candidate reference genes based on the relative expression levels of "gene pairs" in each group of sample comparisons, while the mean standard deviation of gene expression differences (STDEV) is inversely proportional to its stability using the raw Ct value [34]. The stability of the transcript levels of each candidate reference gene was evaluated based on the STDEV value. The gene with the minimum STDEV value was regarded as the most stably expressed gene. The results demonstrated that the optimal reference genes were different in the different experimental sets (Figure 2). In seedling samples, *helicase-15* and *PAB2* showed the lowest ΔCt values (0.43), indicating the most stability (Figure 2A), while *UBC28* (0.31) was the most stably expressed gene in adult trees (Figure 2B). In other tissues, *helicase-15* (0.51) and *UBC28* (0.57) were the most stable reference genes (Figure 2C), which is consistent with the results for seedlings and adult trees. *ACT7* (0.58) and *UBC36* (0.59) were more stable across different leaf developmental stages (Figure 2D). Data analyses from the entire growth cycle indicated that *ubiquinone* (0.67) and *UBC36* (0.68) were the most stable (Figure 2E). *UBC36* showed good stability in all three cold stress sets (Figure 2F–H). *PAB2* (0.69) had the highest stability under heat treatment (Figure 2I). Across all temperature stresses, *PAB2* (0.6) was the most stably expressed gene (Figure 2J). For all samples, ubiquinone (0.74), *EF2* (0.75) and *PAB2* (0.76) showed the most stability (Figure 2K). *TUA* had relatively higher Ct values, indicating that it was the least stable reference gene in most of the experimental sets (Figure 2A–K).

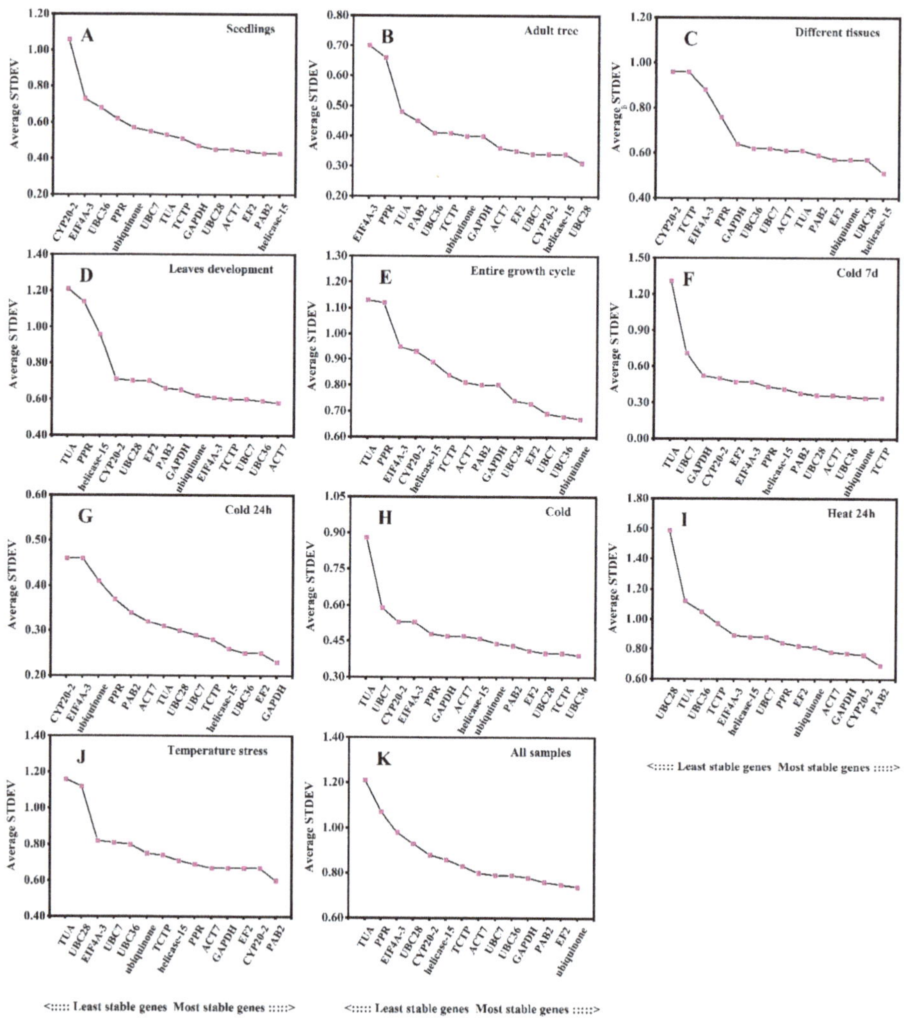

Figure 2. Ranking of expression stability of the 14 candidate reference genes in *L. megaphylla* using ΔCt analysis. Genes are listed across bottom of each plot in order of increasing stability from left to right. Results from (**A**) different seedling tissues; (**B**) different adult tree tissues; (**C**) all seedling and adult tree tissues; (**D**) 16 leaf developmental stages; (**E**) entire growth cycle including all different seedlings and adult tree tissues as well as 16 leaf developmental stages; (**F**) cold treatment for 7 days; (**G**) cold treatment for 24 h; (**H**) cold treatments including 7 days and 24 h; (**I**) heat treatment for 24 h; (**J**) different temperature treatments including cold and heat; (**K**) total samples.

3.5. geNorm Analysis

The stability of the reference genes was ranked by calculating the average expression stability values (M value) using the geNorm program, taking into account only similar intergroup variation [35].

Genes with an M value less than 1.5 were considered stably expressed, with smaller M values indicating a more stable gene [28]. The geNorm analysis results for the 14 candidate genes in the different experimental sets are shown in Figure 3A–K. The M values of the 14 candidate reference genes were less than 1.5 under the different experimental conditions (Figure 3). *EF2* and *PAB2* had the highest stability in seedlings, with M values of 0.012 (Figure 3A), while *CYP26-2* (0.076) and *helicase-15* (0.076) were most stably expressed in the adult tree (Figure 3B). Between the different tissue sets, *PAB2* and *helicase15* were the optimal candidate genes, which is similar to the results in seedlings and adult trees (Figure 3C). The genes *PAB2* and *helicase-15* had the lowest stability values (0.285), which is consistent with the results of the ΔCt analysis. The two most stably expressed genes among the different leaf developmental stages were similar to those for the entire growth cycle, namely, *UBC36* and *UBC7* (Figure 3D,E). These results are also similar to those of the ΔCt analysis. *UBC36* was more stable than the other candidate reference genes under the cold treatments (Figure 3F–H) and was the same regardless of whether the cold treatment lasted for 7 days or 24 h. *GAPDH* (0.269), *CYP20-2* (0.269), and *PPR* (0.306) were more stable than the other candidate genes under heat treatment (Figure 3I). However, *PPR* and *PAB2* (0.343) exhibited the strongest stability under temperature stress (Figure 3J). These results are consistent with the results of the ΔCt analysis. *UBC36* and *UBC7* (0.406) showed the strongest stability in all samples (Figure 3K). In contrast, *TUA* and *PPR* were the least stable across most sets.

Best practices include using multiple reference genes as internal controls for standardization to improve the accuracy of RT-qPCR data [28,49]. The number of optimal genes for standardization of the different datasets from *L. megaphylla* was calculated using the $V_{n/n+1}$ function of geNorm, with a threshold of 0.15 (Figure 4). Interestingly, the values of $V_{2/3}$ were less than 0.15 for most experimental groups (0.033, 0.045, 0.111, 0.112, 0.126, 0.034, 0.044, 0.052, 0.098, and 0.129) except for the 'all samples' group, as shown in Figure 4. This suggested that two was the optimal number of reference genes for each type of samples. However, for the 'all samples' set, the $V_{2/3}$ and $V_{3/4}$ values were greater than 0.15, and it is not until four reference genes are used ($V_{4/5}$ value of 0.117) that the values is less than 0.15. Thus, at least four genes are required to obtain accurate results across many tissues and treatments.

3.6. NormFinder Analysis

To further determine the reliability of the results obtained by the geNorm algorithm, the NormFinder application was used to evaluate both the intra- and inter-group variation to calculate the stability of the candidate reference genes, with lower values of inter- and intra-group variation corresponding to increased stability of the candidate gene [48]. The results of the NormFinder analysis are shown in Figure 5. *UBC28* (0.055) and *helicase-15* (0.138) were the most stable genes in seedlings, while *UBC28* (0.063) and *UBC7* (0.123) were the most stable genes in adult trees (Figure 5A,B). The best combination of reference genes for different tissues was *helicase-15* (0.134) and *ubiquinone* (0.267) (Figure 5C). In the combined leaf developmental stages set, the optimal combination of candidate reference genes was *UBC7* (0.190) and *ACT7* (0.245), while over the entire growth cycle, the most stable candidate genes were *ubiqunone* (0.267) and *UBC36* (0.295) (Figure 5D,E). *TCTP* (0.030) and *ubiqunone* (0.035) were the most stable genes under cold treatment for 7 d, while *GAPDH* (0.062) and *UBC36* (0.088) were the most stable genes under cold treatment for 24 h (Figure 5F,G). Including both cold stress treatments, the candidate reference genes *EF2* (0.168) and *UBC36* (0.193) were the most stable (Figure 5H). Under heat treatment for 24 h, *PAB2* (0.089) and *CYP20-2* (0.414) were the most stable reference genes (Figure 5I). Under all temperature stresses, *PAB2* (0.185) and *EF2* (0.349) showed the highest stability (Figure 5J).

The best combination of reference genes to compare all sample sets was *ubiquinone* (0.375), *EF2* (0.408), *PAB2* (0.413), and *GAPDH* (0.483) (Figure 5K). Notably, *TUA* and *PPR* were the least stable genes in most sets, similar to the results calculated by geNorm. In general, the results of the NormFinder analysis for the 14 candidate reference genes under different experimental conditions were similar to the results of the ΔCt and geNorm analyses.

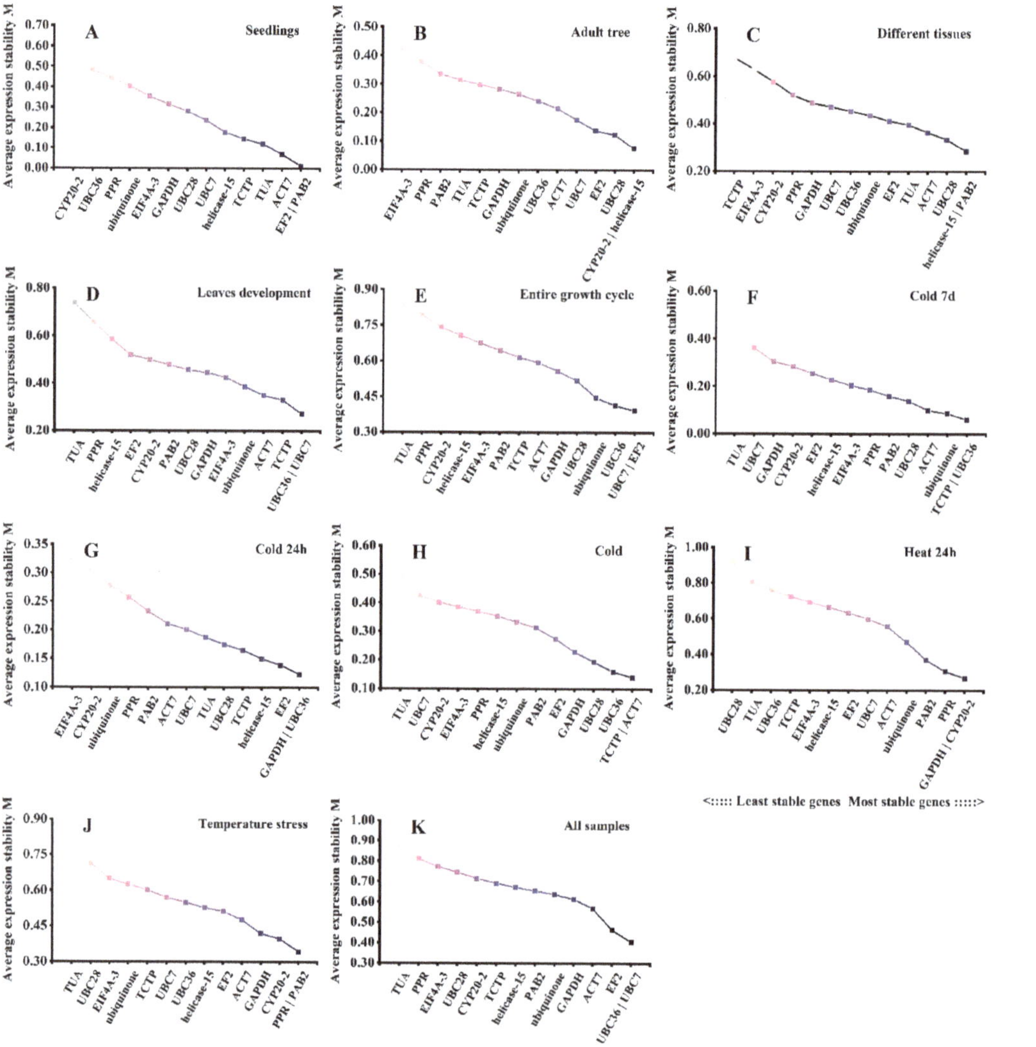

Figure 3. Ranking of expression stability of the 14 candidate reference genes in *L. megaphylla* using geNorm analysis. Results from (**A**) different seedling tissues; (**B**) different adult tree tissues; (**C**) all seedling and adult tree tissues; (**D**) 16 leaf developmental stages; (**E**) entire growth cycle including all different seedling and adult tree tissues as well as 16 leaf developmental stages; (**F**) cold treatment for 7 days; (**G**) cold treatment for 24 h; (**H**) cold treatments including 7 days and 24 h; (**I**) heat treatment for 24 h; (**J**) different temperature treatments including cold and heat; (**K**) total samples.

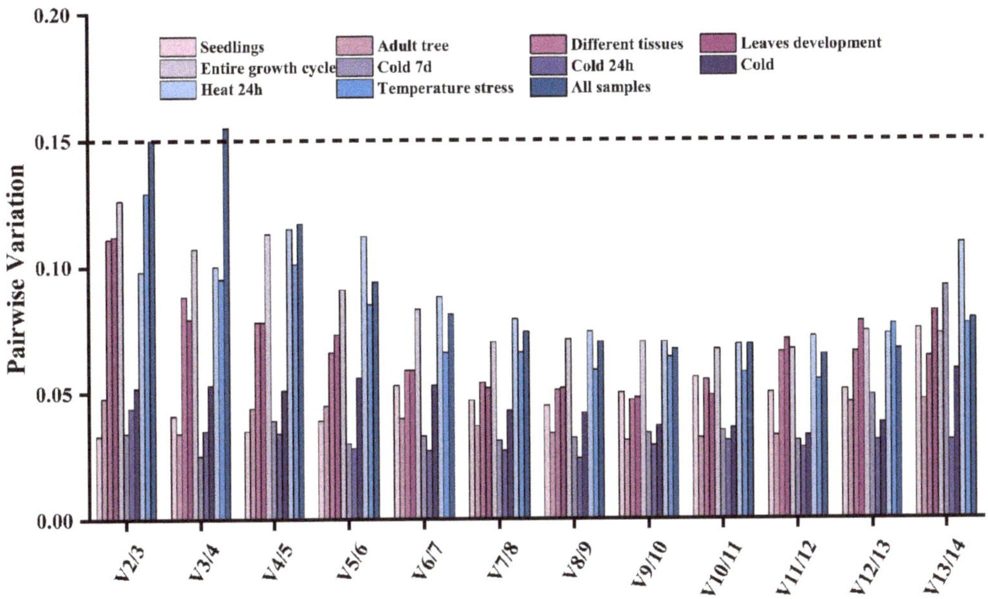

Figure 4. Pairwise variation values calculated by geNorm were used to determine the optimal number of reference genes from the 14 candidates in *L. megaphylla*. The cutoff value of $V_{n/n+1}$ to determine the optimal number of reference genes for RT-qPCR normalization was 0.15.

3.7. BestKeeper Algorithm

The BestKeeper algorithm evaluates the most stable candidate reference genes based on the standard deviation (SD) and coefficient of variance (CV) of the average cycle threshold (Ct) values [16,19,32]. In general, the smaller the SD value is, the more stable the gene is. As shown in Table 3, the genes *UBC7* (0.06 ± 0.24) and *helicase-15* (0.14 ± 0.57) had the lowest SD values in seedlings, while *GAPDH* (0.09 ± 0.48) and *TCTP* (0.18 ± 0.83) were the most stable genes in adult trees. The genes with the most stable expression across different tissues were *ACT7* (0.29 ± 1.36) and *UBC28* (0.31 ± 1.40), while the expression of *UBC28* (0.60 ± 2.76) and *GAPDH* (0.66 ± 3.24) was the most stable across different leaf developmental stages. The genes with the most stable expression across the entire growth cycle were *UBC28* (0.58 ± 2.64) and *TCTP* (0.59 ± 2.73). Under cold treatment, *PPR* (0.18 ± 0.82) and *EIF4A-3* (0.19 ± 0.91) were the most stable genes, while under heat treatment, *UBC28* (0.22 ± 1.03) and *ubiquinone* (0.54 ± 2.35) were the most stable genes. Under both types of temperature stress, the expression of *UBC28* (0.37 ± 1.70) and *PPR* (0.42 ± 1.86) were the most stable. For all samples, the top four most stable reference genes were *UBC28* (0.51 ± 2.32), *TCTP* (0.56 ± 2.57), *GAPDH* (0.60 ± 2.98) and *ubiquinone* (0.64 ± 2.76), while *TUA* showed the highest value and thus the lowest expression stability.

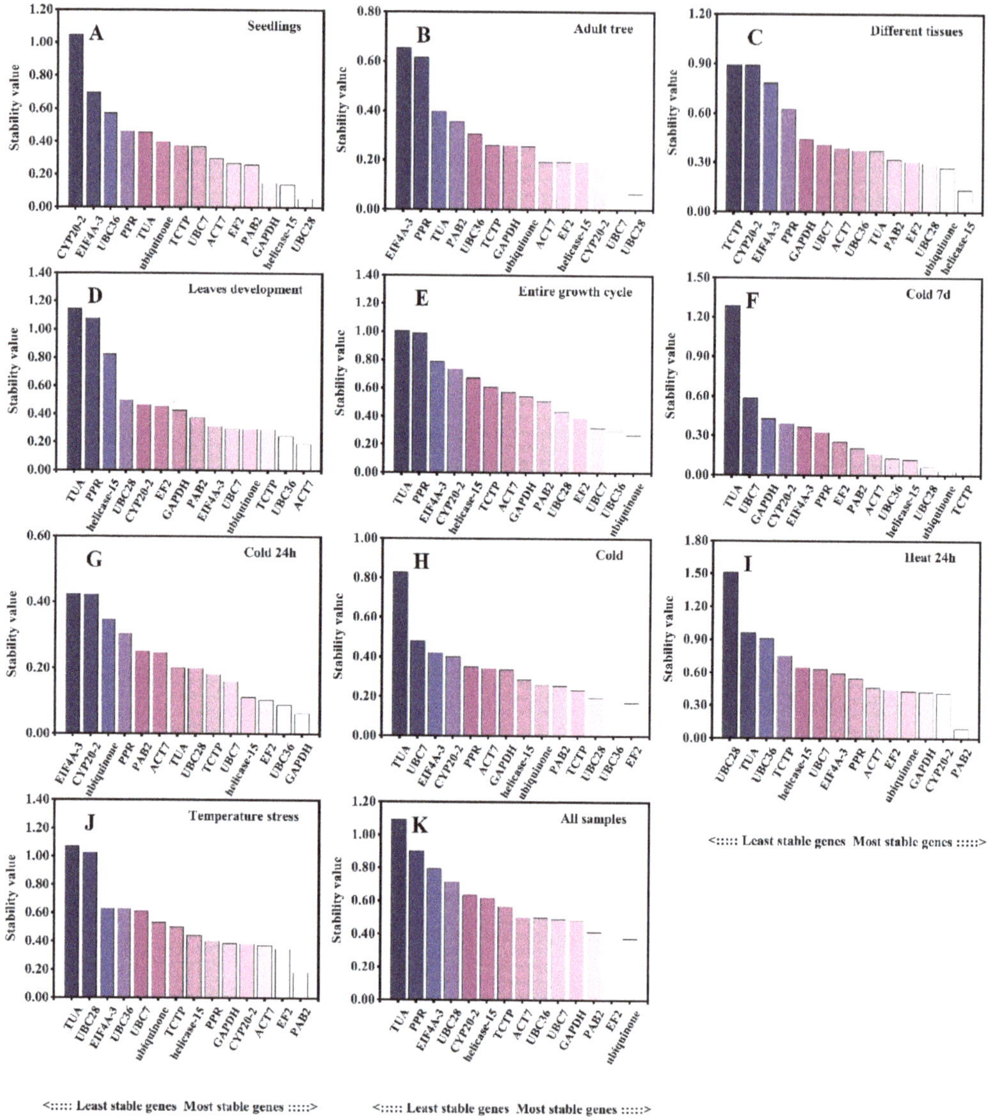

Figure 5. Ranking of expression stability of the 14 candidate reference genes in *L. megaphylla* using NormFinder analysis. Results from (**A**) different seedling tissues; (**B**) different adult tree tissues; (**C**) all seedling and adult tree tissues; (**D**) 16 leaf developmental stages; (**E**) entire growth cycle including all different seedling and adult tree tissues as well as 16 leaf developmental stages; (**F**) cold treatment for 7 days; (**G**) cold treatment for 24 h; (**H**) cold treatments including 7 days and 24 h; (**I**) heat treatment for 24 h; (**J**) different temperature treatments including cold and heat; (**K**) total samples.

Table 3. Ranking of expression stability of the 14 candidate reference genes in *L. megaphylla* using BestKeeper analysis.

Ranking	Seedlings			Adult Tree			Different Tissues			Leaf Development			Entire Growth Cycle			Cold 7 d		
	Gene Name	SD	CV (%)	Gene Name	SD	CV (%)	Gene Name	SD	CV (%)	Gene Name	SD	CV (%)	Gene Name	SD	CV (%)	Gene Name	SD	CV (%)
1	*UBC7*	0.06	0.24	*GAPDH*	0.09	0.48	*ACT7*	0.29	1.36	*UBC28*	0.60	2.76	*UBC28*	0.58	2.64	*PPR*	0.06	0.27
2	*helicase-15*	0.14	0.57	*TCTP*	0.18	0.83	*UBC28*	0.31	1.40	*GAPDH*	0.66	3.24	*TCTP*	0.59	2.73	*PAB2*	0.11	0.55
3	*EF2*	0.22	0.93	*ubiquinone*	0.21	0.91	*TCTP*	0.37	1.71	*TCTP*	0.71	3.27	*UBC7*	0.69	3.01	*EIF4A-3*	0.14	0.68
4	*PAB2*	0.22	1.01	*UBC7*	0.24	1.08	*EIF4α*	0.37	1.80	*UBC7*	0.73	3.21	*ACT7*	0.69	3.20	*GAPDH*	0.21	1.04
5	*ACT7*	0.27	1.28	*ACT7*	0.25	1.20	*ubiquinone*	0.39	1.69	*PPR*	0.80	3.56	*GAPDH*	0.69	3.40	*UBC36*	0.23	1.01
6	*GAPDH*	0.28	1.32	*EIF4A-3*	0.27	1.30	*PTB*	0.39	1.85	*ACT7*	0.80	3.67	*ubiquinone*	0.75	3.27	*TCTP*	0.25	1.15
7	*UBC28*	0.3	1.35	*UBC28*	0.30	1.37	*TUA*	0.45	1.97	*EIF4A-3*	0.88	4.11	*UBC36*	0.78	3.44	*UBC28*	0.25	1.15
8	*PPR*	0.34	1.38	*TUA*	0.35	1.54	*helicase-15*	0.46	1.93	*UBC36*	0.89	3.97	*EIF4A-3*	0.78	3.69	*ACT7*	0.25	1.19
9	*TUA*	0.35	1.51	*UBC36*	0.36	1.58	*UBC36*	0.50	2.17	*CYP20-2*	0.89	4.21	*EF2*	0.89	4.04	*ubiquinone*	0.27	1.16
10	*TCTP*	0.37	1.73	*EF2*	0.36	1.64	*EF2*	0.60	2.66	*ubiquinone*	0.91	3.98	*PPR*	0.95	4.17	*CYP20-2*	0.32	1.47
11	*ubiquinone*	0.46	1.93	*CYP20-2*	0.4	1.85	*PPR*	0.69	2.92	*EF2*	1.05	4.78	*CYP20-2*	0.97	4.51	*helicase-15*	0.34	1.45
12	*EIF4A-3*	0.54	2.63	*helicase-15*	0.42	1.79	*UBC7*	0.69	2.97	*PAB2*	1.06	5.34	*PAB2*	1.02	5.05	*EF2*	0.40	1.80
13	*UBC36*	0.58	2.46	*PAB2*	0.43	2.05	*GAPDH*	0.73	3.62	*helicase-15*	1.38	6.02	*helicase-15*	1.14	4.91	*UBC7*	0.63	2.74
14	*CYP20-2*	0.80	3.40	*PPR*	0.61	2.65	*CYP26-2*	0.96	4.33	*TUA*	1.59	7.25	*TUA*	1.32	5.93	*TUA*	1.10	4.79

Ranking	Cold 24 h			Cold			Heat 24 h			Stress Treatment			All Samples		
	Gene Name	SD	CV (%)	Gene Name	SD	CV (%)	Gene Name	SD	CV (%)	Gene Name	SD	CV (%)	Gene Name	SD	CV (%)
1	*EIF4A-3*	0.1	0.46	*PPR*	0.18	0.82	*UBC28*	0.22	1.03	*UBC28*	0.37	1.7	*UBC28*	0.51	2.32
2	*PPR*	0.14	0.65	*EIF4A-3*	0.19	0.91	*ubiquinone*	0.54	2.35	*PPR*	0.42	1.86	*TCTP*	0.56	2.57
3	*helicase-15*	0.3	1.29	*PAB2*	0.29	1.42	*TCTP*	0.77	3.5	*ubiquinone*	0.45	1.94	*GAPDH*	0.6	2.98
4	*TUA*	0.32	1.41	*UBC36*	0.3	1.29	*PPR*	0.8	3.5	*EIF4A-3*	0.47	2.24	*ubiquinone*	0.64	2.76
5	*PAB2*	0.33	1.64	*GAPDH*	0.32	1.55	*GAPDH*	0.82	3.98	*GAPDH*	0.49	2.39	*EIF4A-3*	0.66	3.14
6	*GAPDH*	0.35	1.69	*ubiquinone*	0.34	1.45	*EF2*	0.83	3.55	*TCTP*	0.51	2.35	*ACT7*	0.68	3.12
7	*UBC7*	0.36	1.54	*UBC28*	0.36	1.65	*PAB2*	0.91	4.39	*PAB2*	0.51	2.52	*UBC7*	0.71	3.08
8	*ubiquinone*	0.37	1.59	*TCTP*	0.37	1.71	*EIF4A-3*	0.91	4.44	*CYP20-2*	0.63	2.96	*UBC36*	0.72	3.16
9	*UBC36*	0.38	1.65	*helicase-15*	0.39	1.67	*CYP20-2*	0.95	4.38	*ACT7*	0.66	3.04	*EF2*	0.81	3.62
10	*UBC28*	0.42	1.87	*ACT7*	0.4	1.83	*ACT7*	1.03	4.63	*EF2*	0.68	2.99	*PAB2*	0.82	4.05
11	*EF2*	0.43	1.94	*EF2*	0.42	1.86	*UBC7*	1.09	4.5	*UBC7*	0.7	2.97	*PPR*	0.83	3.64
12	*TCTP*	0.51	2.3	*UBC7*	0.43	1.84	*helicase-15*	1.17	4.81	*UBC36*	0.72	3.09	*CYP20-2*	0.85	3.94
13	*ACT7*	0.53	2.43	*CYP20-2*	0.53	2.48	*TUA*	1.3	5.33	*helicase-15*	0.73	3.08	*helicase-15*	0.98	4.18
14	*CYP20-2*	0.66	3.11	*TUA*	0.67	2.93	*UBC36*	1.36	5.64	*TUA*	1.21	5.2	*TUA*	1.22	5.4

3.8. Comprehensive Stability Analysis Using RefFinder

Since the different algorithms determined different stability rankings for the candidate genes, the program RefFinder was used to calculate the geometric mean of the ranking results from the four methods [16]. The comprehensively ranked candidate genes (Table 4) did not present one or two universal reference genes for the normalization of gene expression data for all samples. However, the consensus for the top two genes in adult trees during leaf development, under cold treatment for 7 days or 24 h, and under heat treatment for 24 h was consistent with the results of the ΔCt and NormFinder analyses. The top two reference genes in seedlings or under temperature stress were consistent with the results of the ΔCt or geNorm analysis, respectively. In the other experimental conditions, the top two most stable genes in the overall ranking appeared in either the top two or top three positions in one of the other four algorithms. In addition, all analyses revealed that *TUA* was the most unstable gene. We then analyzed the top five most stable genes as determined using the ΔCt, geNorm, NormFinder, BestKeeper, and RefFinder algorithms (Figure 6). The top two most stable reference genes selected using RefFinder for the various experimental sets were ranked highly by three or four of the other software programs.

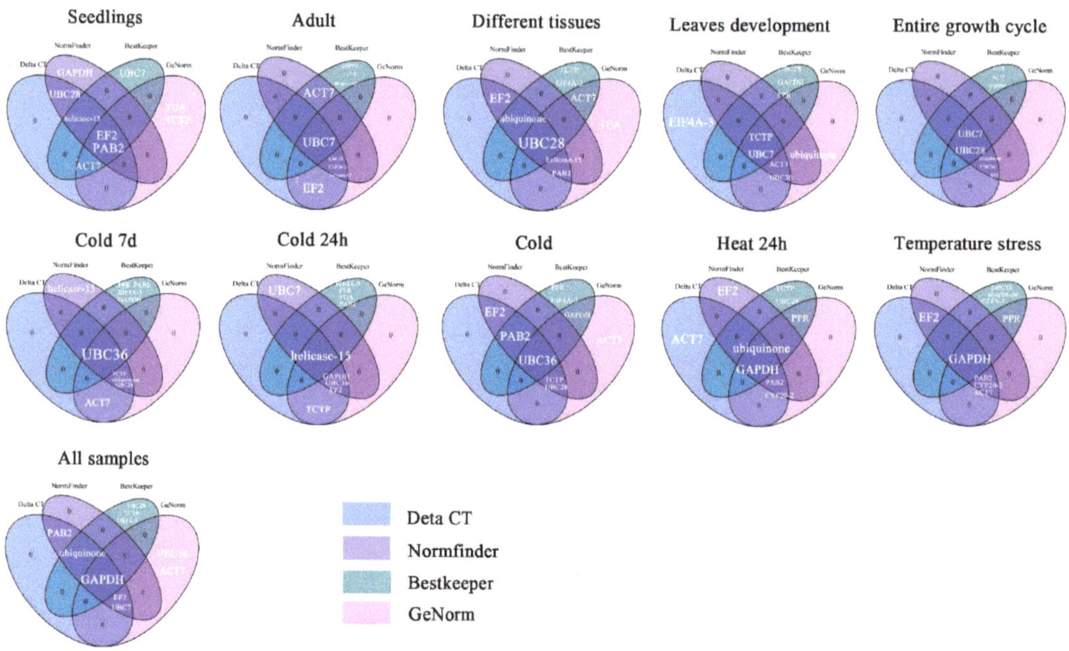

Figure 6. Venn diagrams displaying the overlap between the top five most stable reference genes as ranked by ΔCt, geNorm, NormFinder, and BestKeeper for the different datasets. Genes in the overlapping area were confirmed by more than one algorithm. The light purple, purple, light green and pink ellipses contain the top five most stable reference genes selected by ΔCt, geNorm, NormFinder, and BestKeeper, respectively.

Table 4. Ranking of expression stability of the 14 candidate reference genes in *L. megaphylla* using RefFinder analysis.

Ranking	Seedlings		Adult Tree		Different Tissues		Leaf Development		Entire Growth Cycle		Cold 7 d	
	Gene Name	Geomean of Ranking Values	Gene Name	Geomean of Ranking Values	Gene Name	Geomean of Ranking Values	Gene Name	Geomean of Ranking Values	Gene Name	Geomean of Ranking Values	Gene Name	Geomean of Ranking Values
1	*PAB2*	2.21	*UBC28*	2.21	*helicase-15*	1.68	*ACT7*	2.11	*ubiquinone*	2.21	*TCTP*	1.86
2	*helicase-15*	2.21	*UBC7*	2.21	*UBC28*	2.91	*UBC36*	2.45	*UBC7*	2.45	*ubiquinone*	2.71
3	*EF2*	2.78	*CYP20-2*	2.78	*PAB2*	3.34	*UBC7*	2.99	*UBC36*	3.13	*UBC36*	2.94
4	*UBC28*	4.09	*helicase-15*	4.09	*ACT7*	3.72	*TCTP*	3.00	*UBC28*	3.34	*UBC28*	4.53
5	*ACT7*	4.36	*GAPDH*	4.36	*ubiquinone*	3.98	*GAPDH*	5.29	*EF2*	3.46	*PPR*	4.74
6	*UBC7*	4.58	*ACT7*	4.58	*EF2*	4.68	*UBC28*	5.30	*GAPDH*	5.45	*PAB2*	4.74
7	*GAPDH*	5.58	*ubiquinone*	5.58	*TUA*	6.19	*ubiquinone*	5.89	*TCTP*	6.00	*ACT7*	5.57
8	*TCTP*	7.27	*EF2*	7.27	*UBC36*	7.97	*EIF4A-3*	5.96	*ACT7*	6.88	*EIF4A-3*	7.00
9	*TUA*	7.33	*TCTP*	7.33	*EIF4A-3*	9.30	*PAB2*	8.82	*PAB2*	7.90	*helicase-15*	7.26
10	*ubiquinone*	10.22	*UBC36*	10.22	*UBC7*	9.46	*CYP20-2*	9.69	*EIF4A-3*	10.02	*GAPDH*	9.12
11	*PPR*	10.38	*TUA*	10.38	*TCTP*	9.53	*EF2*	10.22	*helicase-15*	10.94	*EF2*	9.64
12	*EIF4A-3*	11.93	*EIF4A-3*	11.93	*GAPDH*	10.68	*PPR*	10.72	*CYP20-2*	11.24	*CYP20-2*	10.74
13	*UBC36*	12.49	*PAB2*	12.49	*PPR*	11.24	*helicase-15*	12.24	*PPR*	12.17	*UBC7*	13.00
14	*CYP20-2*	14.00	*PPR*	14.00	*CYP20-2*	12.98	*TUA*	14.00	*TUA*	14.00	*TUA*	14.00

Ranking	Cold 24 h		Cold		Heat 24 h		Stress Treatment		All Samples	
	Gene Name	Geomean of Ranking Values	Gene Name	Geomean of Ranking Values	Gene Name	Geomean of Ranking Values	Gene Name	Geomean of Ranking Values	Gene Name	Geomean of Ranking Values
1	*GAPDH*	1.57	*UBC36*	2.21	*PAB2*	2.30	*PAB2*	1.57	*ubiquinone*	2.21
2	*UBC36*	2.45	*TCTP*	3.13	*CYP20-2*	2.45	*PPR*	2.91	*EF2*	3.22
3	*helicase-15*	3.72	*UBC28*	3.60	*GAPDH*	2.59	*ACT7*	4.05	*UBC7*	3.64
4	*EF2*	4.15	*EF2*	4.03	*ubiquinone*	3.76	*GAPDH*	4.16	*GAPDH*	3.94
5	*UBC7*	6.40	*PAB2*	4.79	*PPR*	4.92	*EF2*	4.68	*UBC36*	4.12
6	*TUA*	6.51	*ACT7*	5.33	*ACT7*	6.16	*CYP20-2*	4.68	*PAB2*	5.01
7	*TCTP*	6.51	*PPR*	5.62	*EF2*	6.16	*UBC28*	6.85	*TCTP*	5.83
8	*PPR*	7.18	*GAPDH*	6.32	*UBC28*	7.24	*ubiquinone*	7.19	*ACT7*	5.86
9	*EIF4A-3*	7.24	*ubiquinone*	6.45	*TCTP*	7.95	*helicase-15*	8.17	*UBC28*	6.04
10	*UBC28*	7.36	*EIF4A-3*	7.50	*UBC7*	8.89	*TCTP*	8.18	*helicase-15*	9.58
11	*PAB2*	8.41	*helicase-15*	7.94	*EIF4A-3*	8.94	*EIF4A-3*	8.92	*EIF4A-3*	9.64
12	*ACT7*	9.87	*CYP20-2*	11.72	*helicase-15*	9.64	*UBC7*	10.22	*CYP20-2*	10.47
13	*ubiquinone*	10.84	*UBC7*	12.74	*UBC36*	12.47	*UBC36*	10.36	*PPR*	12.47
14	*CYP20-2*	13.24	*TUA*	14.00	*TUA*	13.00	*TUA*	14.00	*TUA*	14.00

3.9. Validation of Reference Genes

To verify the reliability of the selected reference genes, the transcripts of two genes were quantified using either different combinations of the two most stable genes, using a single reference gene or using the relatively unstable reference gene *TUA* under six different experimental conditions. The test genes were *NAC* and *ERF*, both of which show relatively high abundance levels, with fragments per kilobase of transcript per million fragments mapped (FPKM) values of approximately 22–223. The relative expression patterns and levels of the *LmNAC83* gene showed similar trends when the most stable genes were used alone or in combination as reference genes for standardization (Figure 7A–F). In contrast, the relative expression levels of *LmNAC83* showed significant fluctuations when the relatively unstable gene *TUA* was used for relative quantification. For example, under heat treatment for 24 h, the expression levels of *LmNAC83* in leaves was the highest at 30 °C when using the most stable genes as the reference genes. However, the relative expression of *LmNAC83* was low when using the unstable gene (*TUA*) as the reference gene (Figure 7F). In addition, the expression levels and trends of the *LmERF60* gene (Figure 8) were very similar to those found in the analysis of *LmNAC83*. It is evident that the use of unstable internal reference genes for gene expression analysis in *L. megaphylla* can lead to unreliable results. This test illustrates the importance of screening stable internal reference genes under different experimental conditions.

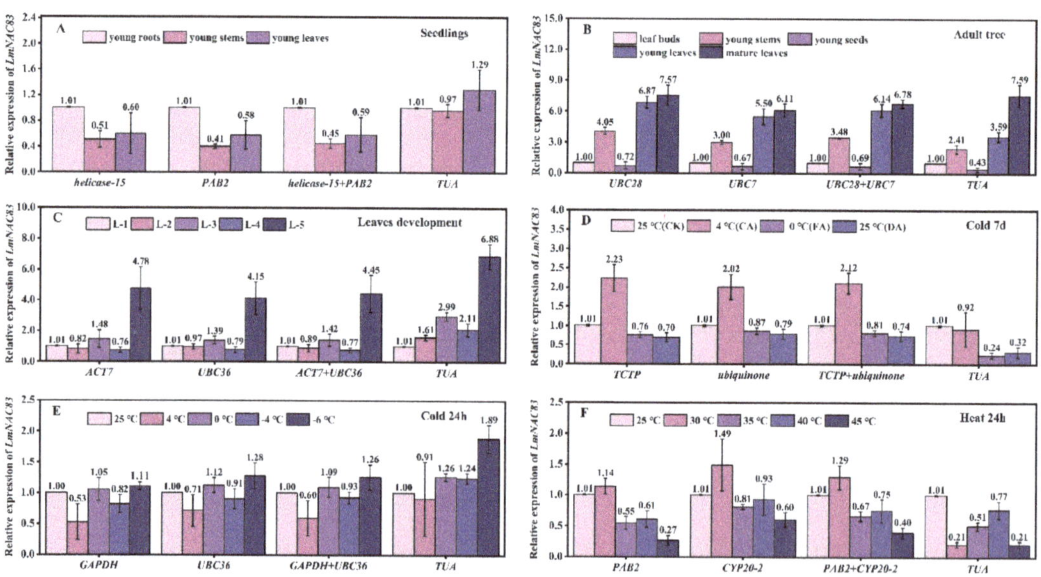

Figure 7. Relative expression of *LmNAC83* in six experimental sample sets. The results were normalized using the stable reference genes (alone or in combination) or an unstable gene in six sample sets, including (**A**) different seedling tissues; (**B**) different adult tree tissues; (**C**) different leaf developmental stages; (**D**) cold treatment for 7 days; (**E**) cold treatment for 24 h; and (**F**) heat treatment for 24 h. The bars indicate the standard deviation (±SD) evaluated from three biological replicates.

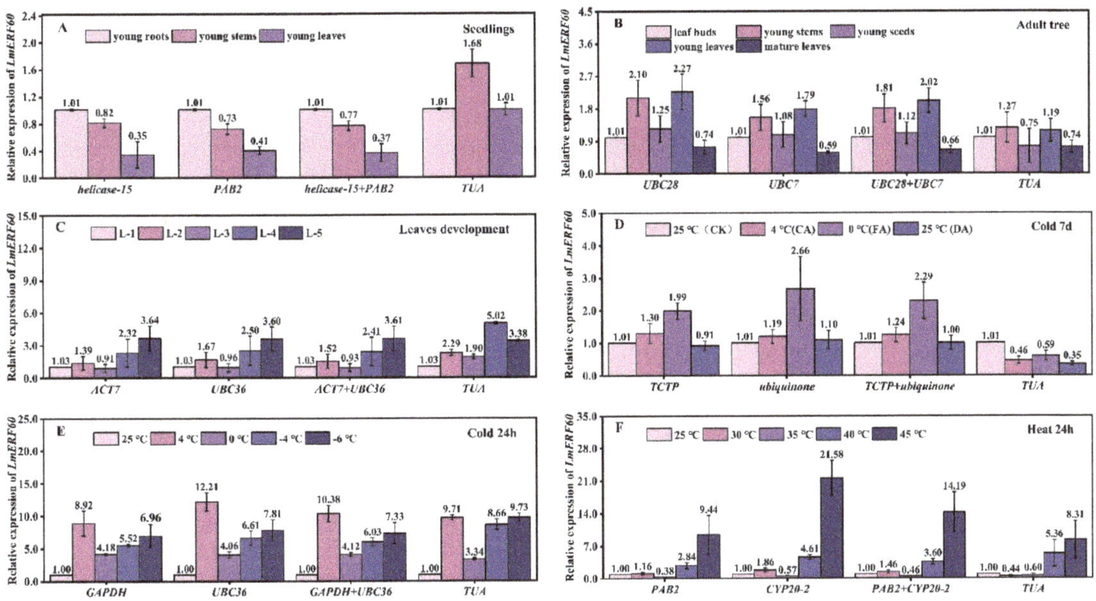

Figure 8. Relative expression of *LmERF60* in six experimental sample sets. The results were normalized using the stable reference genes (alone or in combination) or an unstable gene in six sample sets, including (**A**) different seedling tissues; (**B**) different adult tree tissues; (**C**) different leaf developmental stages; (**D**) cold treatment for 7 days; (**E**) cold treatment for 24 h; and (**F**) heat treatment for 24 h. The bars indicate the standard deviation (±SD) evaluated from three biological replicates.

4. Discussion

Changes in plant secondary metabolism are closely related to the transcriptional activities of key genes, and gene expression analysis is a key technique for understanding the mechanisms involved in these processes [32,49]. RT-qPCR is the most accurate technique to obtain gene expression profiles that relate to biological function and regulatory networks [50–52]. However, the accuracy of RT-qPCR results mainly depends on using optimal internal reference genes that are stably expressed in samples across different experimental conditions. Reference genes are crucial for normalization of gene expression data and avoiding experimental errors by minimizing non-biological variation between different samples [24,33,35]. To ensure the accuracy of experiments, it is important to select suitable reference genes for each species that similar transcript levels under different experimental conditions.

Lindera megaphylla is an ecologically important and dominant broad-leaved evergreen tree species that is naturally distributed in the warm-temperate and subtropical zones of China [1]. This tree contributes to the seasonal landscape and produces volatile compounds with strong effects on bacteria and toxic gases. For example, the terpenes produced have a strong antibacterial effect [6,10,53]. *L. megaphylla* is also used as a medicinal plant [2]. However, few studies have focused on the molecular biology of *L. megaphylla* due to limited genomic information, and to date, no reference genes have been reported. In this study, we obtained transcriptome databases of different tissues of *L. megaphylla* and identified appropriate internal reference genes for use when studying the expression of genes. We initially used 40 internal reference genes from published model plants such as *Arabidopsis thaliana* and other species in the Lauraceae family to screen for constitutively expressed reference genes from the transcriptome database of *L. megaphylla*. We obtained 20 candidate genes based on an FPKM value > 50. As shown in Table 2, the final 14 candidate genes were selected based on their E-values in the range of 91.04–107.17% and R2 values in the range of 0.991–0.999, which indicated that the primer pairs for standardized evaluation by RT-qPCR had high sensitivity and accuracy. In addition, the average Ct values of the candidate genes ranged from 17.94 (*PAB2*) to 27.51 (*UBC36*), indicating different expression levels (Figure 1). The results obtained are similar to those of many previous studies, such as on *Cryptomeria fortune* [33], *Gerbera hybrid* [19], and *Piper species* [48]. The results indicate that none of the reference genes had constant expression levels under all tested experimental conditions or in different species. Thus, it is necessary to carefully select the most appropriate reference gene to ensure the accuracy of RT-qPCR analysis for specific conditions and with specific materials. In this study, we combined four statistical algorithms (ΔCt, geNorm, NormFinder, and BestKeeper) to assess the expression stability of 14 candidate genes (*TUA, PPR, EIF4A-3, CYP26-2, helicase-15, TCTP, ACT7, PAB2, GAPDH, UBC28, EF2, UBC7, UBC36,* and *ubiquinone*) in different tissues across 16 different leaf developmental stages, and under different temperature stresses. The results demonstrated that the optimal reference genes were not the same under different conditions (Table 4).

There were slight differences in the rankings of candidate reference genes between the different algorithms. However, analysis by the ΔCt and NormFinder algorithms consistently identified the most stable or unstable candidate reference genes for most experimental sets, while in a few experimental groups, the expression patterns of similar genes were the most or least stable in geNorm and NormFinder. The ranking of candidate genes by BestKeeper suggested some differences compared to the other algorithms. For example, for different leaf developmental stages, the ΔCt and NormFinder platforms indicated that the *ACT7* and *UBC36* genes were the most stable. geNorm placed these genes in fourth and first place, while the BestKeeper program placed these genes in sixth and seventh place. Although the rankings of candidate genes produced by the different algorithms were slightly different, the top five stable candidate genes selected by the algorithms were similar for each group of experimental conditions (Table 3). For instance, *ACT7, UBC36, TCTP, UBC7,* and *ubiquinone* were the top five most stable genes based on the geNorm and NormFinder analyses across the leaf development stages, and the ΔCt analysis showed similar results, except for *ubiquinone*. BestKeeper analysis identified two stable genes: *UBC7* and *TCTP*. Numerous other studies have found similar differences between the outputs of geNorm and NormFinder [32,54], and many studies also demonstrated that these subtle differences result from the use of different algorithm models [33,34,55].

To comprehensively synthesize the results of the four algorithms, RefFinder was utilized to rank the identified candidate genes in *L. megaphylla*. This analysis plays an important role in integrating the screening results of reference genes from other algorithms by assigning an appropriate weight to each gene and calculating the geometric mean of its weights to produce a final ranking [32,34,35]. Fortunately, we found that the results from RefFinder were similar to those of the different algorithms in each experimental set, proving that RefFinder can assess and screen the optimal reference genes [56], as shown in Figure 6 and Table 4.

The results also indicate that we screened and identified the optimal reference gene combinations for use in *L. megaphylla* samples generated under different experimental conditions. To compare expression in different seedling tissues, *helicase-15* and *PAB2* were the most suitable, whereas *UBC28* and *UBC7* were most stably expressed in different adult trees tissues. When two different groups of tissues were analyzed, *helicase-15* and *UBC28* emerged as the most stable gene combination. For different leaf developmental stages, *ACT7* and *UBC36* were best, whereas *ubiquinone* and *UBC7* were best when analyzing samples over the entire growth cycle. Interestingly, for cold stress of 7 days and 24 h, the optimal reference genes were *TCTP* + *ubiquinone* and *GAPDH* + *UBC36*, respectively. For heat treatment, the best reference genes were *PAB2* and *CYP20-2*, while for overall temperature stress, *PAB2* and *PPR* were most stable. When all samples were tested, *ubiquinone, EF2, UBC7,* and *GAPDH* were the optimal candidate reference genes overall for the normalization of gene expression in *L. megaphylla* (Table 4). These analyses are sufficient to demonstrate the necessity of screening suitable internal reference genes under different experimental conditions for each species.

With increasing demand for accurate scientific data, it has become important to screen for the best internal reference genes in a greater number of plant species and for different experimental treatments [19,32,49]. Some housekeeping genes involved in cytoskeleton structure or primary metabolism, including *ACT*, *TUA*, and *EIF4α*, are extensively used as reference genes in many plant species. For example, *ACT2/7* and *TUA* are the three most stable genes across different developmental stages of *Glycine max* [57]. *AhyACT*, *AhyMDH*, and *AhyEF-1a* are the most stable genes in different tissues of amaranth [56]. Research on bamboo revealed that *eIF4α* was most stable in different organs, while *CYP*, *eEF1α*, and *UBQ5* were found to be the optimal reference genes for different developmental stages of *Bambusa tulda* [18]. In *Litsea cubeba*, *F-BOX*, *EF1α*, and *EIF4α* were the most stable reference genes across different tissues and developmental stages [40]. In the present study, *TUA* was the least stable candidate gene in *L. megaphylla*, as calculated by the different programs, while the best combinations of genes were *helicase-15* + *UBC28* and *ACT7* + *UBC36* in different tissues and developmental stages.

Zhong et al. [24] found that *ACT* was stably expressed in high-temperature-stressed *Psoralea corylifolia*. Chen et al. [22] indicated that *EF-1α* was the most stably expressed and suitable reference gene under heat and cold treatments. In this study, *PAB2* + *CYP20-2* and *UBC36* +*TCTP* were identified as the most stable reference genes under heat and cold treatments, respectively. The experimental results once again show the importance of screening the best reference genes under different experimental conditions for different species.

5. Conclusions

helicase-15 and *UBC28* can be used as internal reference genes when detecting gene expression pattern in different tissues of *L. megaphylla*. *ACT7* and *UBC36* were considered to be the most suitable internal reference genes at different leaf developmental stages of *L. megaphylla*. *UBC36*, *TCTP*, *PAB2*, and *CYP20* were optimum reference genes when *L. megaphylla* was suffering from abiotic stress. Among them, *UBC36* and *TCTP* were the best under cold treatment, while *PAB2* and *CYP20-2* were the best under heat treatment. We recommend that two internal reference genes be used to normalize RT-qPCR data in future experiments. In short, our results lay a foundation for future molecular biology research on *L. megaphylla*.

Supplementary Materials: The following supporting information can be downloaded at: https://www.mdpi.com/article/10.3390/plants12112185/s1, Figure S1: The PCR products electrophoresis result of the 20 candidate reference genes; Figure S2: Melting curves of the 20 candidate reference genes in *L. megaphylla*. Table S1: The cDNA sequences of the 20 candidate reference genes from *L. megaphylla*; Table S2: Primer sequences of candidate reference genes in *L. megaphylla*.

Author Contributions: H.L. and Y.L. conceived and designed the research; H.L., J.L. (Jing Liu), P.C., X.Z. and K.W. conducted the experiments; H.L. drafted the manuscript; Y.L., P.C. and J.L. (Jiuxing Lu) revised the manuscript. All authors have read and agreed to the published version of the manuscript.

Funding: The research was supported by the National Natural Science Foundation of China (no. 32201615), the Science and Technology Invigorating Forestry Foundation of He'nan Province (no. YLK202211), and the Key Technology R&D Program of He'nan Province (No. 232102111106).

Data Availability Statement: All data needed to evaluate the conclusions in this paper are present in the paper and/or the Supplementary Materials.

Acknowledgments: We thank the anonymous reviewers for their comments. The authors specifically thank Anita K. Snyder for language editing; their efforts greatly improved the manuscript. We also thank Ruimin Li for revising the manuscript.

Conflicts of Interest: The authors declare no conflict of interest.

References

1. Jiang, K.; Wang, Z.W.; Huang, W.C.; Hu, Y.H. Characterization of two complete chloroplast genomes of *Lindera megaphylla* (Lauraceae). *Mitochondrial DNA B Resour.* **2019**, *4*, 2851–2852. [CrossRef] [PubMed]
2. Cao, Y.; Xuan, B.; Peng, B.; Li, C.; Chai, X.; Tu, P. The genus Lindera: A source of structurally diverse molecules having pharmacological significance. *Phytochem. Rev.* **2015**, *15*, 869–906. [CrossRef]
3. Lin, H.F.; Huang, H.L.; Liao, J.F.; Shen, C.C.; Huang, R.L. Dicentrine analogue-induced G2/M arrest and apoptosis through inhibition of topoisomerase II activity in human cancer cells. *Planta Med.* **2015**, *81*, 830–837. [CrossRef] [PubMed]
4. Huang, R.L.; Chen, C.C.; Huang, Y.L.; Ou, J.C.; Hu, C.P.; Chen, C.F.; Chang, C. Anti-tumor effects of d-dicentrine from the root of Lindera megaphylla. *Planta Med.* **1998**, *64*, 212–215. [CrossRef] [PubMed]
5. Chou, C.J.; Lin, L.C.; Chen, K.T.; Chen, C.F. Northalifoline, a new isoquinolone alkaloid from the pedicels of *Lindera megaphylla*. *J. Nat. Prod.* **1994**, *57*, 689–694. [CrossRef]
6. Wei, S. Effects of Three Plant Volatiles on Negative Air Ions and Air Microorganisms. Master's Thesis, Zhejiang A&F University, Hangzhou, China, 2021. [CrossRef]
7. Dong, J.; Zhu, S.; Zhao, H.; Wang, J. The research progress on *Lindera megaphylla*. *J. Henan For. Sci. Technol.* **2021**, *41*, 12–15.
8. Wu, X.; Yuan, J.; Zhou, M.; Ding, Y.; Liu, M. Research on the overwintering survival of *Lindera megaphylla* in central China. *Anhui Agric. Sci. Bull.* **2017**, *23*, 106–108. [CrossRef]
9. Yang, J.; Ning, L.; Yang, L.; Wang, J.; Qian, Y.; Chen, Y. Structural characteristics of *Lindera megaphylla* wood and its volatile organic compounds. *J. Zhejiang AF Univ.* **2018**, *35*, 927–934. [CrossRef]
10. Bian, J.; Cheng, M.; Luo, S.; Chen, S.; Liu, S.; Bai, Z. GC/MS Analysis of volatile substances in essential oilof *Lindera megaphylla* blade and its application. *J. Southwest Univ. (Nat. Sci. Ed.)* **2014**, *36*, 82–88. [CrossRef]
11. Ding, Q.; Zhang, Y.; Liu, G.; Xu, Z.; Fei, Y. Effects of water stress on photosynthesis characteristics of cinnamomum camphora and *Lindera megaphylla* Seedlings. *J. Southwest For. Univ.* **2015**, *35*, 14–20. [CrossRef]
12. Teng, K.; Yang, Q.; Dai, G.; Ji, Y. The effect of low temperature stress on physiological and membrtine lipid peroxidatic reaction of *Lindera megaphylla* seedling. *J. Henan Agric. Univ.* **2002**, *36*, 151–154. [CrossRef]
13. Zhang, L.; Zhang, Q.; Jiang, Y.; Li, Y.; Zhang, H.; Li, R. Reference genes identification for normalization of qPCR under multiple stresses in *Hordeum brevisubulatum*. *Plant Methods* **2018**, *14*, 110. [CrossRef]
14. Lin, Y.; Zhang, A.; Yang, S.; Huang, L. Reference gene selection for real-time quantitative PCR normalization in *Hemarthria compressa* and *Hemarthria altissima* leaf tissue. *Mol. Biol. Rep.* **2019**, *46*, 4763–4769. [CrossRef] [PubMed]
15. Fleige, S.; Pfaffl, M.W. RNA integrity and the effect on the real-time qRT-PCR performance. *Mol. Asp. Med.* **2006**, *27*, 126–139. [CrossRef] [PubMed]
16. Song, H.; Mao, W.; Duan, Z.; Que, Q.; Zhou, W.; Chen, X.; Li, P. Selection and validation of reference genes for measuring gene expression in Toona ciliata under different experimental conditions by quantitative real-time PCR analysis. *BMC Plant Biol.* **2020**, *20*, 450. [CrossRef]
17. Czechowski, T.; Stitt, M.; Altmann, T.; Udvardi, M.K.; Scheible, W.R. Genome-wide identification and testing of superior reference genes for transcript normalization in Arabidopsis. *Plant Physiol.* **2005**, *139*, 5–17. [CrossRef]
18. Chakraborty, S.; Dutta, S.; Biswas, P.; Das, M. Identification of candidate reference genes in tropical bamboos stable across species, tissues, and developmental stages. *Biol. Plant.* **2019**, *63*, 253–261. [CrossRef]
19. Li, F.; Cheng, Y.; Ma, L.; Li, S.; Wang, J. Identification of reference genes provides functional insights into meiotic recombination suppressors in Gerbera hybrida. *Hortic. Plant J.* **2022**, *8*, 123–132. [CrossRef]
20. Qu, R.; Miao, Y.; Cui, Y.; Cao, Y.; Zhou, Y.; Tang, X.; Yang, J.; Wang, F. Selection of reference genes for the quantitative real-time PCR normalization of gene expression in *Isatis indigotica* fortune. *BMC Mol. Biol.* **2019**, *20*, 9. [CrossRef]
21. Ragni, E.; Vigano, M.; Rebulla, P.; Giordano, R.; Lazzari, L. What is beyond a qRT-PCR study on mesenchymal stem cell differentiation properties: How to choose the most reliable housekeeping genes. *J. Cell Mol. Med.* **2013**, *17*, 168–180. [CrossRef]
22. Chen, M.; Wang, B.; Li, Y.; Zeng, M.; Liu, J.; Ye, X.; Zhu, H.; Wen, Q. Reference gene selection for qRT-PCR analyses of luffa (*Luffa cylindrica*) plants under abiotic stress conditions. *Sci. Rep.* **2021**, *11*, 3161. [CrossRef]
23. Umadevi, P.; Suraby, E.J.; Anandaraj, M.; Nepolean, T. Identification of stable reference gene for transcript normalization in black pepper-Phytophthora capsici pathosystem. *Physiol. Mol. Biol. Plants* **2019**, *25*, 945–952. [CrossRef] [PubMed]
24. Zhong, Y.; Gai, Y.; Gao, J.; Nie, W.; Bao, Z.; Wang, W.; Xu, X.; Wu, J.; He, Y. Selection and validation of reference genes for quantitative real-time PCR normalization in *Psoralea corylifolia* (Babchi) under various abiotic stress. *J. Plant Physiol.* **2022**, *274*, 153722. [CrossRef] [PubMed]
25. Sheng, Z.; Luan, Y.; Xu, C.; Tao, J.; Zhao, D. Selection of Stable Reference Genes for Quantitative Real-Time PCR on Herbaceous Peony (*Paeonia lactiflora* Pall.) in Response to Drought Stress. *J. Oceanol. Limnol.* **2023**, *3*, 801–814. [CrossRef]
26. Yao, Y.; Wang, X.; Chen, B.; Zheng, S.; Wang-Pruski, G.; Chen, X.; Guo, R. Evaluation of Reference Genes Suitable for Gene Expression during Root Enlargement in Cherry Radish Based on Transcriptomic Data. *Horticulturae* **2023**, *9*, 20. [CrossRef]
27. Silver, N.; Best, S.; Jiang, J.; Thein, S.L. Selection of housekeeping genes for gene expression studies in human reticulocytes using real-time PCR. *BMC Mol. Biol.* **2006**, *7*, 33. [CrossRef]
28. Vandesompele, J.; De Preter, K.; Pattyn, F.; Poppe, B.; Van Roy, N.; De Paepe, A.; Speleman, F. Accurate normalization of real-time quantitative RT-PCR data by geometric averaging of multiple internal control genes. *Genome Biol.* **2002**, *3*, research0034.1. [CrossRef]

29. Andersen, C.L.; Jensen, J.L.; Orntoft, T.F. Normalization of real-time quantitative RT-PCR data: A model based variance estimation approach to identify genes suited for normalization applied to bladder and colon cancer data sets. *Cancer Res.* **2004**, *64*, 5245–5250. [CrossRef]
30. Pfaffl, M.W.; Tichopad, A.; Prgomet, C.; Neuvians, T.P. Determination of stable housekeeping genes, differentially regulated target genes and sample integrity: BestKeeper—Excelbased tool using pair-wise correlations. *Biotechnol. Lett.* **2004**, *26*, 509–515. [CrossRef]
31. Xie, F.; Xiao, P.; Chen, D.; Xu, L.; Zhang, B. miRDeepFinder: A miRNA analysis tool for deep sequencing of plant small RNAs. *Plant Mol. Biol.* **2012**, *80*, 75–84. [CrossRef]
32. Dudziak, K.; Sozoniuk, M.; Szczerba, H.; Kuzdralinski, A.; Kowalczyk, K.; Borner, A.; Nowak, M. Identification of stable reference genes for qPCR studies in common wheat (*Triticum aestivum* L.) seedlings under short-term drought stress. *Plant Methods* **2020**, *16*, 58. [CrossRef] [PubMed]
33. Zhang, Y.; Xue, J.; Zhu, L.; Hu, H.; Yang, J.; Cui, J.; Xu, J. Selection and optimization of reference genes for microRNA expression normalization by qRT-PCR in chinese cedar (*Cryptomeria fortunei*) under multiple stresses. *Int. J. Mol. Sci.* **2021**, *22*, 7246. [CrossRef] [PubMed]
34. Zhang, Y.; Zhang, Z.; Ren, M.; Liu, X.; Zhou, X.; Yang, J. Selection of reference genes for RT-qPCR analysis in the hawthorn spider mite, amphitetranychus viennensis (Acarina: Tetranychidae), under acaricide treatments. *J. Econ. Entomol.* **2022**, *115*, 662–670. [CrossRef] [PubMed]
35. Wu, Y.; Zhang, C.; Yang, H.; Lyu, L.; Li, W.; Wu, W. Selection and validation of candidate reference genes for gene expression analysis by RT-qPCR in rubus. *Int. J. Mol. Sci.* **2021**, *22*, 10533. [CrossRef] [PubMed]
36. Chen, X.; Chen, H.; Yuan, J.S.; Kollner, T.G.; Chen, Y.; Guo, Y.; Zhuang, X.; Chen, X.; Zhang, Y.; Fu, J.; et al. The rice terpene synthase gene *OsTPS19* functions as an (S)-limonene synthase in planta, and its overexpression leads to enhanced resistance to the blast fungus *Magnaporthe oryzae*. *Plant Biotechnol. J.* **2018**, *16*, 1778–1787. [CrossRef]
37. Qi, C.; Jiang, H.; Zhao, X.; Mao, K.; Liu, H.; Li, Y.; Hao, Y. The characterization, authentication, and gene expression pattern of the mdcer family in malus domestica. *Hortic. Plant J.* **2019**, *5*, 1–9. [CrossRef]
38. Mergaert, P.; Kereszt, A.; Kondorosi, E. Gene expression in nitrogen-fixing symbiotic nodule cells in *Medicago truncatula* and other nodulating plants. *Plant Cell* **2020**, *32*, 42–68. [CrossRef]
39. Kiseleva, A.A.; Bragina, M.K.; Muterko, A.F.; Salina, E.A. Functional characterization of genes with daily expression patterns in common wheat. *Plant Mol. Biol.* **2022**, *109*, 135–146. [CrossRef]
40. Lin, L.; Han, X.; Chen, Y.; Wu, Q.; Wang, Y. Identification of appropriate reference genes for normalizing transcript expression by quantitative real-time PCR in *Litsea cubeba*. *Mol. Genet. Genom.* **2013**, *288*, 727–737. [CrossRef]
41. Zhao, M.; Jin, J.; Gao, T.; Zhang, N.; Jing, T.; Wang, J.; Ban, Q.; Schwab, W.; Song, C. Glucosyltransferase CsUGT78A14 regulates flavonols accumulation and reactive oxygen species scavenging in response to cold stress in *Camellia sinensis*. *Front Plant Sci.* **2019**, *10*, 1675. [CrossRef]
42. Yang, N.; Liu, J.; Gao, Q.; Gui, S.; Chen, L.; Yang, L.; Huang, J.; Deng, T.; Luo, J.; He, L.; et al. Genome assembly of a tropical maize inbred line provides insights into structural variation and crop improvement. *Nat. Genet.* **2019**, *51*, 1052–1059. [CrossRef] [PubMed]
43. Nicot, N.; Hausman, J.F.; Hoffmann, L.; Evers, D. Housekeeping gene selection for real-time RT-PCR normalization in potato during biotic and abiotic stress. *J. Exp. Bot.* **2005**, *56*, 2907–2914. [CrossRef] [PubMed]
44. Hu, X.; Xie, F.; Liang, W.; Liang, Y.; Zhang, Z.; Zhao, J.; Hu, G.; Qin, Y. *HuNAC20* and *HuNAC25*, two novel NAC genes from pitaya, confer cold tolerance in transgenic *Arabidopsis*. *Int. J. Mol. Sci.* **2022**, *23*, 2189. [CrossRef] [PubMed]
45. Mijiti, M.; Wang, Y.; Wang, L.; Habuding, X. *Tamarix hispida* NAC transcription factor *ThNAC4* confers salt and drought stress tolerance to transgenic *Tamarix* and *Arabidopsis*. *Plants* **2022**, *11*, 2647. [CrossRef]
46. Srivastava, R.; Kobayashi, Y.; Koyama, H.; Sahoo, L. Cowpea NAC1/NAC2 transcription factors improve growth and tolerance to drought and heat in transgenic cowpea through combined activation of photosynthetic and antioxidant mechanisms. *J. Integr. Plant Biol.* **2022**, *65*, 25–44. [CrossRef]
47. Yu, Y.; Yu, M.; Zhang, S.; Song, T.; Zhang, M.; Zhou, H.; Wang, Y.; Xiang, J.; Zhang, X. Transcriptomic identification of wheat AP2/ERF transcription factors and functional characterization of *TaERF-6-3A* in response to drought and salinity stresses. *Int. J. Mol. Sci.* **2022**, *23*, 3272. [CrossRef]
48. Huang, Y.; Liu, L.; Hu, H.; Tang, N.; Shi, L.; Xu, F.; Wang, S. *Arabidopsis ERF012* is a versatile regulator of plant growth, development and abiotic stress responses. *Int. J. Mol. Sci.* **2022**, *23*, 6841. [CrossRef]
49. De Oliveira, L.F.; Piovezani, A.R.; Ivanov, D.A.; Yoshida, L.; Segal Floh, E.I.S.; Kato, M.J. Selection and validation of reference genes for measuring gene expression in Piper species at different life stages using RT-qPCR analysis. *Plant Physiol. Biochem.* **2022**, *171*, 201–212. [CrossRef]
50. Zhang, L.; Chen, F.; Zhang, X.; Li, Z.; Zhao, Y.; Lohaus, R.; Chang, X.; Dong, W.; Ho, S.Y.W.; Liu, X.; et al. The water lily genome the early evolution of flowering plants. *Nature* **2020**, *577*, 79–84. [CrossRef]
51. Exelby, K.; Herrera-Delgado, E.; Perez, L.G.; Perez-Carrasco, R.; Sagner, A.; Metzis, V.; Sollich, P.; Briscoe, J. Precision of tissue patterning is controlled by dynamical properties of gene regulatory networks. *Development* **2021**, *148*, dev197566. [CrossRef] [PubMed]

52. Pierandrei, S.; Truglio, G.; Ceci, F.; Del Porto, P.; Bruno, S.M.; Castellani, S.; Conese, M.; Ascenzioni, F.; Lucarelli, M. DNA methylation patterns correlate with the expression of *SCNN1A*, *SCNN1B*, and *SCNN1G* (Epithelial Sodium Channel, ENaC) genes. *Int. J. Mol. Sci.* **2021**, *22*, 3754. [CrossRef] [PubMed]
53. Abbas, F.; Ke, Y.; Yu, R.; Yue, Y.; Amanullah, S.; Jahangir, M.M.; Fan, Y. Volatile terpenoids: Multiple functions, biosynthesis, modulation and manipulation by genetic engineering. *Planta* **2017**, *246*, 803–816. [CrossRef]
54. Elbl, P.; Navarro, B.V.; de Oliveira, L.F.; Almeida, J.; Mosini, C.; dos Santos, A.L.W.; Rossi, M.; Floh, E.I.S. Identification and evaluation of reference genes for quantitative analysis of Brazilian pine (*Araucaria angustifolia* Bertol. Kuntze) gene expression. *PLoS ONE* **2015**, *10*, e0136714. [CrossRef] [PubMed]
55. Khan, M.M.; Guo, C.F.; Peng, J.; Fan, Z.Y.; Hafeez, M.; Ali, D.; Wang, K.; Almarzoug, M.H.A.; Qiu, B.L. Screening and validation of reference genes using in RT-qPCR for gene expression studies in *Paederus fuscipes*, a medically and agriculturally important insect. *J. King Saud Univ. Sci.* **2022**, *34*, 101654. [CrossRef]
56. Hernandez, F.P.V.; Nunez, M.M.; Rivas, M.R.; Portillo, R.E.V.; Martinez, M.D.B.; Suarez, S.L.; Cardenas, F.D.R. Reference genes for RT-qPCR normalisation in different tissues, developmental stages and stress conditions of amaranth. *Plant Biol.* **2018**, *20*, 713–721. [CrossRef] [PubMed]
57. Jian, B.; Liu, B.; Bi, Y.; Hou, W.; Wu, C.; Han, T. Validation of internal control for gene expression study in soybean by quantitative real-time PCR. *BMC Mol. Biol.* **2008**, *9*, 59. [CrossRef]

plants

MDPI

Article

Establishing a Virus-Induced Gene Silencing System in *Lycoris chinensis*

Guanghao Cheng [1,2,3,†], Xiaochun Shu [1,2,3,†], Zhong Wang [1,2,3], Ning Wang [1,2,3] and Fengjiao Zhang [1,2,3,*]

1 Institute of Botany, Jiangsu Province and Chinese Academy of Sciences, Nanjing 210014, China; chengghao@yeah.net (G.C.); sxc@cnbg.net (X.S.); wangzhong@cnbg.net (Z.W.); wangning813@njau.edu.cn (N.W.)
2 Jiangsu Key Laboratory for the Research and Utilization of Plant Resources, Nanjing 210014, China
3 Nanjing Botanical Garden Mem. Sun Yat-Sen, Nanjing 210014, China
* Correspondence: fengjiao@cnbg.net
† These authors contributed equally to this work.

Abstract: *Lycoris* is an important plant with both medicinal and ornamental values. However, it does not have an efficient genetic transformation system, which makes it difficult to study gene function of the genus. Virus-induced gene silencing (VIGS) is an effective technique for studying gene functions in plants. In this study, we develop an efficient virus-induced gene-silencing (VIGS) system using the leaf tip needle injection method. The widely used TRV vector is constructed, and the *Cloroplastos Alterados 1* (*CLA1*) and *Phytoene Desaturase* (*PDS*) genes are selected as visual indicators in the VIGS system. As a result, it is observed that leaves infected with TRV-*LcCLA1* and TRV-*LcPDS* both show a yellowing phenotype (loss of green), and the chlorosis range of TRV-*LcCLA1* was larger and deeper than that of TRV-*LcPDS*. qRT-PCR results show that the expression levels of *LcCLA1* and *LcPDS* are significantly reduced, and the silencing efficiency of *LcCLA1* is higher than that of *LcPDS*. These results indicate that the VIGS system of *L. chinensis* was preliminarily established, and *LcCLA1* is more suitable as a gene-silencing indicator. For the monocotyledonous plant leaves with a waxy surface, the leaf tip injection method greatly improves the infiltration efficiency. The newly established VIGS system will contribute to gene functional research in *Lycoris* species.

Keywords: *Lycoris*; VIGS system; leaf tip injection; infiltration efficiency; indicator genes

Citation: Cheng, G.; Shu, X.; Wang, Z.; Wang, N.; Zhang, F. Establishing a Virus-Induced Gene Silencing System in *Lycoris chinensis*. *Plants* **2023**, *12*, 2458. https://doi.org/10.3390/plants12132458

Academic Editors: Qian-Hao Zhu, Aiping Song and Yu Chen

Received: 1 June 2023
Revised: 21 June 2023
Accepted: 25 June 2023
Published: 27 June 2023

1. Introduction

Lycoris belongs to the Amaryllidaceae family and consists of around 20 distinct species worldwide. They are native to Eastern Asia and distributed in moist warm temperature areas, especially in China, Japan and Korea [1]. Most species are valued for their stunningly vibrant and distinct coloration and striking blooms. As an important class of bulbous flowers, they also have low environmental requirements and strong adaptability, making them a good choice for landscaping [2]. In addition, the bulbs contain a large number of alkaloid compounds, which have anti-malarial and anti-tumor effects and are a treatment for senile dementia [3,4], so the plants also have important medicinal value. Due to the great ornamental and medicinal value of *Lycoris*, more and more transcriptomic data have been provided, and functional genes related to excellent breeding traits have been identified [5–8]. However, due to the lack of efficient genetic transformation, there has been no report on the gene function of *Lycoris* thus far. Moreover, *Lycoris* is a perennial bulbous flower; it takes 3–5 years to grow from a small bulb to a flowering plant [9]. In the study of the functional genes of flowering traits, a long period is always needed to observe the phenotype after conventional genetic transformation. Therefore, it is necessary to establish an efficient, rapid and appropriate transformation system to promote research on the molecular regulation mechanism of this genus.

Virus-induced gene silencing (VIGS) is an excellent alternative to obtaining information about gene function by transiently knocking out the gene of interest [10]. It is a post-transcriptional gene silencing (PTGS)-based technology that utilizes natural defense mechanisms employed by plants to protect against invading viruses [11]. Virus-infected plants induce double-stranded RNA-mediated PTGS, which degrades viral RNA [12]. Recombinant viruses carrying partial sequences of host target genes are used to infect and spread throughout the plant [13]. Viral gene transcripts and plant target genes are recognized and degraded by endogenous PTGS, resulting in reduced target gene expression [14]. Compared with other transgenic technologies, VIGS technology can avoid plant transformation and has the advantages of short cycle, low cost and simple operation. At present, it has been applied to a variety of plants for gene function verification, such as Arabidopsis [15], tomato [11], pepper [16], *Lilium* [17], etc. VIGS has great advantages, especially in perennial woody plants and perennial herbaceous plants. In *Vernicia fordii*, VIGS can shorten the time to phenotype observation and identify phenotypes after loss-of-function of a gene of interest within a single generation [14]. For herbaceous plants, the transformation operation of VIGS is simple and fast and can function in different genetic backgrounds [12].

Many viruses have been used to develop VIGS vectors, such as tobacco mosaic virus (TMV) [18], potato virus X (PVX) [19], tomato golden mosaic virus (TGMV) [20], tobacco rattle virus (TRV) [21] and apple latent spherical virus (ALSV) [22]. Among these viral vectors, TRV has been widely used to construct VIGS vectors for silencing target genes in various bulbous perennial flowers; for example, VIGS experiments using TRV vectors in lily petals showed that anthocyanin accumulation was reduced when *LvMYB5* was silenced [23]. After silencing *NtPDS* with TRV vector in Chinese narcissus, extensive chlorosis of leaves was found [24]. However, it is unclear whether TRV vector-based VIGS can be used to reveal gene function in *Lycoris*, and there is no suitable VIGS system established by now.

Cloroplastos Alterados 1 (*CLA1*) and *Phytoene Desaturase* (*PDS*) are the most commonly used indicator genes for VIGS system establishment. The *CLA1* gene is involved in chloroplast development and has shown a highly pronounced albino phenotype, so the silenced *CLA1* serves as a useful marker to determine silencing efficiency [25]. *CLA1* has mainly been used in cotton as a positive control for VIGS [26]. When the albino phenotype is observed, it indicates that the gene has been silenced. To determine whether *HyPRP1* is required for cotton resistance to Verticillium wilt, the marker gene *CLA1*-VIGS plants are used to determine the silencing efficiency of *HyPRP1* [27]. In upland cotton, using TRV-*CLA1* as a control, it was found that silencing of *GhCLCg-1* resulted in impaired salt tolerance [28]. It has also been successfully used in other species, such as *Arabidopsis* [14] and *Vernicia fordii* [29]. The gene *PDS* also produces an albino phenotype and is widely used as an indicator gene for the VIGS system [30]. The TRV-*PDS* system has been successfully applied in many ornamental plants. For example, in tree peonies, a typical albino phenotype was found in the newly sprouted top leaves of TRV-*PoPDS*-infected triennial tree peony seedlings using leaf syringe infiltration and seedling vacuum infiltration [31]. This shows that TRV-based VIGS technology can be applied to the high-throughput functional characterization of tree peony genes. In *Lilium* × *formolongi* [17], using the inoculation method of rubbing plus injection, albino was observed in newly developing leaves of TRV-*LhPDS*-infected lily seedlings 56 days after infiltration. In *Solanum pseudocapsicum* L. [32], the TRV-*SpPDS* system was used to infect leaves, and obvious albino was found. In *Catharanthus roseus* [33], the TRV-*CrPDS* system was used to infect roots, stems, leaves, and flowers, and it was found that all tissues showed obvious albino, and the phenotypes of leaves and flowers were more obvious. Nishii et al. used a TRV vector with a wide host range to silence the *SrPDS* gene in *Streptocarpus rexii* through *Agrobacterium* inoculation, and finally obtained silenced plants with albino phenotypes [34]. These examples demonstrate the wide application of the TRV-PDS system. However, the *PDS* gene is not only involved in chlorophyll content but also in carotenoid biosynthesis [35]. For example, in highbush blueberry, which is rich in polyphenols and anthocyanins, a phenotype could occur without chlorophyll, but with red coloration after knockout of the *PDS* gene [36]. It suggests that the chlorophyll-lacking phenotype is varied in different species.

The leaves of *Lycoris* plants are linear and have the characteristics of longitudinal and orderly growth of vascular tissue. The surface of the leaves is covered with a thin waxy layer, and thus it is not easy to infiltrate the bacterial solution with the traditional infiltration method. In this study, we developed a leaf tip injection method, constructed TRV vectors using *LcCLA1* and *LcPDS* as reporter genes, and tested the feasibility of the TRV-VIGS system in spring-leafed *L. chinensis*, in which the young leaves emerge from the bulb in early spring when they are suitable for injection. Two weeks after infection, the leaves of *L. chinensis* injected with *LcCLA1* and *LcPDS* showed chlorosis, but the phenotype of leaves injected with *LcPDS* was not as obvious as that injected with *LcCLA1*. The expression level of *LcCLA1* in the chlorotic leaves was significantly lower than that in non-injected leaves. In terms of gene expression, the expression levels of *LcCLA1* and *LcPDS* in chlorotic leaves were significantly lower than those in uninfected leaves, while the gene expression of *LcCLA1* was lower, which indicated that *LcCLA1* was more suitable as an indicator gene in gene function studies. Therefore, TRV-*LcCLA1* can silence the genes in *L. chinensis* more effectively than the TRV-*LcPDS* VIGS system, which lays a good foundation for the gene function verification of *L. chinensis* in the future.

2. Results

2.1. Comparison of Agrobacterium Infection Efficiency of Different Infiltration Methods

In this study, we used the tip needle injection method, which requires 1–2 mL bacterial solution and takes 15–20 s to infiltrate a whole leaf (Figure 1a), compared with the conventional leaf infiltration method, which commonly requires at least 5 mL of bacterial solution and takes 1–2 min to infiltrate a leaf (Figure 1b). The reason for such a large difference is the waxy layer on the surface of the leaves of the *Lycoris*. It is difficult to inject *Agrobacterium* solution into the entire leaf when infiltrating. To infiltrate the whole leaf, multiple wounds need to be created. In addition, the infiltration process is very easy to cause the loss of *Agrobacterium* solution, and more wounds are not conducive to normal plant growth and development, thus affecting the experimental results. The tip needle injection method is an easier way to infiltrate the whole leaf, as well as being more efficient, simple to operate, and able to save more solution.

Figure 1. Comparison of tip needle injection method and traditional injection infiltration method. (**a**) Tip needle injection method to infect leaves of *L. chinensis*. (**b**) Traditional infiltration method to infect leaves of *L. chinensis*. The red oval area shows the infection solution that was successfully injected into the leaf.

2.2. Silencing Efficiency of LcCLA1 and LcPDS in L. chinensis after Infection

Two weeks after injection, an obvious yellow leaf phenotype was observed in the leaves of *L. chinensis* after injection of pTRV2-*LcCLA1*. The difference is that the leaves injected with pTRV2-*LcPDS* infection solution had less yellow leaf phenotype. Injection efficiency was detected using PCR, and leaves not injected with *Agrobacterium* (CK), pTRV1+pTRV2 (Mock), pTRV1+pTRV2-*LcCLA1* and pTRV1+pTRV2-*LcPDS* were selected for PCR detection. The results showed that no band was seen in CK, while pTRV1+pTRV2 and pTRV1+pTRV2-*LcCLA1* had a clear band at 260 bp (Figure 2a). Similarly, for the injection of *LcPDS*, no band was observed in the CK region, but about 240 bp bands were observed in pTRV1+pTRV2 and pTRV1+pTRV2-*LcPDS* (Figure 2b). This indicates that the tip needle injection method was successful in achieving infection of the leaves of *L. chinensis*.

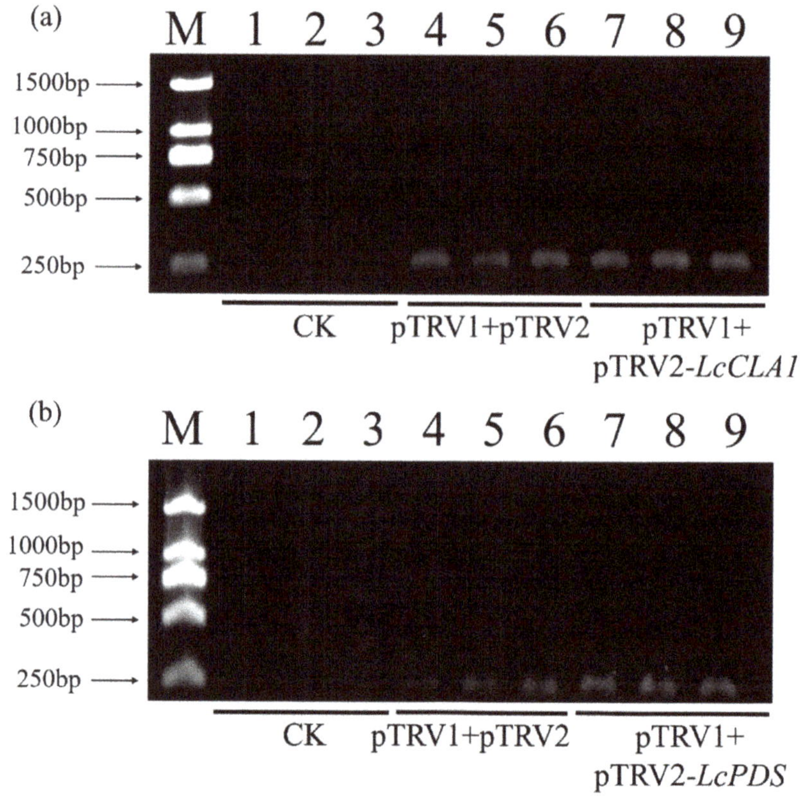

Figure 2. Testing of *LcCLA1* and *LcPDS* injection efficiency. M: marker; 1–3: leaves without injection; 4–6: leaves injected with pTRV1 and pTRV2; 7–9 (**a**): leaves injected with pTRV1 and pTRV2-*LcCLA1*: 7–9 (**b**): leaves injected with pTRV1 and pTRV2-*LcPDS*.

From the perspective of leaf phenotype, plants inoculated with pTRV1 and pTRV2 showed no significant difference in leaf morphology compared to CK (Figure 3a,b). In comparison with CK, leaves injected with pTRV1+pTRV2-*LcCLA1* were found to show a distinct etiolation phenotype (Figure 3c). However, leaves injected with pTRV1+pTRV2-*LcPDS* showed a weaker etiolation phenotype, which was less pronounced than that of leaves injected with pTRV1+pTRV2-*LcCLA1*. (Figure 3d). Phenotypic differences suggest that the expression of the gene *LcCLA1* may be repressed in leaves infiltrated by pTRV2-*LcCLA1*. The expression level of *LcPDS* might not be completely suppressed because the leaves showed incomplete chlorotic phenotype. The reason for this discrepancy is unclear, but it is speculated that *LcPDS* may have a lower density of *Agrobacterium* and less function in leaves.

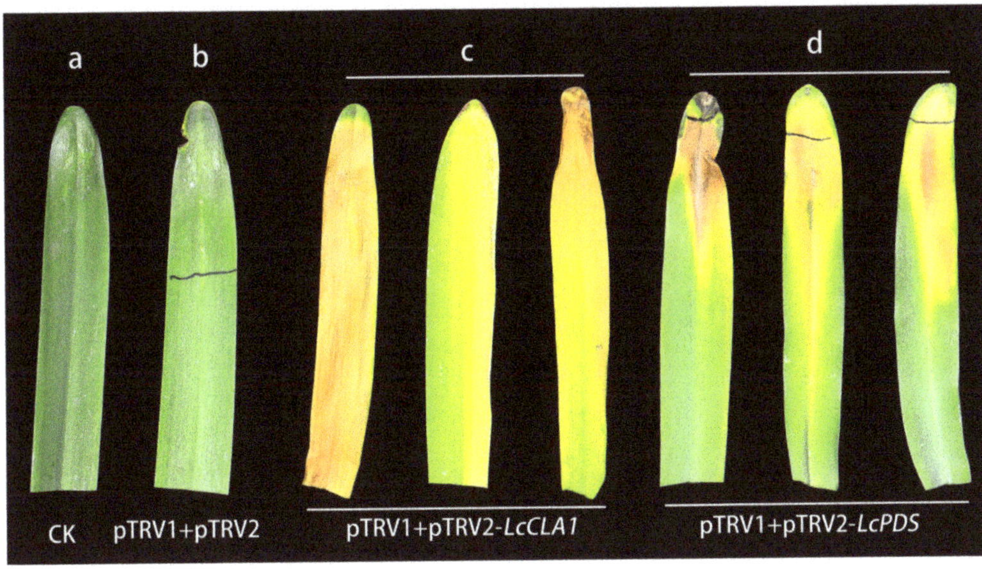

Figure 3. Silencing of *LcCLA1* and *LcPDS* in *L. chinensis* using TRV VIGS vector. (**a**) Leaf of *L. chinensis* without *Agrobacterium* injected as a control (CK). (**b**) Leaf of *L. chinensis* injected with pTRV1 and pTRV2. (**c**) Leaves of *L. chinensis* injected with pTRV1 and pTRV2-*LcCLA1* showed etiolation phenotype after two weeks. (**d**) Leaves of *L. chinensis* injected with pTRV1 and pTRV2-*LcPDS* showed less etiolation phenotype after two weeks.

2.3. qRT-PCR Analysis of *LcCLA1* and *LcPDS* Silencing Level

After detection of viral infection in *L. chinensis* leaves, we next analyzed gene expression after infection using qRT-PCR. The gene expression of *LcCLA1* and *LcPDS* in untreated wild-type, empty-vector and virus-infected plants was compared to assess gene-silencing efficiency. For both genes, gene expression was significantly reduced in infected *Lycoris* leaves compared to wild-type and empty TRV vector plants (Figure 4). The efficiency of gene silencing was analyzed by monitoring the expression levels of *LcCLA1* in plants showing leaf bleaching phenotype. According to the results, it was found that the expression level of *LcCLA1* in pTRV1- and pTRV2-injected leaves was not significantly changed compared with CK. However, the expression level of *LcCLA1* in leaves injected with pTRV2-*LcCLA1* was about 75% lower than that in CK (Figure 4a). The etiolation phenotype of the leaves was consistent with the expression level of *LcCLA1* after silencing. The expression level of *LcPDS* was also analyzed in leaves injected with pTRV2-*LcPDS* but with less etiolation phenotype. From the results, it was found that the expression level of *LcPDS* in CK leaves was approximately the same as that in leaves injected with pTRV1 and pTRV2. However, the expression level of *LcPDS* in leaves injected with pTRV2-*LcPDS* was only about 50% of that in CK (Figure 4b). Compared with the silencing efficiency of *LcCLA1*, the silencing efficiency of *LcPDS* is much lower.

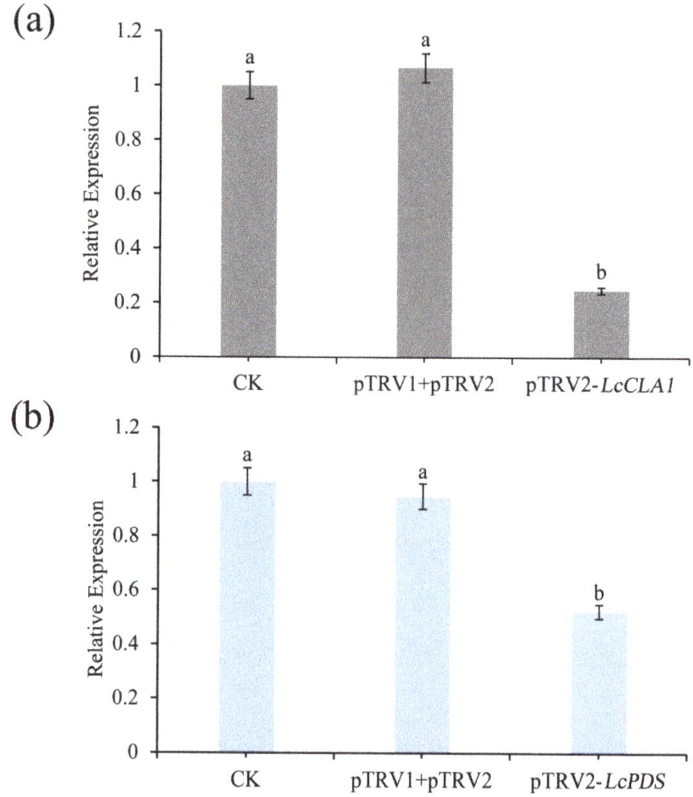

Figure 4. Relative expression levels of *LcCLA1* and *LcPDS* in not injected (CK), TRV vector-injected (pTRV1+pTRV2), pTRV2-*LcCLA1*- and pTRV2-*LcPDS*-injected leaves of *L. chinensis*. (**a**) Relative expression level of *LcCLA1*. (**b**) Relative expression level of *LcPDS*. Error bars represent standard errors, and different lowercase letters indicate significant differences at $p \leq 0.05$.

3. Discussion

In this study, we demonstrated that the endogenous genes in *L. chinensis* can be effectively down-regulated using the TRV-VIGS system. Since the current genetic transformation system of *Lycoris* is very immature, it is difficult to rapidly verify the gene function of this genus. Breaking through this barrier would not only take a lot of time and effort but would also need to overcome a lot of technical challenges. Therefore, effective and low-cost technologies need to be developed to temporarily replace genetic transformation systems. We developed the TRV-VIGS system to contribute to the gene regulation research of *Lycoris*, especially in its flowering traits.

Agrobacterium infection is the most effective method of virus-based vector infestation of plants [37]. The silencing efficiency varies greatly with different injection methods for the same plant [32]. The silencing efficiency is related to the fluid volume of *Agrobacterium* entering the plant. Since the leaf surface of *Lycoris* has a waxy layer, it is difficult to get the *Agrobacterium* solution into the leaf using the normal method of injection infiltration. We inserted the solution by injection with the needle of a syringe from the tip of the leaf to fully infiltrate the leaf, thus greatly improving the injection efficiency. To visualize whether genes were being knocked down, we used two different reporter genes, *CLA1* and *PDS*, which are commonly used in VIGS and which produce a visible phenotype after successful silencing.

The *CLA1* gene is involved in chloroplast development and is a potent marker in the TRV-VIGS system [38]. The *CLA1* gene was selected as a reporter, and it was tested out to determine if the TRV-VIGS system can be well applied to the study of gene function in *Solanum melongena* [39]. In *Arabidopsis*, the bleaching phenotype after silencing *CLA1* was used as a visible indicator of the silencing efficiency of the TRV-VIGS system [40]. Up to now, *CLA1* has been used as a reporter gene for gene silencing many times in cotton studies. For example, when weather cold-induced changes in non-coding gene expression had a significant effect on the cold tolerance of cotton seedlings, TRV2-*CLA1* was used as a positive control to determine that silencing of lincRNA *XH123* resulted in increased sensitivity to cold injury [41]. When studying the function of the *CLE* gene family in cotton, *CLA* was again used as a positive control, and it was found that the silencing of *GhCLE5* in cotton resulted in dwarf seedlings [42]. Inhibition of *GauGRAS1* by VIGS resulted in glandless stems and petioles of *Gossypium australe* compared with positive control *TRV-CLA* and negative control TRV empty vector [43]. We isolated a homologue of *CLA1* from *L. chinensis* leaves, named *LcCLA1*. Plants were infiltrated with *Agrobacterium tumefaciens* expressing partial sequences of *LcCLA1*, and etiolation phenotype was observed on the injected leaves approximately two weeks after infiltration. qRT-PCR analysis revealed low levels of *LcCLA1* transcript in *L. chinensis* leaves inoculated with the TRV-VIGS system, which confirmed the silencing of this gene. This suggests that the expression of *LcCLA1* is significantly down-regulated in *L. chinensis* through the TRV-VIGS system, resulting in etiolation phenotype of the leaves.

The gene *PDS* encodes an enzyme in carotenoid biosynthesis whose silencing leads to the albino phenotype of plant tissues [18], so it is often used as a marker in VIGS experiments. Up to now, VIGS has been reported to silence *PDS* in many plants, including *Solanum melongena* [39], *Capsicum annuum* [16], tomato [11], *Nicotiana benthamiana* [44], *Sorghum bicolor* [45], soybean [46], etc. In this study, the endogenous *PDS* gene also was selected as the reporter gene in *L. chinensis*. Injected leaves were significantly altered compared to controls, exhibiting a pronounced chlorotic phenotype, especially in the upper and middle sections of leaves. It indicated that the *PDS* gene could be selected as the indicator gene for VIGS system in *L. chinensis*. However, the chlorotic phenotype of *LcCLA1* was more obvious than that of *LcPDS*. qRT-PCR validation also showed that the expression of *LcCLA1* was lower than that of *LcPDS*. The reason may be that the *PDS* gene is not only involved in chlorophyll content but also carotenoid biosynthesis [18,35]. For example, in highbush blueberry that knock out the *PDS* gene, the leaves exhibited a phenotype without chlorophyll, but with red coloration [36]. Thus, our study suggested that the *PDS* gene may also be involved in carotenoid biosynthesis in *Lycoris*, leading to less albino phenotype in leaves.

There are many factors that affect silencing efficiency, such as the growth temperature after inoculation, the growth stage of the inoculated plants, the type of *Agrobacterium* strain, the inoculation method and the inoculum concentration. Different plants require different temperatures to produce a good silencing phenotype after VIGS. In the tomato, the optimal silencing phenotype was obtained when the growth temperature was 22 °C after inoculation [47]. In *S. pseudocapsicum*, silencing efficiency decreased when inoculated seedlings were grown at 18 °C or 30 °C [32]. We need to improve the growth temperature of the VIGS system between 20 °C and 25 °C. The growth stage of the target plant affects silencing efficiency. Gerbera seedlings at the early stages of vegetative development were much more sensitive to TRV VIGS than those at the middle and late stages [48]. We can try to inoculate between one and two weeks after leaf emergence. The optimal *Agrobacterium* strain for VIGS varies from plant to plant and has a strong impact on gene silencing efficiency. Studies have shown that both *Agrobacterium* strains GV2260 and GV3101 can be used in *N. benthamiana*, with GV2260 working best [18]. LBA4404 and GV2260 can be used on the tomato, but the silencing efficiency is very low, and GV3101 has the best silencing effect [49]. We will select LBA4404 and GV2260 to compare with GV3101 and select the strain with the best silencing efficiency. Due to the different inoculation methods and characteristics of different infected plants, the concentration of the infection solution strongly affects the gene silencing efficiency of VIGS experiments. In the TRV-VIGS experiment of *Arabidopsis*, the most suitable concentration of *Agrobacterium* infection solution was OD600 = 1.5, and the silencing efficiency was almost 100% [15]. In addition, naked cotton seeds soaked in *Agrobacterium* inoculum with OD600 of 1.5 for 90 min showed the best silencing efficiency [50]. For vacuum infiltration, some studies have found that the optimal conditions for VIGS are vacuum treatment with *Agrobacterium* liquid with OD600 of 0.3 for 30–60 s, and co-cultivation with the same concentration of *Agrobacterium* for 15 h [51]. The above experiments show that different species require different concentrations of *Agrobacterium* inoculation; different inoculation methods also require different concentrations of *Agrobacterium*. In this study, the silencing of *LcCLA1* and *LcPDS* were successfully achieved by *Agrobacterium* infection at OD600 = 2.0 and cultured at 18–26 °C for 2 weeks. In addition, *L. chinensis* has an obvious chlorosis phenotype. Therefore, this can be used as the recommended concentration for the VIGS system of *Lycoris* plants.

4. Materials and Methods

4.1. Plant Materials and Infection Methods

L. chinensis bulbs were planted in the National Germplasm Bank of *Lycoris*, located in the Institute of Botany, Jiangsu Province and Chinese Academy of Sciences (Nanjing Botanical Garden Mem. Sun Yat-Sen), Nanjing, China. The plants grow in conditions of 12 h/12 h day/night periods and 26 °C/18 °C day/night temperatures. The experiment was performed in March when the leaves were lush. The leaf tip injection method (Figure 1a) was used for bacterial infection, which makes it easier to infiltrate the whole leaf, showing more efficiency than the conventional leaf infiltration method (Figure 1b). After the infection was completed, the plant material was cultured in the dark at 18 °C for 2 d and then grown in a normal environment. After tip injection, when the leaves presented the chlorotic phenotype, the injected leaves and normal growing leaves were collected, labeled as treatment group (T) and control group (CK). All samples were immediately frozen in liquid nitrogen after collection and stored at −80 °C until RNA extraction. The collected leaves were used to analyze the expression of the *LcCLA1* and *LcPDS* genes.

4.2. Cloning of the LcCLA1 and LcPDS Genes and VIGS Vector Construction

The gene sequences of *LcCLA1* and *LcPDS* were obtained from transcriptome data of *L. chinensis* [6]. Raw reads were deposited in the NCBI database (https://www.ncbi.nlm.nih.gov/, accessed on 8 June 2022) under BioProject number PRJNA847051. The fresh leaves of *L. chinensis* were collected and put into liquid nitrogen for RNA isolation. Total RNA was extracted using the Plant RNA Extraction Kit (Huayueyang, Beijing, China) according to the manufacturer's instructions. The RNA concentration and purity were detected using the OneDrop OD1000 system (Wuyi Technology, Nanjing, China) and gel electrophoresis. The cDNA was synthesized using the PrimeScriptTM RT kit with the gDNA Eraser kit (TaKaRa, Dalian, China) according to the manufacturer's instructions. Primers of *LcCLA1* and *LcPDS* genes (Table 1) were designed using Primier 5 software. The fragments of *LcCLA1* and *LcPDS* were amplified using the designed primers *LcCLA1*-F/R and *LcPDS*-F/R, respectively. The Xba I restriction site was added to the 5′ end of the upstream primer, and the Kpn I restriction site was added to the 5′ end of the downstream primer. The PCR reaction was performed in a 50 μL volume containing 4 μL cDNA, 1 μL dNTP Mix (10 mM), 25 μL 2 × Phanta Max Buffer, 2 μL each of the forward and reverse primers (10 μM), 1 μL Phanta Max Super-Fidelity DNA Polymerase and 15 μL ddH2O. The PCR reaction was performed in the following conditions: denaturation at 95 °C for 3 min, followed by 35 cycles of 95 °C for 15 s, 56 °C for 15 s, 72 °C for 1 min, and finally extension at 72 °C for 5 min.

Table 1. List of primers used in this study.

Primer Name	Primer Sequence
EXP1-RT-F	TTGATGTTGACAAGGTAAGGTGC
EXP1-RT-R	AGGCAGGAAATCTCCAAAGC
LcCLA1-F	TTGACGGGCAGGAGGGAC
LcCLA1-R	GCGGAGGAAGCAGTTTAGGC
LcCLA1-XbaI-F	CTAGTCTAGATTGACGGGCAGGAGGGAC
LcCLA1-KpnI-R	CGGGGTACCGCGGAGGAAGCAGTTTAGGC
LcCLA1-RT-F	GTTGCTCATTTCTTGGCACTCA
LcCLA1-RT-R	CAGCACCAACGGTCTCCACT
LcPDS-F	GTTAGGTCAGTTTCTGCTGTTTGTC
LcPDS-R	GTTGTTCCTCAAGATAGCCCATA
LcPDS-XbaI-F	CTAGTCTAGAGTTAGGTCAGTTTCTGCTGTTTGTC
LcPDS-KpnI-R	CGGGGTACCGTTGTTCCTCAAGATAGCCCATA
LcPDS-RT-F	AAAACCGTACCCGACTGTGAG
LcPDS-RT-R	CGGCTGTAGACACTTTCTTGCT

The TRV-VIGS vectors pTRV1 and pTRV2 were used in this study as reported in previous studies [52]. AxyPrep DNA Gel Extraction Kit (Axygen, Nanjing, China) was used to purify PCR products. After purification, *LcCLA1* and *LcPDS* fragments were assembled into the pTRV2 vector using ClonExpress II one-step cloning kit (Vazyme, Nanjing, China). The specific fragments were respectively attached to pTRV2, and then the silent vectors pTRV2-*LcCLA1* and pTRV2-*LcPDS* were obtained (Figure 5).

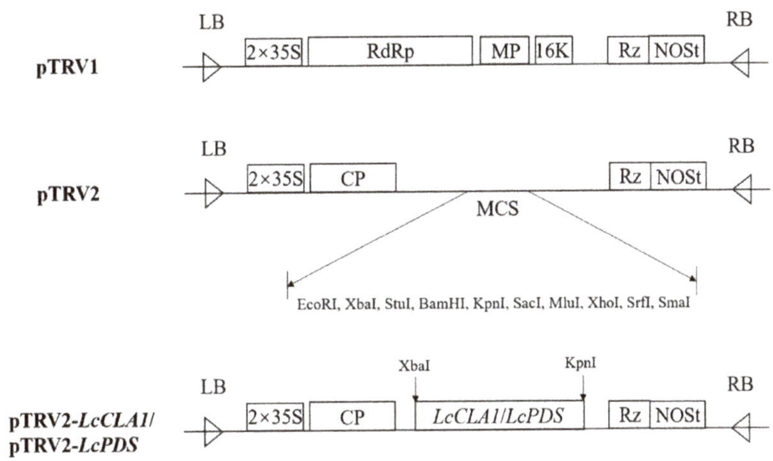

Figure 5. TRV-based VIGS vectors and construction. LB and RB: left and right borders of T-DNA; 2 × 35 S: the duplicated CaMV 35 S promoter; RdRp: RNA-dependent RNA polymerase; MP: movement protein; CP: coat protein; 16 K: 16 kDa cysteine-rich protein; MCS: multiple cloning sites; Rz: self-cleaving ribozyme; NOSt: nopaline synthase terminator.

4.3. *Preparation of Agroinfiltration Infection Solution*

The plasmids of TRV2-*LcCLA1*, TRV2-*LcPDS*, TRV2 and TRV1 were transformed into *Agrobacterium tumefaciens* strain GV3101 using the freeze-thaw method [53], separately. The transformed plasmids of TRV1, TRV2, TRV2-*LcCLA1* and TRV2-*LcPDS* were cultured in LB solid medium (with 50 µg/mL kanamycin; 25 µg/mL rifampicin). Freshly cultured *A. tumefaciens* GV3101 colonies containing pTRV1, pTRV2, pTRV2-*LcCLA1* and pTRV2-*LcPDS* were selected and added to 3 mL YEB liquid medium (with 50 µg/mL kanamycin; 25 µg/mL rifampicin). Then, they were cultured in a shaker at 28 °C and 170 rpm/min for 16 h. Add 1 mL of the cultured solution to 100 mL of YEB liquid medium (with

50 μg/mL kanamycin; 25 μg/mL rifampicin), and incubate at 28 °C and 170 rpm/min for 20–24 h. When the OD600 of the detected solution is about 1.0, take 40 mL and centrifuge at 4000 rpm for 10 min to collect the cells, and add an appropriate volume of infection buffer to resuspend until the final concentration of OD600 is 2.0. The configuration of the infection buffer was 100 mL of permeate containing 1 mL of $MgCl_2$ (1 mol/L), 1 mL of MES (1 mol/L), and 200 μL of acetosyringone (1 mol/L), and the pH was adjusted to 5.6. Make up to volume with sterilized distilled water, which needs to be prepared and used immediately. Two kinds of infection solutions were prepared by mixing pTRV1 with pTRV2 and pTRV2-*LcCLA1* resuspension at a volume ratio of 1:1 [54], respectively. Another infection solution also mixed pTRV1 and pTRV2-*LcPDS* at the same ratio. The mixed infecting solution was placed in a 25 °C shaker at 100 rpm/min and cultured in the dark for 4 h to infect the leaves.

4.4. Expression Analysis of *LcCLA1* and *LcPDS* Using qRT-PCR

Leaf chlorosis was observed after 2 weeks, and the leaves of *L. chinensis* that were not injected with infection solution, injected with pTRV1+pTRV2 and injected with pTRV1+pTRV2-*LcCLA1*, pTRV1+pTRV2-*LcPDS* were collected. The collected leaves were immediately frozen in liquid nitrogen and stored at −80 °C for qRT-PCR experiments. The *EXP1* gene was used as an endogenous control to normalize the results. To determine the relative levels of endogenous *LcCLA1* and *LcPDS* transcripts in infected leaves, qRT-PCR was performed using the primer pair EXP1-RT-F, EXP1-RT-R, *LcCLA1*-RT-F, *LcCLA1*-RT-R and *LcPDS*-RT-F, *LcPDS*-RT-R (Table 1). The steps of extracting total RNA, detecting RNA concentration, purity, integrity and reverse transcription are the same as in Section 4.2. qRT-PCR was performed using a StepOnePlus real-time PCR system (Applied Biosystems, Beijing, China) and a 20 μL reaction mixture. Each 20 μL reaction mixture contained 10 μL SYBR Premix Ex TaqTMII (TaKaRa, Dalian, China), 5 μL diluted cDNA, 0.8 μL forward and reverse primer (10 μM), 0.4 μL ROX Reference Dye and 3 μL ddH_2O. The experiment was repeated three times, and the data were analyzed using the $2^{-\Delta\Delta Ct}$ method to calculate relative gene expression levels [55].

5. Conclusions

For species where plant transformation, regeneration and genetic backgrounds are challenging, VIGS provides a powerful tool for transient gene function studies. We demonstrated that the TRV-based VIGS system can effectively silence genes in *L. chinensis* and established a TRV-based VIGS system that could successfully induce yellow leaf phenotypes by silencing *LcCLA1* and *LcPDS*, respectively. According to the influencing factors such as *Agrobacterium* strain GV3101, tip needle injection method and inoculated plant growth temperature of 26 °C, the relative expression level of *LcCLA1* and *LcPDS* can reach about 25–50%, respectively. The TRV-mediated VIGS system developed and used in this study provides an alternative tool for functional gene studies in *L. chinensis*. This study also could provide a reference for the development of a rapid, transient and stable transformation system for *Lycoris*. In addition, we have improved the injection method for blades with waxy surfaces, which greatly improves the injection efficiency and ease of operation. This can provide new ideas for other plants that are not easily penetrated. These results help to lay the foundation for gene expression analysis in *Lycoris* and better studying of the molecular mechanism of related traits.

Author Contributions: F.Z. conceived the project and designed the experiments; G.C. and X.S. performed the experiments and analyzed the data; N.W. and Z.W. prepared and collected the samples; G.C. and F.Z. wrote and revised the manuscript; Z.W. and N.W. reviewed the manuscript. All authors have read and agreed to the published version of the manuscript.

Funding: This research was funded by the National Natural Science Foundation of China (31801900); the Natural Science Foundation of Jiangsu Province (BK20180310); the Open Fund of Jiangsu Key Laboratory for the Research and Utilization of Plant Resources (JSPKLB202205); the Jiangsu Provincial Crop Germplasm Resource Bank for Conservation (2022-SJ-015); Jiangsu Provincial Forestry Science and Technology Promotion Project (LYKJ[2022]08).

Data Availability Statement: The data are available from the corresponding author upon request.

Conflicts of Interest: The authors declare no conflict of interest.

References

1. Kurita, S. Variation and evolution on the karyotype of *Lycoris*, Amaryllidaceae II. Karyotype analysis of ten taxa among which seven are native in China. *Cytologia* **1987**, *52*, 19–40. [CrossRef]
2. Tae, K.H.; Ko, S.C. A taxonomic study on the genus *Lycoris* (Amaryllidaceae). *Kor. J. Plant Tax.* **1996**, *26*, 19–35. [CrossRef]
3. Cahlikova, L.; Breiterova, K.; Opletal, L. Chemistry and Biological Activity of Alkaloids from the Genus *Lycoris* (Amaryllidaceae). *Molecules* **2020**, *25*, 4797. [CrossRef]
4. Yeo, H.J.; Kim, Y.J.; Nguyen, B.V.; Park, Y.E.; Park, C.H.; Kim, H.H.; Kim, J.K.; Park, S.U. Comparison of Secondary Metabolite Contents and Metabolic Profiles of Six *Lycoris* Species. *Horticulturae* **2021**, *7*, 5. [CrossRef]
5. Sun, B.; Wang, P.; Wang, R.; Li, Y.; Xu, S. Molecular Cloning and Characterization of a *meta/para-O*-Methyltransferase from *Lycoris aurea*. *Int. J. Mol. Sci.* **2018**, *19*, 1911. [CrossRef]
6. Zhang, F.; Cheng, G.; Shu, X.; Wang, N.; Wang, Z. Transcriptome Analysis of *Lycoris chinensis* Bulbs Reveals Flowering in the Age-Mediated Pathway. *Biomolecules* **2022**, *12*, 899. [CrossRef]
7. Wang, N.; Song, G.; Zhang, F.; Shu, X.; Cheng, G.; Zhuang, W.; Wang, T.; Li, Y.; Wang, Z. Characterization of the *WRKY* Gene Family Related to Anthocyanin Biosynthesis and the Regulation Mechanism under Drought Stress and Methyl Jasmonate Treatment in *Lycoris radiata*. *Int. J. Mol. Sci.* **2023**, *24*, 2423. [CrossRef]
8. Cheng, G.; Zhang, F.; Shu, X.; Wang, N.; Wang, T.; Zhuang, W.; Wang, Z. Identification of Differentially Expressed Genes Related to Floral Bud Differentiation and Flowering Time in Three Populations of *Lycoris radiata*. *Int. J. Mol. Sci.* **2022**, *23*, 14036. [CrossRef]
9. Cai, J.; Fan, J.; Wei, X.; Zhang, D.; Ren, J.; Zhang, L. Differences in floral development between *Lycoris radiata* and *Lycoris sprengeri*. *Scienceasia* **2020**, *46*, 271. [CrossRef]
10. Senthil-Kumar, M.; Mysore, K.S. New dimensions for VIGS in plant functional genomics. *Trends Plant Sci.* **2011**, *16*, 656–665. [CrossRef]
11. Liu, Y.; Schiff, M.; Dinesh-Kumar, S.P. Virus-induced gene silencing in tomato. *Plant J.* **2002**, *31*, 777–786. [CrossRef] [PubMed]
12. Burch-Smith, T.M.; Anderson, J.C.; Martin, G.B.; Dinesh-Kumar, S.P. Applications and advantages of virus-induced gene silencing for gene function studies in plants. *Plant J.* **2004**, *39*, 734–746. [CrossRef] [PubMed]
13. Waterhouse, P.M.; Wang, M.B.; Lough, T. Gene silencing as an adaptive defence against viruses. *Nature* **2001**, *411*, 834–842. [CrossRef] [PubMed]
14. Jiang, Y.; Ye, S.; Wang, L.; Duan, Y.; Lu, W.; Liu, H.; Fan, D.; Zhang, F.; Luo, K. Heterologous gene silencing induced by tobacco rattle virus (TRV) is efficient for pursuing functional genomics studies in woody plants. *Plant Cell Tiss. Org.* **2014**, *116*, 163–174. [CrossRef]
15. Burch-Smith, T.M.; Schiff, M.; Liu, Y.; Dinesh-Kumar, S.P. Efficient virus-induced gene silencing in Arabidopsis. *Plant Physiol.* **2006**, *142*, 21–27. [CrossRef]
16. Chung, E.; Seong, E.; Kim, Y.C.; Chung, E.J.; Oh, S.K.; Lee, S.; Park, J.M.; Joung, Y.H.; Choi, D. A Method of High Frequency Virus-induced Gene Silencing in Chili Pepper (*Capsicum annuum* L. cv. Bukang). *Mol. Cells* **2004**, *17*, 377–380.
17. Xu, H.; Xu, L.; Yang, P.; Cao, Y.; Tang, Y.; He, G.; Yuan, S.; Lei, J.; Ming, J. Virus-induced *Phytoene Desaturase* (*PDS*) Gene Silencing Using *Tobacco Rattle Virus* in *Lilium × formolongi*. *Hortic. Plant J.* **2019**, *5*, 31–38. [CrossRef]
18. Kumagai, M.H.; Donson, J.; Della-Cioppa, G.; Harvey, D.; Hanley, K.; Grill, L. Cytoplasmic inhibition of carotenoid biosynthesis with virus-derived RNA. *Proc. Natl. Acad. Sci. USA* **1995**, *92*, 1679–1683. [CrossRef]
19. Ruiz, M.T.; Voinnet, O.; Baulcombe, D.C. Initiation and Maintenance of Virus-Induced Gene Silencing. *Plant Cell* **1998**, *10*, 937–946. [CrossRef]
20. Peele, C.; Jordan, C.V.; Muangsan, N.; Turnage, M.; Egelkrout, E.; Eagle, P.; Robertson, D. Silencing of a meristematic gene using geminivirus-derived vectors. *Plant J.* **2001**, *27*, 357–366. [CrossRef]
21. Liu, Y.; Schiff, M.; Marathe, R.; Dinesh-Kumar, S.P. Tobacco *Rar1*, *EDS1* and *NPR1/NIM1* like genes are required for *N*-mediated resistance to tobacco mosaic virus. *Plant J.* **2002**, *30*, 415–429. [CrossRef]
22. Igarashi, A.; Yamagata, K.; Sugai, T.; Takahashi, Y.; Sugawara, E.; Tamura, A.; Yaegashi, H.; Yamagishi, N.; Takahashi, T.; Isogai, M.; et al. *Apple latent spherical virus* vectors for reliable and effective virus-induced gene silencing among a broad range of plants including tobacco, tomato, *Arabidopsis thaliana*, cucurbits, and legumes. *Virology* **2009**, *386*, 407–416. [CrossRef] [PubMed]
23. Yin, X.; Zhang, Y.; Zhang, L.; Wang, B.; Zhao, Y.; Irfan, M.; Chen, L.; Feng, Y. Regulation of MYB Transcription Factors of Anthocyanin Synthesis in Lily Flowers. *Front. Plant Sci.* **2021**, *12*, 761668. [CrossRef]
24. Zhou, P.; Peng, J.; Zeng, M.; Wu, L.; Fan, Y.; Zeng, L. Virus-induced gene silencing (VIGS) in Chinese narcissus and its use in functional analysis of *NtMYB3*. *Hortic. Plant J.* **2021**, *7*, 565–572. [CrossRef]
25. Zhao, Y.; Yang, Z.; Ding, Y.; Liu, L.; Han, X.; Zhan, J.; Wei, X.; Diao, Y.; Qin, W.; Wang, P.; et al. Over-expression of an R2R3 MYB Gene, *GhMYB73*, increases tolerance to salt stress in transgenic Arabidopsis. *Plant Sci.* **2019**, *286*, 28–36. [CrossRef] [PubMed]
26. Hu, Q.; Zhu, L.; Zhang, X.; Guan, Q.; Xiao, S.; Min, L.; Zhang, X. GhCPK33 Negatively Regulates Defense against *Verticillium dahliae* by Phosphorylating GhOPR3. *Plant Physiol.* **2018**, *178*, 876–889. [CrossRef]

27. Yang, J.; Zhang, Y.; Wang, X.; Wang, W.; Li, Z.; Wu, J.; Wang, G.; Wu, L.; Zhang, G.; Ma, Z. *HyPRP1* performs a role in negatively regulating cotton resistance to *V. dahliae* via the thickening of cell walls and ROS accumulation. *BMC Plant Biol.* **2018**, *18*, 339. [CrossRef] [PubMed]
28. Liu, W.; Feng, J.; Ma, W.; Zhou, Y.; Ma, Z. *GhCLCg-1*, a Vacuolar Chloride Channel, Contributes to Salt Tolerance by Regulating Ion Accumulation in Upland Cotton. *Front. Plant Sci.* **2021**, *12*, 765173. [CrossRef] [PubMed]
29. Manhães, A.M.E.D.A.; de Oliveira, M.V.V.; Shan, L. Establishment of an Efficient Virus-Induced Gene Silencing (VIGS) Assay in *Arabidopsis* by *Agrobacterium*-Mediated Rubbing Infection. In *Plant Gene Silencing: Methods and Protocols*; Mysore, K.S., Senthil-Kumar, M., Eds.; Springer: New York, NY, USA, 2015; pp. 235–241.
30. Kirigia, D.; Runo, S.; Alakonya, A. A virus-induced gene silencing (VIGS) system for functional genomics in the parasitic plant *Striga hermonthica*. *Plant Methods* **2014**, *10*, 16. [CrossRef]
31. Xie, L.; Zhang, Q.; Sun, D.; Yang, W.; Hu, J.; Niu, L.; Zhang, Y. Virus-induced gene silencing in the perennial woody *Paeonia ostii*. *PeerJ* **2019**, *7*, e7001. [CrossRef]
32. Xu, H.; Xu, L.; Yang, P.; Cao, Y.; Tang, Y.; He, G.; Yuan, S.; Ming, J. Tobacco rattle virus-induced *PHYTOENE DESATURASE* (*PDS*) and *Mg-chelatase H* subunit (*ChlH*) gene silencing in *Solanum pseudocapsicum* L. *PeerJ* **2018**, *6*, e7001. [CrossRef] [PubMed]
33. Sung, Y.C.; Lin, C.P.; Chen, J.C. Optimization of virus-induced gene silencing in *Catharanthus roseus*. *Plant Pathol.* **2014**, *63*, 1159–1167. [CrossRef]
34. Nishii, K.; Fei, Y.; Hudson, A.; Moller, M.; Molnar, A. Virus-induced Gene Silencing in *Streptocarpus rexii* (Gesneriaceae). *Mol. Biotechnol.* **2020**, *62*, 317–325. [CrossRef]
35. Zhou, J.; Hunter, D.A.; Lewis, D.H.; McManus, M.T.; Zhang, H. Insights into carotenoid accumulation using VIGS to block different steps of carotenoid biosynthesis in petals of California poppy. *Plant Cell Rep.* **2018**, *37*, 1311–1323. [CrossRef]
36. Vaia, G.; Pavese, V.; Moglia, A.; Cristofori, V.; Silvestri, C. Knockout of phytoene desaturase gene using CRISPR/Cas9 in highbush blueberry. *Front. Plant Sci.* **2022**, *13*, 1074541. [CrossRef] [PubMed]
37. Lindbo, J.A. High-efficiency protein expression in plants from agroinfection-compatible *Tobacco mosaic virus* expression vectors. *BMC Biotechnol.* **2007**, *7*, 52. [CrossRef]
38. Pang, J.; Zhu, Y.; Li, Q.; Liu, J.; Tian, Y.; Liu, Y.; Wu, J. Development of Agrobacterium-mediated virus-induced gene silencing and performance evaluation of four marker genes in *Gossypium barbadense*. *PLoS ONE* **2013**, *8*, e73211. [CrossRef]
39. Liu, H.; Fu, D.; Zhu, B.; Yan, H.; Shen, X.; Zuo, J.; Zhu, Y.; Luo, Y. Virus-induced Gene Silencing in Eggplant (*Solanum melongena*). *J. Integr. Plant Biol.* **2012**, *54*, 422–429. [CrossRef]
40. Liu, J.; Huang, Y.; Kong, L.; Yu, X.; Feng, B.; Liu, D.; Zhao, B.; Mendes, G.C.; Yuan, P.; Ge, D.; et al. The malectin-like receptor-like kinase LETUM1 modulates NLR protein SUMM2 activation via MEKK2 scaffolding. *Nat. Plants* **2020**, *6*, 1106–1115. [CrossRef]
41. Cao, Z.; Zhao, T.; Wang, L.; Han, J.; Chen, J.; Hao, Y.; Guan, X. The lincRNA *XH123* is involved in cotton cold-stress regulation. *Plant Mol. Biol.* **2021**, *106*, 521–531. [CrossRef]
42. Wan, K.; Lu, K.; Gao, M.; Zhao, T.; He, Y.; Yang, D.L.; Tao, X.; Xiong, G.; Guan, X. Functional analysis of the cotton *CLE* polypeptide signaling gene family in plant growth and development. *Sci. Rep.* **2021**, *11*, 5060. [CrossRef]
43. Cai, Y.; Cai, X.; Wang, Q.; Wang, P.; Zhang, Y.; Cai, C.; Xu, Y.; Wang, K.; Zhou, Z.; Wang, C.; et al. Genome sequencing of the Australian wild diploid species *Gossypium australe* highlights disease resistance and delayed gland morphogenesis. *Plant Biotechnol. J.* **2020**, *18*, 814–828. [CrossRef]
44. Liu, E.; Page, J.E. Optimized cDNA libraries for virus-induced gene silencing (VIGS) using tobacco rattle virus. *Plant Methods* **2008**, *4*, 5–18. [CrossRef] [PubMed]
45. Bredow, M.; Natukunda, M.I.; Beernink, B.M.; Chicowski, A.S.; Salas-Fernandez, M.G.; Whitham, S.A. Characterization of a foxtail mosaic virus vector for gene silencing and analysis of innate immune responses in *Sorghum bicolor*. *Mol. Plant Pathol.* **2023**, *24*, 71–79. [CrossRef] [PubMed]
46. Luo, Y.; Na, R.; Nowak, J.S.; Qiu, Y.; Lu, Q.S.; Yang, C.; Marsolais, F.; Tian, L. Development of a Csy4-processed guide RNA delivery system with soybean-infecting virus ALSV for genome editing. *BMC Plant Biol.* **2021**, *21*, 419. [CrossRef]
47. Yan, H.X.; Fu, D.Q.; Zhu, B.Z.; Liu, H.P.; Shen, X.Y.; Luo, Y.B. Sprout vacuum-infiltration: A simple and efficient agroinoculation method for virus-induced gene silencing in diverse solanaceous species. *Plant Cell Rep.* **2012**, *31*, 1713–1722. [CrossRef]
48. Deng, X.; Elomaa, P.; Nguyen, C.X.; Hytonen, T.; Valkonen, J.P.; Teeri, T.H. Virus-induced gene silencing for Asteraceae--a reverse genetics approach for functional genomics in *Gerbera hybrida*. *Plant Biotechnol. J.* **2012**, *10*, 970–978. [CrossRef] [PubMed]
49. Dinesh-Kumar, S.; Anandalakshmi, R.; Marathe, R.; Schiff, M.; Liu, Y. *Virus-Induced Gene Silencing*; Humana Press: Totowa, NJ, USA, 2003; pp. 287–293.
50. Zhang, J.; Wang, F.; Zhang, C.; Zhang, J.; Chen, Y.; Liu, G.; Zhao, Y.; Hao, F.; Zhang, J. A novel VIGS method by agroinoculation of cotton seeds and application for elucidating functions of *GhBI-1* in salt-stress response. *Plant Cell Rep.* **2018**, *37*, 1091–1100. [CrossRef]
51. Zhang, J.; Yu, D.; Zhang, Y.; Liu, K.; Xu, K.; Zhang, F.; Wang, J.; Tan, G.; Nie, X.; Ji, Q.; et al. Vacuum and Co-cultivation Agroinfiltration of (Germinated) Seeds Results in Tobacco Rattle Virus (TRV) Mediated Whole-Plant Virus-Induced Gene Silencing (VIGS) in Wheat and Maize. *Front. Plant Sci.* **2017**, *8*, 393. [CrossRef]
52. Ratcliff, F.; Martin-Hernandez, A.M.; Baulcombe, D.C. Technical advance: Tobacco rattle virus as a vector for analysis of gene function by silencing. *Plant J.* **2001**, *25*, 237–245. [CrossRef]

53. Höfgen, R.; Willmitzer, L. Storage of competent cells for Agrobacterium transformation. *Nucleic Acids Res.* **1988**, *16*, 9877. [CrossRef] [PubMed]
54. Chong, X.; Wang, Y.; Xu, X.; Zhang, F.; Wang, C.; Zhou, Y.; Zhou, T.; Li, Y.; Lu, X.; Chen, H. Efficient Virus-Induced Gene Silencing in *Ilex dabieshanensis* Using Tobacco Rattle Virus. *Forests* **2023**, *14*, 488. [CrossRef]
55. Livak, K.J.; Schmittgen, T.D. Analysis of relative gene expression data using real-time quantitative PCR and the $2^{-\Delta\Delta CT}$ method. *Methods* **2001**, *25*, 402–408. [CrossRef] [PubMed]

MDPI
St. Alban-Anlage 66
4052 Basel
Switzerland
www.mdpi.com

Plants Editorial Office
E-mail: plants@mdpi.com
www.mdpi.com/journal/plants